ADVANCES IN
MULTIUSER DETECTION

ADVANCES IN MULTIUSER DETECTION

Edited by

Michael L. Honig

Northwestern University

A JOHN WILEY & SONS, INC., PUBLICATION

Published by John Wiley & Sons, Inc., Hoboken, New Jersey.
Published simultaneously in Canada.

For general information on our other products and services or for technical support, please contact our
Customer Care Department within the United States at (800) 762-2974, outside the United States at (317)
572-3993 or fax (317) 572-4002.

Wiley also publishes its books in variety of electronic formats. Some content that appears in print
may not be available in electronic format. For more information about Wiley products, visit our
web site at www.wiley.com.

Library of Congress Cataloging-in-Publication Data:

Advances in multiuser detection / Michael L. Honig, editor.
    p. cm.
  Includes bibliographical references and index.
  ISBN 978-0-471-77971-1 (cloth)
  1. Multiuser detection (Telecommunication)   I. Honig, Michael L.
  TK5103.2.A3886 2009
  621.382--dc22

                                      2009013356

Printed in the United States of America.

10 9 8 7 6 5 4 3 2 1

CONTENTS

3   Blind Multiuser Detection in Fading Channels   **127**

Daryl Reynolds, H. Vincent Poor, and Xiaodong Wang

5 Generic Multiuser Detection and Statistical Physics　　　　**251**

Dongning Guo and Toshiyuki Tanaka

6   Joint Detection for Multi-Antenna Channels   311

*Antonia Tulino, Matthew R. McKay, Jeffrey G. Andrews,
Iain B. Collings, and Robert W. Heath, Jr.*

8   Capacity-Approaching Multiuser Communications Over Multiple Input/Multiple Output Broadcast Channels   417

Uri Erez and Stephan ten Brink

PREFACE

The proliferation of telecommunications systems and services over the past couple of decades has been accompanied by numerous advances in physical layer communications and associated signal processing techniques. Many of these services have in fact been enabled by the increase in spectral efficiency provided by improved modulation, coding, and reception capability. Furthermore, the evolution of communications networks continues to stimulate efforts to push the performance and reliability of these networks to their fundamental limits.

For the most part, many of the recent advances in signal processing methods for communications have been motivated by the evolution of mobile cellular systems, i.e., from first-generation analog systems, introduced in the 1980s, to next (fourth)-generation systems currently being designed. Additional motivation has been provided by other wireless systems and standards, such as wireless local and metropolitan area networks, and also the desire to provide broadband services over existing copper subscriber lines in the telephone network. Although wired channels, such as subscriber lines, do not experience the time variations in received signal strength associated with mobile cellular channels, other challenges remain, such as overcoming frequency-selectivity and efficient spectrum sharing among multiple users with different channel characteristics.

This book reviews recent advances in multiuser detection, which generally refers to methods for detecting digital data associated with multiple interfering signals. These advances comprise some of the signal processing techniques just mentioned, and have been an active area of research and development over the past couple of decades. In the title of this book, "recent" generally refers to advances made over the past ten years, i.e., since the publication of the first book on multiuser detection [87]. Except for the first chapter, which gives a general introduction and overview, each of the eight chapters is contributed by a different set of authors, and is meant to be a self-contained discussion of a particular topic. Namely, Chapter 2 discusses iterative techniques for combined multiuser detection and decoding of error control codes; Chapter 3 dicusses multiuser detection in the presence of linear channel impairments, such multipath and intersymbol interference; Chapters 4 and 5 present performance analysis of multiuser detection methods with random signatures and channels; Chapter 6 discusses the application of joint detection methods to Multi-Input/Multi-Output (MIMO) channels, corresponding to wireless links with multiple antennas at the transmitter and/or receiver; Chapter 7 discusses interference avoidance methods at the transmitter (i.e., through the choice of signatures assuming spread spectrum signaling); and Chapter 8 discusses transmitter precoding methods for the MIMO downlink (broadcast channel). A more detailed overview of the chapters is given at the end of Chapter 1.

These topics represent a sampling of major advances that have been made in multiuser detection over the past ten years. Because this research area has been quite active, comprehensive coverage of recent progress would be quite difficult. This book therefore serves as an entry point for exploring ongoing research in multiuser detection and for learning about existing unsolved problems and issues. The intended audience is therefore graduate students in communications, as well as practicing engineers and researchers who are familiar with digital communications at the level of [87] and [60], and wish to gain a deeper understanding of multiuser detection techniques. This area continues to progress and it is our hope that these contributions will stimulate further advances.

MICHAEL HONIG

Evanston, Illinois
July 2009

CONTRIBUTORS

Jeffrey G. Andrews, Department of Electrical and Computer Engineering, University of Texas, Austin, Texas

Iain B. Collings, Wireless Technologies Lab, CSIRO ICT Centre, Sydney, Australia

Uri Erez, Department of Electrical Engineering-Systems, Tel Aviv University, Tel Aviv-Yafo, Israel

Alex Grant, Institute for Telecommunications Research, University of South Australia, Mawson Lakes, South Australia

Dongning Guo, Department of Electrical Engineering and Computer Science, Northwestern University, Evanston, Illinois

Robert W. Heath, Jr., Department of Electrical and Computer Engineering, University of Texas, Austin, Texas

Michael L. Honig, Department of Electrical Engineering and Computer Science, Northwestern University, Evanston, Illinois

Matthew R. McKay, Department of Electronic and Computer Engineering, Hong Kong University of Science and Technology, Kowloon, Hong Kong

Matthew J. M. Peacock, Credit-Suisse Bank, New York, New York

H. Vincent Poor, Department of Electrical Engineering, Princeton University, Princeton, New Jersey

Dimitrie C. Popescu, Department of Electrical and Computer Engineering, Old Dominion University, Norfolk, Virginia

Lars K. Rasmussen, KTH, Royal Institute of Technology, School of Electrical Engineering, Stockholm, Sweden

Daryl Reynolds, Department of Computer Science and Electrical Engineering, West Virginia University, Morgantown, West Virginia

Christopher Rose, Wireless Information Network Laboratory (WINLAB), Rutgers University, North Brunswick, New Jersey

Toshiyuki Tanaka, Department of Systems Science, Kyoto University, Kyoto, Japan

Stephan ten Brink, Wionics Technologies, Irvine, California

Antonia Tulino, Dip. Di Ing. Elettronica e delle Telecomunicazioni, Università degli Studi di Napoli, Naples, Italy

Sennur Ulukus, Department of Electrical and Computer Engineering, University of Maryland, College Park, Maryland

Xiaodong Wang, Electrical Engineering Department, Columbia University, New York, New York

Roy Yates, Wireless Information Network Laboratory (WINLAB), Rutgers University, North Brunswick, New Jersey

1

OVERVIEW OF MULTIUSER DETECTION

Michael L. Honig

... But what is the use of counterpoint when, if played, one imagines that four different orchestras are playing at the same time four different tunes in four different keys and measures? A veritable nightmare!

—Music critic Arthur Bird, writing about R. Strauss's tone
poem *Ein Heldenleben*, May 1, 1899 [69, p. 186]

1.1 INTRODUCTION

One of the key challenges in designing multiuser communications systems is mitigating interference. This challenge is apparent for modern wireless networks, such as mobile cellular, and wireless local and metropolitan area networks, where achieving high spectral efficiencies requires aggressive frequency reuse. Interference also limits the performance of many wired channels, such as the digital subscriber line (DSL). Although each DSL is typically associated with a single user, capacitive coupling between pairs of DSLs in close physical proximity causes cross-talk interference, which degrades performance (e.g., see [72]).

Interference encompasses *self*-interference, due to reflections of the same signal, in addition to *multi-user*-interference associated with other signals sharing the same bandwidth. Self-interference arises from multipath in a wireless channel, a bridge tap in a Digital Subscriber Line (DSL), and bandwidth constraints, which cause

Advances in Multiuser Detection. Edited by Michael L. Honig
Copyright © 2009 John Wiley & Sons, Inc.

intersymbol interference. Throughput this book, interference typically refers to signals from other users (or more generally, data streams) associated with the *same* system. Clearly, techniques for reducing, or mitigating interference lead directly to improved performance, either in terms of reduced error rate, increased data rate, or number of users that can be served.

Effective techniques for mitigating interference must depend, of course, on what is known at the transmitter and receiver. For example, in a multiple-access channel the receivers for all transmitters are co-located. Hence instead of detecting the transmitted bits from a particular user in isolation, treating the other signals as background noise, the receiver can *jointly* detect *all* of the transmitted information bits, or symbols. In that case, the structure of the multiple access interference can be used to aid the detection of the desired symbol. For example, it may be possible to use estimates of interfering symbols to cancel, or at least reduce, the level of the interfering signals. In contrast, if the receivers are not co-located, then interference cancellation, which relies on accurate interference estimates, may not be practical.

Multiuser detection refers to the scenario in which a single receiver jointly detects multiple simultaneous transmissions. Examples include the uplink of a single cell in a cellular system, and a group of twisted-pair copper subscriber lines that terminate at the same central switching office. More generally, multiuser detection techniques apply to the joint detection of different signals transmitted over any multi-input/multi-output (MIMO) channel. In addition to the preceding examples, others include channels in which multiple transmitted information streams are multiplexed over multiple transmit antennas. In that scenario, the multiple "users" refer to the multiple information streams, even though the transmitted signal may originate from a single user.

Closely related to multiuser detection is *interference suppression*. The key distinction is that a multiuser detector attempts to retrieve *multiple* (i.e., at least two) transmitted signals, or information streams, whereas interference suppression implies that the receiver is interested in only one signal among the received superposition of transmitted signals. For example, this is typically the case for the downlink of a cellular system, in which a mobile wishes to demodulate a single transmitted information stream in the presence of interfering signals from the associated base station and from nearby base stations. Of course, in general a receiver may wish to demodulate a subset of two or more signals from among a larger mix of signals. In that case, the receiver jointly detects the subset of desired signals while suppressing the interfering signals. An example of this is the uplink of a cellular system, in which the receiver is interested in demodulating the transmitted signals from users within the cell in the presence of interference from other cells.

1.1.1 Applications

Much of the work on multiuser detection and interference suppression over the past couple of decades has been motivated by the commercial success of Code-Division Multiple Access (CDMA) in mobile cellular systems. Namely, second-generation CDMA cellular systems were introduced in the early 1990s, and CDMA is currently used in third-generation cellular systems. The performance of CDMA is generally limited by multi-user interference. In particular, the performance of the uplink, which refers to the multiple access channel from users to base station, is sensitive to received power variations across users. The classic example of this is the near–far problem, in which a user close to the base station causes excessive interference to a user far from the base station. Commercial CDMA systems generally use closed-loop power control to minimize received power variations (both across time to mitigate fading and across users). It was recognized early on, however, that the sensitivity to interference power is not inherent to CDMA, but rather is a property of the conventional matched filter detector. That motivated studies of multiuser detection techniques for uplink CDMA, starting with the work of Verdú [86].[1]

Prior to the introduction of CDMA for mobile cellular systems, some specific multiuser detectors had been derived for linear MIMO channels. See, for example, [80], which discusses the Maximum Likelihood (ML) detector, and [1,45,46,63], which derive linear and decision feedback detectors.[2]

In recent years, the primary motivation for multiuser detection has shifted from CDMA to other applications (in particular, links with multiple antennas). There are several reasons for this. First, as shown in the initial studies on the achievable rates for wireless links with multiple antennas [19,77], substantial gains in spectral efficiency and reliability can be achieved by adding antennas to the transmitter and/or receiver of a single-user wireless channel. That stimulated an enormous amount of activity on coding and reception techniques for MIMO channels, so that nearly all evolving wireless systems and standards include provisions for multiple transmit antennas. In those scenarios, multiuser detection techniques are useful for mitigating interference among the different transmit antennas [6].

The second reason for the shift away from CDMA applications is that current designs for next-generation cellular and wide-area wireless networks are based primarily on Orthogonal Frequency Division Multiplexing (OFDM) and Orthogonal Frequency Division Multiple Access (OFDMA). That is due in part to the more substantial role played by data services, as opposed to voice, in evolving wireless networks. Namely, current CDMA cellular systems rely on "interference averaging" and power control to achieve robustness with respect to variations in the active user set and associated channels. That requires a relatively large number of low-rate (e.g., voice) users. In contrast, data traffic associated with internet services is bursty, and depends on the rates provided. Higher rates enable shorter bursts, but typically generate more interference due to higher transmit power. Hence without further coordination among users, the interference seen at a base station or mobile is likely to vary substantially as high-rate users enter and leave, degrading performance.[3]

---

[1]Although [86] stimulated much of the subsequent work on multiuser detection, the structure of the optimal detector for CDMA had been previously derived in [67].

[2]Although that work predates the development of mobile cellular and DSL, similar types of "multi-terminal" applications are mentioned, including multi-cable and diversity channels. The MIMO models in [1,45] were originally motivated by dual polarization radio transmission with crosstalk.

[3]In principle, fluctuations in interference in a CDMA system can be mitigated through the application of multiuser detection, although the challenges subsequently discussed still apply.

To mitigate interference associated with bursty users and high data rates, trans-missions can be scheduled over different time slots (scheduling intervals). Because data is delay-tolerant, that also allows the scheduler to select users with favorable channel conditions, thereby exploiting multiuser diversity [3,5,49]. Hence current (third-generation) cellular data systems typically rely on scheduling, as opposed to spreading (as in CDMA), to reduce interference. Emerging systems based on OFDMA provide further flexibility in allocating both time *and* frequency resources among requests to mitigate interference and exploit multiuser diversity. Interference mitigation techniques are potentially useful for OFDMA, although the interference originates from other cells (or sectors), so that associated channel conditions may be more difficult to estimate.

1.1.2 Mobile Cellular Challenges

Cellular systems pose several challenges that prevent a straightforward application of most multiuser detection techniques. This is especially true for uplink asynchronous CDMA with full spreading, or frequency reuse. The main difficulty is that for interfer-ence averaging, it is desirable to extend spreading sequences over many symbols, so that interference from a particular user is averaged over many (randomly chosen) signatures. In that way, the performance is not limited by the possibility of choosing a particular set of signatures with undesirable correlation properties. These "long" signatures, however, greatly complicate the implementation of a multiuser detector, which exploits properties of the particular set of assigned signatures for each symbol. That is, the structure of the multiuser detector must change from symbol to symbol, which may require excessive computation. Long signatures also prevent the application of standard adaptive filtering methods (used to equalize single-user channels), which require linear modulation with short signatures (i.e., the signature for a particular user is repeated from symbol to symbol).[4]

Another challenge for multiuser detection posed by mobile cellular systems is fading. The multiuser detector must track or adapt to channel variations in time and frequency caused by mobility. That includes channel variations associated with the interferers as well as the desired user. This may be feasible when the channel variations are slowly varying (e.g., over a few hundred symbols), or when the number of interfer-ers is relatively small, so that the number of channel parameters to track is manageable. However, in a CDMA cellular system there may be a large number of mobile users, several of which experience rapid fading. The inability to track all of these channel variations can significantly compromise performance. Furthermore, the complexity of the multiuser detector generally increases with the system size (number of users and processing gain). For this reason, application of multiuser detection to uplink CDMA has been mostly limited to relatively simple interference cancellation techniques.

---

[4]At a basic level, one might argue that interference averaging as a design objective for cellular systems runs contrary to a design based on multiuser detection. Namely, interference averaging is associated with a large number of weak interferers, whereas multiuser detection becomes most attractive when there are a small number of strong interferers.

One way to address the preceding challenges is to redesign the CDMA system with multiuser detection in mind. In addition to using short signatures, that generally means reducing the number of users, or transmitted information streams, which are jointly detected at the base station, and also slowing down the channel variations (fade rate). Reducing the number of users for joint detection can be accomplished by sub-dividing the channel resources in time or frequency. For example, the Time-Division Duplex (TDD) version of UMTS[5] is based on a combination of Time-Division Multiple Access (TDMA) and CDMA [30]. The uplink channel is divided into 10 msec frames with 15 time slots, and multiple users can be assigned to a particular time slot through the use of direct-sequence spread spectrum signaling (i.e., direct-sequence CDMA). For example, if the original CDMA system supports 75 users in a cell, then introducing 15 time slots means that the users can be divided into 15 groups, each containing five users, which are assigned to the different time slots. That reduces the required spreading (processing gain) by approximately a factor of 15, and eases the burden on a multiuser detector. Furthermore, the introduction of time slots increases the symbol rate, and hence decreases the fade rate (rate of channel variations normalized by the symbol rate). That simplifies the associated channel estimation.

As previously discussed, the desire to integrate voice services with different data services having variable Quality of Service (e.g., delay) requirements has motivated the trend towards OFDM and OFDMA. The channel is then divided into both time and frequency slots, and each slot is designated for a *single* user. Hence in general the transmitted signals are not spread. That obviates the need for multiuser detection on the uplink, although interference from other cells still limits performance. Hence interference suppression techniques may still be beneficial. Nevertheless, multiuser detection and interference suppression for OFDM systems are primarily focused on mitigating interference among multiple antennas, which creates MIMO channels across the sub-carriers.

1.1.3 Chapter Outline

In the next section we discuss the linear (matrix) channel model, which is the basis for the multiuser detection methods discussed in this book. We then give a brief overview of a few different multiuser detectors along with performance comparisons for a simple version of this channel model in which the mixing matrix has independent, identically distributed (*i.i.d.*) elements. This discussion is meant to provide some background and motivation for the topics discussed in subsequent chapters. We also discuss and compare some additional multiuser detection techniques. A comprehensive treatment of some of the detectors presented here (e.g., optimal and linear) is given in the book by Verdú [87].

---

[5]Universal Mobile Telecommunications System is an air interface standard for third generation cellular systems. The standard was approved in 1998, and consists of two modes corresponding to Frequency-Division Duplex (FDD) and TDD operation [30,57].

1.2 MATRIX CHANNEL MODEL

In its simplest and most general form the matrix channel model is given by:

$$\mathbf{y}(i) = \mathbf{M}(i)\mathbf{b}(i) + \mathbf{n}(i) \tag{1.1}$$

where $\mathbf{y}$ is an $N \times 1$ vector of received samples, $\mathbf{b}$ is a $K \times 1$ vector of transmitted symbols, $\mathbf{M}$ is an $N \times K$ *mixing* matrix, and $\mathbf{n}$ is an $N \times 1$ vector of additive noise samples. Also, i is the discrete time index. Hence this discrete-time model includes the combination of any filtering (analog or digital) at the transmitters and receivers. Assuming quadrature modulation, all of the variables in (1.1) are complex valued.

For synchronous CDMA the kth entry of $\mathbf{b}(i)$, denoted $b_k(i)$, is the ith symbol transmitted by user k, and the kth column of $\mathbf{M}$ is the signature for user k. In that case the elements of $\mathbf{M}$ are typically chosen from a discrete set, e.g., ± 1 for real codes, or $\pm 1 \pm j$ for complex signatures. In practice, the signature elements are often chosen randomly, although it is also desirable if the signatures (columns of $\mathbf{M}$) are orthogonal. To represent different powers across the users, we can write $\mathbf{M} = \mathbf{SA}$, where $\mathbf{S}$ is the signature matrix, and $\mathbf{A}$ is a diagonal matrix of amplitudes.

For asynchronous CDMA, the model (1.1) still applies, but with different interpretations for $\mathbf{b}(i)$ and $\mathbf{M}$. Namely, assuming each signature spans a single symbol interval, we can write:

$$\mathbf{y}(i) = \mathbf{M}_1\mathbf{b}(i-1) + \mathbf{M}_2\mathbf{b}(i) + \mathbf{M}_3\mathbf{b}(i+1) + \mathbf{n} = \mathbf{M}\underline{\mathbf{b}}(i) + \mathbf{n} \tag{1.2}$$

where the kth columns of $\mathbf{M}_1$, $\mathbf{M}_2$, and $\mathbf{M}_3$ are the corresponding segments of signatures (padded with zeros) associated with $b_k(i-1)$, $b_k(i)$, and $b_k(i+1)$, respectively, $\underline{\mathbf{b}}(i)$ contains $\mathbf{b}(i-1)$, $\mathbf{b}(i)$, and $\mathbf{b}(i+1)$ stacked on top of each other, and $\mathbf{M}$ is the corresponding block-diagonal matrix.

For a single-user link with K transmit antennas and N receive antennas, if $\mathbf{b}(i)$ contains K symbols (i.e., is $K \times 1$), then b_k in (1.1) is transmitted by the kth antenna, and the (n, k)th entry of $\mathbf{M}$ is the complex channel gain from the kth transmit antenna to the nth receive antenna. More generally, the number of transmitted symbols K' [dimension of $\mathbf{b}(i)$] may be less than the number of transmit antennas K, in which case an additional $K \times K'$ *precoding matrix* $\mathbf{V}$ is needed to map the symbols to antennas. That is, $\mathbf{M} = \mathbf{HV}$, where $\mathbf{H}$ is the $N \times K$ channel matrix. The channel gains are often modeled as *i.i.d.* random variables (typically complex Gaussian corresponding to flat Rayleigh fading). Other statistical models for the matrix $\mathbf{M}$, corresponding to correlated fading across antennas, are discussed in Chapter 6.

The *interference channel* can also be modeled by (1.1), where $\mathbf{y}$ corresponds to the signal at a particular receiver, which estimates a *subset* of $\mathbf{b}$. The remaining symbols are presumably transmitted to other receivers at different locations, and are treated as interference. That applies, for example, to the uplink where the estimated symbols are transmitted by users within the cell and the interfering symbols are transmitted by users in other cells.

The model (1.1) also applies to a single-input/single-output channel with inter-symbol interference (ISI). Namely, suppose a single transmitter transmits the sequence of symbols $\{b(i)\}$ through a linear, dispersive channel with impulse response:

$$h(-m), h(-m + 1), \ldots, h(0), h(1), \ldots, h(m),$$

where the length of the impulse response is assumed to be $2m + 1$, so that the output:

$$y(i) = \sum_{k=-m}^{m} h(k)b(i - k) \tag{1.3}$$

We can represent this in the form (1.1) by defining the vector of $N = 2n + 1$ received samples, corresponding to time i, as:

$$\mathbf{y}(i) = [y(i + n), \ldots, y(i), \ldots, y(i - n)]^T. \tag{1.4}$$

where $[\cdot]^T$ denotes transpose, and the vector of $K = 2(n + m) + 1$ transmitted symbols as $\mathbf{b}(i) = [b(i + n + m), b(i + n + m - 1), \ldots, b(i - n - m)]^T$. The channel matrix $\mathbf{M}$ is then the $N \times K$ Toeplitz matrix:

$$\mathbf{M} = \begin{bmatrix} h(-m) & \ldots & h(m) & 0 & \ldots & 0 \\ 0 & h(-m) & \ldots & h(m) & \ldots & 0 \\ & & \ddots & & \ddots & \\ 0 & \ldots & 0 & h(-m) & \ldots & h(m) \end{bmatrix}. \tag{1.5}$$

In practice, the receiver would presumably use $\mathbf{y}(i)$ to detect the subset of symbols $b(i + n - m), \ldots, b(i - n + m)$, which forms the reduced symbol vector $\bar{\mathbf{b}}(i)$. That is, the receiver estimates only the subset of transmitted symbols, which correspond to columns of $\mathbf{M}$ containing the *entire* impulse response. (That maximizes the received energy associated with each detected symbol.) Assuming that symbols are transmitted continuously, the trailing $2m$ symbols of $\mathbf{b}$, namely, $b(i - n + m - 1), \ldots, b(i - n - m)$, therefore experience intersymbol interference from the preceding vector $\mathbf{b}(i - 1)$, and the leading $2m$ symbols $b(i + n + m), \ldots,$ $b(i + n - m + 1)$ interfere with $\mathbf{b}(i + 1)$.

To model an OFDM system, the channel matrix $\mathbf{M}$ is first converted to a *circulant* matrix by setting the first m symbols of $\mathbf{b}$ equal to the trailing m symbols of $\bar{\mathbf{b}}(i)$, and the trailing m symbols of $\mathbf{b}$ equal to the first m symbols of $\bar{\mathbf{b}}(i)$ [i.e., $b(i + n + m) = b(i - n + 2m - 1), \ldots, b(i + n + 1) = b(i - n + m)$ and $b(i - n - 1) = b(i + n - m), \ldots, b(i - n - m) = b(i + n - 2m + 1)$]. If we rewrite (1.1) in terms of $\bar{\mathbf{b}}(i)$, i.e.:

$$\mathbf{y}(i) = \bar{\mathbf{M}}\bar{\mathbf{b}}(i) + \mathbf{n}(i), \tag{1.6}$$

then $\bar{\mathbf{M}}$ is circulant, i.e., its rows are cyclic shifts of the first row[6]

$$h(0), \ldots, h(m), 0, \ldots, 0, h(-m), \ldots, h(-1).$$

The matrix $\bar{\mathbf{M}}$ can be diagonalized by pre-multiplying $\bar{\mathbf{b}}$ by the inverse DFT matrix $\mathbf{W}$, and post-multiplying $\mathbf{y}(i)$ by the DFT matrix $\mathbf{W}^{\dagger}$. Hence the conclusion is that (1.1) models a single-user OFDM channel, where the mixing matrix $\mathbf{M}$ can be assumed to be diagonal. The diagonal entries are the complex channel gains across sub-channels.

Other interpretations of the model (1.1) that pertain to uplink and downlink CDMA with dispersive channels are discussed in Chapter 4. In those scenarios, the mixing matrix $\mathbf{M}$ is the product of a Toeplitz, circulant, or diagonal channel matrix, depending on the implementation, and a signature matrix. It becomes apparent that other network configurations with multiple users, multiple antennas, dispersive channels, and with or without spreading can be modeled by (1.1), where $\mathbf{M}$ and $\mathbf{b}(i)$ take on the appropriate forms. For purposes of the following overview, we will focus on the preceding interpretations of the model (1.1), which apply to synchronous CDMA, multi-antenna links (MIMO channels), and dispersive Single-Input/Single-Output (SISO) channels.

1.3 OPTIMAL MULTIUSER DETECTION

We now provide a brief overview of multiuser detection techniques that have been proposed for the matrix model (1.1), starting in this section with optimal detectors. The intent is to provide some background and points of reference for the advances and performance results presented in subsequent chapters. We also discuss some additional related techniques.

1.3.1 Maximum Likelihood (ML)

Referring to (1.1), the ML detector chooses the vector of estimated symbols as:

$$\hat{\mathbf{b}} = \arg \max_{\mathbf{b}} \Pr(\mathbf{y} \text{ received} \mid \mathbf{b} \text{ transmitted}) \tag{1.7}$$

where the dependence on i has been dropped for convenience. If the additive noise $\mathbf{n}$ is Gaussian, then this is equivalent to selecting:

$$\hat{\mathbf{b}} = \arg \min_{\mathbf{b}} \|\mathbf{y} - \mathbf{Mb}\| \tag{1.8}$$

where the norm is the regular Euclidean norm and the elements of $\mathbf{b}$ are constrained to be constellation points.

---

[6]Alternatively, the transmitted packet can be rearranged so that the first $2m$ symbols are the same as the trailing $2m$ symbols. The first $2m$ symbols are then the "cyclic prefix" [24, Ch. 12].

The introduction of the ML detector for CDMA in [86] essentially launched the field of multiuser detection. The main reason for this is that it was shown that the performance of the ML detector is insensitive to power variations among the users, unlike the matched filter detector. Hence this suggested that improving the CDMA receiver could relax the requirements on closed-loop power control, which was one of the most complex aspects of CDMA system design.

Although the ML detector can nearly eliminate the degradation in performance due to multiuser interference for low to moderate loads K/N (users per degree of freedom), it has two main drawbacks: complexity and required side information. Because **b** takes on discrete values, computing the estimate $\hat{\mathbf{b}}$ is an optimization over a discrete set, which is known to be computationally difficult. Specifically, the computation associated with known algorithms for determining the minimum in (1.1), assuming that **b** and **y** can be chosen arbitrarily, grows exponentially with K (the size of the vector **b**), corresponding to the exponential growth in the size of the set over which the minimization is taken. This is not a major issue if the dimension of **b** is small (say, $K<10$ with binary signaling), so that ML detection may be appropriate for single multiantenna links, or for a TDMA system in which there are relatively few co-channel users per time slot. However, the ML search complexity clearly becomes impractical for a CDMA system with much more than ten users per sector.

The second issue with the ML detector is that (1.1) implicitly assumes that **M** is known. Again, this is probably not a major issue for multi-antenna links, which experience slow fading, so that the channel can be accurately estimated. Even so, for some wireless applications, ML multiuser detection may not be as attractive as other simpler decision feedback techniques, to be discussed. The reason is that when combined with error control coding, the combined ML detector for the error control code concatenated with the channel becomes prohibitively complex. In that case, it is desirable to produce soft estimates of the transmitted symbols, i.e., with reliability information. This is further discussed later in this chapter and in Chapter 2.

1.3.2 Optimal (Maximum *a Posteriori*) Detection

The Maximum *a Posteriori* (MAP) detector selects:

$$\hat{\mathbf{b}} = \arg \max_{\mathbf{b}} \Pr(\mathbf{b} \text{ transmitted} \mid \mathbf{y} \text{ received}), \qquad (1.9)$$

which minimizes the probability of error. This is the same as the ML estimate if the symbols are equally likely. However, when combined with error control coding and iterative soft decoding, the decoder can pass reliability information to the multiuser detector in the form of the *a priori* distribution, or likelihood ratio for each transmitted symbol. Hence, in that scenario the MAP estimate generally differs from the ML estimate. Furthermore, the MAP detector itself computes soft estimates of each symbol (e.g., likelihood ratios), although the final (hard) estimates are obtained from the soft estimates by thresholding.

If the receiver detects a subset of the vector **b**, then the MAP estimate maximizes the corresponding marginal distribution [e.g., $\Pr(b_k \,|\, \mathbf{y})$ for a particular user k]. In general, this differs from the estimate in (1.9) and requires less computation.

For asynchronous CDMA, the multiuser MAP detector can be implemented using the standard *forward-backward algorithm* (e.g., see [87, Sec. 4.2]). The MAP detector suffers from the same drawbacks as the ML detector, namely, the complexity grows exponentially with the size of **b**, and it requires knowledge of **M**. However, in some applications where the system size is relatively small, the complexity may be manageable.

1.3.3 Sphere Decoder

Because of the high complexity of the optimal detector, it is generally desirable to incur some performance loss in order to simplify the receiver. That trade off motivates the linear and decision feedback detectors to be discussed. Other reduced-complexity detectors have been proposed that essentially approximate the optimal detector (e.g., see [51,70,75]), or rely on specific properties of the signatures (e.g., [64,66]).

Although the *worst-case* complexity of the ML search increases exponentially with K(e.g., see [87, Ch. 4]),[7] the *average* complexity, taking into account the model (1.1), may be much less. An example of a search algorithm with relatively low average complexity, assuming the Gaussian noise model, is the *sphere decoder*, discussed in [10,33].

To describe the sphere decoder, we first observe that if each component of **b** is selected from a rectangular (e.g., QAM) constellation, then each **b** corresponds to a point in a rectangular lattice. The ML estimate in (1.8) can then be computed by performing an exhaustive search over that lattice. To reduce the search complexity, the sphere decoder restricts **b** to lie in a hypersphere of radius r centered at **y**. If the hypersphere contains at least one lattice point, then this restricted search still gives the ML estimate.

Given r, an algorithm for finding all points in a lattice within the hypersphere defined by:

$$(\mathbf{b} - \hat{\mathbf{b}})^{\dagger}\mathbf{M}^{\dagger}\mathbf{M}(\mathbf{b} - \hat{\mathbf{b}}) \le r^2, \qquad (1.10)$$

where $\hat{\mathbf{b}}$ is the center of the sphere, was presented in [16]. Application of this sphere decoding algorithm to decoding a lattice code was presented in [11], and was subsequently proposed for MIMO channels in [10,33,90]. (See also [76], which considers a somewhat more general search constraint.) For convenience, we assume that all variables are real-valued (hence Hermitian transpose † will be replaced by transpose T). The extension to rectangular QAM constellations is obtained by stacking the real and imaginary components of each complex vector (see [10]). The extension to circular (PSK) constellations is discussed in [33].

---

[7]"Worst-case" refers to the scenario in which **y** and **b** can be chosen to maximize the search time, given a particular search algorithm.

We first note that if **b** is not constrained to take on discrete values (i.e., lie in a lattice), then the ML solution in (1.8) is given by $\hat{\mathbf{b}} = (\mathbf{M}^{\dagger}\mathbf{M})^{-1}\mathbf{M}^{\dagger}\mathbf{y}$. It is then easy to show that minimizing $(\mathbf{b} - \hat{\mathbf{b}})^{T}\mathbf{M}^{T}\mathbf{M}(\mathbf{b} - \hat{\mathbf{b}})$ over the discrete set of **b**'s is equivalent to minimizing $\|\mathbf{y} - \mathbf{Mb}\|^{2}$. The sphere decoder first performs the Cholesky factorization $\mathbf{M}^{T}\mathbf{M} = \mathbf{U}^{T}\mathbf{U}$, where $\mathbf{U}$ is upper triangular. Substituting for $\mathbf{M}^{T}\mathbf{M}$ in (1.10) then gives the equivalent condition:

$$\sum_{k=1}^{K} u_{kk}^{2} \left[b_k - \hat{b}_k + \sum_{j=k+1}^{K} \frac{u_{kj}}{u_{kk}}(b_j - \hat{b}_j) \right]^{2} \leq r^{2}. \tag{1.11}$$

This inequality is then used to generate upper and lower bounds for each component b_k, starting with b_K and repeating for b_{K-1}, b_{K-2}, and so forth. The lattice points within the sphere can then be enumerated by searching through all combinations of component values that lie within the bounding intervals.

More specifically, because all terms in (1.11) are positive, we can discard the first $K-1$ terms in the sum to write:

$$u_{KK}^{2}(b_K - \hat{b}_K)^{2} \leq r^{2}, \tag{1.12}$$

which implies:

$$\underline{Q}\left(\hat{b}_K - \frac{r}{u_{KK}} \right) \leq b_K \leq \bar{Q}\left(\hat{b}_K + \frac{r}{u_{KK}} \right), \tag{1.13}$$

where $\bar{Q}(\cdot)$ and $\underline{Q}(\cdot)$ denote, respectively, the largest and smallest constellation values closest to the argument. Once a candidate value for b_K is selected, this is used in a similar way to compute upper and lower bounds on b_{K-1}. Namely, discarding the first $K-2$ terms in the sum in (1.10) gives bounds on b_{K-1} in terms of the candidate value b_K. A candidate value for b_{K-1} is then selected, which is used with the candidate value b_K to compute bounds for b_{K-2}. A candidate value for b_{K-2} is then selected, and the algorithm continues in this way to select candidate values for b_{K-3}, b_{K-4}, and so forth.

If the bounding intervals computed for each b_k, $k = K, K-1, \ldots, 1$, contain at least one constellation value, then the preceding procedure finds a candidate value for **b**, i.e., the corresponding **b** satisfies (1.10). However, it can also happen that for some k', the bounding interval does not contain a constellation point, which implies that any **b** with the corresponding candidate values $b_{k'+1}, \ldots, b_K$ must lie outside the sphere defined by (1.10). In that case, the algorithm must backtrack to find alternative candidate values that lead to a candidate $b_{k'}$. That is, the algorithm chooses another candidate value $b_{k'+1}$, if available, and continues as before. If such a $b_{k'+1}$ is not available, or still leads to an empty bounding interval for $b_{k'}$, then the algorithm backtracks to index $k' + 2$. All candidate points **b** are then obtained by searching through all possible candidate values at each index k, either including the resulting

b if the procedure continues to $k = 1$, or terminating the procedure and backtracking when the bounding interval is empty.[8]

To reduce the search time for the ML estimate, each time a candidate point **b** is found, the radius r can be reduced to the corresponding value. That way, the preceding procedure terminates (i.e., continues to $k = 1$) only if a better estimate is found. The complexity of the algorithm is determined by the initial choice of r. As r increases, so does the number of candidate values, and hence the search time. As r decreases, the probability that the sphere does not contain any lattice points increases. In that case, the search would likely start over with a larger value of r. For the model (1.1) in which the noise is Gaussian, r can be chosen to ensure that this probability is sufficiently small (e.g., see [33,90]).

The expected complexity of the sphere decoder has been shown in [31] to be roughly cubic in the number of data streams K for constellations of interest.[9] This is a big improvement over the worst-case exponential complexity. This cubic complexity order is comparable to that of the decision feedback detectors to be discussed. Still, decision feedback generally has somewhat lower complexity overall, since it can make intermediate decisions for a subset of symbols (e.g., $b_{k+1}, \ldots, b_K$) before estimating b_k, thereby avoiding a search through all possible candidates.

The sphere decoder can be extended to MAP as well as ML detection with similar benefits. This extension and the application to iterative detection are discussed in [33,90]. Other related applications are discussed in [89].

1.4 LINEAR DETECTORS

Linear detectors are generally much less complex than the optimal detector, making them practical for most applications. A linear multiuser detector takes a linear combination of channel outputs to obtain an estimate of $\mathbf{b}(i)$. It is therefore represented by the $N \times K$ matrix C which forms the (soft) estimate:

$$\tilde{\mathbf{b}}(i) = \mathbf{C}^\dagger(i)\mathbf{y}(i) \tag{1.14}$$

As indicated, **C** may depend on the time index i, e.g., when estimated in an adaptive mode. In principle, **C** can be selected to minimize the probability of error; however, it is more convenient to select **C** to minimize the Mean Squared Error (MSE):

$$\xi = E\left[\|\mathbf{b}(i) - \tilde{\mathbf{b}}(i)\|^2\right], \tag{1.15}$$

---

[8]This algorithm has been described in terms of a tree in which the nodes at level k correspond to the candidate values of b_k [10]. The sphere decoder finds all leaves in the tree by tracing paths from different starting points.
[9]Strictly speaking, the expected complexity still grows exponentially with K when K becomes large [40]. This is because the radius r must grow with K to maintain a target detection probability. However, the complexity is more accurately described as being polynomial with K when K is sufficiently small, or the SNR is sufficiently large.

in which case the solution is:

$$C = M(M^\dagger M + \sigma^2 I_K)^{-1} \qquad (1.16)$$

where the noise samples in (1.1) are assumed to be uncorrelated with variance σ^2, and I_K is the $K \times K$ identity matrix.

It is apparent that a linear detector can detect any subset of the elements of b without having to estimate the remaining symbols. In other words, we have:

$$\tilde{b}_k(i) = c_k^\dagger(i) y(i) \qquad (1.17)$$

where c_k is the kth column of C. The linear multiuser detector is therefore a bank of uncoupled single-user (or SISO) filters, each of which suppresses interference. This makes the linear multiuser detector suitable for interference suppression applications, such as downlink cellular, where the objective is to detect a single transmitted data stream in the presence of multiuser interference.

Selecting c_k to minimize the MSE $\xi_k = E[|b_k - c_k^\dagger y|^2]$ gives:

$$c_k = R^{-1} m_k \qquad (1.18)$$

where $R = E[yy^\dagger] = MM^\dagger + \sigma^2 I_N$ is the received signal covariance matrix and m_k is the kth column of M. Hence we can write $C = R^{-1}M$, which is an alternate expression to that given in (1.16). The signal power is $E[|c_k^\dagger m_k|^2]$ and the noise plus interference power is obtained by subtracting this from the total power $E[|c_k^\dagger y|^2]$. Evaluating these for the optimal c_k gives the Signal-to-Interference Plus Noise Ratio (SINR):

$$\rho_k = m_k^\dagger R_k^{-1} m_k \qquad (1.19)$$

where $R_k = R - m_k m_k^\dagger$ is the interference-plus-noise covariance matrix for user k.

1.4.1 Comparison with Optimal Detection

For a particular mixing matrix M the SINR can be computed from (1.19). To gain additional insight, we can consider a random model for M, and evaluate the performance averaged over M. This approach is discussed in Chapter 4, where performance results are presented for MMSE filters with random inputs and different assumptions concerning the form of M. Here we assume that M contains *i.i.d.* elements, which applies to CDMA with randomly assigned signatures and equal received powers, and to a MIMO channel with random channel gains. The results in this section are taken primarily from [88].

Figure 1.1 shows plots of sum spectral efficiency versus load $\beta = K/N$ (users per dimension), and versus energy per bit over noise density (E_b/N_0). In Figure 1.1a, $E_b/N_0 = 5$ dB and in Figure 1.1b the load $\beta = 0.5$. The sum spectral efficiency is the Shannon capacity summed over all users, and normalized by the degrees of

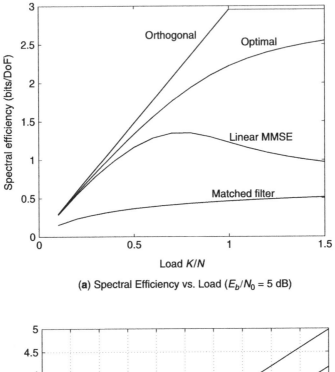

(a) Spectral Efficiency vs. Load ($E_b/N_0 = 5$ dB)

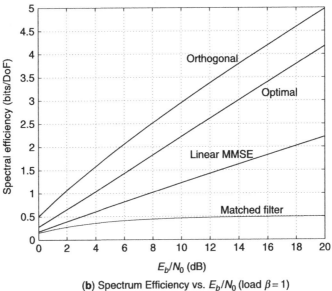

(b) Spectrum Efficiency vs. E_b/N_0 (load $\beta = 1$)

Figure 1.1. Spectral Efficiency vs. Load E_b/N_0 for the matched filter, linear **MMSE**, and optimal (**ML**) receivers with a random matrix model (1.1) (i.e., **M** has *i.i.d.* elements). The results are asymptotic as K and N tend to infinity with fixed load $\beta = K/N$. Also shown is the spectral efficiency with orthogonal signatures.

freedom N. With the optimal (ML) receiver the sum spectral efficiency is:

$$C = \frac{1}{N} \log \det(\mathbf{I}_N + \sigma^{-2}\mathbf{M}\mathbf{M}^{\dagger}). \tag{1.20}$$

in bits per channel use per degree of freedom. The Signal-to-Noise Ratio (SNR) is $1/\sigma^2$, so that $E_b/N_0 = 1/(C\sigma^2)$.

With the optimal linear receiver, treating the residual interference plus noise as Gaussian, the spectral efficiency is:

$$C_{\text{lin}} = \frac{1}{N} \sum_{k=1}^{K} \log(1 + \rho_k) \tag{1.21}$$

where ρ_k is the SINR defined in (1.19). Also shown in the figure is the spectral efficiency for the matched filter $\mathbf{c}_k = \mathbf{m}_k$.

The curves shown in Figure 1.1 correspond to the large system limit in which K and N tend to infinity with fixed K/N. Explicit expressions for C and C_{lin} in this limit are given in Chapter 4,[10] and it is shown there that this limit accurately predicts the performance of small finite-size systems averaged over $\mathbf{M}$. These results show that the spectral efficiency achieved by the optimal linear receiver is close to the maximum (achieved with the ML detector) at relatively small loads; however, there is a significant performance gap at large loads (e.g., $\beta > 1/2$), which increases with SNR. This is because as the load increases, the linear receiver has fewer degrees of freedom with which to suppress the interference, and becomes more closely aligned with the matched filter.

1.4.2 Properties of Linear Multiuser Detection

Linear detection is discussed in more detail in subsequent chapters. Here we state some of the important properties and limitations of linear multiuser detection.

1. It can be implemented as an adaptive digital filter, which requires little side information. Specifically, either a training sequence is needed at the start of each transmission, or the receiver must know the desired user's signature, channel, and associated timing. Amplitudes, phases, and signatures of interferers are not required for adaptation.

2. As the noise variance $\sigma^2 \to 0$, the MMSE detector completely suppresses a limited number of interferers. Specifically, an $N \times 1$ MMSE filter can completely suppress $N-1$ synchronous interfering streams. If the streams are asynchronous, then a digital filter spanning a single symbol interval can suppress $N/2$ interferers. By increasing the observation window, the filter can suppress up to $N-1$ interferers [47]; however, adaptation becomes more difficult.

---

[10]Both C and C_{lin} converge to *deterministic* limits.

3. The MMSE solution coherently combines all multipath within the window spanned by the filter. This is evident from (1.16), which shows that the MMSE filter is the concatenation of the matched filter matrix $\mathbf{M}$ with an interference suppression matrix. (The matched filter $\mathbf{M}$ combines the multipath.)

4. The performance of the MMSE detector degrades gracefully with the number of (equal power) interferers, although for K/N close to one, the performance of the MMSE solution approaches that of the matched filter.

These properties make linear MMSE estimation attractive for interference suppression applications where little or no information is available about the interferers. Still, mobile cellular applications present major challenges for implementation, as discussed in Section 1.1 Namely, current CDMA systems typically use long signatures, which span a large number of symbols. Although this precludes the use of adaptive filtering for interference suppression, an adaptive filter can still be used for *equalization*. This is important for downlink CDMA, where transmission of orthogonal signatures at the base station does not generally produce an orthogonal set at each mobile receiver. That is, a frequency-selective channel introduces inter-chip interference, which alters the received signature from what was transmitted. Hence, equalizing the channel at the mobile can restore orthogonality, thereby eliminating intra-cell interference [20].

A second challenge to adaptive interference suppression is filter estimation in the presence of time variations due to fading.[11] Various noncoherent techniques have been proposed to enhance performance in the presence of fading, including differential detection [36,39,52,82], generalized likelihood ratios [62,85,91,92], and subspace, or reduced-rank techniques, to be discussed in the next section and in Chapter 3. Those techniques help to make performance more robust with respect to the desired user's channel variations. However, fast time- and frequency-selective fading can substantially degrade the performance of adaptive interference suppression. This is primarily due to the fact that frequency-selective fading changes the interference subspace, making it difficult to track.[12]

The preceding discussion implies that the linear multiuser detector is most appropriate with a relatively small number of users in slow fading. In that case, it is insensitive to power variations across interferers. However, to accommodate more users and provide higher spectral efficiencies, it is necessary to use nonlinear techniques such as ML and decision-feedback, to be discussed.

1.5 REDUCED-RANK ESTIMATION

Estimation of a linear MMSE interference suppression filter generally requires a training sequence, which increases linearly with the filter length. This may present

---

[11]Bursty interference is also a challenge with asynchronous transmissions, e.g., see [34].

[12]This becomes important as the number of interferers increases. Namely, with a small number of interferers, the filter may be able to suppress individual multipath components from all interferers. However, that is no longer possible once the total number of multipath components summed over all users becomes comparable with the filter length.

a problem with long filters (e.g., associated with a large processing gain) due to the associated overhead, or due to channel variations, which may occur within the training period. In those scenarios, to reduce training it may be desirable to *approximate* the linear MMSE detector with another detector, which is suboptimal, but which can be rapidly estimated. An extreme example of this is the matched filter ($c_k = m_k$), which may require little or no training, but is not robust with respect to power variations among interferers.

A *reduced-rank* estimate of the linear MMSE filter selects c_k to minimize the MSE ξ_k subject to the constraint that c_k lies within a pre-defined subspace. Here we assume that the subspace is spanned by the columns of the $N \times D$ matrix S_k, where the subspace dimension $D < N$. We can therefore write $c_k = S_k v_k$, where v_k is the $D \times 1$ vector of combining coefficients. Selecting v_k to minimize the MSE ξ_k, given S_k, gives:

$$v_k = (S_k^\dagger R S_k)^{-1}(S_k^\dagger m_k) \tag{1.22}$$

and the associated SINR for the rank-D filter is:

$$\rho_{k,D} = m_k^\dagger S_k^\dagger (S_k^\dagger R_k S_k)^{-1} S_k m_k. \tag{1.23}$$

The advantage of reduced-rank estimation is that given S_k, the number of parameters to estimate is reduced from N to D. The amount of training overhead needed for filter estimation is therefore reduced in proportion. This also reduces the complexity associated with filter estimation. Of course, this generally comes at the cost of an increase in MSE.

Assuming that S_k has rank D (i.e., is full column rank), as D increases, the MSE decreases, and when $D = N$ the MSE becomes the MMSE. That is because c_k is no longer constrained to lie in a lower dimensional subspace.

Of course, the performance (MSE) depends on the choice of S_k. A few different methods for constructing S_k have been proposed in the literature, and offer different tradeoffs between performance and complexity. Some of these methods are described next. Here we discuss MMSE versions of the reduced-rank filters in which the subspace S_k is given, and the reduced-rank filter v_k is optimized. Adaptive versions of these filters can also be developed in which the subspace is estimated along with v_k (i.e., from the sample covariance matrix).

1.5.1 Subspaces from the Matched Filter

Some of the initial reduced-rank schemes for CDMA interference suppression were based on simple manipulations of the desired user's signature m_k. Namely, the columns of S_k can be taken to be cyclic shifts of the matched filter [53]. Another possibility, proposed in [68] is to let the columns of S_k be nonoverlapping segments of m_k padded with zeros. That corresponds to taking linear combinations of the outputs of "partial" matched filters, which are matched to different segments of the matched filter (so-called "partial despreading").

The main advantage of these schemes is that S_k is easily computed, and a modest performance improvement over the matched filter can be obtained with relatively small values of D. Furthermore, D can be varied between $D = 1$, corresponding to the matched filter, and $D = N$, corresponding to the MMSE filter. For partial despreading the corresponding MSE in dB varies almost linearly with D.

1.5.2 Eigen-Space Methods

Some of the initial work on reduced-rank filtering was motivated by the array (spatial) processing application [44]. The vector c_k in that scenario again consists of weights across antenna elements, as for the MIMO (multi-antenna) interpretation of the model (1.1), although the weights are generally adjusted to steer the spatial array in a given "look" direction (as opposed to being optimized for data detection). In those schemes, the columns of S_k (or some subset) are taken to be eigen-vectors of the covariance matrix R, or the interference-plus-noise covariance matrix $R_k = R - m_k m_k^\dagger$. In some situations (specifically, when the dimension of the signal space K is small relative to the number of dimensions N), these methods can achieve MMSE performance with a small rank D. Some different eigen-space methods are briefly described below.

1.5.2.1 Principal Components (PC) [65] In this scheme, the columns of S_k are the D eigen-vectors of R corresponding to the largest D eigen-values. In other words, the filter c_k is constrained to lie in the D-dimensional subspace, which contains the maximum signal energy. The main motivation for this approach is that the MMSE filter c_k must lie in the K-dimensional subspace spanned by the columns of M. Hence the PC reduced-rank filter is the MMSE filter when $D \geq K$. This can be advantageous when K is small relative to N, since a low-rank filter gives optimal performance.

For larger values of K, a low-rank PC filter can perform quite poorly (worse than the matched filter). This is especially true when the interfering signals are strong relative to the desired signal. In that case, the filter c_k is constrained to lie in a subspace that is primarily aligned with the interfering signatures, as opposed to the desired signature. That reduces the signal component at the filter output.

1.5.2.2 Generalized Side-lobe Canceller (GSC) [81] A block diagram of the GSC is shown in Figure 1.2. The received vector y is passed through a matched

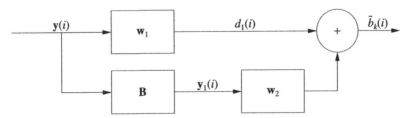

Figure 1.2. Block diagram of the Generalized Sidelobe Canceller (GSC).

filter $\mathbf{w}_1 = \mathbf{m}_k$, and also a *blocking matrix* $\mathbf{B}$, which *blocks* the desired signature, i.e., $\mathbf{B}\mathbf{w}_1 = 0$. For example, we can take $\mathbf{B} = \mathbf{I} - \mathbf{w}_1\mathbf{w}_1^\dagger$. The filter output is then $\tilde{b}_k = (\mathbf{w}_1 + \mathbf{B}\mathbf{w}_2)^\dagger\mathbf{y}$. In the full-rank version of the GSC the filter $\mathbf{w}_2$ is selected to minimize the MSE $E[|d_1 - \mathbf{w}_2^\dagger\mathbf{y}_1|^2]$, where $\mathbf{y}_1 = \mathbf{B}\mathbf{y}$ is the output of the blocking matrix. (Equivalently, we can choose $\mathbf{w}_2$ to minimize the output variance $E[|\tilde{b}_k|^2]$.) With this choice, the GSC is equivalent to a scaled MMSE filter. To obtain the MMSE filter, we simply multiply the output of the GSC by an appropriate constant. (Note that this scaling does not change the output SINR, but does affect the MSE.)

In the reduced-rank version of the GSC, $\mathbf{w}_2$ is constrained to lie in a lower-dimensional subspace. Specifically, we can constrain $\mathbf{w}_2$ to lie in an *interference* subspace, spanned by eigen-vectors of the input covariance matrix $\tilde{\mathbf{R}}_1 = E[\mathbf{y}_1\mathbf{y}_1^\dagger] = \mathbf{B}\mathbf{R}\mathbf{B}^\dagger$. To suppress the most interference, and thereby maximize the SINR, we choose the subspace associated with the largest eigen-values of $\tilde{\mathbf{R}}_1$.

The filter subspace for the rank-D GSC is therefore spanned by the desired signature $\mathbf{s}_k$ along with $D - 1$ eigen-vectors of $\tilde{\mathbf{R}}_1$. (Those are the columns of the projection matrix $\mathbf{S}_k$.) Note that this subspace is equivalent to that spanned by $\mathbf{m}_k$ and eigen-vectors of the interference-plus-noise covariance matrix $\mathbf{R}_k = \mathbf{R} - \mathbf{m}_k\mathbf{m}_k^\dagger$. Consequently, the reduced-rank GSC can perform no worse than the matched filter, and is optimal in the MMSE sense when D exceeds the dimension of the *interference* subspace. Substantial performance gains, relative to the matched filter, can be achieved with small values of D.

1.5.2.3 Cross-Spectral Method [25]

In this method, the columns of $\mathbf{S}_k$ are the eigen-vectors of $\mathbf{R}$, which minimize the MSE objective:

$$\xi_k = 1 - \mathbf{s}_k^\dagger\mathbf{R}^{-1}\mathbf{s}_k = 1 - \sum_{i=1}^{D}\frac{|\mathbf{s}_k^\dagger\mathbf{v}_i|^2}{\lambda_i} \tag{1.24}$$

where $\mathbf{v}_1, \ldots, \mathbf{v}_N$ and $\lambda_1, \ldots, \lambda_N$ are the eigen-vectors and corresponding eigen-values of $\mathbf{R}$, and the eigen-values are sorted in decreasing order. Hence, to minimize the MSE we choose the D eigen-vectors associated with the largest values of $|\mathbf{s}_k^\dagger\mathbf{v}_i|^2/\lambda_i$. This metric clearly takes into account how closely $\mathbf{v}_i$ is aligned with the desired signature $\mathbf{s}_k$. It is also interesting that this metric decreases with λ_i. Hence, unlike PC, a larger value of λ_i *decreases* the likelihood that the corresponding eigen-vector will be chosen. Although this method clearly must perform better than PC (with respect to MSE), it generally performs worse than the GSC for small D.

1.5.2.4 Comparison

All of the preceding methods achieve the MMSE if the filter rank D exceeds the dimension of the signal subspace K. This can be beneficial when $K \ll N$. Although for given D, principal components performs worse than the other eigen-space methods, the projection matrix $\mathbf{S}_k$ does not depend on the signature $\mathbf{m}_k$. This saves computation when computing multiple reduced-rank filters (e.g., for the multiple-access channel). Also, the performance of the other methods can be substantially degraded when $\mathbf{m}_k$ is not precisely known (e.g., due to channel impairments).

Computing $\mathbf{S}_k$ for eigen-space methods is generally more complex than for the other reduced-rank methods described here. This may be especially challenging in a scenario where the mixing matrix $\mathbf{M}$ is time-varying due to fading and changing interference. Eigen-space methods will be discussed further in Chapter 3 with more elaborate channel models that explicitly account for frequency-selective fading and multiple antennas.

1.5.3 Krylov Subspace Methods

Given the covariance matrix $\mathbf{R}$ and the signature $\mathbf{m}_k$, the associated D-dimensional Krylov subspace is defined as:[13]

$$\mathcal{K}_k = \mathrm{span}\{\mathbf{m}_k, \mathbf{R}\mathbf{m}_k, \mathbf{R}^2\mathbf{m}_k, \ldots, \mathbf{R}^{D-1}\mathbf{m}_k\}. \tag{1.25}$$

Constraining $\mathbf{c}_k$ to lie in this subspace gives the optimal reduced-rank filter in (1.22), which can be rewritten as:

$$\mathbf{v}_k = \mathbf{\Gamma}^{-1}\boldsymbol{\gamma} \tag{1.26}$$

where $\mathbf{\Gamma}$ is a $D \times D$ matrix with (l, m)th element $\gamma_{l+m-1} = \mathbf{m}_k^\dagger \mathbf{R}^{l+m-1} \mathbf{m}_k$, and $\boldsymbol{\gamma} = [\gamma_0, \gamma_1, \ldots, \gamma_{D-1}]^T$.

Attractive properties of this reduced-rank technique are: (i) There are efficient (low complexity) algorithms for computing the reduced-rank filter $\mathbf{v}_k$ given by (1.26), and (ii) the rank needed to achieve near-MMSE (full-rank) performance is typically quite small. In fact, full-rank performance can often be achieved with a D much smaller than the dimension of the signal space.

This type of reduced-rank filter has taken a few different forms in the literature[14], and is closely related to the Lancosz and conjugate gradient algorithms for solving sets of linear equations. An overview of those techniques is given in [12]. Here we emphasize the connection to the Multi-Stage Wiener Filter (MSWF), presented in [26]. We also present a rank-recursive algorithm, which recursively generates filters of increasing rank. That is useful for determining an appropriate filter rank in an adaptive mode. A further discussion of iterative methods for linear interference suppression is given in Chapter 2.

1.5.3.1 Multi-Stage Wiener Filter (MSWF) To explain the MSWF, we refer to the block diagram of the GSC in Figure 1.2. We observe that $\mathbf{w}_2$ in this block diagram can be viewed as an estimation filter for d_1 given the input $\mathbf{y}_1$. Replacing $\mathbf{w}_2$ by the corresponding MMSE filter gives the full-rank (MMSE) solution for the overall filter $\mathbf{c}_k$.

---

[13]This subspace does not change if $\mathbf{R}$ is replaced by the interference-plus-noise covariance matrix $\mathbf{R}_k$.

[14]For example, see [7,26,27,29,54,55,58,93] and the discussion of iterative linear techniques in Chapter 2.

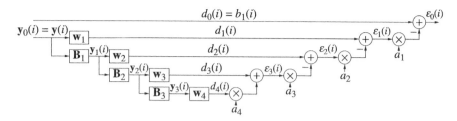

Figure 1.3. Block diagram of the Multi-Stage Wiener Filter (MSWF).

We can decompose $\mathbf{w}_2$ in Figure 1.2 in the same way as the GSC decomposition of $\mathbf{c}_k$. This is illustrated in Figure 1.3, where $\mathbf{w}_1$ and $\mathbf{B}_1$ are the same as in Figure 1.2. Also, Figure 1.3 shows the filter output error $\varepsilon_0(i) = b_k(i) - \mathbf{c}_k^\dagger \mathbf{y}(i)$. (The scale factors a_k, $k = 1, \ldots, D$, are needed to obtain the MMSE estimate, as discussed in Section 1.5.2.2.) Since $\mathbf{w}_2$ estimates d_1 given $\mathbf{y}_1$, the associated "matched filter" is defined as $\mathbf{w}_2 = E[d_1^* \mathbf{y}_1]$, and the blocking matrix $\mathbf{B}_2$ in Figure 1.3 is chosen so that $\mathbf{B}_2 \mathbf{w}_2 = \mathbf{0}$.

Similarly, the filter $\mathbf{w}_3$, which follows $\mathbf{B}_2$ in Figure 1.3, can be viewed as estimating $d_2(i)$ from $\mathbf{y}_2(i)$. Selecting $\mathbf{w}_3$ as the corresponding MMSE filter gives the full-rank MMSE filter. However, Figure 1.3 again shows a GSC decomposition of the filter after $\mathbf{B}_2$. Namely, we take $\mathbf{w}_3 = E[d_2^* \mathbf{y}_2(i)]$, which is the corresponding matched filter, and $\mathbf{B}_3 \mathbf{w}_3 = \mathbf{0}$. Iterating in this fashion, replacing each filter $\mathbf{w}_k$ with the corresponding GSC structure, expands the MMSE filter in terms of the N "matched filters" $\mathbf{w}_1, \ldots, \mathbf{w}_N$. Truncating this process after $D < N$ iterations, setting $\mathbf{w}_D = E[d_{D-1}^* \mathbf{y}_{D-1}]$, gives a MSWF with rank D. Figure 1.3 shows an example with $D = 4$. (If $\mathbf{w}_4$ is instead the MMSE filter for estimating $d_3(i)$ given $\mathbf{y}_3(i)$, then we obtain the full-rank MMSE filter.)

From this description it is straightforward to show that the reduced-rank MSWF lies in the subspace spanned by the basis vectors $\mathbf{w}_1, \mathbf{B}_1 \mathbf{w}_2, \mathbf{B}_1 \mathbf{B}_2 \mathbf{w}_3, \ldots,$ $\left(\prod_{i=1}^{D-1} \mathbf{B}_i \right) \mathbf{w}_D$. (These are the columns of the projection matrix $\mathbf{S}_k$.) Furthermore, it can be shown that this basis is an orthogonal basis for the D-dimensional Krylov subspace (1.25) [38]. The combining coefficients in the MSWF are determined by the scale factors $a_0, \ldots, a_{D-1}$ shown in Figure 1.3. The MSWF recursions are given by (1.27)–(1.33).

ALGORITHM 1: Recursions for Multi-stage Wiener Filter (MSWF)

Initialization:

$$d_0(i) = b_k(i), \qquad \mathbf{y}_0(i) = \mathbf{y}(i) \tag{1.27}$$

For $n = 1, \ldots, D$ (*Forward Recursion*):

$$\mathbf{w}_n = E[d_{n-1}^* \mathbf{y}_{n-1}(i)] / \|E[d_{n-1}^* \mathbf{y}_{n-1}]\| \qquad (1.28)$$

$$d_n(i) = \mathbf{w}_n^\dagger \mathbf{y}_{n-1}(i) \qquad (1.29)$$

$$\mathbf{B}_n = \text{null}(\mathbf{w}_n) \qquad (\text{if } n < D) \qquad (1.30)$$

$$\mathbf{y}_n = \mathbf{B}_n^\dagger \mathbf{y}_{n-1} \qquad (\text{if } n < D) \qquad (1.31)$$

Decrement $n = D, \ldots, 1$ (*Backward Recursion*):

$$a_n = E[d_{n-1}^*(i)\varepsilon_n(i)] / E[|\varepsilon_n(i)|^2] \qquad (1.32)$$

$$\varepsilon_{n-1}(i) = d_{n-1}(i) - a_n^* \varepsilon_n(i) \qquad (1.33)$$

Where $\varepsilon_D(i) = d_D(i)$. The estimate of d_0 is $a_1^* \varepsilon_1$.

1.5.3.2 Rank-Recursive (Conjugate Gradient) Algorithm An alternative method for computing the Krylov reduced-rank filter can be obtained by expressing the rank D filter in terms of the rank $D - 1$ filter. To state the algorithm, it will be convenient to denote the basis vectors for the subspace associated with the rank D MSWF as the columns of $\mathbf{V}_D = [\mathbf{v}_1, \ldots, \mathbf{v}_D]$. (That is, $\mathbf{V}_D = \mathbf{S}_k$ for user k.) The MSWF has the property that the projected input $\tilde{\mathbf{y}}(i) = \mathbf{V}_D^\dagger \mathbf{y}(i)$ has a real-valued, *tri-diagonal* covariance matrix $\tilde{\mathbf{R}}_D = \mathbf{V}_D^\dagger \mathbf{R} \mathbf{V}_D$ [26]. Let $\tilde{\mathbf{R}}_D$ have diagonal elements $\alpha_1, \ldots, \alpha_D$, and off-diagonal elements $\delta_2, \ldots, \delta_D$. That is, δ_n is the $(n, n - 1)$st, or $(n - 1, n)$th element of $\tilde{\mathbf{R}}_D$. (All other elements are zero.)

The conjugate gradient (or Lanczos) version of the Krylov reduced-rank filter is given by the recursions (1.35)–(1.45) (for derivations see [43] or [12]). The reduced-rank filter is $\mathbf{c} = \mathbf{V}_D \tilde{\mathbf{c}}_D$, where $\tilde{\mathbf{c}}_D$ is the $D \times 1$ vector of combining coefficients, and $\mathbf{v}_{D,n}$ denotes the nth element of $\mathbf{v}_D$. This algorithm can be used to increment the filter rank to any desired rank starting with the rank-one (matched) filter. The MMSE for the rank-D filter is given by:

$$\xi_D = 1 - \mathbf{m}_k^\dagger \mathbf{V}_D \tilde{\mathbf{R}}_D^{-1} \mathbf{V}_D^\dagger \mathbf{m}_k = 1 - \beta_1 \tilde{c}_{D,1} \qquad (1.34)$$

The recursions (1.35)–(1.45) are an efficient way to compute the reduced-rank filter. What's more, as will be discussed further, this method is easily combined with a stopping rule, which determines when to terminate the recursions (i.e., selects the desired filter rank). Of course, the algorithm must terminate if $\beta_D = 0$, in which case the filter rank is taken to be $D - 1$ (assuming $\beta_{D-1} \neq 0$).

1.5.3.3 Performance For a particular mixing matrix $\mathbf{M}$ the reduced-rank SINR as a function of D can be computed from (1.23). To gain additional insight, we again consider a random model for $\mathbf{M}$ (i.e., $\mathbf{M}$ has *i.i.d.* elements). The SINR can be explicitly evaluated in the large system limit as $K \to \infty$ and $N \to \infty$ with fixed ratio $\beta = K/N$ [38]. In this limit the SINR for the rank-D filter converges to a *deterministic* value $\bar{\rho}_D$ (independent of the user), which is given by:

$$\bar{\rho}_D = \frac{1}{\sigma^2 + \dfrac{\beta}{1 + \bar{\rho}_{D-1}}} \tag{1.46}$$

with $\bar{\rho}_0 = 0$. Hence, the SINR in this limit converges to the full-rank MMSE as a *continued fraction*. It takes only a few iterations for ρ_D to converge to the full-rank MMSE (e.g., $D \leq 8$), independent of the load β and the noise level σ^2.

ALGORITHM 2: Rank-Recursive (Conjugate Gradient) Version of a Krylov Subspace Reduced-Rank Filter

---

Initialization (rank-one filter):

$$\beta_1 = \|\mathbf{m}_k\|, \quad \mathbf{v}_1 = \mathbf{m}_k/\beta_1 \tag{1.35}$$
$$\mathbf{t}_1 = \mathbf{R}\mathbf{v}_1 \tag{1.36}$$
$$\alpha_1 = \mathbf{v}_1^\dagger \mathbf{t}_1 \tag{1.37}$$
$$\tilde{c}_1 = \beta_1/\alpha_1, \quad \mathbf{q}_1 = [] \; (null) \tag{1.38}$$

Increment rank D and perform the following recursions until the stopping rule is satisfied:

$$\mathbf{u}_D = \mathbf{t}_{D-1} - \alpha_{D-1}\mathbf{v}_{D-1} - \beta_{D-1}\mathbf{v}_{D-2} \tag{1.39}$$
$$\beta_D = \|\mathbf{u}_D\| \tag{1.40}$$
$$\mathbf{v}_D = \mathbf{u}_D/\beta_D \tag{1.41}$$
$$\mathbf{t}_D = \mathbf{R}\mathbf{v}_D \tag{1.42}$$
$$\alpha_D = \mathbf{v}_D^\dagger \mathbf{t}_i \tag{1.43}$$
$$\mathbf{q}_D = \begin{bmatrix} \mathbf{q}_{D-1} \\ -1 \end{bmatrix} \frac{\beta_D}{\beta_{D-1}\mathbf{q}_{D-1,D-2} - \alpha_{D-1}} \tag{1.44}$$
$$\tilde{\mathbf{c}}_D = \begin{bmatrix} \tilde{\mathbf{c}}_{D-1} \\ 0 \end{bmatrix} + \frac{\beta_D c_{D-1,D-1}}{\alpha_D - \beta_D \mathbf{q}_{D,D-1}} \begin{bmatrix} \mathbf{q}_D \\ -1 \end{bmatrix} \tag{1.45}$$

We emphasize that to obtain the result in (1.46), the rank D is *fixed* while both K and N become large. Hence the D needed to achieve full-rank performance[15] does

---

[15]More precisely, "full-rank" performance refers to a target MSE, which can be arbitrary close but not equal to the full-rank MSE.

not depend on the dimension of the signal subspace, unlike the eigen-space methods. The extension of this analysis to non-uniform power distributions over the transmitted symbols is presented in [48,50]. Related results, which apply to iterative methods for computing the linear MMSE filter, are discussed in [78]. (See also Chapter 2.)

If we instead consider the single-user intersymbol interference model where $\mathbf{M}$ is given by (1.5), then the performance of the reduced-rank filter depends on the channel impulse response. Although there is currently no expression for the SINR analogous to (1.46) in this case, the SINR can be computed from the moments of the squared magnitude of the channel frequency response [74]. Again the full-rank performance can typically be achieved with a relatively low-rank filter.

1.5.3.4 Adaptive Rank Selection   In practice, the parameters of the reduced-rank filter can be estimated with an adaptive algorithm. For example, we can directly estimate the matrix $\mathbf{S}_k$ associated with the Krylov space by replacing $\mathbf{R}$ with the sample covariance matrix $\hat{\mathbf{R}} = \sum_{i=0}^{N} \mathbf{y}(i)\mathbf{y}^{\dagger}(i)$. An estimate for the signature $\mathbf{m}_k$ can also be obtained given a training sequence. For the MSWF, the filters $\mathbf{w}_0, \ldots, \mathbf{w}_{D-1}$ and the combining coefficients $a_0, \ldots, a_{D-1}$ can be estimated by similar types of time-averages (e.g., see [35]).

With limited training or data samples for filter estimation, the choice of filter rank becomes important. Decreasing the rank means fewer parameters must be estimated, which increases estimation accuracy (assuming a fixed training length), but reduces the degrees of freedom available to suppress interference. Choosing the rank is also complicated by the fact that as D increases, the basis vectors spanning $\mathcal{K}_k$ in (1.25) become nearly linearly dependent, even when D is relatively small (i.e., $<K$). Hence computing $\mathbf{v}_k$ via (1.22) becomes ill-conditioned.

As mentioned earlier, the rank-recursive method for computing the reduced-rank filter presented in Section 1.5.3.2 is easily combined with a rank selection, or stopping criterion. For a nonadaptive filter in which $\mathbf{R}$ and the steering vector $\mathbf{m}_k$ are known *a priori*, the MSE can be computed directly from (1.34). That could be used to select D according to a target performance objective. For an adaptive reduced-rank filter in which $\mathbf{R}$ and $\mathbf{m}_k$ are unknown *a priori*, the following stopping criteria can be used.

Estimated MSE.   Given a training sequence $\{b_k(i)\}$, which is known to the receiver, the filter error for a rank D filter is $e_{k,D}(i) = b_k(i) - \tilde{\mathbf{c}}_D^{\dagger} \mathbf{V}_D^{\dagger} \mathbf{y}(i)$, and the MSE can be estimated as $\hat{\xi}_D = \sum_i |e_{k,D}(i)|^2$, where the summation index corresponds to the training symbols. This estimate, however, cannot be used to select the rank D, since $\tilde{\mathbf{c}}_D$ is computed from the data in the sum, so that the corresponding estimated MSE decreases monotonically with D. The minimizing rank would therefore be the largest, which can give relatively poor performance. One way to circumvent this problem is to compute the error over non-training symbols, for example, using an estimate for the symbol (e.g., the sign of the filter output if $b_k(i)$ is ± 1). To determine the optimal rank, the cost function must be computed for all possible ranks.

A more efficient method for selecting the termination rank is to use the *relative incremental* estimated MSE as a metric [73, Ch. 3]:

$$f(D) = \frac{\hat{\xi}_{D-1} - \hat{\xi}_D}{\hat{\xi}_{D-1}}. \tag{1.47}$$

That is, we terminate the rank-update algorithm when $f(D) < \varepsilon$, where ε is a stopping threshold (e.g., between 0.01 and 0.1). The metric $f(D)$ can be efficiently computed in terms of the algorithm parameters, namely:

$$\hat{\xi}_{D-1} - \hat{\xi}_D = \frac{1}{\Delta_D \Delta_{D-1}} \prod_{i=1}^{D} \beta_i^2 \tag{1.48}$$

where $\Delta_D = \det \tilde{\mathbf{R}}_D$, and satisfies:

$$\Delta_D = \alpha_D \Delta_{D-1} - \beta_D^2 \Delta_{D-2}. \tag{1.49}$$

The preceding recursions are initialized as:

$$\hat{\xi}_1 = \frac{1}{T} \sum_{i=1}^{T} |b_k(i)|^2 - \frac{\beta_1^2}{\alpha_1} \tag{1.50}$$

$$\Delta_0 = 1 \qquad \Delta_1 = \alpha_1. \tag{1.51}$$

Projection Norm. As the dimension of the Krylov subspace increases, the basis vectors become nearly linearly dependent. As a consequence, the matrix $\tilde{\mathbf{R}} = \mathbf{V}_D^\dagger \mathbf{R} \mathbf{V}_D$ becomes ill-conditioned. One way to prevent this is to increment the rank from D to $D + 1$ only if the new basis vector $\mathbf{R}^D \mathbf{m}_k$ is far enough outside the space spanned by $\mathcal{K}_D$. This is equivalent to ensuring that the norm of the orthogonal projection of $\mathbf{R}^D \mathbf{m}_k$ onto $\mathcal{K}_D$ is sufficiently large (i.e., greater than some stopping threshold ε). This norm can be interpreted as a measure of the "new information" provided by the new basis vector. An efficient method for computing this norm in terms of the β_i parameters is given in [73, Ch. 3]. This method tends to overestimate the rank with the best performance.

1.5.4 Performance Comparison

To illustrate some of the previous discussion, Figure 1.4 (taken from [35]) shows performance (error rate) versus filter rank for the GSC, cross-spectral filter, the MSWF, and a matched filter. Here a CDMA scenario is assumed in which asynchronous users transmit streams of binary symbols with randomly chosen signatures. The filter is estimated from 200 training symbols (i.e., using the sample covariance and an estimate for $\mathbf{m}_k$), the signature length is 128, and there are 42 users with a log-normal

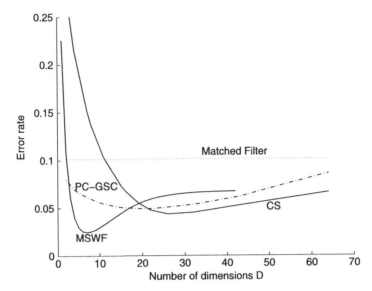

Figure 1.4. Error rate versus filter rank for reduced-rank algorithms after training with 200 symbols (taken from [35]). The signature length is 128, there are 42 asynchronous users, and the received powers are log-normal with standard deviation 6 dB. The desired user's SNR is 10 dB.

distribution over the received power (variance 6 dB). The results show that the optimal rank for the MSWF is quite small (around eight), and that the associated performance is better than that of the other (more complicated) eigen-space methods.

1.6 DECISION-FEEDBACK DETECTION

One of the main drawbacks of linear multiuser detection is that the performance degrades significantly when the dimension of the space containing strong interference approaches the filter length N. This is illustrated in Figure 1.1 which shows that a significantly larger spectral efficiency can be achieved with an optimal (nonlinear) receiver when the load β becomes large. However, the optimal receiver may be prohibitively complex when K and N are moderate to large. This motivates the study of relatively simple nonlinear decision feedback (DF), or interference cancellation schemes, which can perform better than the linear receiver.

A block diagram of a DF multiuser detector (DFD) is shown in Figure 1.5. It consists of an $N \times K$ linear feedforward filter (matrix) $\mathbf{F}$ followed by a feedback loop with a $K \times K$ feedback filter (matrix) $\mathbf{B}$. The input to the decision device is:

$$\mathbf{x}(i) = \mathbf{F}^{\dagger}\mathbf{y}(i) - \mathbf{B}^{\dagger}\hat{\mathbf{b}}(i) \tag{1.52}$$

where $\hat{\mathbf{b}}(i)$ is the $K \times 1$ estimate of the transmitted symbol vector $\mathbf{b}(i)$. The purpose of the feedback filter is therefore to *cancel* interference associated with other transmitted symbols. Typically the decision device is nonlinear (i.e., $\hat{\mathbf{b}}$ is a nonlinear

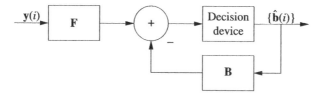

Figure 1.5. Block diagram of a decision feedback detector.

function of **b**), which makes the receiver nonlinear.[16] We assume that **B** has zeros along the diagonal to avoid cancelling the desired symbol.

Successive DF means that the users are detected in successive order $k = 1, \ldots, K$, so that the estimate $\hat{b}_k(i)$ depends only upon the *previous* estimated symbols $\hat{b}_1(i), \ldots, \hat{b}_{k-1}(i)$. The feedback filter **B** in the Successive-DFD (S-DFD) is therefore lower diagonal. In contrast, *parallel* DF means that estimates of *all* other symbols $\hat{b}_m(i)$, $m \neq k$, are available for cancellation. In that case, **B** is generally full except for zeros along the diagonal. The P-DFD is generally associated with iterative cancellation methods, where the estimated vector $\hat{b}(i)$, which is fed back for cancellation, is computed in the preceding iteration.

Successive interference cancellation has its origins in earlier work on the Gaussian multiple access channel. Namely, successive decoding and interference cancellation achieves the associated capacity region [21], [9, Ch. 14]. The basic idea is that by coding a user's stream at a rate just below capacity, treating the other users as noise, that user's data can be reliably decoded and then used to cancel the associated interference to other users. The S-DFD was proposed for the CDMA application in [14,15], and is related to earlier work on decision-feedback equalization for MIMO channels [13,45]. It is shown in [28] that the S-DFD achieves the sum capacity at corner points of the CDMA capacity region. (Hence the sum spectral efficiency of the S-DFD receiver coincides with the curve in Figure 1.1 corresponding to the optimal detector.) The S-DFD has been proposed and analyzed for multi-antenna applications in [17,18].

Following the discussion in [94], we can jointly optimize the filters **F** and **B**. Our performance objective is again MSE, where the error is defined as:

$$\mathbf{e}_{dfd}(i) = \mathbf{b}(i) - \mathbf{x}(i), \qquad (1.53)$$

where **x** is defined by (1.52). In general, the decision device either makes a hard decision on **b** (e.g., ± 1 for binary signaling), or outputs a soft estimate of **b** (e.g., the conditional expectation $E[\mathbf{b} \,|\, \mathbf{y}]$), which uses knowledge about the prior distribution (likelihoods of the different hypotheses) and the noise statistics. Soft estimates are especially desirable when combined with error control coding, and will be discussed in more detail in Chapter 2. For purposes of optimizing the filters **F** and **B**, here we make the simplifying assumption of perfect feedback, i.e., $\hat{\mathbf{b}} = \mathbf{b}$,

---

[16]A *linear* decision device [e.g., taking $\hat{b}(i) = \mathbf{x}(i)$] leads to iterative solution methods for the optimal *linear* receiver, which are discussed in Chapter 2.

corresponding to high SNRs. Of course, in practice the feedback contains errors, which can propagate across users. That is much more difficult to analyze, and furthermore, the ideal feedback assumption gives significant insight into the roles of $\mathbf{F}$ and $\mathbf{B}$ and associated performance.

We divide the users into two sets:

$$\mathcal{D} = \{ j : \hat{b}_j \text{ is fed back} \} \tag{1.54}$$

$$\mathcal{U} = \{ j : j \notin \mathcal{D} \} \tag{1.55}$$

i.e., "detected" and "undetected" users. In general, these two sets depend on the particular user being detected. We also define the $N \times |\mathcal{D}|$ matrix of signatures for the detected users as $\mathbf{M}_{\mathcal{D}}$, and similarly, $\mathbf{M}_{\mathcal{U}}$ contains the signatures for $k \in \mathcal{U}$.

We observe that the MSE for user k, $\varepsilon_{\mathrm{dfd},k} = E[|\mathbf{e}_{\mathrm{dfd},k}|^2]$, depends only on $\mathbf{F}_k$ and $\mathbf{B}_k$, i.e., the kth column of $\mathbf{F}$ and $\mathbf{B}$, respectively. Optimizing $\mathbf{F}_k$ and $\mathbf{B}_k$ gives:

$$\mathbf{F}_k = \mathbf{R}_{\mathcal{U}}^{-1} \mathbf{p}_k, \qquad \mathbf{B}_k = \mathbf{P}_{\mathcal{D}}^{\mathrm{H}} \mathbf{F}_k \tag{1.56}$$

where:

$$\mathbf{R}_{\mathcal{U}} \triangleq \mathbf{M}_{\mathcal{U}} \mathbf{M}_{\mathcal{U}}^{\mathrm{H}} + \sigma^2 \mathbf{I} \tag{1.57}$$

$$= \mathbf{R} - \mathbf{M}_{\mathcal{D}} \mathbf{M}_{\mathcal{D}}^{\dagger} \tag{1.58}$$

is the covariance matrix for the undetected users.

The feedforward filter for user k is therefore the linear MMSE filter assuming that only users in $\mathcal{U}$ are present. The resulting MMSE is:

$$\mathcal{E}_{\mathrm{dfd},kk} = 1 - \mathbf{m}_k^{\dagger} \mathbf{R}_{\mathcal{U}}^{-1} \mathbf{m}_k, \tag{1.59}$$

where $\mathcal{E}_{\mathrm{dfd}} = E[\mathbf{e}_{\mathrm{dfd}} \mathbf{e}_{\mathrm{dfd}}^{\dagger}]$ is the error covariance matrix, which has the same form as the MMSE for a linear receiver, except that $\mathbf{R}$ is replaced by $\mathbf{R}_{\mathcal{U}}$. In the absence of error propagation, interference from all users in set $\mathcal{D}$ is therefore eliminated, and user k is affected only by the users in set $\mathcal{U}$. That is, the MMSE DFD cancels interference from the users in set $\mathcal{D}$, while suppressing interference from users in $\mathcal{U}$ in an MMSE sense.[17]

An alternative interpretation for the DFD filters can be obtained by minimizing the sum MSE:

$$\xi_{\mathrm{dfd}} = \sum_{k=1}^{K} E\left[\left| \mathbf{e}_{\mathrm{dfd},k} \right|^2 \right] = tr[\mathcal{E}_{\mathrm{dfd}}], \tag{1.60}$$

with respect to $\mathbf{F}$ and $\mathbf{B}$. Minimizing with respect to $\mathbf{F}$ with fixed $\mathbf{B}$ gives:

$$\mathbf{F} = \mathbf{F}_{\mathrm{lin}}(\mathbf{I} + \mathbf{B}). \tag{1.61}$$

where $\mathbf{F}_{\mathrm{lin}} = \mathbf{R}^{-1} \mathbf{M}$ is the MMSE linear filter. The feedforward filter is therefore a concatenation of the linear MMSE filter with the filter $\mathbf{I} + \mathbf{B}$, as shown in Figure 1.6.

---

[17]This result is analogous to earlier results for decision-feedback, or "data-aided," equalization [22,56]. It is shown there that conditioned on perfect feedback, the feedback filter cancels the associated intersymbol interference (ISI), and the feedforward filter suppresses the remaining ISI.

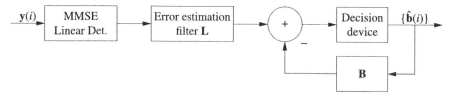

Figure 1.6. DFD with an error estimation filter.

From (1.52), (1.53), and (1.61) the error covariance matrix can be expressed as:

$$\mathcal{E}_{\text{dfd}} = (\mathbf{I} + \mathbf{B})^{\dagger} \mathcal{E}_{\text{lin}} (\mathbf{I} + \mathbf{B}), \tag{1.62}$$

where $\mathcal{E}_{\text{lin}}$ is the error covariance matrix for the *linear* MMSE filter. That is:

$$\mathcal{E}_{\text{lin}} = E[\mathbf{e}_{\text{lin}} \mathbf{e}_{\text{lin}}^{\dagger}] = \mathbf{I} - \mathbf{M}^{\dagger} \mathbf{R}^{-1} \mathbf{M} = (\sigma^{-2} \mathbf{I} + \mathbf{M}^{\dagger} \mathbf{M})^{-1} \tag{1.63}$$

where $\mathbf{e}_{\text{lin}} = \mathbf{b}(i) - \mathbf{F}_{\text{lin}}^{\dagger} \mathbf{y}(i)$ is the error for the linear MMSE filter, and the last equality follows from the matrix inversion lemma.

We therefore wish to determine the $\mathbf{B}_k$, which minimizes $\text{tr}[\mathcal{E}_{\text{dfd}}]$ (equivalently, $[\mathcal{E}_{\text{dfd}}]_{kk}$), for a given set $\mathcal{D}$. Let $\mathbf{B}_{k,\mathcal{D}}$ be the $D \times 1$ vector containing the elements of $\mathbf{B}_k$ with indices in $\mathcal{D}$. By definition, all other elements of $\mathbf{B}_k$ are zero. Minimizing (1.62) over $\mathbf{B}_k$ gives:

$$\mathbf{B}_{k,\mathcal{D}} = (\mathbf{I} - \mathbf{M}_{\mathcal{D}}^{\dagger} \mathbf{R}^{-1} \mathbf{M}_{\mathcal{D}})^{-1} \mathbf{M}_{\mathcal{D}}^{\dagger} \mathbf{R}^{-1} \mathbf{m}_k \tag{1.64}$$

$$= -[\mathcal{E}_{\text{lin}}]_{\mathcal{D},\mathcal{D}}^{-1} [\mathcal{E}_{\text{lin}}]_{k,\mathcal{D}} \tag{1.65}$$

where $[\mathcal{E}_{\text{lin}}]_{k,\mathcal{D}}$ is the kth column of $\mathcal{E}_{\text{lin}}$ taking only rows in set $\mathcal{D}$ and $[\mathcal{E}_{\text{lin}}]_{\mathcal{D},\mathcal{D}}$ is the matrix formed from only those rows and columns of $\mathcal{E}_{\text{lin}}$ in $\mathcal{D}$.

The expression for $\mathbf{B}_k$, given by (1.65), is the MMSE estimation filter for the error $\mathbf{e}_{\text{lin},k}$ (kth component of $\mathbf{e}_{\text{lin}}$) given $\mathbf{e}_{\text{lin},\mathcal{D}}$ (i.e., $\mathbf{e}_{\text{lin},m}$ for $m \in \mathcal{D}$). That is, combining (1.52) and (1.61) with (1.53) gives:

$$\mathbf{e}_{\text{dfd}} = (\mathbf{I} + \mathbf{B})^{\dagger} \mathbf{e}_{\text{lin}} \tag{1.66}$$

so that:

$$\mathbf{e}_{\text{dfd},k} = \mathbf{e}_{\text{lin},k} + \mathbf{B}_{k,\mathcal{D}}^{\dagger} \mathbf{e}_{\text{lin},\mathcal{D}}. \tag{1.67}$$

Selecting $\mathbf{B}_{k,\mathcal{D}}$ to minimize $E[|\mathbf{e}_{\text{dfd},k}|^2]$ gives (1.65). Note that the orthogonality principle implies:

$$E[\mathbf{e}_{\text{dfd},k} \mathbf{e}_{\text{lin},m}^*] = 0 \tag{1.68}$$

where $m \in \mathcal{D}$. We also remark that the expression for $\mathbf{B}_k$ in (1.56) follows directly from (1.65) by applying the Matrix Inversion Lemma [32, Sec. 13.2].

1.6.1 Successive Decision Feedback

For the S-DFD, we have for user k:

$$\mathcal{D} = \{1, \ldots, k-1\}, \qquad \mathcal{U} = \{k, \ldots, K\}. \tag{1.69}$$

From (1.68) and (1.67), it is easily shown that:

$$E[\mathbf{e}_{\text{dfd},k} \mathbf{e}_{\text{dfd},m}^*] = 0, \quad k \neq m, \tag{1.70}$$

which from (1.66) implies that:

$$\mathcal{E}_{\text{dfd}} = (\mathbf{I} + \mathbf{B})^{\text{H}} \mathcal{E}_{\text{lin}} (\mathbf{I} + \mathbf{B}) \tag{1.71}$$

is *diagonal*. That is, $\mathbf{I} + \mathbf{B}$ can be interpreted as an *error whitening filter* [15]. In this case, $\mathbf{B}$ can be computed via a Cholesky factorization of the error covariance matrix $\mathcal{E}_{\text{lin}}$ in (1.63).

1.6.2 Parallel Decision Feedback

For the P-DFD, assuming that all K streams are detected, we have for user k:

$$\mathcal{U} = \{k\}, \qquad \mathcal{D} = \{1, \ldots, k-1, k+1, \ldots, K\} \tag{1.72}$$

The initial symbol estimates for feedback might be obtained from the linear MMSE filter, or from an S-DFD. From (1.56) we have:

$$\mathbf{F}_k = \mathbf{R}_{\mathcal{U}}^{-1} \mathbf{m}_k = a \mathbf{m}_k \tag{1.73}$$

where $a = 1/(\|\mathbf{m}_k\|^2 + \sigma^2)$, and combining (1.61) with (1.73) gives:

$$\mathbf{I} + \mathbf{B} = a(\mathbf{M}^{\dagger} \mathbf{M} + \sigma^2 \mathbf{I}) \tag{1.74}$$

That is, the feedforward filter is a bank of scaled matched filters, and the off-diagonal components of the feedback matrix are cross-correlations between columns of $\mathbf{M}$. Ideal interference cancellation is therefore achieved with correct feedback estimates. Substituting (1.58) into (1.59) gives the corresponding MMSE:

$$[\mathcal{E}_{\text{p-dfd}}]_{kk} = 1 - \mathbf{m}_k^{\dagger} (\mathbf{m}_k \mathbf{m}_k^{\dagger} + \sigma^2 \mathbf{I})^{-1} \mathbf{m}_k = a\sigma^2, \tag{1.75}$$

which applies to a single user in the absence of interference.

For the general case in which $\mathcal{D}$ is an arbitrary subset of transmitted streams, from (1.56) it is apparent that the feedforward filter uses the available degrees of freedom to suppress *only* interference from symbols in $\mathcal{U}$. This applies, for example, to a CDMA system in which $\mathcal{D}$ contains users within the cell and $\mathcal{U}$ contains users in other cells.

From (1.68) and (1.67), it can be shown that the error covariance matrix for the P-DFD is given by:

$$\mathcal{E}_{\text{p-dfd}} = (\mathbf{I} - \mathbf{B})\mathbf{D}_p \tag{1.76}$$

where $\mathbf{D}_p$ is the diagonal matrix with $[\mathbf{D}_p]_{k,k} = E[|e_{\text{p-dfd},k}|^2]$. In contrast with the S-DFD, the error covariance matrix for the P-DFD is generally a full matrix, and the matrix $\mathbf{I} + \mathbf{B}$ no longer has the interpretation of an error whitening filter.

1.6.3 Filter Adaptation

According to the previous discussion, the optimal feedforward and feedback filters depend on knowledge of user signatures, channel characteristics, and the noise variance. However, as with a linear multiuser detector, the DFD filters can be estimated adaptively with a training sequence. For example, if the transmitter transmits a training sequence $\{\mathbf{b}(i)\}$ known to the receiver, then the DFD filters $\mathbf{F}$ and $\mathbf{B}$ can be selected to minimize the Least Squares cost function:

$$\mathcal{C}(i) = \sum_{m=1}^{i} \| \mathbf{e}_{\text{dfd}}(m) \|^2 \tag{1.77}$$

where $\mathbf{e}_{\text{dfd}(m)}$ is given by (1.66). Note that in the multiuser scenario this requires all users to train at the same time (although they may be symbol-asynchronous).

The filters, which minimize $\mathcal{C}(i)$, have the same form as the MMSE filters given in (1.56), and (1.61) and (1.65), where associated expectations are replaced by sums over the time index. Hence, the filters can be computed from either the least squares version of (1.56), or (1.61) and (1.65), where $\mathbf{F}_{\text{lin}}$ is replaced by the linear least squares filter. We also remark that as for the linear receiver, reduced-rank versions of the DFD can be derived to reduce training. Namely, referring to the decomposition of the feedforward filter $\mathbf{F} = \mathbf{F}_{\text{lin}}(\mathbf{I} + \mathbf{B})$, we can replace $\mathbf{F}_{\text{lin}}$ by a reduced-rank filter. The error estimation filter (1.65) can also be approximated as a reduced-rank filter, although the order of this filter may not be large enough to benefit from rank reduction (e.g., see [74]).

1.6.4 Error Propagation and Iterative Decision Feedback

The performance of the DFD can be substantially degraded by decision errors in the feedback estimates. This is especially problematic at low SNRs when the detection (DFD) and decoding for error control are separated. (That is, the decoder generates hard decisions based on DFD outputs alone.) In that case, the accumulation of feedback errors can degrade the performance of the DFD relative to a linear detector.[18]

The effects of error propagation can be mitigated by using successive cancellation and demodulation rather than parallel cancellation. Namely, provided that the rate for a particular stream (e.g., user) is less than the capacity, treating the undecoded streams as

---

[18]Computing an exact expression for the error probability associated with a particular transmitted stream, accounting for error propagation, is generally difficult. However, it is possible to determine the probability that *at least one element* of $\hat{\mathbf{b}}$ is incorrect [59,83].

noise, that stream can be reliably decoded and then cancelled for purposes of detecting the remaining streams. This type of successive cancellation achieves the sum capacity associated with the CDMA or MIMO channel [17,28], but requires that the transmitted streams be coded at the appropriate rates. (Those would have to be determined at the receiver, given the mixing matrix $\mathbf{M}$, and fed back to the transmitters.) Also, the decoding order can affect performance (e.g., see [83]). An associated disadvantage of the S-DFD, relative to the P-DFD for some applications, is that it generally does not provide uniform performance over the multiplexed data streams. Also, since each data stream must be decoded before the next stream can be detected, streams towards the bottom of the decoding order may experience substantial latency.

DFD performance can be improved by combining *soft* estimates for feedback with *iterative* refinements of those estimates. Soft estimation typically refers to weighting the feedback symbols by their "reliabilities." For example, the hard estimate of the symbol can be replaced with its expected value, given some assumptions about the error statistics (e.g., Gaussian). This mitigates the effect of hard decision errors, since symbols in error are likely to have lower reliabilities than correct decisions. (A review of related non-iterative interference cancellation techniques is given in [2].)

Iterative versions of the DFD are obtained by updating and feeding back the estimates $\hat{\mathbf{b}}$. That is, $\hat{\mathbf{b}}$ in iteration i is used to generate the input to the decision device $\mathbf{x}$ in (1.52), which is used to generate the estimate $\hat{\mathbf{b}}$ at iteration $i + 1$. This type of iterative cancellation approach to multiuser detection for CDMA was initially proposed in [84], assuming hard decisions for $\hat{\mathbf{b}}$.

When the decision device in Figure 1.5 generates hard estimates, it has been observed that additional DFD iterations typically provide a modest reduction in the error probability [37,94]. Feeding back soft decisions for cancellation further improves performance somewhat; however, much more substantial performance gains are obtained by iterating the soft decision DFD with a soft decision decoder (assuming the presence of error control coding), analogous to iterative decoding of a concatenated (turbo) code. This type of joint iterative multiuser detection and decoding is discussed in detail in the next chapter.

1.6.5 Application to MIMO Channels

Both the S- and P-DFD can be directly applied to single-user MIMO channels, where the matrix $\mathbf{M}$ is the channel matrix, and $\mathbf{b}$ is the vector of symbols across transmit antennas. From the preceding discussion, either DFD can be applied independently to each received vector $\mathbf{y}(i)$ to obtain the estimate $\hat{\mathbf{b}}(i)$. The application of the S-DFD to estimate K transmitted streams (one per transmit antenna), where $K \leq N$, was proposed in [18], and further evaluated in [17].[19] An underlying assumption is that each stream is individually coded, hence each transmitted stream corresponds to a user in a CDMA system, where the signature for user (stream) k is the kth column of the channel

---

[19]This combined transmission/reception scheme without coding is given the acronymn "Vertical (V)-BLAST (Bell Laboratories Layered Space-Time)" in [18], since each received vector $\mathbf{y}$ is processed individually as a column (vertically). With coding the acronymn is changed in [17] to "Horizontal (H)-BLAST," since each stream is coded "horizontally" across time.

matrix $\mathbf{M}$ (array of channel gains from transmit antenna k to the receive antennas). The DFD output can then be passed to a bank of K decoders, one for each stream.[20]

Processing each received vector individually by the DFD, as just described, and subsequently decoding each stream is suboptimal in the sense that it cannot achieve the sum capacity of the MIMO channel. This is due to error propagation associated with the DFD. As discussed in the preceding section, two ways to improve performance are (1) successively decode each entire *stream* (i.e., corresponding to a frame or packet) before using the estimates to cancel interference for the next stream; and (2) use iterative methods to refine the DFD estimates. An important feature of the MIMO channel, as opposed to CDMA, is that the transmitter can coordinate the transmission of the K codewords across space and time (e.g., see [79, Ch. 8], [17]). The pattern can therefore be chosen to facilitate successive interference cancellation, and also to equalize the code rates across streams, so that feeding back a set of K rates is no longer required. Alternatively, the transmitter can interleave a single codeword across space and time, and the receiver can apply iterative detection methods with either an S- or P-DFD.

1.7 INTERFERENCE MITIGATION AT THE TRANSMITTER

In addition to interference suppression and cancellation at the receiver, in some situations it may be possible to mitigate, or *avoid* interference at the transmitter. In general, this is often accomplished by transmitting on different frequencies or time slots. For the model (1.1), this can also be accomplished by optimizing a precoding matrix for a MIMO channel, or by optimizing signatures assigned to different users.[21]

To illustrate, here we assume that the mixing matrix $\mathbf{M} = \mathbf{HS}$, where $\mathbf{H}$ is a given $N \times K$ channel matrix, and $\mathbf{S}$ is a $K \times K'$ precoding or signature matrix, where K' is the number of multiplexed data streams. That is, the columns of $\mathbf{S}$ can be interpreted as signatures across antennas (for a MIMO channel), or across time (for CDMA). In the case of a MIMO channel, we can choose the matrix $\mathbf{S}$ to optimize a performance metric over all data streams (e.g., the sum rate). In that case we refer to the data streams as being *coordinated*. In contrast, if each column of $\mathbf{S}$ is optimized independently to maximize a performance metric for the associated data stream, then the data streams are *uncoordinated*.

1.7.1 Precoding for Coordinated Data Streams

Precoding is illustrated in Figure 1.7, where the channel is decomposed according to the eigen-decomposition:

$$\mathbf{H}^{\dagger}\mathbf{H} = \mathbf{V}\Lambda\mathbf{V}^{\dagger} \tag{1.78}$$

---

[20]The successive canceller for V-BLAST, presented in [18], corresponds to the form of the S-DFD given in (1.56). The equivalence between this form and the S-DFD described in Section 1.6.1 and shown in Figure 1.6, where the "error estimation filter" is a whitening filter, was pointed out in [23].

[21]Note that assigning channels and time slots can be viewed as a special case of "signature" assignment, since the signatures can be defined across frequency and/or time.

Figure 1.7. Linear precoding at the transmitter for a MIMO channel with coordinated data streams.

where $\mathbf{V}$ is a $K \times K$ unitary matrix, the columns of which are the eigen-vectors of $\mathbf{H}^\dagger \mathbf{H}$, and Λ is the $K \times K$ diagonal matrix of nonnegative eigen-values. In Figure 1.7, the precoding matrix $\mathbf{S} = \mathbf{V}_{K'}\mathbf{A}$, where $\mathbf{A}$ is a real $K' \times K'$ diagonal matrix, which allocates power across the K' streams, and the columns of $\mathbf{V}_{K'}$ are the eigen-vectors (columns of $\mathbf{V}$) corresponding to the K' largest eigen-values (i.e., $\mathbf{V}_{K'}$ is $K \times K'$). The transmitted symbol vector is then $\mathbf{b}' = \mathbf{V}_{K'}\mathbf{A}\mathbf{b}$, and the receiver multiplies the channel output $\mathbf{y}$ with the receive filter $\mathbf{C}^\dagger = \mathbf{A}\mathbf{V}_{K'}^\dagger \mathbf{H}^\dagger$ to obtain:

$$\mathbf{y}' = \mathbf{A}\mathbf{V}_{K'}^\dagger \mathbf{H}^\dagger \mathbf{H}\mathbf{V}_{K'}\mathbf{A}\mathbf{b} + \mathbf{A}\mathbf{V}_{K'}^\dagger \mathbf{H}^\dagger \mathbf{n} \tag{1.79}$$

$$= \Lambda \mathbf{A}^2 \mathbf{b} + \mathbf{n}' \tag{1.80}$$

where $\mathbf{n}' = \mathbf{A}\mathbf{V}_{K'}^\dagger \mathbf{H}^\dagger \mathbf{n}$ and $E[\mathbf{n}'\mathbf{n}'^\dagger] = \sigma^2 \mathbf{A}^2 \Lambda$. Hence choosing "signatures" (columns of $\mathbf{V}_{K'}$) aligned with the eigen-vectors of $\mathbf{M}$ diagonalizes the channel matrix, thereby eliminating interference.

With additive white Gaussian noise, the corresponding maximum achievable rate, or sum capacity, with precoding is given by [79, Ch. 7]:

$$C = \log \det(\mathbf{I}_{K'} + \sigma^{-2}\mathbf{A}\mathbf{V}^\dagger \mathbf{H}^\dagger \mathbf{H}\mathbf{V}\mathbf{A}) \tag{1.81}$$

$$= \sum_{i=1}^{K'} \log(1 + A_i^2 \lambda_i / \sigma^2) \tag{1.82}$$

where A_i and λ_i are the corresponding diagonal elements of $\mathbf{A}$ and Λ. To maximize C, the amplitudes are chosen according to the water-pouring distribution $A_i^2 = \max\{\eta - \sigma^2/\lambda_i, 0\}$, where η is the water-pouring level selected to enforce the power constraint $\sum_{i=1}^{K'} A_i^2 \leq P_{av}$.

Jointly optimizing the precoding matrix with the receiver clearly must perform better than interference mitigation at the receiver alone. The performance improvement can be substantial at high loads when the receiver is constrained to be suboptimal (e.g., a linear receiver).[22]

With an optimal receiver, the additional gain due to optimal precoding may be modest (e.g., see [79, Ch. 8]); however, precoding generally simplifies the coding, decoding, and detection. Namely, the preceding discussion for coordinated data

---

[22]The load refers to the number of independent data streams relative to the number of degrees of freedom (e.g., number of antennas or processing gain).

streams implies that with optimal precoding a linear matched filter achieves the sum capacity. Furthermore, codes designed for the single-input/single-output AWGN channel can be applied directly by coding over the channel modes (eigen-vectors). This approach to combined error control coding and linear precoding avoids the use of more complicated space-time codes, and achieves the associated channel capacity.

The main drawback of precoding is that the channel or precoding matrix must be estimated at the receiver, and then relayed back to the transmitter through a feedback channel. This must be done frequently in situations where the data streams and channels are time-varying (e.g., due to mobility).

1.7.1.1 Precoding for Equalizing SNR Performance

The capacity given by (1.82) can be achieved by transmitting random Gaussian codewords, where the elements of the codewords are spread across the different channel eigen-vectors (columns of $\mathbf{V}_{K'}$). The variance of the elements are chosen according to the water pouring distribution. Practical codes, however, choose the codeword elements from discrete constellations. Hence in practice, coded bits can be mapped to a set of constellation points, one for each eigen-mode, where the size of a constellation is matched to the associated channel gain (eigen-value). This leads to "bit loading" schemes in which the rate for the ith eigen-mode (i.e., information bits per symbol) is selected to approximate $\log(1 + A_i^2 \lambda_i^2 / \sigma^2)$, where A_i^2 is the optimized power [8].[23]

Constraining the set of constellations, e.g., restricting the codeword elements to be selected from the *same* constellation over all modes, limits the ability to do bit loading. In that situation it is desirable to decompose the channel into effective modes or "sub-channels" having the *same* SNR [i.e., as opposed to the eigen-decomposition (1.78), which gives the set of SNRs $\{A_i^2 \lambda_i^2 / \sigma^2\}$]. This type of *uniform* channel decomposition is presented in [42,95], and relies on the following observations:

1. Replacing $\mathbf{S}$ in Figure 1.7 with $\mathbf{S}\Phi$, where Φ is any unitary matrix, does not change the sum capacity.

2. Any $N \times K$ matrix $\mathbf{M}$ with rank r can be written as:

$$\mathbf{M} = \mathbf{QUP}, \tag{1.83}$$

where $\mathbf{Q}$ is $N \times r$, $\mathbf{P}$ is $K \times r$, both $\mathbf{P}$ and $\mathbf{Q}$ have orthonormal columns, and $\mathbf{U}$ is an $r \times r$ upper triangular matrix with equal diagonal elements.

The first observation gives the additional degrees of freedom needed to vary the set of received SNRs per transmitted stream without sacrificing total rate. The second observation is used to compute the precoder and the receiver. (The proof of (1.83) is given in [41,96].)

---

[23]In practice, the same power can be used on each active channel without sacrificing much rate.

Replacing the optimal precoding matrix in Figure 1.7, $\mathbf{S} = \mathbf{V}_{K'}\mathbf{A}$, by $\mathbf{V}_{K'}\mathbf{A}\Phi$, where Φ is orthonormal, introduces interference among the transmitted substreams, hence the linear receiver $\mathbf{C}$ shown in Figure 1.7, is no longer optimal. For the uniform channel decomposition, we therefore replace $\mathbf{C}$ by an S-DFD, which has front-end filter $\mathbf{F}$ given by (1.61). Let $\tilde{\mathbf{M}} = \mathbf{HS} = \mathbf{HV}_{K'}\mathbf{A}\Phi$ denote the combined channel and precoder matrix. Then from (1.71) and (1.63) we have:

$$\tilde{\mathbf{R}} = \tilde{\mathbf{M}}^{\dagger}\tilde{\mathbf{M}} + \sigma^{-2}\mathbf{I}_{K'} \qquad (1.84)$$

$$= (\mathbf{I}_{K'} + \mathbf{B})\mathbf{D}^{-1}(\mathbf{I}_{K'} + \mathbf{B})^{\dagger} \qquad (1.85)$$

where $\mathbf{D}$ is diagonal.

As discussed in Section 1.6.1, assuming perfect cancellation, the MSE for the kth substream is $\mathbf{D}_{kk}$. Although the sum capacity $C = \log \det[\tilde{\mathbf{R}}] = \sum_{k=1}^{K'} \log \mathbf{D}_{kk}^{-1}$ does not depend on Φ, it can be used to change the MSEs (or SNRs) across substreams. This can be achieved by factoring out Φ in (1.84), and writing $\tilde{\mathbf{R}} = \Phi^{\dagger}\mathbf{U}^{\dagger}\mathbf{U}\Phi$, where $\mathbf{U}$ is upper triangular. The preceding decomposition (1.83) can then be applied to $\tilde{\mathbf{R}}^{1/2} = \mathbf{QU}\Phi$, where $\mathbf{Q}$ is selected along with $\Phi = \mathbf{P}$. Algorithms for determining Φ and $\mathbf{Q}$ are presented in [42,95]. The resulting precoding matrix $\mathbf{S}\Phi$ combined with an S-DFD at the receiver then gives the same SNR across all K' data streams.

1.7.2 Signature Optimization with Uncoordinated Data Streams

In some scenarios, such as a peer-to-peer network, the data streams correspond to different users, who may not be able to coordinate transmissions. In that case, a particular user may adjust his own signature (for spread spectrum transmission), or precoding matrix (with multiple antennas), to optimize his own performance, ignoring the effect on other users.

This is illustrated with the system model:

$$\mathbf{y}_k = \mathbf{H}_k\mathbf{s}_k A_k b_k + \sum_{i \neq k} \mathbf{H}_i\mathbf{s}_i A_i b_i + \mathbf{n} \qquad (1.86)$$

where $\mathbf{y}_k$ is the received signal for user k, and $\mathbf{H}_i$ is the channel from transmitter i to receiver k. Suppose that user k has a linear MMSE receiver, and can select the signature $\mathbf{s}_k$ to minimize the MSE $E[|\mathbf{c}^{\dagger}\mathbf{y}_k - b_k|^2]$. It can be shown that, assuming all other signatures and channels are fixed, the optimized signature $\mathbf{s}_k$ is the eigenvector of the matrix $\mathbf{H}_k^{\dagger}\mathbf{R}_k^{-1}\mathbf{H}_k$ corresponding to the maximum eigen-value, where $\mathbf{R}_k = \sum_{i \neq k} A_i^2 \mathbf{H}_i\mathbf{s}_i\mathbf{s}_i^{\dagger}\mathbf{H}_i^{\dagger} + \sigma^2\mathbf{I}$ is the interference-plus-noise covariance matrix at receiver k.

The optimal signature depends on the channels, so that significant feedback may be required with mobile users. The model (1.86) also applies to the multiple access channel (i.e., with co-located receivers), and in that case it is desirable to select the set of signatures $\{s_1, \ldots, s_K\}$ to maximize a global performance objective, such as sum rate over the users. A basic question is whether or not the individual (or "greedy") optimization described here, which does not account for interference to other users, can maximize such a global performance objective. This is discussed in detail in Chapter 7.[24]

1.7.3 Network Configurations

The ability to coordinate different data streams depends on the network configuration. Coordination is relatively easy when the transmitted data streams are co-located, as in a single-user MIMO channel. Another example in which coordinated precoding can provide substantial benefits is the downlink of a single cell in a cellular system. There the optimal precoding, which maximizes the total sum rate over the users, is *nonlinear*. In addition, the transmission to a particular user depends on the messages being transmitted to other users [9, Ch. 14]. This optimal precoder and associated performance are discussed in Chapter 8.

For the uplink the base station can, in principle, compute a set of optimized signatures for the users given knowledge of the uplink channels [e.g., $\mathbf{H}_i, i = 1, \ldots, K$ in (1.86)]. Coordination in that scenario mainly consists of synchronizing the users. Unlike the downlink, users are not likely to be able to coordinate codewords. The uplink sum capacity is achieved by the dual operation of successive decoding and cancellation of interfering users at the base station. (See [79, Ch. 10] and the discussion in Chapter 8.)

As previously mentioned, coordination of data streams may be difficult in an ad hoc, or peer-to-peer network in which transmitter-receiver pairs are not co-colocated. In those scenarios each transmitter-receiver pair may optimize a local performance objective, ignoring the effect on other transmitter-receiver pairs. The performance of such a network can be studied within the framework of game theory [71]. Namely, the transmitter-receiver pairs can be viewed as *non-cooperative agents*, each acting according to their own self-interest. The users can then be expected to adjust their parameters (e.g., powers, signatures) until they reach an equilibrium in which no further improvements to local performance are possible.[25] This game-theoretic framework is currently being used to study distributed resource allocation in a variety of communications scenarios (e.g., see [4]).

In general, a wireless network may consist of a combination of coordinated and uncoordinated nodes. For example, it is generally easier to coordinate users within a cell than it is to coordinate users across cells. Optimization of signaling methods along with signatures, or precoding matrices, in such a general environment is a challenging direction for future work.

---

[24]For the model considered here (uncoordinated users each with a single signature) greedy optimization generally does not maximize the sum rate, although it typically comes close [61].

[25]More precisely, a *Nash equilibrium* is achieved when each user has no incentive to change its current power and/or signature given a fixed set of choices for powers and signatures across the other users.

1.8   OVERVIEW OF REMAINING CHAPTERS

We conclude this chapter by relating the previous discussion and overview of multiuser detection techniques to the advances discussed in subsequent chapters. Generally speaking, these advances fall into two main categories: advances in detection techniques, and advances in performance analysis. With the exception of iterative techniques, discussed in Chapter 2, the advances in detection techniques are motivated by more elaborate, or different modeling assumptions (i.e., that account for multipath, fading, and downlink transmissions). The advances in performance analysis are motivated by parallel advances in relevant mathematical methods.

Referring to the fundamental performance limits shown in Figure 1.1, optimal multiuser-detection offers a significant increase in spectral efficiency relative to the linear detector at high SNRs and high loads. As discussed in Section 1.6, that has motivated the development of relatively simple nonlinear decision feedback and iterative techniques that can close this gap. When combined with error control coding, the multiuser detector can exploit reliability information generated by the decoder. Iterating between the (soft output) multiuser detector and the soft decoder is analogous to iterative (Turbo) decoding, and the performance approaches that corresponding to a single isolated user (i.e., without interference) over a wide range of loads and SNRs. An overview of these types of iterative techniques is presented in Chapter 2.

The results in Figure 1.1 are based on the model (1.1), where $\mathbf{M}$ has *i.i.d.* elements. Although this model has been quite useful for developing and analyzing multiuser detection techniques, it does not account for channel impairments encountered in practice, e.g., associated with wireless channels. In particular, multipath introduces fading and intersymbol interference, which pose additional challenges for multiuser receiver design. Linear receivers for multiuser channels with multipath are developed in Chapter 3, including variants of reduced-rank receivers, which are based on eigen-decompositions of appropriate signal subspaces.

For the basic multiuser model (1.1), where $\mathbf{M}$ is random, it is often desirable to obtain performance measures that are averaged over realizations of $\mathbf{M}$. This can be difficult in general, but sometimes becomes tractable in the large system limit in which the dimensions of $\mathbf{M}$ are scaled with a fixed ratio. This has already been suggested by the performance results shown in Figure 1.1. This type of large system analysis is discussed Chapter 4, where derivations of some key multiuser performance measures are presented for a few variations on the channel model (1.1). For the random matrix model (1.1) the techniques in Chapter 4 cannot be applied in a straightforward way to determine the performance (error probability) of the *optimal* detector. To evaluate that performance measure in the large system limit, it has been necessary to apply techniques from the statistical physics literature. That analysis is introduced in Chapter 5, along with a general estimation framework for interpreting the performance of different (suboptimal) multiuser detectors.

As previously mentioned, the model (1.1) applies to both channels with multiple antennas and multi-user systems. This connection is developed in more detail in Chapter 6, where fundamental (capacity) limits associated with multi-antenna systems are presented along with performance results for linear receivers. In particular, results for the random matrix model (1.1) are presented for finite dimensions, which complement the large system results in Chapter 4.

Chapters 7 and 8 discuss methods for *jointly optimizing* transmitters and receivers, as previously discussed in Section 1.7. The discussion in Chapter 7 is focused on *avoiding* interference through selection of signatures at a transmitter. As previously pointed out in Section 1.7, that can be viewed as a linear precoding technique. In contrast, Chapter 8 discusses optimal multiuser precoding for the downlink, which processes the input symbols in a nonlinear fashion. There the optimization criterion is the rate summed over users. The discussion is generalized to the situation in which there are multiple antennas at the base station and at each mobile (i.e., the MIMO downlink). In those scenarios substantial performance gains are achievable through optimal precoding.

Finally, while there have been many advances in multiuser detection over the past few years, there are a number of important related directions for future work. These include extensions and generalizations of the techniques and analysis presented here to other channel models (e.g., combinations of wideband, MIMO, slow or fast fading) and network configurations (e.g., multi-cell, peer-to-peer, multi-hop). Additional modifications may be motivated by the presence (or lack) of limited information about channels and interferers, which may (or may not) be available in practice.

From a practical point of view, numerous challenges arise when trying to incorporate many of the techniques discussed here in actual systems. These include adaptive channel estimation in the presence of time- and frequency-selective fading (for interferers as well as for the desired data stream), allocation of system (computing) resources within a receiver (e.g., between decoding and detection), power management (consumed by the device as well as transmitted power), limited feedback schemes for relaying channel (or decoder state) information from the receiver to a transmitter, in addition to managing system cost. Some of these issues, such as channel estimation, become more complicated as the number of degrees of freedom increases (e.g., by increasing bandwidth or number of antennas). It is our hope that the advances described in this collection will serve as a springboard for future investigations into all of these areas.

REFERENCES

1. N. Amitay and J. Salz. Linear equalization theory in digital data transmission over dually polarized fading radio channels. *Bell System Technical Journal*, pages 2215–2259, December 1984.

2. J. G. Andrews. Interference cancellation for cellular systems: a contemporary overview. *IEEE Wireless Communications*, 12(2):19–29, April 2005.

3. M. Andrews, K. Kumaran, K. Ramanan, A. L. Stolyar, R. Vijayakumar, and P. Whiting. Providing quality of service over a shared wireless link. *IEEE Communications Magazine*, 39(2):150–154, 2001.

4. T. Basar, J. Huang, N. B. Mandayam, V. P. Palomar, J. Walrand, and S. B. Wicker, editors. *IEEE Journal on Selected Areas in Communications*, 2008. Special Issue on Game Theory in Communication Systems.

5. P. Bender, P. Black, M. Grob, R. Padovani, N. Sindhushayana, and A. Viterbi. CDMA/ HDR: a bandwidth-efficient high-speed wireless data service for nomadic users. *IEEE Communications Magazine*, pages 70–77, July 2000.

6. E. Biglieri, R. Calderbank, A. Constantinides, A. Goldsmith, A. Paulraj, and H. V. Poor. *MIMO Wireless Communications*. Cambridge University Press, 2007.

7. W. Chen, U. Mitra, and P. Schniter. On the equivalence of three reduced rank linear estimators with applications to DS-CDMA. *IEEE Transactions on Information Theory*, 48(9):2609–2614, 2002.

8. P. S. Chow, J. M. Cioffi, and J. A. C. Bingham. A practical discrete multitone transceiver loading algorithm for data transmission over spectrally shaped channels. *IEEE Transactions on Communications*, 43(234):773–775, April 1995.

9. T. Cover and J. Thomas. *Elements of Information Theory*. John Wiley and Sons, 2nd edition, 2006.

10. M. O. Damen, H. El Gamal, and G. Caire. On maximum-likelihood detection and the search for the closest lattice point. *IEEE Transactions on Information Theory*, 49(10):2389–2402, 2003.

11. O. Damen, A. Chkeif, and J. C. Belfiore. Lattice code decoder for space-time codes. *IEEE Communications Letters*, 4(5):161–163, 2000.

12. G. K. E. Dietl. Linear estimation and detection in Krylov subspaces. *Foundations in Signal Processing, Communications, and Networking*, 1, 2007.

13. A. Duel-Hallen. Equalizers for multiple input/multiple output channels and PAM systems with cyclostationary input sequences. *IEEE Journal on Selected Areas in Communications*, 10(3):630–639, April 1992.

14. A. Duel-Hallen. Decorrelating decision-feedback multiuser detector for synchronous code-division multiple-access channel. *IEEE Transactions on Communications*, 41(2):285–290, February 1993.

15. A. Duel-Hallen. A family of multiuser decision-feedback detectors for asynchronous code-division multiple-access channels. *IEEE Transactions on Communications*, 43(2/3/4):421–434, February/Mar/Apr 1995.

16. U. Fincke and M. Pohst. Improved methods for calculating vectors of short length in a lattice, including a complexity analysis. *Mathematics of Computation*, 44(170):463–471, 1985.

17. G. J. Foschini, D. Chizhik, M. J. Gans, C. Papadias, and R. A. Valenzuela. Analysis and performance of some basic space-time architectures. *IEEE Journal on Selected Areas in Communications*, 21(3):303–320, April 2003.

18. G. J. Foschini, G. D. Golden, R. A. Valenzuela, and P. W. Wolniansky. Simplified processing for high spectral efficiency wireless communication employing multi-element arrays. *IEEE Journal on Selected Areas in Communications*, 17(11):1841–1852, November 1999.

19. G. J. Foschini and M. J. Gans. On limits of wireless communications in a fading environment when using multiple antennas. *Wireless Personal Communications (Kluwer)*, 6:311–335, 1998.

20. C. D. Frank, E. Visotsky, and U. Madhow. Adaptive interference suppression for the downlink of a direct sequence CDMA system with long spreading sequences. *Journal of VLSI Signal Processing*, 30:273–291, 2002.

21. R. Gallager. A perspective on multiaccess channels. *IEEE Transactions on Information Theory*, 31(2):124–142, March 1985.

22. A. Gersho and T. L. Lim. Adaptive cancellation of intersymbol interference for data transmission. *Bell System Technical Journal*, 60:1997–2021, November 1981.

23. G. Ginis and J. M. Cioffi. On the relation between V-BLAST and the GDFE. *Communication Letters*, 5(9):364–366, September 2001.

24. A. Goldsmith. *Wireless Communication*. Cambridge University Press, 2005.

25. J. S. Goldstein and I. S. Reed. Reduced rank adaptive filtering. *IEEE Trans. Signal Processing*, 45(2):492–496, February 1997.

26. J. S. Goldstein, I. S. Reed, and L. L. Scharf. A multistage representation of the Wiener filter based on orthogonal projections. *IEEE Trans. Inform. Theory*, 44(7), November 1998.

27. A. Grant and C. Schlegel. Convergence of linear interference cancellation multiuser receivers. *IEEE Transactions on Communications*, 10(49):1824–1834, October 2001.

28. T. Guess and M. K. Varanasi. An information-theoretic framework for deriving canonical decision-feedback receivers in Gaussian channels. *IEEE Transactions on Information Theory*, 51(1):173–187, January 2005.

29. D. Guo, L. K. Rasmussen, S. Sun, and T. J. Lim. A matrix-algebraic approach to linear parallel interference cancellation in CDMA. *IEEE Transactions on Communications*, 48:152–161, January 2000.

30. M. Haardt, A. Klein, R. Koehn, S. Oestreich, M. Purat, V. Sommer, and T. Ulrich. The TD-CDMA based UTRA TDD mode. *Selected Areas in Communications, IEEE Journal on*, 18(8):1375–1385, 2000.

31. B. Hassibi and H. Vikalo. On the sphere-decoding algorithm I. Expected complexity. *IEEE Transactions on Signal Processing*, 53(8 Part 1):2806–2818, 2005.

32. S. Haykin. *Adaptive Filter Theory*. Prentice Hall, 4th edition, 2002.

33. B. M. Hochwald and S. Ten Brink. Achieving near-capacity on a multiple-antenna channel. *IEEE Transactions on Communications*, 51(3):389–399, 2003.

34. M. L. Honig. Adaptive linear interference suppression for packet DS-CDMA. *European Transactions on Telecommunications and Related Technologies*, 9(2):173–181, March/apr 1998.

35. M. L. Honig and J. S. Goldstein. Adaptive reduced-rank interference suppression based on the multistage Wiener filter. *IEEE Transactions on Communications*, 50(6):986–994, June 2002.

36. M. L. Honig, S. L. Miller, M. J. Shensa, and L. B. Milstein. Performance of adaptive linear interference suppression in thepresence of dynamic fading. *IEEE Transactions on Communications*, 49(4): 635–645, 2001.

37. M. L. Honig and R. Ratasuk. Large-system performance of iterative multiuser decision-feedback detection. *IEEE Transactions on Communications*, 51(8):1368–1377, August 2003.

38. M. L. Honig and W. Xiao. Performance of reduced-rank linear interference suppression. *IEEE Transactions on Information Theory*, 47(5):1928–1946, July 2001.

39. H. Huang and S. Verdú. Linear differentially coherent multiuser detection for multipath channels. *Wireless Personal Communications*, 6(1/2):113–136, 1998.

40. J. Jalden and B. Ottersten. On the Complexity of Sphere Decoding in Digital Communications. *IEEE Transactions on Signal Processing*, 53(4):1474–1484, 2005.

41. Y. Jiang, J. Li, and W. W. Hager. Joint transceiver design for MIMO communications using geometric mean decomposition. *IEEE Transactions on Signal Processing*, 53(10):3791–3803, October 2005.

42. Y. Jiang, J. Li, and W. W. Hager. Uniform Channel Decomposition for MIMO Communications. *IEEE Transactions on Signal Processing*, 53(11):4283–4294, November 2005.

43. M. Joham, Y. Sun, M. D. Zoltowski, M. Honig, and S. Goldstein. A new backward recursion for the multi-stage nested Wiener filter employing Krylov subspace methods. In *Proc. MILCOM 2001*, pages 1210–1213, McLean, Va., October 2001.

44. D. H. Johnson and D. E. Dudgeon. *Array Signal Processing*. Prentice Hall, Signal Processing Series, 1992.

45. M. Kavehrad and J. Salz. Cross-polarization cancellation and equalization in digital transmission over dually polarized multipath fading channels. *Bell System. Technical Journal*, pages 2211–2245, December 1985.

46. A. R. Kaye and D. A. George. Transmission of multiplexed PAM signals over multiple channel and diversity systems. *IEEE Trans. Commun. Techn.*, pages 520–526, October 1970.

47. Kiran and D. N. C. Tse. Effective interference and effective bandwidth of linear multiuser receivers in asynchronous CDMA systems. *IEEE Transactions on Information Theory*, 46(4):1426–1447, July 2000.

48. L. Li, A. M. Tulino, and S. Verdú. Design of reduced-rank MMSE multiuser detectors using random matrix methods. *IEEE Transactions on Information Theory*, 50(6):986–1008, 2004.

49. X. Liu, E. K. P̄. Chong, and N. Shroff. Opportunistic transmission scheduling with resource sharing constraints in wireless networks. *IEEE Journal on Selected Areas in Communications*, 19(10), October 2001.

50. P. Loubaton and W. Hachem. Asymptotic analysis of reduced rank Wiener filters. *Proc. 2003 Information Theory Workshop*, pages 328–331, March 2003.

51. W.-K. Ma, T. N. Davidson, K. M. Wong, Z.-Q. Luo, and P.-C. Ching. Quasi-maximum-likelihood multiuser detection using semi-definite relaxation with application to synchronous CDMA. *IEEE Transactions on Signal Processing*, 50(4):912–922, April 2002.

52. U. Madhow, K. Bruvold, and L. J. Zhu. Differential MMSE: A Framework for Robust Adaptive Interference Suppression for DS-CDMA Over Fading Channels. *IEEE Transactions on Communications*, 53(8):1377–1390, 2005.

53. U. Madhow and M. L. Honig. MMSE interference suppression for direct-sequence spread spectrum CDMA. *IEEE Transactions on Communications*, 42(12):3178–3188, December 1994.

54. N. B. Mandayam and S. Verdú. Analysis of an approximate decorrelating detector. *Wireless Personal Communications: An International Journal*, 6(1–2):97–111, January 1998.

55. S. Moshavi, E. G. Kanterakis, and D. L. Schilling. Multistage linear receivers for DS-CDMA systems. *International Journal of Wireless Information*, 3:1–17, January 1996.

56. M. S. Mueller and J. Salz. A unified theory of data-aided equalization. *Bell System Technical Journal*, 60(9):2023–2038, November 1981.

57. T. Ojanpera and R. Prasad. An overview of air interface multiple access for UMTS/IMT-2000. *IEEE Commun. Mag*, 36(9):82–95, 1998.

58. D. A. Pados and S. N. Batalama. Low-complexity blind detection of DS/CDMA signals: Auxiliary-vector receivers. *IEEE Transactions on Communications*, 45(12):1586–1594, December 1997.

59. N. Prasad and M. K. Varanasi. Analysis of decision feedback detection for MIMO Rayleigh-fading channels and the optimization of power and rate allocations. *IEEE Transactions on Information Theory*, 50(6):1009–1025, June 2004.

60. J. G. Proakis and M. Salehi. *Digital Communications*. McGraw-Hill, 5th edition, 2008.

61. G. S. Rajappan and M. L. Honig. Signature sequence adaptation for DS-CDMA with multipath. *IEEE Journal on Selected Areas in Communications*, 20(2):384–395, February 2002.

62. D. J. Ryan, I. B. Collings, and I. V. L. Clarkson. GLRT-Optimal noncoherent lattice decoding. *IEEE Transactions on Signal Processing*, 55(7 Part 2):3773–3786, 2007.

63. J. Salz. Digital transmission over cross-coupled linear channels. *AT&T Technical Journal*, 64(6):1147–1159, July 1985.

64. C. Sankaran and A. Ephremides. Solving a class of optimum multiuser detection problems with polynomial complexity. *IEEE Transactions on Information Theory*, 44(5): 1958–1961, 1998.

65. L. L. Scharf. The SVD and reduced rank signal processing. *Signal Processing*, 25(2):113–133, 1991.

66. C. Schlegel and A. Grant. Polynomial complexity optimal detection of certain multiple-access systems. *IEEE Transactions on Information Theory*, 46(6):2246–2248, 2000.

67. K. S. Schneider. Optimum detection of code division multiplexed signals. *IEEE Transactions on Aerospace and Electronic Systems*, 15(1):181–185, January 1979.

68. R. Singh and L. B. Milstein. Interference suppression for DS/CDMA. *IEEE Transactions on Communications*, 47(3):446–453, March 1999.

69. N. Slonimsky. *Lexicon of Musical Invective: Critical Assaults on Composers since Beethoven's Time*. University of Washington Press, 1987.

70. P. Spasojevic and C. N. Georghiades. The slowest descent method and its application to sequence estimation. *IEEE Transactions on Communications*, 49(9):1592–1604, 2001.

71. V. Srivastava, J. Neel, A. B. Mackenzie, R. Menon, L. A. Dasilva, J. E. Hicks, J. H. Reed, and R. P. Gilles. Using game theory to analyze wireless ad hoc networks. *IEEE Communications Surveys & Tutorials*, 7(4):46–56, 2005.

72. T. Starr, J. M. Cioffi, and P. J. Silverman. Understanding digital subscriber line technology. *Prentice Hall Communications Engineering and Emerging Technologies Series*, 1999.

73. Y. Sun. *Transmitter and Receiver Techniques for Wireless Fading Channels*. PhD thesis, Northwestern University, Evanston, IL, December 2004.

74. Y. Sun and M. L. Honig. Performance of reduced-rank equalization. *IEEE Transactions on Information Theory*, 52(10):4548–4562, October 2005.

75. P. H. Tan and L. K. Rasmussen. Multiuser detection in CDMA—a comparison of relaxations, exact, and heuristic search methods. *IEEE Transactions on Wireless Communications*, 3(5):1802–1809, 2004.

76. P. H. Tan, L. K. Rasmussen, and T. J. Lim. Constrained maximum-likelihood detection in CDMA. *IEEE Transactions on Communications*, 49(1):142–153, 2001.

77. E. Telatar. Capacity of multi-antenna gaussian channels. *Euro. Trans. Telecommun.*, 10:585–595, Nov.–Dec. 1999.

78. L. F. Trichard, J. S. Evans, and I. B. Collings. Large system analysis of linear multistage parallel interference cancellation. *IEEE Transactions on Communications*, 50(11):1778–1786, 2002.

79. D. Tse and P. Viswanath. *Fundamentals of Wireless Communication*. Cambridge University Press, 2005.

80. W. Van Etten. Maximum likelihood receiver for multiple channel transmission systems. *IEEE Transactions on Communications*, pages 276–283, February 1976.

81. B. D. Van Veen and K. M. Buckley. Beamforming: a versatile approach to spatial filtering. *IEEE Signal Processing Magazine*, 5(2):4–24, 1988.

82. M. K. Varanasi. Noncoherent detection in asynchronous multiuser channels. *IEEE Transactions on Information Theory*, 39(1):157–176, 1993.

83. M. K. Varanasi. Decision feedback multiuser detection: a systematic approach. *IEEE Transactions on Information Theory*, 45(1):219–240, January 1999.

84. M. K. Varanasi and B. Aazhang. Multistage detection in asynchronous code-division multiple-access communications. *IEEE Transactions on Communications*, 38(4): 509–519, April 1990.

85. M. K. Varanasi and A. Russ. Noncoherent decorrelative detection for nonorthogonal multi-pulsemodulation over the multiuser Gaussian channel. *IEEE Transactions on Communications*, 46(12):1675–1684, 1998.

86. S. Verdri. Minimum probability of error for asynchronous Gaussian multiple-access channels. *IEEE Transactions on Information Theory*, 32(1):85–96, January 1986.

87. S. Verdú. *Multiuser Detection*. Cambridge University Press, 1998.

88. S. Verdú and S. Shamai. Spectral efficiency of CDMA with random spreading. *IEEE Transactions on Information Theory*, 45(2):622–640, March 1999.

89. H. Vikalo and B. Hassibi. On the sphere-decoding algorithm II. Generalizations, second-order statistics, and applications to communications. *IEEE Transactions on Signal Processing, [see also IEEE Transactions on Acoustics, Speech, and Signal Processing]*, 53(8 Part 1):2819–2834, 2005.

90. H. Vikalo, B. Hassibi, and T. Kailath. Iterative decoding for MIMO channels via modified sphere decoding. *IEEE Transactions on Wireless Communications*, 3(6):2299–2311, 2004.

91. E. Visotsky and U. Madhow. Noncoherent multiuser detection for CDMA systems with nonlinearmodulation: a non-Bayesian approach. *IEEE Transactions on Information Theory*, 47(4):1352–1367, 2001.

92. D. Warrier and U. Madhow. Noncoherent communication in space and time. *IEEE Transactions on Information Theory*, 48:651–668, 2002.

93. M. E. Weippert, J. D. Hiemstra, J. S. Goldstein, and M. D. Zoltowski. Insights from the relationship between the multistage Wiener filter and the method of conjugate gradients. *Sensor Array and Multichannel Signal Processing Workshop Proceedings, 2002*, pages 388–392, August 2002.

94. G. Woodward, R. Ratasuk, M. L. Honig, and P. B. Rapajic. Minimum mean-squared error multiuser decision-feedback detectors for DS-CDMA. *IEEE Transactions on Communications*, 50(12):2104–2112, December 2002.

95. J. K. Zhang, T. N. Davidson, and K. M. Wong. Uniform decomposition of mutual information using MMSE decision feedback detection. *IEEE International Symposium on Information Theory*, pages 714–718, September 2005.

96. J. K. Zhang, A. Kavtic, and K. M. Wong. Equal-diagonal QR decomposition and its application to precoder design for successive-cancellation detection. *IEEE Transactions on Information Theory*, 51(1):154–172, January 2005.

2

ITERATIVE TECHNIQUES

Alex Grant and Lars K. Rasmussen

2.1 INTRODUCTION

Theoretical analysis promises performance improvements for multiple-access communications through the use of joint detection (in the case of uncoded transmission) or joint decoding (for encoded data). These gains in bit error probability and/or achievable reliable information transmission rate can be significant in nonorthogonal multiple-access channels. Except in a few special circumstances [1–3] however, the optimal joint detection/decoding problem is prohibitively complex [4,5]. Typically, the implementation complexity scales exponentially with the number of independent transmitters. This adds an additional layer of complexity, on top of that required for the optimal detection or decoding of single-user transmissions.

The ensuing engineering challenge is therefore to find practical encoders and decoders yielding performance approaching the theoretical limits. Since the 1993 "turbo revolution," modern coding practice has been dominated by iteratively decoded codes such as parallel [6] and serial [7] turbo codes, low-density parity-check codes [8] and repeat-accumulate codes [9]. The success of these codes is largely due to their iterative decoding algorithms, which approximate optimal decoding with manageable computational complexity. As a result, iterative processing has emerged as the framework of choice for the design of near-optimal systems in a variety of communications scenarios.

This chapter describes iterative processing techniques for both uncoded and coded multiple-access channels. For completeness and to introduce notation, we establish a system model in Section 2.1.1. Linear multiuser filters are an important component of

Advances in Multiuser Detection. Edited by Michael L. Honig
Copyright © 2009 John Wiley & Sons, Inc.

the methods to be discussed in this chapter. For reference, we give an overview of some well-known linear filters in Section 2.1.2, with generalizations to accommodate non-uniform prior probabilities. The latter becomes important when the linear filters are used as components of iterative multiuser decoders.

Section 2.2 describes iterative methods for the implementation of multiuser detectors. Both linear and non-linear iterations are considered. Linear iterative detectors are developed within the framework of well-known iterative methods for solving linear systems (whose exact solutions correspond to given linear filters, such as the decorrelator). This general set-up reveals the commonality between many engineering approaches reported in the literature. It also allows the system designer to specify an arbitrary linear filter (e.g., decorrelator, or linear minimum mean squared error filter) as the goal of the iteration. Convergence analysis is reduced to determination of eigen-values of certain iteration matrices. The gradient method for solution of linear systems emerges as the common underlying iterative method, and in particular the conjugate gradient method results in fast convergence, finite termination and a convenient interference cancellation structure.

Section 2.3 describes iterative multiuser decoders, based on the turbo principle. We describe a general framework, which iterates between linear multiuser filtering and single-user decoding. Variance transfer methods are used for convergence analysis.

The following notation will be used. Column vectors will be represented as lower-case bold symbols, $\mathbf{x} \in \mathbb{R}^n$ with real elements $x_i \in \mathbb{R}$, $i = 1, 2, \ldots, n$. Similarly, matrices will be uppercase bold symbols, $\mathbf{X} \in \mathbb{R}^{n \times m}$, with elements $X_{ij} \in \mathbb{R}$, $i = 1, 2, \ldots, n$, $j = 1, 2, \ldots, m$. The superscript ^{t} denotes matrix transpose, $^{-1}$ denotes matrix inverse when applied to matrices, while $\mathbf{X} = \text{diag}(x_1, x_2, \ldots, x_n)$ is a diagonal matrix with diagonal entries $X_{ii} = x_i$. Similarly, $\text{diag}(\mathbf{X})$ is the diagonal matrix with the same principal diagonal as $\mathbf{X}$. The standard inner product between two vectors is denoted $(\mathbf{x}, \mathbf{y}) = \mathbf{x}^t\mathbf{y}$. For a function $f : \mathbb{R}^n \mapsto \mathbb{R}$, $\partial f/\partial \mathbf{x} = (\partial f/\partial x_1, \ldots, \partial f/\partial x_n)^t$.

The expectation operator is denoted $\mathsf{E}\{\cdot\}$, and:

$$\text{Cov}\{\mathbf{x}, \mathbf{y}\} = \mathsf{E}\{(\mathbf{x} - \mathsf{E}\{\mathbf{x}\})(\mathbf{y} - \mathsf{E}\{\mathbf{y}\})^t\}$$

denotes the covariance matrix of random vectors $\mathbf{x}$ and $\mathbf{y}$. $N(\mathbf{m}, \boldsymbol{\Sigma})$ denotes a multivariate normal distribution with mean vector $\mathbf{m}$ and covariance matrix $\boldsymbol{\Sigma}$.

2.1.1 System Model

In this chapter, we consider a K user symbol-synchronous multiple-access system with error control coding, transmitting over a common additive white Gaussian noise (AWGN) channel. We adopt the well-known chip/symbol synchronous discrete-time model [10], for notational and conceptual simplicity. We further assume real, binary modulation. Extentions of this model to asynchronous multipath fading channels with multiple antennae are relatively straightforward but notationally cumbersome. More detailed models, including the development from continuous-time to discrete-time can be found in [11,12].

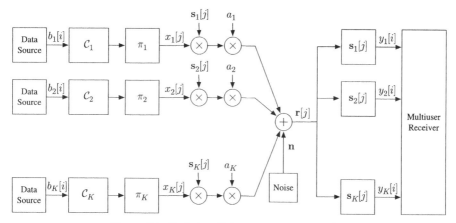

Figure 2.1. Coded multiple-access system model.

A block diagram of the system including the transmission side and the receiver front-end is shown in Figure 2.1. The source for user $k = 1, 2, \ldots, K$ produces a sequence of $R_k L$ information bits, $b_k[i]$, $i = 1, 2, \ldots, R_k L$. These sequences are mutually independent. Each user's sequence is encoded by a binary code, $\mathcal{C}_k$, of rate R_k, passed through an interleaver, π_k, and modulated onto an antipodal, e.g., binary phase shift keying (BPSK) signal constellation to produce the interleaved and modulated code bit sequence $x_k[j], j = 1, 2, \ldots, L$.

With the assumption of chip synchronism, the output of the chip-match-filtered multiple-access channel is characterized by a set of length N modulation, or spreading vectors $\mathbf{s}_k[i] = (s_{k,1}[i], s_{k,2}[i], \ldots, s_{k,N}[i])^t$, for user $k = 1, 2, \ldots, K$, at symbol interval $i = 1, 2, \ldots, L$. The $s_{k,n}[i]$ are referred to as chip amplitudes. With the further assumption of symbol synchronism, the length N received vector $\mathbf{r}[i]$ at symbol i can be written:

$$\mathbf{r}[i] = \sum_{k=1}^{K} \mathbf{s}_k[i] a_k x_k[i] + \mathbf{n}[i] \tag{2.1}$$

$$= \mathbf{S}[i]\mathbf{A}\mathbf{x}[i] + \mathbf{n}[i] \tag{2.2}$$

where $\mathbf{S}[i] = (\mathbf{s}_1[i], \mathbf{s}_2[i], \ldots, \mathbf{s}_K[i])$ is a $N \times K$ matrix which has as column k the modulation vector $\mathbf{s}_k[i]$ for user k, symbol i. The scalars a_k are the received user amplitudes, combining the effect of different transmit levels and any channel attenuation (for simplicity we assume that these amplitudes are fixed for the duration of the transmission). These are collected into a diagonal matrix $\mathbf{A} = \text{diag}(a_1, a_2, \ldots, a_K)$. The vector $\mathbf{x}[i] = (x_1[i], x_2[i], \ldots, x_K[i])^t$ consists of the transmitted symbols at interval i, and $\mathbf{n}$ is a vector of zero-mean additive white Gaussian noise samples with $\mathsf{E}\{\mathbf{n}[i]\mathbf{n}'[i]\} = \sigma^2 \mathbf{I}$. For this real-valued channel, the noise variance is related to the underlying thermal noise density via $\sigma^2 = N_0/2$.

We will assume binary unit norm modulation vectors, $s_k[i] \in \{-1/\sqrt{N}, +1/\sqrt{N}\}^N$ and real amplitudes $a_k \in \mathbb{R}$. This models direct-sequence code-division multiple-access, and we will use CDMA terminology throughout the chapter. Extension to complex values accommodates a wider range of linear multiple-access channel including TDMA, FDMA, MC-CDMA, and OFDM.

One useful analytical model for the modulation vectors is to let the $s_{k,m}[i]$ be random, chosen uniformly *i.i.d.* from $\{-1/\sqrt{N}, +1/\sqrt{N}\}$. We shall refer to this model as random spreading. Otherwise, in certain circumstances, the modulation vectors for each user may be fixed for all time, $s_k[i] = s_k$. We shall refer to this as fixed spreading.

Equation (2.2) is a chip-level discrete-time model. A symbol-level model is obtained via matched filtering of the modulating spreading sequences, $s_k[i]$, for all k and i. Specifically for symbol interval i, we get:

$$\mathbf{y}[i] = \mathbf{S}^t[i]\mathbf{r}[i] \tag{2.3}$$

$$= \mathbf{S}^t[i]\mathbf{S}[i]\mathbf{A}\mathbf{x}[i] + \mathbf{S}^t[i]\mathbf{n}[i] \tag{2.4}$$

$$= \mathbf{R}[i]\mathbf{A}\mathbf{x}[i] + \mathbf{z}[i] \tag{2.5}$$

where $\mathbf{R}[i] = \mathbf{S}^t[i]\mathbf{S}[i]$ is the correlation matrix[1] at interval i, and $\mathbf{z}[i]$ is a vector of colored Gaussian noise samples with $\mathsf{E}\{\mathbf{z}[i]\mathbf{z}^t[i]\} = \sigma^2\mathbf{R}[i]$. Note that according to our assumption of unit norm modulation vectors, $R_{ii} = 1$ and $|R_{ij}| \leq 1$ for $i \neq j$.

The vector $\mathbf{y}[i]$ is referred to as the symbol-matched-filter output. The single-user matched-filter-detector outputs:

$$\hat{\mathbf{x}}_{\mathrm{MF}}[i] = \mathrm{sgn}(\mathbf{y}[i]) \tag{2.6}$$

i.e., independent hard decisions for each user, based on the symbol matched filter output for each user.

The discrete-time models are developed for a given arbitrary time interval i. Throughout the chapter, the symbol interval index will only be included when conceptually required. For example, the symbol index is omitted for cases where an arbitrary symbol interval is considered.

2.1.2 Multiuser Detectors

The minimum probability of error detector for linear multiple-access channels is the maximum likelihood (ML) detector, considering all users jointly [13]. Optimal multiuser detection is, however, an NP-complete problem [4], where the brute-force implementation of the ML detector is in effect a Viterbi algorithm applied to the received matched-filtered signal. Such a system has an inherent complexity of the order $\mathcal{O}(2^K)$ for binary modulation.

---

[1]$\mathbf{R}$ is also commonly denoted the symbol-level channel matrix.

As alternatives to the ML detector, an abundance of sub-optimal reduced-complexity receiver structures exist in the literature. For an introduction, see [10,12,14]. Most of these structures are sub-optimal approximations to classic design criteria. Some approaches, however, satisfy relaxed optimality criteria. The decorrelator [15] and the linear minimum mean-squared error (LMMSE) detector [16–18] are two examples of linear detectors that are optimal with respect to modified criteria. Such linear detection schemes are, however, fundamentally limited in the spectral efficiency that they can offer in conjunction with independent decoding of each user's data [19,20].

LMMSE filters, incorporating *a-priori* information on the data bits, have been described for iterative multiuser decoding in [21–25]. Similar filters have been derived based on probabilistic data association [26], the nonlinear MMSE criterion [27], and neural networks [28].

In the following subsections, the optimal multiuser detector, the decorrelator and a set of MMSE detectors are briefly described. Iterative implementations of the decorrelator and the linear MMSE detectors will subsequently be considered in Section 2.2 for uncoded transmission, and in Section 2.3 for coded transmission.

2.1.2.1 Optimal Multiuser Detectors
The optimal multiuser detector first appeared in [29] and was subsequently developed in [13,30,31]. Minimum error probability optimal multiuser detection results from maximum *a posteriori* probability (MAP) decisions:

$$\hat{x}_{\mathrm{MAP}} = \arg \max_{\mathbf{x} \in \{-1,1\}^K} \Pr(\mathbf{x}|\mathbf{r}) \tag{2.7}$$

$$= \arg \max_{\mathbf{x} \in \{-1,1\}^K} p(\mathbf{r}|\mathbf{x})\Pr(\mathbf{x}) \tag{2.8}$$

Assuming that $\Pr(\mathbf{x}) = 2^{-K}$ for all data symbol vectors, we get the maximum-likelihood (ML) multiuser detector:

$$\hat{x}_{\mathrm{ML}} = \arg \max_{\mathbf{x} \in \{-1,1\}^K} p(\mathbf{r}|\mathbf{x}) \tag{2.9}$$

$$= \arg \max_{\mathbf{x} \in \{-1,1\}^K} \exp\left(-\frac{1}{2}||\mathbf{r} - \mathbf{SAx}||^2\right) \tag{2.10}$$

$$= \arg \max_{\mathbf{x} \in \{-1,1\}^K} \exp\left(-\frac{1}{2}(\mathbf{y} - \mathbf{RAx})\mathbf{R}^{-1}(\mathbf{y} - \mathbf{RAx})^t\right) \tag{2.11}$$

In general, MAP and ML multiuser detectors suffer from complexity of the order $\mathcal{O}(2^K)$, rendering such approaches impractical for large K.

Linear multiuser detectors such as the decorrelator or the linear minimum mean-squared error filter can significantly reduce the complexity, while still improving the bit error rate of uncoded transmission as compared to the single-user matched filter (2.6).

2.1.2.2 Decorrelator Detector   The decorrelator approach to multiuser detection introduced in [15] is closely related to the zero-forcing equalizer for intersymbol interference (ISI) channels [32].

Suppose the correlation matrix $\mathbf{R} = \mathbf{S}'\mathbf{S}$ is invertible, and let:

$$\hat{\mathbf{x}} = (\mathbf{S}'\mathbf{S})^{-1}\mathbf{S}'\mathbf{r} \tag{2.12}$$

$$= \mathbf{R}^{-1}\mathbf{y} \tag{2.13}$$

$$= \mathbf{A}\mathbf{x} + \mathbf{R}^{-1}\mathbf{z} \tag{2.14}$$

where $\mathsf{E}\{\mathbf{R}^{-1}\mathbf{z}\mathbf{z}'\mathbf{R}^{-1}\} = \sigma^2\mathbf{R}^{-1}$. Whereas the matched filter output $\mathbf{y}$ (2.6) consists of correlated data $\mathbf{R}\mathbf{A}\mathbf{x}$ and correlated noise $\mathbf{z}$, $\hat{\mathbf{x}}$ consists of independent data $\mathbf{A}\mathbf{x}$ and correlated noise $\mathbf{R}^{-1}\mathbf{z}$. The inverse correlation matrix $\mathbf{R}^{-1}$ can increase the effect of the noise when $\mathbf{R}$ is badly conditioned, leading to a trade-off between multiple-access interference (MAI) elimination and noise enhancement.

The decorrelator is the maximum likelihood detector for two different modifications of the original detection problem (2.9). Firstly, it is the ML detector resulting from relaxing the integer constraints $\mathbf{x} \in \{-1, 1\}^K$ to $\mathbf{x} \in \mathbb{R}^K$.

Secondly, the decorrelator does not require knowledge of the user amplitudes $\mathbf{A}$, which is one implementation advantage of this detector. The decorrelator is the ML detector in the absence of knowledge of $\mathbf{A}$. It provides an estimate of $\mathbf{A}\mathbf{x}$ rather than of $\mathbf{x}$. Note that the amplitudes $\mathbf{A}$ in (2.14) do not affect hard decisions.

2.1.2.3 Linear Minimum Mean-Squared Error Detectors   The LMMSE filter for multiuser detection [16] is closely related to the LMMSE equalizer for ISI channels [32]. Adaptive methods for LMMSE filtering were first suggested in [17,18].

For jointly random vectors $\mathbf{x}$ and $\mathbf{y}$, with:

$$\bar{\mathbf{x}} = \mathsf{E}\{\mathbf{x}\}$$

$$\bar{\mathbf{y}} = \mathsf{E}\{\mathbf{y}\}$$

$$\mathbf{G}_{xy} = (\mathsf{Cov}\{\mathbf{y}, \mathbf{y}\})^{-1}\mathsf{Cov}\{\mathbf{y}, \mathbf{x}\}$$

the LMMSE estimate of $\mathbf{x}$ given $\mathbf{y}$ is [33, V.C.19]:

$$\bar{\mathbf{x}} + \mathbf{G}'_{xy}(\mathbf{y} - \bar{\mathbf{y}})$$

For jointly Gaussian $\mathbf{x}, \mathbf{y}$ this linear estimate in fact minimizes the mean squared error.

The LMMSE estimate can be found for either the chip-level model (2.2) considering $\mathbf{x}, \mathbf{r}$ or the symbol-level model (2.5), considering $\mathbf{x}, \mathbf{y}$:

$$\hat{\mathbf{x}}^r = \mathsf{E}\{\mathbf{x}\} + \mathbf{G}'_{xr}(\mathbf{r} - \bar{\mathbf{r}}) \tag{2.15}$$

$$\hat{\mathbf{x}}^y = \mathsf{E}\{\mathbf{x}\} + \mathbf{G}'_{xy}(\mathbf{y} - \bar{\mathbf{y}}) \tag{2.16}$$

Suppose the data symbols $\mathbf{x} \in \{-1, 1\}^K$ are independent with:

$$E\{\mathbf{x}\} = \bar{\mathbf{x}} \tag{2.17}$$

in turn leading to:

$$\text{Cov}\{\mathbf{x}, \mathbf{x}\} = \mathbf{I} - \text{diag}(\bar{\mathbf{x}}\bar{\mathbf{x}}^t) = \mathbf{V} \tag{2.18}$$

The corresponding LMMSE estimates are then:

$$\hat{\mathbf{x}}^r = \bar{\mathbf{x}} + \mathbf{VAS}^t(\mathbf{SAVAS}^t + \sigma^2\mathbf{I})^{-1}(\mathbf{r} - \mathbf{SA}\bar{\mathbf{x}}) \tag{2.19}$$

$$\hat{\mathbf{x}}^y = \bar{\mathbf{x}} + \mathbf{VAR}(\mathbf{RAVAR} + \sigma^2\mathbf{R})^{-1}(\mathbf{y} - \mathbf{RA}\bar{\mathbf{x}}) \tag{2.20}$$

$$= \bar{\mathbf{x}} + \mathbf{A}^{-1}(\mathbf{R} + \sigma^2(\mathbf{AVA})^{-1})^{-1}(\mathbf{y} - \mathbf{RA}\bar{\mathbf{x}}) \tag{2.21}$$

In case $E\{\mathbf{x}\} = 0$, we have $\text{Cov}\{\mathbf{x}, \mathbf{x}\} = \mathbf{I}$, and thus, (2.19) and (2.21) simplify into the well-known LMMSE estimates [16][2]:

$$\hat{\mathbf{x}}^r = \mathbf{AS}^t(\mathbf{SA}^2\mathbf{S}^t + \sigma^2\mathbf{I})^{-1}\mathbf{r} \tag{2.22}$$

$$\hat{\mathbf{x}}^y = \mathbf{A}^{-1}(\mathbf{R} + \sigma^2\mathbf{A}^{-2})^{-1}\mathbf{y} \tag{2.23}$$

It should be emphasized that the two LMMSE filter representations above are equivalent. Also if σ^2 is set to zero in (2.22) and (2.23), the filters simplify into the decorrelator filters in (2.12) and (2.13), respectively.

Since for binary random variables, the mean and probability mass function are uniquely related, $\bar{\mathbf{x}}$ represents the *a-priori* information available to the detector. Interference cancellation and iterative multiuser decoding are two examples where such prior information can be easily exploited. The case $E\{\mathbf{x}\} = \mathbf{0}$ corresponds to uniform prior probabilities.

2.1.2.4 *Per-User Linear Minimum Mean-Squared Error Detectors* In some applications where prior information is available, it may still be desirable to ignore information pertaining to the user of interest. This is the case for joint iterative multiuser decoding, discussed in Section 2.3. In such cases, a different filter needs to be derived for each user. The filters presented here were first proposed in [21,22] for iterative multiuser decoding.

Assuming user k is the user of interest, we set $E\{x_k\} = 0$ in (2.17) for derivation purposes, even though prior information for user k may be available. The corresponding covariance matrix can be expressed as a function of $\bar{x}_k$ and $\mathbf{V}$ in (2.17) and (2.18),

---

[2]For the symbol-level estimate in (2.23), the scaling by $\mathbf{A}^{-1}$ can be ignored as subsequent hard decisions are not affected by scaling.

respectively:

$$\text{Cov}\{\mathbf{x}, \mathbf{x}\} = \mathbf{V}_k = \mathbf{V} - \mathbf{e}_k \mathbf{e}_k^t \bar{x}_k^2 \qquad (2.24)$$

where $\mathbf{e}_k$ is a length K column vector with a one in position k and zeros elsewhere. The per-user LMMSE estimates for user k are now:

$$\hat{x}_k^r = a_k \mathbf{s}_k^t \left(\mathbf{SAV}_k \mathbf{AS}^t + \sigma^2 \mathbf{I}\right)^{-1} \left(\mathbf{r} - \sum_{j \neq k} \mathbf{s}_j a_j \bar{x}_j\right) \qquad (2.25)$$

$$\hat{x}_k^y = a_k \mathbf{s}_k^t \mathbf{S} \left(\mathbf{RAV}_k \mathbf{AR} + \sigma^2 \mathbf{R}\right)^{-1} \left(\mathbf{y} - \sum_{l=1}^{K} \sum_{j \neq k} \mathbf{e}_l R_{lj} a_j \bar{x}_j\right) \qquad (2.26)$$

2.1.2.5 *Per-User Approximate Nonlinear MMSE Detector* Suppose the multiple-access interference experienced by user k is a multivariate Gaussian random vector, i.e.:

$$\mathbf{r} = \sum_{j=1}^{K} \mathbf{s}_j a_j \mathbf{x}_j + \mathbf{n} \qquad (2.27)$$

$$= \mathbf{s}_k a_k x_k + \underbrace{\sum_{j \neq k} \mathbf{s}_j a_j x_j}_{\text{MAI}} + \mathbf{n} \qquad (2.28)$$

where:

$$\sum_{j \neq k} \mathbf{s}_j a_j x_j \sim N\left(\sum_{j \neq k} \mathbf{s}_j a_j \bar{x}_j, \sum_{j \neq k} \mathbf{s}_j \mathbf{s}_j^t a_j^2 (1 - \bar{x}_j^2)\right) \qquad (2.29)$$

$$= N\left(\sum_{j \neq k} \mathbf{s}_j a_j \bar{x}_j, \mathbf{SAW}_k \mathbf{AS}^t\right) \qquad (2.30)$$

with:

$$\mathbf{W}_k = \mathbf{V} - \mathbf{e}_k \mathbf{e}_k^t \left(1 - \bar{x}_k^2\right) \qquad (2.31)$$

This approach was suggested in [26] as an application of probabilistic data association to multiuser detection, and later derived based on approximations to the nonlinear MMSE (NMMSE) criterion in [27]. The principle has previously been proposed for feedback equalization of ISI channels in [34].

From (2.30) it follows that $\mathbf{r}$ given x_k is a multivariate Gaussian random vector with distribution:

$$\mathbf{r} \sim N\left(\mathbf{s}_k a_k x_k + \sum_{j \neq k} \mathbf{s}_j a_j \bar{x}_j, \ \mathbf{SAW}_k \mathbf{AS}^t + \sigma^2 \mathbf{I} \right) \tag{2.32}$$

and the corresponding conditional probability distribution function for $\mathbf{r}$ given x_k is thus proportional to:

$$p(\mathbf{r}|x_k) \propto \exp\left(x_k a_k \mathbf{s}_k^t (\mathbf{SAW}_k \mathbf{AS}^t + \sigma^2 \mathbf{I})^{-1} \left(\mathbf{r} - \sum_{j \neq k} \mathbf{s}_j a_j \bar{x}_j \right) \right) \tag{2.33}$$

Given these Gaussian assumptions, we can now find the approximate nonlinear MMSE estimate as:

$$\hat{x}_k^r = \tanh\left(a_k \mathbf{s}_k^t (\mathbf{SAW}_k \mathbf{AS}^t + \sigma^2 \mathbf{I})^{-1} \left(\mathbf{r} - \sum_{j \neq k} \mathbf{s}_j a_j \bar{x}_j \right) \right) \tag{2.34}$$

The corresponding symbol-level estimate is determined as:

$$\hat{x}_k^y = \tanh\left(a_k \mathbf{s}_k^t \mathbf{S} (\mathbf{RAW}_k \mathbf{AR} + \sigma^2 \mathbf{R})^{-1} \left(\mathbf{y} - \mathbf{s}_k^t \sum_{j \neq k} \mathbf{s}_j a_j \bar{x}_j \right) \right) \tag{2.35}$$

As the number of users K grows large the diagonal elements of the matrix $\mathbf{SAW}_k \mathbf{AS}^t + \sigma^2 \mathbf{I}$ in (2.34) become dominant, encouraging the approximation [27]:

$$\mathbf{SAW}_k \mathbf{AS}^t + \sigma^2 \mathbf{I} = \sum_{j \neq k} \mathbf{s}_j \mathbf{s}_j^t a_j^2 (1 - \bar{x}_j^2) + \sigma^2 \mathbf{I} \tag{2.36}$$

$$\simeq \left(\frac{1}{N} \sum_{j \neq k} a_j^2 (1 - \bar{x}_j^2) + \sigma^2 \right) \mathbf{I} \tag{2.37}$$

Applying this approximation, the matrix inversion is avoided and we arrive at a simplified estimate:

$$\hat{x}_k^r = \tanh\left(\frac{a_k \mathbf{s}_k^t \left(\mathbf{r} - \sum_{j \neq k} \mathbf{s}_j a_j \bar{x}_j \right)}{\frac{1}{N} \sum_{j \neq k} a_j^2 (1 - \bar{x}_j^2) + \sigma^2} \right) \tag{2.38}$$

Closely related schemes are found in [35] based on direct approximation of the NMMSE detector and in [28] based on neural networks arguments.

The linear filter applied within the argument to the hyperbolic tangent function in (2.34) has been proposed in [23,24] as the *nominal* LMMSE detector. The per-user nominal LMMSE estimate is found as:

$$\hat{x}_k = a_k \mathbf{s}_k^t \left(\mathbf{SAW}_k \mathbf{AS}^t + \sigma^2 \mathbf{I} \right)^{-1} \left(\mathbf{r} - \sum_{j \neq k} \mathbf{s}_j a_j \bar{x}_j \right) \tag{2.39}$$

and the corresponding approximation, similar to (2.38), is found as:

$$\hat{x}_k = \frac{a_k \mathbf{s}_k^t \left(\mathbf{r} - \sum_{j \neq k} \mathbf{s}_j a_j \bar{x}_j \right)}{\dfrac{1}{N} \sum_{j \neq k} a_j^2 (1 - \bar{x}_j^2) + \sigma^2} \tag{2.40}$$

The detector in (2.40) is in fact the matched filter detector proposed for iterative multi-user decoding in [36], exploiting prior information.

2.2 ITERATIVE JOINT DETECTION FOR UNCODED DATA

Optimal multiuser detection is prohibitively complex for a large number of active users. Lower complexity solutions can be provided by iterative or multi-stage detection structures corresponding to serial or parallel interference cancellation. This class of detectors forms new decisions for each user by subtracting estimates of other user interference obtained from previous iterations. The original multistage detector [37,38] is a parallel cancellation device, while the first serial canceller was suggested in [39]. In both cases, data estimates of the interfering users are based on hard tentative decisions. As an alternative, linear multistage cancellation was suggested in [40,41] for parallel and serial scheduling. Linear cancellation schemes were subsequently recognized as iterative methods for solving linear systems [42,43]. Nonlinearities such as the hyperbolic tangent [28,44,45] and the hard limiter (also know as the clipper) [46–49] were later introduced at the output of each cancellation stage, as a compromise between hard decision (sign test) and linear soft decisions. In fact, the clipped soft decision interference canceller was shown in [48] to be an iterative solution to the original optimization problem with modified constraints. We shall pursue this idea further in Section 2.2.3. For an overview of linear and nonlinear interference cancellation techniques, see [50].

In Section 2.2.1 below the concepts of interference cancellation are introduced based on linear parallel processing. The principles are subsequently extended to arbitrary cancellation schedules and nonlinear tentative decision functions. In Section 2.2.2 linear methods are discussed in detail, while nonlinear approaches are investigated as solutions to constrained optimization problems in Section 2.2.3.

2.2.1 Interference Cancellation

In a communications system where performance is interference limited, interference cancellation is appealing as a strategy for improving performance. Consider the symbol-level multiple-access channel model (2.5). The matched filter output for user k at an arbitrary symbol interval is:

$$y_k = \mathbf{s}_k^t \mathbf{r}$$

$$= a_k x_k + \mathbf{s}_k^t \left(\sum_{j \neq k} \mathbf{s}_j a_j x_j + \mathbf{n} \right)$$

$$= a_k x_k + \underbrace{\sum_{j \neq k} R_{kj} a_j x_j}_{\text{multiple-access interference}} + z_k$$

The term:

$$\sum_{j \neq k} \mathbf{s}_k^t \mathbf{s}_j a_j x_j = \sum_{j \neq k} R_{kj} a_k x_j$$

is the multiple-access interference experienced by user k and $z_k = \mathbf{s}_k^t \mathbf{n}$ is Gaussian noise, colored across users.

Suppose that user k possesses estimates of the interfering users' symbols. Then an obvious approach for reducing multiple-access interference is to subtract an estimate of the MAI from the received signal. Typically, this estimate will not be perfect, and there will be some residual MAI. This motivates an iterative cancellation approach for recursively improving symbol estimates and removing residual interference. At each iteration, the cancelled signal is used to generate a (hopefully improved) estimate for user k, which can then be used to cancel interference from other user's observations.

Iterations will be denoted by an index n. Let the symbol estimate of user $k = 1, 2, \ldots, K$ at iteration $n = 0, 1, 2, \ldots$ be $\hat{x}_k^{(n)}$. Given the estimates from iteration n, the updated estimate for user k at iteration $n + 1$ is:

$$\hat{x}_k^{(n+1)} = a_k^{-1} \mathbf{s}_k^t \left(\mathbf{r} - \sum_{j \neq k} \mathbf{s}_j a_j \hat{x}_j^{(n)} \right) \tag{2.41}$$

$$= a_k^{-1} \left(y_k - \sum_{j \neq k} R_{jk} a_j \hat{x}_j^{(n)} \right) \tag{2.42}$$

Equation 2.41 describes an interference canceller working at the chip rate, while Equation 2.42 operates at the symbol rate.

One possible choice for the initial estimate for each user is $\hat{x}_k^{(0)} = 0$. This leads to $\hat{x}_k^{(1)} = a_k^{-1} y_k$, which is equivalent to starting with an initial estimate $a_k^{-1} y_k$. Typically, the matched filter output is used as the initial estimate.

Note that if $\hat{\mathbf{x}}^{(n)} = \mathbf{x}$ the cancellation is perfect, and:

$$\hat{x}_k^{(n+1)} = x_k + \frac{z_k}{a_k}$$

which is an interference-free AWGN channel with signal-to-noise ratio a_k^2/σ^2.

2.2.1.1 Schedules for Iterative Cancellation

The iteration (2.41) describes a *parallel* update of each user's estimate. At each new iteration $n + 1$, all users cancel the contributions of other user estimates from the previous iteration n. Writing (2.42) for all users in vector form:

$$\hat{\mathbf{x}}^{(n+1)} = \mathbf{A}^{-1}(\mathbf{y} - (\mathbf{R} - \mathbf{I})\mathbf{A}\hat{\mathbf{x}}^{(n)}) \tag{2.43}$$

This is shown schematically, from the perspective of user k in Figure 2.2.

Another possibility is that the user's estimates are updated serially, for example in order of user number $k = 1, 2, \ldots$. In this case, at iteration $n + 1$ a particular user k has available the estimates of lower index users from the same iteration $n + 1$ (these have already been computed) and higher index users from the previous iteration n. The resulting *serial* update, in both chip-level (2.44) and symbol-level (2.45) form is:

$$\hat{x}_k^{(n+1)} = a_k^{-1}\mathbf{s}_k^t\left(\mathbf{r} - \sum_{j=1}^{k-1}\mathbf{s}_j a_j \hat{x}_j^{(n+1)} - \sum_{j=k+1}^{K}\mathbf{s}_j a_j \hat{x}_j^{(n)}\right) \tag{2.44}$$

$$= a_k^{-1}\left(y_k - \sum_{j=1}^{k-1}R_{jk} a_j \hat{x}_j^{(n+1)} - \sum_{j=k+1}^{K}R_{jk} a_j \hat{x}_j^{(n)}\right) \tag{2.45}$$

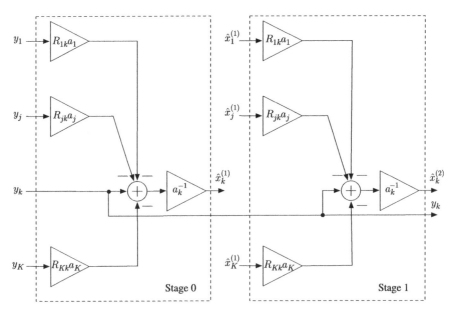

Figure 2.2. Parallel interference cancellation.

Since this approach uses updated estimates as soon as they are available, one might expect serial cancellation to perform better than parallel cancellation. Let $\mathbf{L}$ be the strictly lower triangular part of $\mathbf{R} = \mathbf{L} + \mathbf{L}^t + \mathbf{I}$. Then the vector form of (2.45) is:

$$\hat{\mathbf{x}}^{(n+1)} = \mathbf{A}^{-1}\left(\mathbf{y} - \mathbf{L}\mathbf{A}\hat{\mathbf{x}}^{(n+1)} - \mathbf{L}^t\mathbf{A}\hat{\mathbf{x}}^{(n)}\right) \tag{2.46}$$

This is shown schematically in Figure 2.3.

Both the parallel and serial cancellation structures pass two signals from iteration to iteration, namely the user estimates $\hat{\mathbf{x}}^{(n)}$, and the received vector $\mathbf{y}$.

2.2.1.2 *Implementation via Residual Error Update* The output of the chip-level parallel interference canceller (2.41) can be re-written as:

$$\hat{x}_k^{(n+1)} = a_k^{-1}\mathbf{s}_k^t\left(\mathbf{r} - \sum_{j=1}^{K}\mathbf{s}_j a_j \hat{x}_j^{(n)} + \mathbf{s}_k a_k \hat{x}_k^{(n)}\right)$$

$$= \hat{x}_k^{(n)} + a_k^{-1}\mathbf{s}_k^t\left(\mathbf{r} - \sum_{j=1}^{K}\mathbf{s}_j a_j \hat{x}_j^{(n)}\right) \tag{2.47}$$

$$= \hat{x}_k^{(n)} + a_k^{-1}\mathbf{s}_k^t \boldsymbol{\eta}^{(n)} \tag{2.48}$$

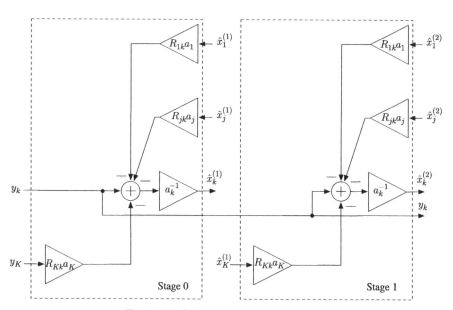

Figure 2.3. Serial interference cancellation.

where:

$$\boldsymbol{\eta}^{(n)} = \mathbf{r} - \sum_{j=1}^{K} \mathbf{s}_j a_j \hat{x}_j^{(n)} \tag{2.49}$$

is the residual error, or noise hypothesis remaining after iteration n. With perfect cancellation, it consists only of the thermal noise, i.e., $\boldsymbol{\eta}^{(n)} = \mathbf{n}$.

Similarly, the output of the chip-level serial canceller (2.44) can be re-written as:

$$\hat{x}_k^{(n+1)} = \hat{x}_k^{(n)} + a_k^{-1} \mathbf{s}_k^t \boldsymbol{\eta}_k^{(n)} \tag{2.50}$$

where:

$$\boldsymbol{\eta}_k^{(n)} = \mathbf{r} - \sum_{j=1}^{k-1} \mathbf{s}_j a_j \hat{x}_j^{(n+1)} - \sum_{j=k}^{K} \mathbf{s}_j a_j \hat{x}_j^{(n)} \tag{2.51}$$

is the residual error seen by user k after iteration n.

Let:

$$\Delta \boldsymbol{\eta}_k^{(n)} = \mathbf{s}_k a_k \left(\hat{x}_k^{(n-1)} - \hat{x}_k^{(n)} \right) \tag{2.52}$$

which simplifies to:

$$\Delta \boldsymbol{\eta}_k^{(n)} = -\mathbf{s}_k \mathbf{s}_k^t \boldsymbol{\eta}_k^{(n-1)} \tag{2.53}$$

Then, the residual error for the parallel canceller can be written recursively as:

$$\boldsymbol{\eta}^{(n)} = \boldsymbol{\eta}^{(n-1)} + \sum_{j=1}^{K} \Delta \boldsymbol{\eta}_j^{(n)} \tag{2.54}$$

while the residual error for user k in the serial canceller can be written recursively as:

$$\boldsymbol{\eta}_k^{(n)} = \begin{cases} \boldsymbol{\eta}_K^{(n-1)} + \Delta \boldsymbol{\eta}_K^{(n)} & k = 1 \\ \boldsymbol{\eta}_{k-1}^{(n)} + \Delta \boldsymbol{\eta}_{k-1}^{(n+1)} & k > 1 \end{cases} \tag{2.55}$$

Rather than passing the received vector from stage to stage, as was described in the previous section, iterative interference cancellation can be accomplished by passing the continually updated residual error. Summarizing this development, parallel interference cancellation is implemented via (2.48), (2.52), and (2.54) as shown in Algorithm 1 below.

ALGORITHM 1: Paralled Cancellation

Initialize $\hat{\mathbf{x}}^{(0)} = 0$, $\boldsymbol{\eta}^{(0)} = \mathbf{r}$
for $n = 1, 2, \ldots$ **do**
   for $k = 1, 2, \ldots, K$ **do**

$$\hat{x}_k^{(n)} = \hat{x}_k^{(n-1)} + a_k^{-1} \mathbf{s}_k^t \boldsymbol{\eta}^{(n-1)}$$

$$\Delta \boldsymbol{\eta}_k^{(n)} = -\mathbf{s}_k \mathbf{s}_k^t \boldsymbol{\eta}_k^{(n-1)}$$

   end for

$$\boldsymbol{\eta}^{(n)} = \boldsymbol{\eta}^{(n-1)} + \sum_{j=1}^{K} \Delta \boldsymbol{\eta}_j^{(n)}.$$

end for

Serial cancellation differs only in the frequency with which the residual error is updated. The serial cancellation scheme is implemented via (2.48), (2.52), and (2.55), which is summarized in Algorithm 2.

ALGORITHM 2: Serial Cancellation

Initialize $\hat{\mathbf{x}}^{(0)} = 0$, $\boldsymbol{\eta}_1^{(0)} = \mathbf{r}$
for $n = 1, 2, \ldots$ **do**
   for $k = 1, 2, \ldots, K$ **do**

$$\hat{x}_k^{(n)} = \hat{x}_k^{(n-1)} + a_k^{-1} \mathbf{s}_k^t \boldsymbol{\eta}_k^{(n-1)}$$

$$\Delta \boldsymbol{\eta}_k^{(n)} = -\mathbf{s}_k \mathbf{s}_k^t \boldsymbol{\eta}_k^{(n-1)}$$

$$\boldsymbol{\eta}_k^{(n)} = \begin{cases} \boldsymbol{\eta}_K^{(n-1)} + \Delta \boldsymbol{\eta}_K^{(n)} & k = 1 \\ \boldsymbol{\eta}_{k-1}^{(n)} + \Delta \boldsymbol{\eta}_{k-1}^{(n+1)} & k > 1. \end{cases} \tag{2.56}$$

   end for
end for

The residual error approach has some desirable features for implementation. Cancellation schemes can be constructed based on a basic interference cancellation

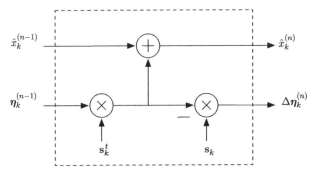

Figure 2.4. Interference cancellation module.

module, shown in Figure 2.4. The module for user k takes as inputs the previous estimate for user k, $\hat{x}_k^{(n-1)}$, and the most recent residual error, $\boldsymbol{\eta}_k^{(n-1)}$. Note that in the parallel case, the user index can be ignored as the residual error at iteration n is the same for all users $\boldsymbol{\eta}_k^{(n-1)} = \boldsymbol{\eta}^{(n-1)}$. From these inputs the module determines as outputs an updated estimate for user k, $\hat{x}_k^{(n)}$, according to (2.48) or (2.50) and a corresponding update, $\Delta\boldsymbol{\eta}_k^{(n)}$, to the residual error according to (2.52).

Using the module as a basic building block, arbitrary cancellation schemes can be systematically constructed. In Figure 2.5 a parallel cancellation scheme is shown, while a serial cancellation scheme is depicted in Figure 2.6.

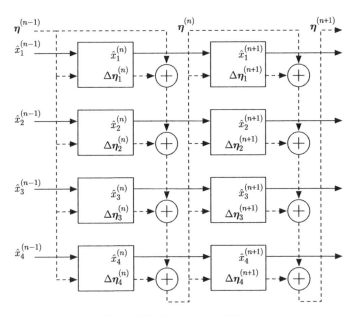

Figure 2.5. Parallel cancellation.

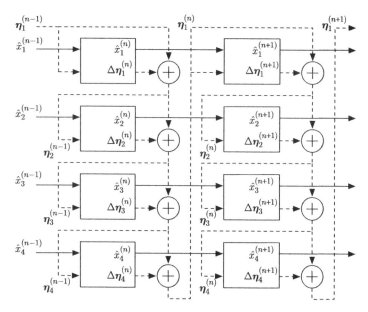

Figure 2.6. Serial cancellation.

2.2.1.3 Tentative Decision Functions

The cancellation strategies so far have been linear. Both the symbol estimates $\hat{x}_k^{(n)}$ and noise hypotheses $\boldsymbol{\eta}_k^{(n)}$ undergo recursive linear updates. The transmitted symbols belong to a discrete set, assumed to be $\{-1, +1\}$ in this chapter, yet the estimates $\hat{x}_k^{(n)}$ could be any real number. This could have a negative effect on the cancellation process. For example, the situation could arise where $|\hat{x}_k^{(n)}| \gg 1$, resulting in cancellation of an impossibly large interference estimate. Conversely, in certain circumstances, the cancellation process might be accelerated by making hard decisions.

This motivates the use of a non-linear *tentative decision function* $\zeta : \mathbb{R} \mapsto [-1, +1]$, which aims to limit the output of the canceller to lie in the interval $[-1, 1]$:

$$\hat{x}_k^{(n+1)} = \zeta(\hat{x}_k^{(n)} + a_k^{-1}\mathbf{s}_k^t\boldsymbol{\eta}_k^{(n)}) \tag{2.57}$$

In this case, however, (2.52) does not simplify into (2.53) due to the non-linearity. The corresponding non-linear interference cancellation module is shown in Figure 2.7.

A wide range of tentative decision functions have been described in the literature. A linear decision function $\zeta(x) = x$ results in linear cancellation which is discussed in detail in Section 2.2.2. Non-linear choices of the decision function obviously result in non-linear cancellation, discussed in detail in Section 2.2.3.

Three prominent examples of non-linear tentative decision functions are the hard limiter, or clipper, [48,49]:

$$\zeta(x) = \text{clip}(x) \triangleq \begin{cases} -1 & x \leq -1 \\ x & |x| < 1 \\ 1 & x \geq 1 \end{cases} \tag{2.58}$$

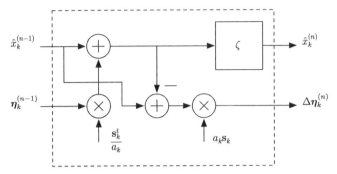

Figure 2.7. Interference cancellation module with tentative decision function.

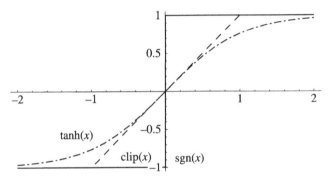

Figure 2.8. Tentative decision functions.

the soft limiter [26,28,35,44,45]:

$$\zeta(x) = \tanh(x) \tag{2.59}$$

and the hard decision [37,39]:

$$\zeta = \text{sgn}(x) \tag{2.60}$$

Figure 2.8 shows each of these non-linear decision functions. The effect of each of these non-linearities is to restrict the output of the canceller to be in the range $[-1, 1]$. This is investigated in more detail in Section 2.2.3.

2.2.2 Linear Methods

Implementation of linear detectors such as the decorrelator and the LMMSE filter require matrix inversion, which has cubic complexity in the number of users. This may be a problem if the modulating waveform set changes for every symbol interval.

A variety of iterative techniques are available for matrix inversion[3], reducing the required complexity. Techniques such as series expansion (polynomial detectors) [56–59], iterations (interference cancellation) [40–43,57,58,60–67], multi-state Wiener filtering [68,69], and gradient decent algorithms [57,58,60,62,67] have been widely applied for linear multiuser detection. Here, we investigate these techniques and show they can all be implemented as interference cancellation.

Taylor series expansion techniques for multiuser detection was first proposed in [56] as the polynomial expansion detector, leading to a multistage implementation. The concept was further developed in [57,67] where series expansion was related to parallel interference cancellation. Detectors with a finite number of stages were designed based on minimizing the output mean squared error. A more comprehensive presentation of polynomial expansion detectors is found in [58,59], where the Cayley Hamilton theorem is applied to obtain the exact matrix inverse in a finite number of stages. Reduced-rank multi-stage Wiener filtering, and related Krylov sub-space techniques, lead to closely related iterative polynomial expansion algorithms [68,70]. The Taylor and Cayley Hamilton series expansions are presented in detail in Section 2.2.2.3. below.

In [42] interference cancellation schemes were recognized as iterative solution methods for solving linear systems. This relationship was further developed in [57,61–63], where weighted (or partial) cancellation suggested in [71] was recognized as relaxation techniques for the well-known Jacobi and Gauss-Seidel iterations. A comprehensive study of iterative solution methods as interference cancellation schemes is found in [43] where convergence aspects were investigated in detail. Optimization of relaxation factors have been investigated in [65,66] based on large system analysis bounds on extreme eigen-values of the correlation matrix.

Iterative solution methods are defined in Section 2.2.2.4, with separate subsections dedicated to the Jacobi and Gauss-Seidel iterations in Sections 2.2.2.5 and 2.2.2.6, respectively. Relaxed iterations are also presented and shown to improve convergence behavior.

Gradient decent schemes, such as the steepest decent and conjugate gradient algorithms, were first applied for multiuser detection in [60]. Independently, partial cancellation with changing relaxation factors was recognized as the steepest decent algorithm in [57,67]. A method for determining the relaxation factors (step sizes) for reaching the minimum mean squared error possible in a given finite number of steps was developed. Cancellation schemes based on the steepest decent algorithm and the conjugate gradient algorithm were considered in [62], and similar to [60] the conjugate gradient approach was found to provide superior performance. The conjugate gradient algorithm and the multi-stage Wiener filter have been shown to be algebraically equivalent in [70,72]. Multiuser interference rejection algorithms based on the multi-stage Wiener filter was investigated in [69].

Section 2.2.2.7 includes a comprehensive presentation of gradient decent algorithms applied for multiuser detection. In particular, the conjugate gradient algorithm

[3]See [51–55] for comprehensive treatments of such techniques.

is found to compare favorably with parallel cancellation in terms of performance versus complexity.

2.2.2.1 Solutions to Linear Systems

The output of either the decorrelator (2.13) or the LMMSE (2.23) filter (represented as $\hat{\mathbf{x}}$ when the filter is left unspecified) can be written as the solution to a least-squares optimization problem, namely:

$$\hat{\mathbf{x}} = \arg \min_{\mathbf{x} \in \mathbb{R}^K} \|\mathbf{M}\mathbf{x} - \mathbf{y}\|_2^2 \qquad (2.61)$$

where $\mathbf{M} = \mathbf{R}$ for the decorrelator, $\mathbf{M} = \mathbf{R}\mathbf{A}$ for the normalized decorrelator, and $\mathbf{M} = \mathbf{R} + \sigma^2 \mathbf{A}^{-2}$ for the LMMSE. In Section 2.2.3, several useful non-linear receiver structures will result by introducing additional constraints to the optimization problem (2.61). The solution to the *unconstrained* optimization problem (2.61) is, however, obtained via solution of a system of linear equations:

$$\mathbf{M}\hat{\mathbf{x}} = \mathbf{y} \qquad (2.62)$$

Efficient solution of (possibly large) linear systems such as (2.62) has been the subject of much study in numerical linear algebra. This viewpoint will prove useful in the design of low-complexity approximations to the decorrelator and LMMSE filters.

2.2.2.2 Direct Solution

Direct solution of a system of linear equations can be accomplished using Gaussian elimination followed by back-substitution. In the case of symmetric $\mathbf{M}$ (which is true for both the decorrelator and LMMSE), this is equivalent to Cholesky factorization of the matrix $\mathbf{M}$ into a lower triangular matrix $\mathbf{F}$:

$$\mathbf{M} = \mathbf{F}\mathbf{F}^t \qquad (2.63)$$

followed by forward and backward substitution,[4]

$$\mathbf{F}\mathbf{z} = \mathbf{y} \quad \text{forward substitution}$$
$$\mathbf{F}^t\hat{\mathbf{x}} = \mathbf{z} \quad \text{backward substitution}$$

In the general case, Cholesky factorization is $\mathcal{O}(K^3/3)$, and each substitution step is $\mathcal{O}(K^2/2)$. Useful references for these methods are [51,52].

If the matrix $\mathbf{M}$ is band-diagonal, i.e., $M_{ij} = 0$ for $|i-j| > b$, where the integer $b \ll K$ is the semibandwidth, then the Cholesky decomposition is $\mathcal{O}(K(b^2 + 3b))$. Likewise, the complexity of the substitution steps decreases to $\mathcal{O}(Kb)$.

Thus, the direct solution approach involves one relatively complex matrix decomposition, followed by two substitution steps.

---

[4]Consider $\mathbf{F}\mathbf{F}^t\hat{\mathbf{x}} = \mathbf{y}$ with $\mathbf{z} = \mathbf{F}^t\hat{\mathbf{x}}$.

2.2.2.3 Series Expansions One strategy for the design of multistage receiver structures is to develop series expansions for $\mathbf{M}^{-1}$, i.e., find coefficients c_n such that:

$$\mathbf{M}^{-1} = \sum_n c_n \mathbf{M}^n \qquad (2.64)$$

Series expansions motivate a multistage structure shown in Figure 2.9. Such series expansion detectors were first proposed in [56] and developed further in [57,58].

A series with K terms involves calculation of $\mathbf{M}^K$, which is $\mathcal{O}(K^3)$. For this approach to be attractive from an implementation point of view, truncation of the series to $n \ll K$ must yield a sufficiently accurate approximation, or the desired level of performance.

Let:

$$p_{\mathbf{M}}(\lambda) = \sum_{n=0}^{K} (-1)^{K-n} c_{K-n} \lambda^n = \det(\mathbf{M} - \lambda \mathbf{I}) = 0 \qquad (2.65)$$

be the characteristic equation of the matrix $\mathbf{M}$. The coefficients c_n, $n = 0, 1, \ldots, K$ in (2.65) are the elementary symmetric functions of the eigen-values $\lambda_1 \geq \lambda_2 \geq \cdots \geq \lambda_K$ of $\mathbf{M}$ [53, p. 41], namely:

$$c_n(\lambda_1, \ldots, \lambda_K) = \sum_{1 \leq i_1 \leq \cdots \leq i_n} \prod_{j=1}^{n} \lambda_{i_j} \qquad (2.66)$$

The c_n are also given as the sum of the $\binom{K}{n}$ different $n \times n$ principal minors of $\mathbf{M}$ [53, Theorem 1.2.12]. In particular, $c_0 = 1$, $c_1 = \mathrm{tr}\,\mathbf{M}$, and $c_K = \det \mathbf{M}$.

The Cayley Hamilton theorem [53, p. 86] says that a matrix satisfies its own characteristic equation.

Theorem 1 (Cayley Hamilton)

$$p_{\mathbf{M}}(\mathbf{M}) = \sum_{n=0}^{K} (-1)^{K-n} c_{K-n} \mathbf{M}^n = 0 \qquad (2.67)$$

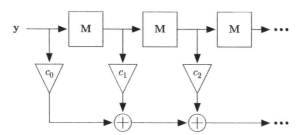

Figure 2.9. Multistage receiver structure motivated by series expansion.

The Cayley Hamilton theorem provides a way to write any power of $\mathbf{M}$ as a linear combination of at most $K + 1$ terms from $\mathbf{M}^n$, $n = 0, 1, \ldots, K$. In particular:

$$\mathbf{M}^{-1} = \frac{1}{(-1)^K \det(\mathbf{M})} \sum_{n=1}^{K} (-1)^{K-n} c_{K-n} \mathbf{M}^{n-1} \tag{2.68}$$

The series (2.68) describes a K-stage multistage implementation. Computation of all the coefficients c_n, however, is just as complex as matrix inversion in the first place. The important point is that there does exist a particular choice of coefficients c_n such that the finite power series (2.68) implements matrix inversion exactly.

An alternative expansion is provided by the following Taylor series:

$$(\mathbf{I} + \mathbf{X})^{-1} = \sum_{n=0}^{\infty} (-\mathbf{X})^n \tag{2.69}$$

which is convergent if the spectral radius of $\mathbf{X}$ satisfies $\rho(\mathbf{X}) < 1$ (recall that the spectral radius of a matrix is the absolute value of its largest eigen-value). Setting $\mathbf{X} = \mathbf{M} - \mathbf{I}$ in (2.69) results in:

$$\mathbf{M}^{-1} = \sum_{n=0}^{\infty} (-1)^n (\mathbf{M} - \mathbf{I})^n \tag{2.70}$$

which is convergent if $\rho(\mathbf{M}) < 2$.

The first order truncation of this series for $\mathbf{M} = \mathbf{R}$ results in the approximate decorrelator [10, Section 5.4] and [73]:

$$\hat{\mathbf{x}}^{(1)} = (2\mathbf{I} - \mathbf{R})\mathbf{y} \tag{2.71}$$

$$= \mathbf{y} - \underbrace{(\mathbf{R} - \mathbf{I})\mathbf{y}}_{\substack{\text{Interference} \\ \text{Estimate}}} \tag{2.72}$$

Equation 2.72 reveals a parallel interference cancellation structure (see Equation 2.75 below). The second term in (2.72) can be regarded as an estimate of the multiple-access interference. In the absence of noise, $\mathbf{y} = \mathbf{R}\mathbf{x} = \mathbf{x} + (\mathbf{R} - \mathbf{I})\mathbf{x}$ and the approximate decorrelator can be seen to be approximating the MAI term $(\mathbf{R} - \mathbf{I})\mathbf{x}$ by $(\mathbf{R} - \mathbf{I})\mathbf{y}$, which corresponds to using the matched filter outputs as direct estimates of the interfering symbols.

Higher-order truncations of the Taylor series result in the following n-th order approximation to the decorrelator [56]:

$$\hat{\mathbf{x}}^{(n)} = \mathbf{y} + (\mathbf{I} - \mathbf{R})\mathbf{y} + (\mathbf{I} - \mathbf{R})^2\mathbf{y} + \cdots + (\mathbf{I} - \mathbf{R})^n\mathbf{y} \tag{2.73}$$

$$= \mathbf{y} - (\mathbf{R} - \mathbf{I})\hat{\mathbf{x}}^{(n-1)} \tag{2.74}$$

which is nothing more than parallel interference cancellation. This is apparent, by writing out the approximation from the perspective of user k as:

$$\hat{x}_k^{(n)} = \underbrace{y_k}_{\text{User } k \text{ matched filter output}} - \underbrace{\sum_{k' \neq k} R_{kk'} \hat{x}_{k'}^{(n-1)}}_{\text{Interference estimate from previous stage}} \qquad (2.75)$$

Figure 2.10a shows stage n of (2.74). This coincides with the structure shown in Figure 2.2 for $\mathbf{A} = \mathbf{I}$. Note that the decorrelator (which does not require knowledge of $\mathbf{A}$) actually estimates $\mathbf{Ax}$, as compared to the normalized decorrelator, which estimates $\mathbf{x}$. Later we shall see how to obtain the structure of Figure 2.2 via iterative implementation of the normalized decorrelator.

In Section 2.2.2.5, the same result will be derived from the viewpoint of iterative solution methods. In particular, (2.73) is known as the Jacobi method for iterative solution of a linear system. The per-iteration performance and convergence characteristics of (2.74) will be evaluated in Section 2.2.2.5.

Similar to (2.72), a n-th order Taylor series approximation may be obtained from the LMMSE filter:

$$\hat{\mathbf{x}}^{(n)} = \mathbf{y} - (\mathbf{R} + \mathbf{A}^{-2}\sigma^2 - \mathbf{I})\hat{\mathbf{x}}^{(n-1)} \qquad (2.76)$$

$$= \mathbf{y} - (\mathbf{R} - \mathbf{I})\hat{\mathbf{x}}^{(n-1)} - \mathbf{A}^{-2}\sigma^2\hat{\mathbf{x}}^{(n-1)} \qquad (2.77)$$

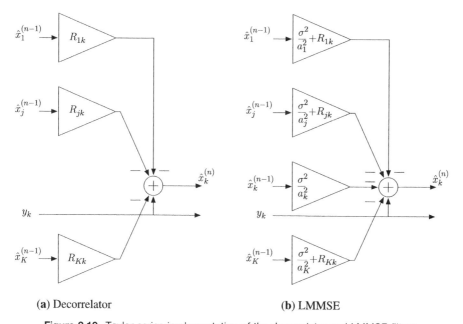

(a) Decorrelator   (b) LMMSE

Figure 2.10. Taylor series implementation of the decorrelator and LMMSE filters.

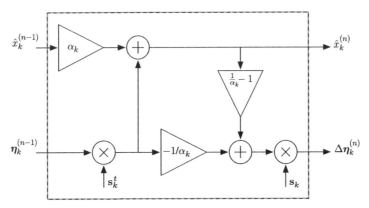

Figure 2.11. Interference cancellation module for decorrelator, $\alpha_k = 1$ and LMMSE, $\alpha_k = 1 - \sigma^2/a_k^2$.

which retains the same parallel interference cancellation structure as (2.72), with the extra cancellation of $\mathbf{A}^{-2}\sigma^2\hat{\mathbf{x}}^{(n-1)}$. Since $\mathbf{A}$ is diagonal, computation of $\mathbf{A}^{-2}$ is easy. Stage n of the Taylor series implementation of the LMMSE is shown in Figure 2.10b. Truncated Taylor series approximation of the LMMSE filter is therefore only marginally more complex than for the decorrelator filter, although it does require knowledge of the signal-to-noise ratio $\mathbf{A}^2/\sigma^2$.

Alternatively, (2.74) and (2.76) can be implemented using the chip-level residual error approach with the interference cancellation module shown in Figure 2.11.

In contrast to the Cayley Hamilton approach (2.68), the Taylor series does not require the computation of any series coefficients (all coefficients are 1). However this comes at a cost. Firstly, the Taylor series requires an infinite number of terms for exact solution. Secondly, the series is only convergent for $\rho(\mathbf{M}) < 2$.

2.2.2.4 Iterative Solution Methods
Motivated by (2.72) and (2.76), it is interesting to consider other iterative approaches for the approximation of solutions to linear systems. This is a well-trodden area of linear algebra and indeed, it is possible to devise many different iterative approaches for solution of linear systems, [54]. The main idea is to devise an easily computable iteration that converges rapidly to the required solution.

A wide class of iterations can be defined based on a linear splitting of the matrix $\mathbf{M}$ into two according to $\mathbf{M} = \mathbf{M}_1 - \mathbf{M}_2$. With such a splitting, (2.62) can be re-written as $(\mathbf{M}_1 - \mathbf{M}_2)\hat{\mathbf{x}} = \mathbf{y}$, resulting in the following fixed point equation:

$$\mathbf{M}_1\hat{\mathbf{x}} = \mathbf{y} + \mathbf{M}_2\hat{\mathbf{x}} \tag{2.78}$$

This motivates an iteration of the form:

$$\mathbf{M}_1\mathbf{x}^{(n+1)} = \mathbf{y} + \mathbf{M}_2\mathbf{x}^{(n)} \tag{2.79}$$

This is a *stationary* iteration, meaning that the same operation is performed each iteration (M_1 and M_2 do not depend on the iteration number n).

There are two main requirements.

1. The matrix M_1 should be chosen such that it is easy to solve systems of the form $M_1 x = z$. Obvious candidates for M_1 would be triangular or diagonal matrices.
2. The matrices M_1 and M_2 should be chosen such that (2.79) converges to the solution of the original system.

Furthermore, (2.79) should converge as quickly as possible, preferably in much less than K iterations.

In order to investigate the convergence properties of (2.79), it is necessary to define an appropriate notion of convergence. Preferably, the convergence would be analyzed in terms of the bit error rate; however, this is not tractable. Instead, one possibility is to consider a norm of the error at each iteration. Let $\hat{x} = M^{-1}y$ be the desired solution, and let $x^{(n)}$ be the output of the n-th iteration of (2.79), where $x^{(0)}$ is defined as the initial input to the iteration. Let $e^{(n)} = \hat{x} - x^{(n)}$ be the associated error vector. Substitution of $x^{(n)} = \hat{x} - e^{(n)}$ into (2.79) results in:

$$e^{(n)} = M_1^{-1}M_2 e^{(n-1)} = (M_1^{-1}M_2)^n e^{(0)} \tag{2.80}$$

Suppose $\|\cdot\|$ is a vector norm (with associated induced matrix norm). According to the defining properties of vector and matrix norms:

$$\|e^{(n)}\| = \|(M_1^{-1}M_2)^n e^{(0)}\| \leq \|(M_1^{-1}M_2)^n\|\|e^{(0)}\| \tag{2.81}$$

Now according to [53, Corollary 5.6.14] matrix norms are in the limit dominated by the spectral radius:

$$\lim_{n \to \infty} \|X^n\| = \rho(X)^n \tag{2.82}$$

and this yields the following result.

Theorem 2 *A necessary and sufficient condition for convergence of (2.79) in any norm is:*

$$\rho(M_1^{-1}M_2) < 1 \tag{2.83}$$

Typically, it is the Euclidean norm that is of interest. In this case, since $M_1^{-1}M_2$ is symmetric:

$$\|(M_1^{-1}M_2)^n\|_2^{1/n} = \rho(M_1^{-1}M_2) \tag{2.84}$$

which means that the total squared error decreases geometrically with rate $\rho(M_1^{-1}M_2)$. For this reason, make the following definitions.

DEFINITION 1 (Iteration Matrix and Convergence Factor).   *For a given linear decomposition* $\mathbf{M} = \mathbf{M}_1 - \mathbf{M}_2$, *define the iteration matrix* $\mathbf{B} = \mathbf{M}_1^{-1}\mathbf{M}_2$. *The quantity* $\|\mathbf{B}^n\|$ *is the convergence factor for n iterations and* $\rho(\mathbf{B})$ *is the asymptotic convergence factor.*

There are several design goals for accelerating convergence of iterative methods. It is advantageous to choose an initial guess $\mathbf{x}^{(0)}$ close to the desired solution (yet with low implementation complexity). For a fixed number of iterations, one could minimize the convergence factor. It remains to find suitable choices of $\mathbf{M}_1$ and $\mathbf{M}_2$.

2.2.2.5 Jacobi Iteration

The iteration (2.79) relies on being able to solve systems of the form $\mathbf{M}_1\mathbf{x} = \mathbf{z}$. This can be performed with very low complexity if $\mathbf{M}_1$ is a diagonal matrix, since inversion of a diagonal matrix requires only element-wise inversion of the diagonal elements.

One possible convenient choice is $\mathbf{M}_1 = \mathbf{I}$ and $\mathbf{M}_2 = \mathbf{I} - \mathbf{M}$, in which case the iteration (2.79) becomes:

$$\mathbf{x}^{(n+1)} = \mathbf{y} - (\mathbf{M} - \mathbf{I})\mathbf{x}^{(n)} \tag{2.85}$$

which is simply the n-th order Taylor series (2.70). The corresponding asymptotic convergence factor is $\rho(\mathbf{M} - \mathbf{I})$, which yields the same convergence criterion given above, namely $\rho(\mathbf{M}) < 2$.

Another possibility for diagonal $\mathbf{M}_1$ is $\mathbf{M}_1 = \mathbf{D} = \text{diag}(\mathbf{M})$, the diagonal matrix with the same diagonal elements as $\mathbf{M}$. More generally, a design parameter ω could be introduced:

$$\mathbf{M}_1 = \omega\mathbf{D} \tag{2.86}$$

$$\mathbf{M}_2 = \omega\mathbf{D} - \mathbf{M} \tag{2.87}$$

Let $\mathbf{x}^{(0)} = \mathbf{y}$ then (2.79) becomes:

$$\mathbf{x}^{(n+1)} = \frac{\mathbf{D}^{-1}}{\omega}(\mathbf{y} - (\mathbf{M} - \omega\mathbf{D})\mathbf{x}^{(n)}) \tag{2.88}$$

This is, in fact, identical to a series expansion of $\mathbf{M}^{-1}$ obtained via the n-th order Taylor series expansion of $(\omega\mathbf{D} + \mathbf{M})^{-1}$, rather than $(\mathbf{I} + \mathbf{M})^{-1}$ as described above. In either case, the resulting receiver structure is multistage parallel interference cancellation.

With $\omega = 1$ and unit energy modulation sequences, $\text{diag}(\mathbf{R}) = \mathbf{I}$ the resulting Jacobi iteration for the decorrelator is the same as (2.74).

Theorem 3   *The Taylor series iterative implementation of the decorrelator is convergent if and only if:*

$$\rho(\mathbf{R}) < 2 \tag{2.89}$$

Since the spectral radius of $\mathbf{R}$ depends upon the choice of spreading sequences, the convergence of the Taylor series expansion is not assured. However, by choosing an appropriate ω, this situation can be rectified.

Theorem 4   *The Jacobi implementation of the decorrelator with $\mathbf{M}_1 = \omega\mathbf{I}$ is convergent for any $\omega > 0$ such that $\rho(\mathbf{R}) < 2\omega$.*

For $\omega = 1$, the LMMSE filter (with unit energy modulation sequences), has $\mathrm{diag}(\mathbf{R} + \mathbf{A}^{-2}/\sigma^2) = \mathbf{I} + \mathbf{A}^{-2}\sigma^2$ and the resulting Jacobi iteration (with $\mathbf{x}^{(0)} = \mathbf{y}$) is:

$$\mathbf{x}^{(n+1)} = \left(\mathbf{I} + \mathbf{A}^{-2}\sigma^2\right)^{-1}\left(\mathbf{y} - (\mathbf{R} - \mathbf{I})\mathbf{x}^{(n)}\right) \tag{2.90}$$

Comparing to (2.74), the only difference is a per-user signal-to-noise ratio scaling each iteration.

Theorem 5   *The Jacobi iterative implementation of the LMMSE filter (2.90) is convergent if and only if:*

$$\rho_J = \rho\left[(\mathbf{I} + \mathbf{A}^{-2}\sigma^2)^{-1}(\mathbf{I} - \mathbf{R})\right] < 1 \tag{2.91}$$

The following theorem gives simpler bounds on convergence for the Jacobi LMMSE.

Theorem 6   *The Jacobi iterative implementation of the LMMSE filter (2.92) is convergent if:*

$$\rho(\mathbf{R} - \mathbf{I}) < 1 + \gamma_{\max}^{-1} \tag{2.92}$$

where $\gamma_{\max} = \max_k A_k^2/\sigma^2$. The iteration (2.90) is not convergent if:

$$\rho(\mathbf{R} - \mathbf{I}) > 1 + \gamma_{\min}^{-1} \tag{2.93}$$

where $\gamma_{\min} = \min_k A_k^2/\sigma^2$. For users with equal powers, (2.92) is also necessary for convergence.

Proof: Convergence occurs if $\rho_J < 1$. Now since both $\mathbf{I} + \mathbf{A}^{-2}\sigma^2$ and $\mathbf{I} - \mathbf{R}$ are symmetric:

$$\rho_J = \|(\mathbf{I} + \mathbf{A}^{-2}\sigma^2)^{-1}(\mathbf{I} - \mathbf{R})\|_2 \tag{2.94}$$

$$\leq \rho\|\left((\mathbf{I} + \mathbf{A}^{-2}\sigma^2)^{-1}\right)\|_2 \|\mathbf{I} - \mathbf{R}\|_2 \tag{2.95}$$

$$= \rho\left((\mathbf{I} + \mathbf{A}^{-2}\sigma^2)^{-1}\right)\rho(\mathbf{I} - \mathbf{R}) \tag{2.96}$$

$$= \max_k \frac{A_k^2}{A_k^2 + \sigma^2}\,\rho(\mathbf{R} - \mathbf{I}). \tag{2.97}$$

The result (2.92) follows from the definition of $\gamma_{\max}$. The negative result (2.93) follows using similar steps, noting that for diagonal $\mathbf{D}$ and symmetric $\mathbf{R}$, $\rho(\mathbf{D}^{-1}(\mathbf{I} - \mathbf{R})) > \min_k D_{kk}^{-1}\rho(\mathbf{R} - \mathbf{I})$.

Comparing Theorems 3 and 5, the Jacobi implementation of the LMMSE admits a wider radius of convergence, increasing that of the decorrelator by at least the inverse of the best user's SNR (Theorem 6).

Results from random matrix theory can be used to gain insight for large systems with random spreading. Of particular interest is the behavior of the extremal eigen-values [74–76].

Lemma 1 *Let $\lambda_{\min}$ and $\lambda_{\max}$ be the smallest and largest eigen-value of $\mathbf{R} = \mathbf{S}^t\mathbf{S}$ where the elements of $\mathbf{S}$ are chosen i.i.d. zero mean, variance $1/N$ and finite higher moments. Then as $K, N \to \infty$ such that $K/N \to \alpha < 1$:*

$$\lambda_{\min} \to (\sqrt{\alpha} - 1)^2$$
$$\lambda_{\max} \to (\sqrt{\alpha} + 1)^2$$

Henceforth, by *large systems*, we mean systems satisfying the assumptions of Lemma 1.

Using Lemma 1 and Theorem 3, it is straightforward to show that for large systems, parallel interference cancellation converges for only lightly loaded systems [77].

Theorem 7 *For large systems, the Taylor series implementation of the decorrelator is convergent only for:*

$$\alpha < \left(\sqrt{2} - 1\right)^2 \approx 0.17$$

The corresponding asymptotic convergence factor is:

$$\alpha + 2\sqrt{\alpha}$$

2.2.2.6 *Gauss-Seidel Iteration*

Another choice of splitting for which the iter-ation (2.79) can be performed with low complexity is if $\mathbf{M}_1$ is a triangular matrix, in which case:

$$\mathbf{M}_1\mathbf{x}^{(n+1)} = \mathbf{y} + \mathbf{M}_2\mathbf{x}^{(n)} \tag{2.98}$$

can be solved for $\mathbf{x}^{(n+1)}$ by back-substitution.

One possible choice for a triangular matrix is $\mathbf{M}_1 = \mathbf{D} + \mathbf{L}$, where $\mathbf{D} = \text{diag}(\mathbf{M})$ and $\mathbf{L}$ is the strictly lower-triangular part of $\mathbf{M} = \mathbf{D} + \mathbf{L} + \mathbf{L}^t$. More generally, one could consider $\mathbf{M}_1 = \mathbf{D}/\omega + \mathbf{L}$, where ω is a design parameter. Since $\omega = 1$ recovers the former choice, it is preferable to proceed with an arbitrary ω. These choices:

$$\mathbf{M}_1 = \frac{1}{\omega}\mathbf{D} + \mathbf{L} \tag{2.99}$$

$$\mathbf{M}_2 = \frac{1 - \omega}{\omega}\mathbf{D} - \mathbf{L}^t \tag{2.100}$$

result in the following iteration (with $\mathbf{x}^{(0)} = \mathbf{y}$):

$$\left(\frac{1}{\omega}\mathbf{D} + \mathbf{L}\right)\mathbf{x}^{(n+1)} = \mathbf{y} + \left(\frac{1-\omega}{\omega}\mathbf{D} - \mathbf{L}^t\right)\mathbf{x}^{(n)} \tag{2.101}$$

Re-arranging to get $\mathbf{x}^{(n+1)}$ on the left hand side gives:

$$\mathbf{x}^{(n+1)} = \mathbf{D}^{-1}\omega(\mathbf{y} - \mathbf{L}^t\mathbf{x}^{(n)} - \mathbf{L}\mathbf{x}^{(n+1)}) + (1-\omega)\mathbf{x}^{(n)} \tag{2.102}$$

Writing out the iteration from the perspective of user k reveals a successive cancellation structure (noting $L_{ij}^t = L_{ij}$ for symmetric $\mathbf{M}$):

$$x_k^{(n+1)} = \frac{\omega}{D_{kk}}\left(y_k - \underbrace{\sum_{i>k}L_{ik}x_i^{(n)}}_{\substack{\text{Cancel higher index} \\ \text{users from previous stage}}} - \underbrace{\sum_{j<k}L_{jk}x_i^{(n+1)}}_{\substack{\text{Cancel lower index} \\ \text{users from current stage}}}\right) + (1-\omega)x_k^{(n)} \tag{2.103}$$

This is the structure of Figure 2.3.

The triangular structure of $\mathbf{L}$ allows cancellation of interference estimates for each user as soon as they are available. This is in contrast to the Jacobi iteration (2.75) where each user's interference estimate is not cancelled until the next iteration.

For $\omega = 1$, (2.103) is known as the Gauss-Seidel iterative method. The parameter ω is a *relaxation* parameter, and in general the method is known as *successive relaxation*. Equation (2.103) shows how for $\omega \neq 1$, each successive estimate is a weighted sum of the previous estimate, and the most up-to-date interference cancelled estimate.

Since successive relaxation uses interference estimates as soon as they are available, it could be expected that this method has superior convergence properties compared to the Jacobi iteration. Indeed, this is the case.

Theorem 8   *Successive relaxation (2.102) is convergent for symmetric positive definite $\mathbf{M}$ and $\omega \in (0, 2)$.*

A proof of this Theorem can be found in [54, p. 232]. Serial cancellation is convergent for any choice of $0 < \omega < 2$, while parallel cancellation can also be made convergent with appropriate choice of ω. Thus the design problem is to select the relaxation parameter ω to accelerate convergence.

2.2.2.7 Descent Algorithms   For symmetric positive definite $\mathbf{M}$, define the vector norm:

$$\|\mathbf{x}\|_{\mathbf{M}^{\frac{1}{2}}} = \|\mathbf{M}^{-\frac{1}{2}}\mathbf{x}\|_2 = \mathbf{x}^t\mathbf{M}^{-1}\mathbf{x} \tag{2.104}$$

i.e., the norm derived from the inner product $(\,\cdot\,,\,\cdot\,)_{\mathbf{M}^{-1}}$. Define:

$$f(\mathbf{x}) = \frac{1}{2}\|\mathbf{M}\mathbf{x} - \mathbf{y}\|_{\mathbf{M}^{-\frac{1}{2}}} \tag{2.105}$$

Then an alternative to the least-squares minimization (2.61), which results in the same solution is:

$$\hat{\mathbf{x}} = \arg\min_{\mathbf{x}\in\mathbb{R}^K} f(\mathbf{x}) \tag{2.106}$$

The equivalence of solutions can be seen by noting that $\mathbf{M}\mathbf{x} = \mathbf{y}$ is the unique stationary point for both problems, since the gradient of $f(\mathbf{x})$ is equal to:

$$\nabla(\mathbf{x}) \triangleq \left(\frac{\partial f(\mathbf{x})}{\partial x_1}, \frac{\partial f(\mathbf{x})}{\partial x_2}, \ldots, \frac{\partial f(\mathbf{x})}{\partial x_K}\right)^t = \mathbf{M}\mathbf{x} - \mathbf{y} \tag{2.107}$$

This reveals that the gradient at each step is equal to the error vector $\mathbf{e} = \mathbf{M}\mathbf{x} - \mathbf{y}$, and the corresponding unique stationary point is $\mathbf{M}\mathbf{x} = \mathbf{y}$.

One well-known approach for numerical solution of optimization problems such as (2.61) is to use a descent algorithm [55]. In general terms, for a convex optimization problem such as (2.106), descent algorithms take the form:

$$\mathbf{x}^{(n+1)} = \mathbf{x}^{(n)} + t_n\mathbf{d}^{(n)} \tag{2.108}$$

where the direction $\mathbf{d}^{(n)}$ is chosen to reduce the value of the objective function and the step size t_n is chosen each iteration to minimize the objective function in the direction of $\mathbf{d}^{(n)}$. Optimization of t_n may be performed in closed form for the problem of interest (this is not always the case for a general convex problem). Let:

$$h(t) = f\left(\mathbf{x}^{(n)}\right) - f\left(\mathbf{x}^{(n+1)}\right) \tag{2.109}$$

$$= f\left(\mathbf{x}^{(n)}\right) - f\left(\mathbf{x}^{(n)} + t_n\mathbf{d}^{(n)}\right) \tag{2.110}$$

be the decrease in the objective function due to step $n + 1$. The objective function, and hence $h(t)$ is quadratic, and the decrease is minimized for:

$$t_n = -\frac{\left(\mathbf{e}^{(n)}, \mathbf{d}^{(n)}\right)}{\left(\mathbf{d}^{(n)}, \mathbf{M}\mathbf{d}^{(n)}\right)} \tag{2.111}$$

Alternatively, a fixed step size may be used, which avoids computation of t_n according to (2.111).

One possible choice for the search direction is the negative gradient of the objective function, evaluated at the current point, $\mathbf{d}^{(n)} = -\nabla(\mathbf{x}^{(n)})$, resulting in:

$$\hat{\mathbf{x}}^{(n+1)} = \hat{\mathbf{x}}^{(n+1)} - t_n\nabla\left(\hat{\mathbf{x}}^{(n)}\right) \tag{2.112}$$

$$= t_n\mathbf{y} - (t_n\mathbf{M} - \mathbf{I})\hat{\mathbf{x}}^{(n)} \tag{2.113}$$

where the optimal step size is:

$$t_n = \frac{\|\mathbf{e}(n)\|_2}{\|\mathbf{M}^{\frac{1}{2}}\mathbf{e}^{(n)}\|_2} \tag{2.114}$$

Considering [52, Section 10.2], we have the following rate of convergence for steepest descent.

Theorem 9 *The error norm of the steepest descent algorithm with optimal step size decreases geometrically with rate at least:*

$$\left(1 - \frac{\lambda_{\min}}{\lambda_{\max}}\right)$$

where $\lambda_{\min}$ and $\lambda_{\max}$ are the smallest and largest eigen-values of **M**.

Theorem 10 *For large systems, the steepest descent implementation of the decorrelator has convergence factor:*

$$\frac{4\sqrt{\alpha}}{(\sqrt{\alpha} + 1)^2}$$

For the sub-optimal choice of step sizes $t_n = 1$, (2.113) is identical to the Taylor series expansion (and of course the Jacobi iteration). The relationship to the Taylor series is not altogether surprising, considering the role of the derivative in the Taylor expansion.

In the earlier discussion of the Gauss-Seidel and successive relaxation methods, a relaxation parameter was used to accelerate convergence. The gradient descent (2.113) is in fact the Jacobi iteration, with a relaxation parameter.

Consider a splitting of **M** into:

$$\mathbf{M}_1 = t\mathbf{I} \tag{2.115}$$

$$\mathbf{M}_2 = t\mathbf{I} - \mathbf{M} \tag{2.116}$$

The resulting iterative method (compare to (2.88)) is exactly the same as (2.113). For fixed t, the asymptotic convergence factor is:

$$\rho(t\mathbf{M} - \mathbf{I}) \tag{2.117}$$

Obviously, t can always be chosen such that $\rho(t\mathbf{M} - \mathbf{I}) < 1$ and the resulting algorithm is always convergent. For this to happen, t must satisfy:

$$t < \frac{2}{\lambda_{\max}(\mathbf{M})} \tag{2.118}$$

where $\lambda_{\max}(\mathbf{M})$ is the largest eigen-value of **M**.

Theorem 11   *There is a choice of relaxation parameter t in (2.113) such that parallel interference cancellation is always convergent.*

Section 2.2.2.6, described the successive relaxation method, in which per-user updates used the most up-to-date interference estimates to achieve faster convergence. The same strategy can be used in the gradient search. In fact, with $\mathbf{D} = \mathbf{I}$ and $\omega = t$, (2.102) is identical to (2.113), with successive updates of each user rather than parallel updates (this amounts to using the coordinate vectors as the directions). This can be seen by noting $\mathbf{M} - \mathbf{I} = \mathbf{L} + \mathbf{L}'$ and re-writing (2.113) using successive updates as:

$$\mathbf{x}^{(n+1)} = t_n\left(\mathbf{y} - \mathbf{L}'\mathbf{x}^{(n)} - \mathbf{L}\mathbf{x}^{(n+1)}\right) + (1 - t_n)\mathbf{x}^{(n)} \tag{2.119}$$

Thus both the parallel and serial interference cancellation are instances of gradient descent. The design goal is to choose the step size t_n at each iteration to yield fastest convergence.

Gradient descent (implemented either as parallel or serial interference cancellation with relaxation) is guaranteed to reduce the value of the objective function (the error norm) at each step. However, the series of search directions that it follows are set by the gradient directions at each stage, and convergence may be slow due to "overlapping" search directions.

A better approach is to ensure that each new search direction $\mathbf{d}^{(n)}$ is *orthogonal* to all previous directions (with respect to the particular inner product under consideration):

$$\left(\mathbf{d}^{(n+1)}, \mathbf{M}\mathbf{d}^{(j)}\right) = 0, \quad j = 0, 1, \ldots, n \tag{2.120}$$

One possible choice of orthogonal directions is a linear combination of the current error vector (direction of steepest descent) and the previous direction [54, p. 462], where the combining coefficient is chosen to ensure orthogonality:

$$\mathbf{d}^{(0)} = -\mathbf{e}^{(0)} \tag{2.121}$$

$$\mathbf{d}^{(n+1)} = -\mathbf{e}^{(n+1)} + \beta_n\mathbf{d}^{(n)} \tag{2.122}$$

where, substituting (2.122) into $\left(\mathbf{d}^{(n+1)}, \mathbf{d}^{(n)}\right)_{\mathbf{M}^{\frac{1}{2}}} = 0$:

$$\beta_n = \frac{\left(\mathbf{e}^{(n+1)}, \mathbf{M}\mathbf{d}^{(n)}\right)}{\left(\mathbf{d}^{(n)}, \mathbf{M}\mathbf{d}^{(n)}\right)} \tag{2.123}$$

The *conjugate gradient* algorithm is defined by (2.108), (2.111), (2.122), and (2.123).

The convergence properties of the conjugate gradient algorithm are characterized as follows [52, Theorem 10.2.6].

Theorem 12   *The error norm for the conjugate gradient algorithm decreases geometrically with rate at least*:

$$\left(\frac{\sqrt{\kappa} - 1}{\sqrt{\kappa} + 1}\right)$$

where $\kappa = \lambda_{\max}/\lambda_{\min}$ is the condition number of $\mathbf{M}$.

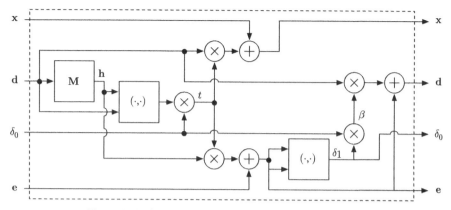

Figure 2.12. One stage of the conjugate gradient algorithm.

Theorem 13 *For large systems, the conjugate implementation of the decorrelator has convergence factor* $1/\sqrt{\alpha}$.

A computationally efficient summary of the algorithm is given as Algorithm 3, [54, p. 470] and a corresponding block diagram is given in Figure 2.12. It requires one matrix multiplication, two vector additions and two inner products per iteration. This compares favorably with the parallel cancellation approach, which requires one matrix multiplication and one vector addition per iteration. Thus, the only computational overhead to ensure both orthogonal search directions and optimal step sizes is two inner products and one vector addition.

The conjugate gradient algorithm has previously been suggested for multiuser detection in [60,62].

ALGORITHM 3: Conjugate Gradient Implementation

Inputs: $\mathbf{M}$, $\mathbf{y}$, $\hat{\mathbf{x}}^{(0)}$, ε
Initialize:

$$\mathbf{x} \leftarrow \hat{\mathbf{x}}^{(0)} \qquad \textit{Initial guess}$$
$$\mathbf{e} \leftarrow \mathbf{Mx} - \mathbf{y} \qquad \textit{Initial residual}$$
$$\delta_0 \leftarrow \mathbf{e}^t \mathbf{e} \qquad \textit{and residual norm}$$
$$\delta_1 \leftarrow \delta_0$$
$$\mathbf{d} \leftarrow -\mathbf{e} \qquad \text{Initial direction}$$

while $\delta_1 > \varepsilon$ **do**

$$\mathbf{h} \leftarrow \mathbf{Md} \qquad \textit{Internal variable}$$
$$t \leftarrow \delta_0 / \mathbf{d}^t \mathbf{h} \qquad \textit{Step size}$$
$$\mathbf{x} \leftarrow \mathbf{x} + t\mathbf{d} \qquad \textit{New estimate}$$

$$\begin{array}{ll}
\mathbf{e} \leftarrow \mathbf{e} + t\mathbf{h} & \textit{Residual} \\
\delta_1 \leftarrow \mathbf{e}^t\mathbf{e} & \textit{Residual norm} \\
\beta \leftarrow \delta_1/\delta_0 & \textit{Combining coefficient} \\
\delta_0 \leftarrow \delta_1 & \textit{Save old residual norm} \\
\mathbf{d} \leftarrow -\mathbf{e} + \beta\mathbf{d} & \textit{Next direction}
\end{array}$$

end while

2.2.3 Non-Linear Methods

Interference cancellation was first developed using hard decisions [37–39]. Although simple in concept, the hard decision canceller is difficult to analyze. Some insights into suboptimal fixed points and limit cycles have been reported in [78], and results from neurodynamics have been applied in [45] providing insight in the large system limit.

In contrast, cancellation based on the clipped soft decision in (2.58) implements the solution to a convex-constrained modification to the least-squares problem in (2.61) [48]. The clipped soft decision function was also used in [44], motivated by expectation-maximization arguments, and later on related to the convex-constrained ML problem [47,48] and nonlinear programming [46,49].

The soft limiter in (2.59) has resulted from several different approaches. The canceller structure in [28] was inspired by Hopfield neural networks, while approximate nonlinear MMSE arguments were used in [27,35] to arrive at interference cancellation with hyperbolic tangent decisions. Statistical physics [45] and probabilistic data association [26] are other approaches that lead to similar canceller structures.

In this section, a general framework based on constrained optimization is developed for deriving and optimizing nonlinear cancellation structures. The nonlinearities discussed above (see also Section 2.2.1.3) are recognized as special cases, and the corresponding optimization constraints are derived. In addition, the framework allows for new and improved nonlinear decision functions to be derived.

2.2.3.1 Constrained Optimization
In Section 2.2.2, we described several linear iterative methods for multiuser detection. These methods could all be viewed as gradient descent algorithms for numerical solution of an unconstrained quadratic optimization problem (2.106):

$$\hat{\mathbf{x}} = \arg\min \frac{1}{2} \|\mathbf{Mx} - \mathbf{y}\|_{\mathbf{M}^{-\frac{1}{2}}} \tag{2.124}$$

These methods converge to the solution of (2.124), namely $\mathbf{M}^{-1}\mathbf{y}$, and appropriate choice of $\mathbf{M}$ results in the decorrelator or LMMSE detector. In contrast, maximum likelihood detection results from solution of (2.124) subject to the constraint $\mathbf{x} \in \mathcal{D} = \{-1, 1\}^K$. Although this provides optimal performance, the required integer constraint causes the problem to be NP-complete in general. The complexity reduction, and associated performance reduction of linear detectors stems from replacing the integer constrained problem $\mathbf{x} \in \mathcal{D}$ with an unconstrained problem, $\mathbf{x} \in \mathbb{R}^K$.

An intermediate approach is to introduce constraints $\mathcal{C} \subset \mathbb{R}^K$ that more closely approximate $\mathcal{D}$. For example, one could require unit energy solutions, $\mathbf{x}'\mathbf{x} \leq 1$, or solutions in which each user's output is individually unit energy, $|x_k|^2 \leq 1$, $k = 1, 2, \ldots, K$.

In general, consider inequality constrained optimization problems of the form:

$$\min f(\mathbf{x}) \quad \text{subject to} \tag{2.125}$$

$$g_i(\mathbf{x}) \leq 0 \quad i = 1, 2, \ldots, m \tag{2.126}$$

where according to (2.105), $f(\mathbf{x}) = \frac{1}{2}\|\mathbf{Mx} - \mathbf{y}\|_{\mathbf{M}^{-\frac{1}{2}}}$ and the $g_i(\mathbf{x})$ are convex, differentiable functions. Clearly, selection of the g_i will have an impact on the performance of a detector that implements (2.125), depending on how well the constraint set $\mathcal{C}$ implied by (2.126), namely:

$$\mathcal{C} = \{\mathbf{x} \ : \ g_i(\mathbf{x}) \leq 0, i = 1, 2, \ldots, m\}$$

approximates $\mathcal{D}$.

The main focus of this section is on implementation of (2.125)–(2.126), rather than selection of $\mathcal{C}$. One approach for numerical solution of an inequality constrained optimization problem is to use barrier functions [55, Section 11.2].

First, we can replace the inequality constrained problem (2.125)–(2.126) by an equivalent unconstrained problem:

$$\min f(\mathbf{x}) + \underbrace{\sum_{i=1}^{m} I(g_i(\mathbf{x}))}_{\text{penalty function}} \tag{2.127}$$

where the ideal barrier function I is defined as:

$$I(u) = \begin{cases} 0 & u \leq 0 \\ \infty & u > 0 \end{cases}$$

The discontinuity of I at $u = 0$ resists analysis, and the main idea of the barrier function approach is to use a differentiable approximation $b(u) \approx I(u)$. For a given barrier function $b(u)$ and inequality constraints g_i, let:

$$\varphi(\mathbf{x}) = \sum_{i=1}^{m} b(g_i(\mathbf{x})) \tag{2.128}$$

be the associated penalty function. Replacing (2.127) with the approximation:

$$\min f(\mathbf{x}) + \varphi(\mathbf{x}) \tag{2.129}$$

we can now consider the class of suboptimal, non-linear multiuser detectors defined by various solution methods for (2.129). In the following development, the choice of barrier function b and constraints g_i is arbitrary. Later, we will see how certain choices of these functions recover several known multiuser detectors.

By way of example however, one common choice for $b(u)$ in the optimization literature is the logarithmic barrier function:

$$l_t(u) = -\frac{1}{t} \log(-u) \tag{2.130}$$

where $t > 0$ tunes the accuracy of the approximation, noting:

$$\lim_{u \to 0^-} l_t(u) = \infty$$

$$\lim_{t \to \infty} l_t(u) = 0, \quad u < 0$$

Returning to our general framework, note that f is linear and hence, if φ is convex, the modified objective function $f + \varphi$ in (2.129) is convex. In that case, a vanishing gradient is a necessary and sufficient condition for optimality. Alternatively, with non-convex φ, there may be local optima and the condition is necessary only.

Now the gradient of the objective function evaluated at a particular point $\hat{\mathbf{x}}$ (2.129) is:

$$\mathbf{M}\hat{\mathbf{x}} - \mathbf{y} + \varphi'(\hat{\mathbf{x}})$$

where:

$$\varphi'(\hat{\mathbf{x}}) = \frac{\partial}{\partial \mathbf{x}} \varphi(\mathbf{x}) \Big|_{\mathbf{x} = \hat{\mathbf{x}}}$$

The resulting gradient descent algorithm is a direct modification of (2.113) to include the gradient of the penalty function:

$$\hat{\mathbf{x}}^{(n+1)} = t_n \mathbf{y} - (t_n \mathbf{M} - \mathbf{I})\hat{\mathbf{x}}^{(n)} - t_n \varphi'\left(\hat{\mathbf{x}}^{(n)}\right) \tag{2.131}$$

In Section 2.2.2.7, with $\varphi = 0$ we saw that setting the step size $t_n = 1$ resulted in parallel interference cancellation. Setting $t_n = 1$ in (2.131) and re-arranging, we obtain:

$$\hat{\mathbf{x}}^{(n+1)} + \varphi'\left(\hat{\mathbf{x}}^{(n)}\right) = \mathbf{y} - (\mathbf{M} - \mathbf{I})\hat{\mathbf{x}}^{(n)} \tag{2.132}$$

Supposing:

$$\xi(\mathbf{x}) = \mathbf{x} + \varphi'(\mathbf{x}) \tag{2.133}$$

has inverse function:

$$\zeta = \xi^{-1} \tag{2.134}$$

Equation (2.132) motivates iterations of the form:

$$\hat{\mathbf{x}}^{(n+1)} = \zeta\left(\mathbf{y} - (\mathbf{M} - \mathbf{I})\,\hat{\mathbf{x}}^{(n)}\right) \tag{2.135}$$

which is simply parallel interference cancellation followed by a non-linearity ζ, which we can identify as a tentative decision function, introduced in Section 2.2.1.

Several non-linear parallel interference cancellation detectors of the form (2.135) have been described in the literature. The framework of constrained optimization that we have just developed gives a means of analysis of such detectors. Given a particular non-linearity ζ, applied to the output of a parallel interference canceller, it may be possible to determine the underlying constraints (approximately) enforced by the detector, via (2.133), (2.128), and (2.129).

The following theorem summarizes this framework for non-linear interference cancellation.

Theorem 14 *The non-linear iteration:*

$$\hat{\mathbf{x}}^{(n+1)} = \zeta\left(\mathbf{y} - (\mathbf{M} - \mathbf{I})\hat{\mathbf{x}}^{(n)}\right)$$

is a gradient method for numerical solution of:

$$\min \frac{1}{2}\|\mathbf{M}\mathbf{x} - \mathbf{y}\|_{\mathbf{M}^{\frac{1}{2}}} + \varphi(\mathbf{x})$$

where φ satisfies:

$$\zeta^{-1}(\mathbf{x}) = \varphi'(\mathbf{x}) + \mathbf{x}$$

If φ is convex, then the iteration is convergent to the unique point $\hat{\mathbf{x}}$ satisfying:

$$\mathbf{M}\hat{\mathbf{x}} + \varphi'(\hat{\mathbf{x}}) = \mathbf{y}$$

In certain cases, it may be possible to decompose φ into a sum of barrier functions and constraint functions (2.128), which further identifies the non-linear iteration as a barrier method for a constrained optimization problem.

The development of the functions $\varphi : \mathbb{R}^K \mapsto \mathbb{R}$ and $\zeta : \mathbb{R}^K \mapsto \mathbb{R}^K$ in this section allows for general mappings with domain $\mathbb{R}^K$. In many cases of interest, the functions φ and ζ operate independently on each coordinate, i.e.:

$$\varphi(\mathbf{x}) = \sum_{k=1}^{K} \varphi_k(x_k) \tag{2.136}$$

$$\zeta(\mathbf{x}) = (\zeta_1(x_1), \zeta_2(x_2), \ldots, \zeta_K(x_K))^t \tag{2.137}$$

Furthermore, it is typical that the functions $\varphi_k : \mathbb{R} \mapsto \mathbb{R}$: and $\zeta_k : \mathbb{R} \mapsto \mathbb{R}$ are identical for each coordinate, $\varphi_i = \varphi_j$ and $\zeta_i = \zeta_j$. In such cases we will write $\varphi(x)$ and $\zeta(x)$ to mean the coordinate-wise functions. In particular, in Sections 2.2.3.2, 2.2.3.3, and 2.2.3.4 below, we demonstrate how parallel interference cancellation followed by either clipped soft decision (hard limiter), hyperbolic tangent (soft limiter) or hard decision fit within this framework of constrained optimization.

Note that in the discussion that follows, the penalty functions are determined only up to a constant of integration, $C \in \mathbb{R}$ [from solution of the differential equation (2.133)]. All functions of the form $\varphi(x) + C$ yield the same optimum point as $\varphi(x)$ alone. Without loss of generality, we will set $C = 0$.

2.2.3.2 Clipped Soft Decision

The maximum likelihood detector requires that the output is a member of $\mathcal{D} = \{-1, 1\}^K$. One simple way of relaxing this constraint is to require instead that the point be constrained to lie in the convex hull of $\mathcal{D}$, i.e. the set:

$$C = \{\mathbf{x} : x_k^2 \leq 1, k = 1, 2, \ldots, K\}$$

Rather than seeking solutions that are the vertexes of a K-dimensional hypercube, we seek the best point that lies inside the hypercube.

Thus we have K inequality constraints $g_k(\mathbf{x}) = x_k^2 - 1 \leq 0$. Using the logarithmic barrier function (2.130) to approximately enforce these constraints results in the following coordinate-wise penalty function:

$$\varphi(x) = -\frac{1}{t} \log (1 - x^2)$$

i.e., according to our coordinate-wise notation convention introduced above:

$$\varphi(\mathbf{x}) = -\frac{1}{t} \sum_{k=1}^{K} \log (1 - x_k^2)$$

The following theorem shows that as $t \rightarrow \infty$, the resulting (coordinate-wise) tentative decision function ζ is the clipped soft decision.

Theorem 15 *Let:*

$$\varphi(x) = -\frac{1}{t} \log \left(1 - x^2\right) \tag{2.138}$$

then

$$\lim_{t \to \infty} \zeta(x) = \text{clip}(x)$$

Proof: Starting from coordinate-wise versions of (2.133) and (2.134):

$$\xi(x) = \varphi'(x) + x$$

$$= x\left(1 + \frac{2}{t(1 - x^2)}\right)$$

The last line results by differentiation of (2.138). Taking $t \to \infty$ completes the proof.

A closed form for $\zeta(x)$ for finite t can be found via solution in x of the cubic equation $\xi t(1 - x^2) = xt(1 - x^2) + 2x$. Figure 2.13 shows the penalty functions and resulting tentative decision functions for $t = 1, 10, 100$.

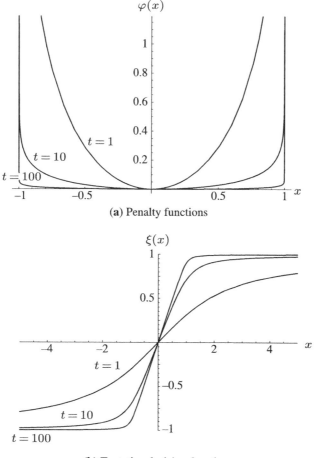

(a) Penalty functions

(b) Tentative decision functions

Figure 2.13. Penalty functions and tentative decision functions for clipped soft decision.

2.2.3.3 *Hyperbolic Tangent*

Theorem 15 showed how use of the standard logarithmic barrier function on the constraints $x_k^2 \leq 1$ resulted in the hard limiter. Naturally, the choice of barrier function is arbitrary, although the particular choice will affect the convergence of the resulting gradient algorithm and the accuracy with which the constraints are enforced. The following theorem identifies the particular choice that leads to the soft limiter.

Theorem 16 *Let*:

$$\varphi(x) = \frac{1}{2}\log(1 - x^2) - \frac{x^2}{2} + \frac{x}{2}\log\left(\frac{1+x}{1-x}\right) \tag{2.139}$$

then:

$$\zeta(x) = \tanh(x)$$

Proof: $\xi(x) = \varphi'(x) + x = \frac{1}{2}\log\left(\frac{1+x}{1-x}\right) = \tanh^{-1}(x)$.

The penalty function (2.139) can be interpreted as a set of constraints, with the standard logarithmic barrier function (2.130). Re-writing (2.139) as:

$$\varphi(x) = -\frac{1}{2}\log\left(\frac{1}{1-x^2}\right) - \frac{1}{2}\log\left(\left(\frac{1+x}{1-x}\right)^{-x}\right) - \frac{1}{2}\log\left(e^{x^2}\right)$$

and we can identify the following three coordinate-wise constraints as the negative arguments of the logarithms, (2.139):

$$\frac{1}{x^2 - 1} \leq 0$$

$$-\left(\frac{1-x}{1+x}\right)^x \leq 0$$

$$-e^{x^2} \leq 0$$

These three constraints are only simultaneously satisfied for $|x| \leq 1$. Thus the soft limiter tentative decision function has the same goal as the hard limiter—to constrain the output for each user to lie in $[-1, 1]$. The two approaches differ only in the implementation of this underlying constraint, and the hence the accuracy to which this constraint is satisfied.

2.2.3.4 *Hard Decision*

The third non-linear tentative decision function mentioned in Section 2.2.1 was the hard decision. The following theorem identifies the associated penalty function, and as a side result, a family of "softened" hard decision functions.

Theorem 17   *Let*:

$$\varphi(x) = -\frac{x^2}{2} - \frac{1}{2t}\log\left(1 - x^2\right) \tag{2.140}$$

then:

$$\xi(x) = \frac{x}{t(1 - x^2)}$$

$$\zeta(x) = \frac{-1 + \sqrt{1 + 4x^2t^2}}{2tx},$$

$$\lim_{t \to \infty} \zeta(x) = \text{sgn}(x)$$

Note that unlike the previous two examples, the penalty function (2.140) is non-convex. For $t \to \infty$ however, it is convex-$\cap$ on the open interval $(-1, 1)$, which is the domain of feasible solutions anyhow.

Figure 2.14 compares the penalty functions that lead to the hard limiter clip($\cdot$), solid line, soft limiter tanh($\cdot$), dashed line and hard decision sgn($\cdot$), dotted line. All of these tentative decision functions ensure that the outputs lie in $[-1, 1]$. The hard limiter correctly introduces this as a constraint to the optimization problem, whereas the soft limiter uses a non-ideal barrier function, which slightly penalizes decisions close to -1 or $+1$. In contrast, the hard decision uses a barrier function that additionally penalizes outputs that are close to zero, while still ensuring that outputs lie in $[-1, 1]$.

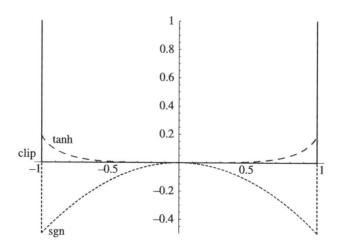

Figure 2.14. Penalty functions corresponding to hard limiter, soft limiter and hard decision.

2.2.4   Numerical Results

Although theoretical results exist concerning the rate of convergence in norm for most linear iterative methods, ultimately we are interested in the bit error rate performance. Furthermore, nonlinear iterations are less amenable to theoretical convergence analysis.

In this section, we present the results of computer simulations. For each scenario, at least 1000 bit errors were collected at the lowest bit error rate reported. For the sake of comparison, we have fixed the number of users at $K = 8$ and targeted uncoded bit error rates of 10^{-3} to 10^{-4}. Antipodal modulation and random spreading sequences were used for all simulations.

2.2.4.1   Parallel Cancellation   Figure 2.15 shows the performance of linear and nonlinear parallel cancellation approximations of the decorrelator versus system load $\alpha = K/N$ at $E_b/N_0 = 7$ dB. All results are for eight iterations, since the goal here is to examine the fundamental convergence limitations of these algorithms. Using more than eight iterations is undesirable from a complexity point of view, since that would require greater computational effort than direct implementation of the decorrelator.

The decorrelator results are shown with long dashed line. Note how linear parallel cancellation departs from decorrelator performance at a system load of about 0.12. This is predicted by the large system result of Theorem 7.

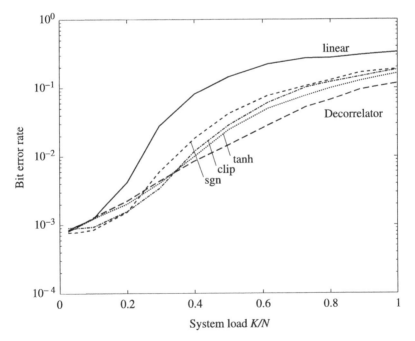

Figure 2.15. Performance versus load for parallel interference cancellation implementation of the decorrelator, $K = 8$, $E_b/N_0 = 7$ dB.

Nonlinear iterations perform better, with hard decisions (sgn) and clipping (clip) outperforming the decorrelator at loads less than $\alpha \approx 0.25$ and $\alpha \approx 0.35$ respectively. The soft decision (tanh) tracks the decorrelator performance up to $\alpha \approx 0.35$. For high loads, all of the nonlinear approaches do not perform as well as the decorrelator. This, however, may not be too concerning, since at these loads the resulting error rate is above 10^{-2} anyway. Figures 2.16 and 2.17 show the performance versus iteration at $E_b/N_0 = 7$ dB and 10 dB respectively. For both sets of results, $K = 8$ and $N = 32$ (resulting in $\alpha = 0.25$ which is above the 0.17 convergence cutoff for linear cancellation).

Figure 2.16 shows how the nonlinear parallel cancellation approaches can not only outperform the decorrelator; they can do so with as few as three iterations. The oscillatory effect evident in these curves has been explained in [43,79]. Figure 2.17 shows that at higher signal-to-noise ratio, the nonlinear approaches fail to deliver decorrelator performance (at least after 8 iterations). This effect is more apparent in Figure 2.18 which shows performance versus SNR after 8 iterations. All of the non-linear parallel cancellation methods suffer from an error floor (which can be reduced at the expense of more iterations).

2.2.4.2 Serial Cancellation

Turning now to serial cancellation methods, Figure 2.19 shows performance versus system load for linear and nonlinear serial

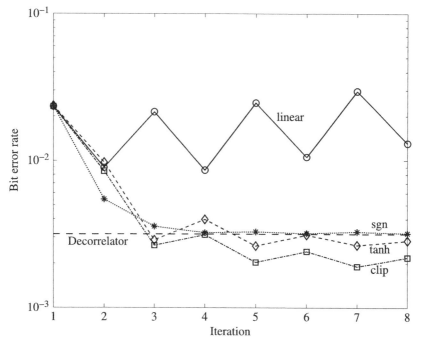

Figure 2.16. Performance versus iteration for non-linear parallel cancellation, $K = 8$, $N = 32$, $E_b/N_0 = 7$ dB.

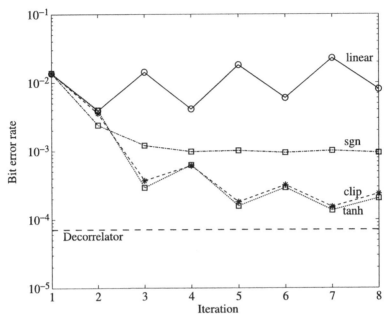

Figure 2.17. Performance versus iteration for non-linear parallel cancellation, $K = 8$, $N = 32$, $E_b/N_0 = 10$ dB.

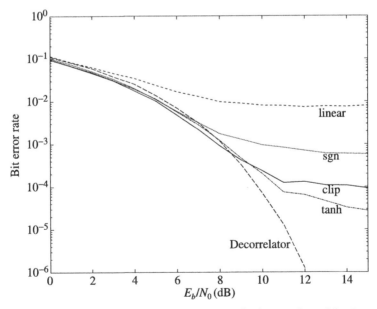

Figure 2.18. Performance versus SNR for parallel iterative implementations of the decorrelator, $K = 8$, $N = 32$, 8 iterations.

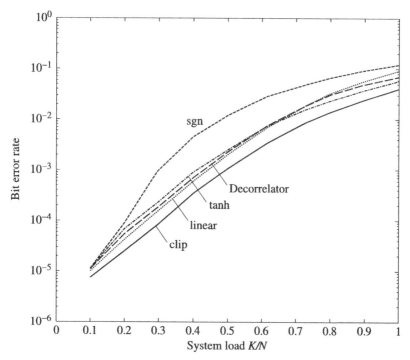

Figure 2.19. Performance versus load for serial interference cancellation implementation of the decorrelator, $K = 8$, $E_b/N_0 = 10$ dB.

cancellation approximations to the decorrelator. The results are for $E_b/N_0 = 10$ dB and 8 iterations. Clearly hard decisions (sgn) are a poor choice with serial cancellation. This makes sense, since hard decisions are made for the first few users without the benefit of much interference cancellation. Linear and soft decisions (tanh) offer similar performance, essentially matching that of the decorrelator for the entire range of system loads. Clipping (clip) outperforms the decorrelator for the entire range of loads tested.

Figure 2.20 shows the corresponding performance versus iteration results for $N = 16$ ($\alpha = 1/2$). Both tanh and clip tentative decision functions result in decorrelator performance after only three iterations. Clipping however continues to improve with more iterations and in fact exceeds the LMMSE performance after four iterations. This is quite remarkable, given the simplicity of the serial cancellation scheme. It is interesting to see that the clip tentative decision function works best, given that it implements an ideal penalty function, as explained in Section 2.2.3.2.

Figure 2.21 confirms that the serial cancellation approaches track the decorrelator performance for a wide range of SNR (these results are for eight iterations—from the previous figure, we know that similar performance is obtained with three or four iterations). The clipping approach is uniformly better than the decorrelator for the range of SNR considered.

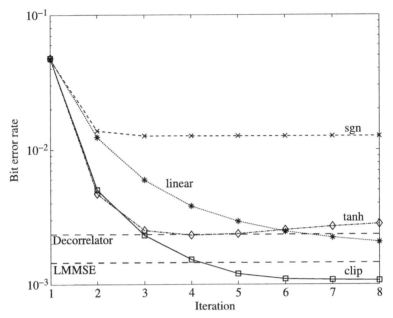

Figure 2.20. Performance versus iteration for non-linear serial cancellation, $K = 8$, $N = 16$, $E_b/N_0 = 10$ dB.

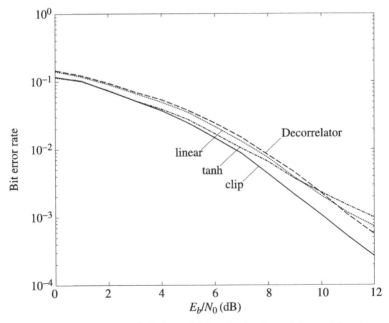

Figure 2.21. Performance versus SNR for serial iterative implementations of the decorrelator, $K = 8$, $N = 16$, 8 iterations.

2.2.4.3 *Gradient Methods* Figures 2.22 and 2.23 report results for linear iterative approximations to the decorrelator and LMMSE respectively. Both sets of results
are at $E_b/N_0 = 10$ dB and $\alpha = 1/2$.

In Figure 2.22 we can observe how both gradient methods "overshoot" the performance of the decorrelator. This figure also demonstrates a useful lesson about convergence. Although the conjugate gradient method converges fastest (as expected),
it offers worse performance than steepest descent for seven and eight iterations.
This is due to the overshoot phenomenon[5] and the slowness of the steepest descent
algorithm to converge. Both gradient methods are significantly faster than linear
serial cancellation.

In Figure 2.23, where the iterative methods are targeting the LMMSE filter, the
overshoot phenomenon disappears. The LMMSE is the best linear filter (in the
sense of an error norm) and not surprisingly outperforms any other linear approach.
In this case, the conjugate gradient method is clearly the best choice amongst the
linear iterative methods.

Figure 2.24 compares linear serial cancellation, conjugate gradient and clipped
serial cancellation implementations of both the decorrelator and LMMSE filter. For
each pair of curves (corresponding to a particular iterative method) the LMMSE
curve is the better one.

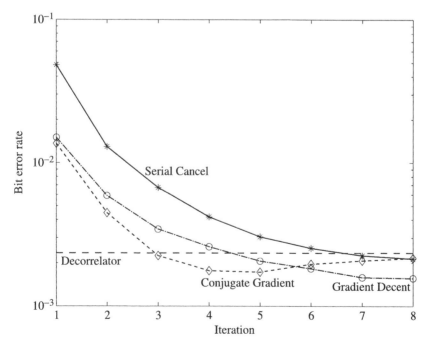

Figure 2.22. Performance versus iteration for iterative implementation of the decorrelator, $K = 8$, $N = 16$, $E_b/N_0 = 10$ dB

---

[5]The convergence behavior of linear cancellation structures is investigated in [79], where the overshoot
phenomenon is also discussed.

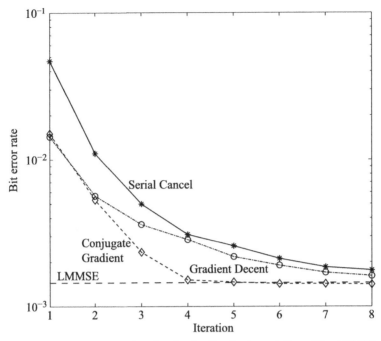

Figure 2.23. Performance versus iteration for iterative implementation of the LMMSE, $K = 8$, $N = 16$, $E_b/N_0 = 10$ dB.

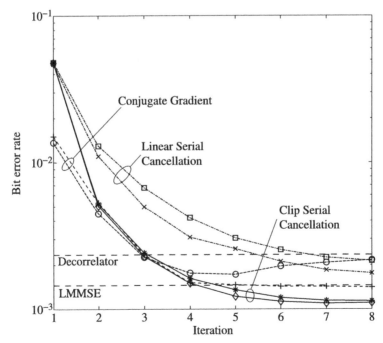

Figure 2.24. Performance versus iteration for iterative implementations of the decorrelator and LMMSE, $K = 8$, $N = 16$, $E_b/N_0 = 10$ dB.

Note how for conjugate gradient, both the decorrelator and LMMSE curve curves yield similar performance up to about three or four iterations, where the decorrelator curve bends up to converge to the exact decorrelator performance. Good performance is obtainable with either approach after four iterations.

The linear serial cancellation is slow to converge for both the decorrelator and the LMMSE as targets. The most interesting result is that of the nonlinear serial cancellation using clip as the tentative decision function. With either the decorrelator or LMMSE as target, similar performance is obtained, and exceeds that of the exact LMMSE with five or more iterations. Thus, clipped serial cancellation is an attractive choice for low-complexity, iterative, multiuser detection.

2.3 ITERATIVE JOINT DECODING FOR CODED DATA

In this section, iterative joint multiuser decoding is considered as a reduced complexity alternative to optimal joint multiuser decoding. ML decoding of coded CDMA was first suggested in [5], where it was shown that the computational complexity grows exponentially with the product of the number of active users and the constraint length of the error control codes used. One simple sub-optimal approach is obtained by separating the multiuser detection and decoding steps. For example, independent decoders for each user can be used at the output of any of the multiuser detectors described in Section 2.2. In [80–82], a linear multiuser detector was designed specifically to generate metrics suitable for input to single-user decoders. The optimal joint multiuser decoder and separate detection and decoding are discussed further in Section 2.3.1 below.

Concatenated codes and iterative decoding [6,7] inspired the canonical iterative joint multiuser decoder, iterating between a multiuser *a posterior* probability (APP) detector, which ignores the constraints due to the error control codes, and individual error control decoders, which ignore any residual multiuser interference [83–87]. Details of the canonical iterative multiuser decoder are described in Section 2.3.2.

Unlike APP decoders for error control codes, which typically exploit code structure (e.g., trellis) for efficient implementation, the multiuser APP detector is in general prohibitively complex. Most work on iterative multiuser decoding replaces this component with lower complexity alternatives.

Linear detectors in general and linear cancellation structures in particular have since been shown to be particularly well suited for iterative joint multiuser decoding. For an overview, see [88]. Inspired by [89], linear interference cancellation as the multiuser detector in an iterative multiuser decoder was further developed in [36]. Linear cancellation followed by instantaneous MMSE filtering was independently proposed in [21] and in [22], providing better performance at the expense of increased complexity. In [90] the same structure is investigated using an adaptive filter that estimates unknown correlations and covariances. Linear cancellation was subsequently recognized in [23] as a simplification of belief propagation applied to the joint multiuser factor graph.

The general structure of the iterative linear multiuser decoder is presented and discussed in Section 2.3.3. Sections 2.3.4 and 2.3.5 describe two of the most common approaches to iterative decoding, based on linear cancellation and LMMSE filtering.

An introduction to convergence analysis based on transfer function analysis is given in Section 2.3.6 and several numerical examples are given in Section 2.3.7.

2.3.1 Joint Optimal Multiuser and Separate Single-User Decoders

Optimal joint multiuser decoding considers detection and decoding *jointly*, as illustrated in Figure 2.25. The joint decoder makes the maximum likelihood decision on the entire set of user information sequences $b_k[i]$, based on the entire sequence of observed vectors $y[i]$, and knowledge of all the encoders C_k and spreading sequences $s_k[i]$.

Joint multiuser ML decoding was first described in [5] for a system where each user applies a memory κ convolutional code. The complexity of the corresponding Viterbi decoder for the resulting super-trellis is $\mathcal{O}(2^{K\kappa})$. As the number of users increases this approach becomes infeasible.

One obvious sub-optimal approach to the problem of multiuser decoding is obtained by separating the multiuser detection and error control decoding steps, as shown in Figure 2.26. The resulting multiuser decoder consists of a multiuser detector which jointly estimates the coded bits $x[i]$, typically in a per-symbol fashion (i.e., ignoring the time dependencies due to the error control codes). The multiuser detector

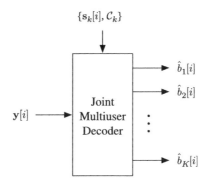

Figure 2.25. Joint multiuser decoding.

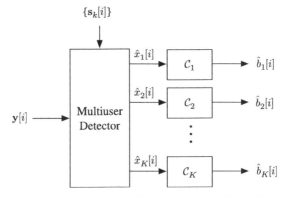

Figure 2.26. Separate multiuser detection and decoding.

is followed by a bank of K error-control decoders, where the decoder for a particular user k takes as input only the corresponding estimates $\hat{x}_k[i]$ and decodes against the constraints of the code C_k. Each user's decoder considers any residual multiple-access interference as noise.

Optimal multiuser detection is $\mathcal{O}(2^K)$ and the subsequent single-user ML decoding for memory κ codes is $\mathcal{O}(K2^\kappa)$. In contrast to the multiuser detector, the complexity of the decoder bank has the desired feature of scaling linearly with the number of users. The complexity of the multiuser detector can be reduced through the use of sub-optimal structures such as the interference cancellation techniques described in Section 2.2. The main drawback of separate detection and decoding is an inherent loss of spectral efficiency as discussed in [19,91,92].

2.3.2 The Canonical Iterative Joint Multiuser Decoder

Iterative joint multiuser decoding, inspired by the development of turbo code, represents a useful compromise between the optimal joint decoder and separate detection and decoding. Iterative decoders can significantly outperform receivers based on separate detection and decoding, at only a small fraction of the complexity required by the optimal joint decoder.

A "turbo decoder" structure for the coded multiple-access channel results directly by recognizing the system to be a bank of K parallel concatenated error control encoders, serially concatenated with the multiuser channel [87,89].

The resulting *canonical iterative joint multiuser decoder* is shown in Figure 2.27. Following the principles for iterative decoding the canonical iterative joint multiuser decoder consists of an *a posteriori* probability multiuser detector, a bank of individual APP error control decoders, together with the appropriate individual interleavers and deinterleavers.

At iteration n, the multiuser APP detector determines the extrinsic log-likelihood ratios (LLRs), $\lambda_k^{(n)}$, $k = 1, 2, \ldots, K$ based on the received signal, $\mathbf{r}$, and the *a priori* LLRs, $\hat{\Lambda}_k^{(n-1)}$, $k = 1, 2, \ldots, K$. The multiuser APP detector operates per symbol and ignores temporal relationships that are due to the error control codes. We therefore omit the symbol interval index.

Assuming independent $x_k \in \{-1, 1\}$ with prior probabilities $\Pr(x_k)$, the posterior probability distribution is:

$$\Pr(x_k|\mathbf{r}) = \frac{\Pr(\mathbf{r}|x_k)\Pr(x_k)}{\sum_{x_j}\Pr(\mathbf{r}|x_j)\Pr(x_j)} \tag{2.141}$$

where:

$$\Pr(\mathbf{r}|x_k) = \sum_{\mathbf{x}_k \in \{-1,1\}^{K-1}} \Pr(\mathbf{r}|\mathbf{x}) \prod_{j \neq k} \Pr(x_j) \tag{2.142}$$

and $\mathbf{x}_k = (x_1, x_2, \ldots, x_{k-1}, x_{k+1}, \ldots, x_K)^t$

The argument that maximizes $\Pr(x_k|\mathbf{r})$ is the per-user marginalized MAP decision. In general, the underlying factor graph admits no simplifications, and the complexity is

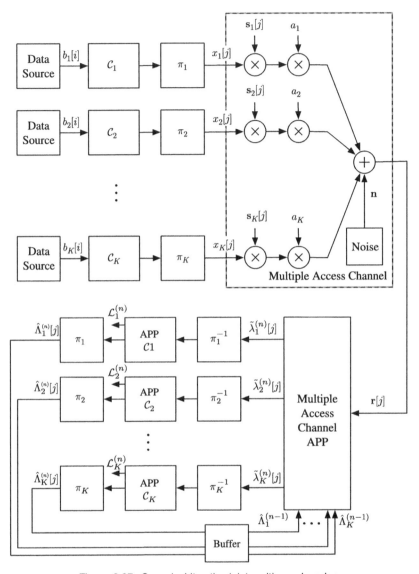

Figure 2.27. Canonical iterative joint multiuser decoder.

dominated by the 2^{K-1} terms in the summation. Extrinsic probabilities are found dividing (2.142) by $\Pr(x_k)$:

$$\frac{\Pr(x_k|\mathbf{r})}{\Pr(x_k)} = \alpha \sum_{\mathbf{x}_{k\!\!/}} \Pr(\mathbf{r}\,|\,x_k, \mathbf{x}_{k\!\!/}) \prod_{j \neq k} \Pr(x_j) \qquad (2.143)$$

where:

$$\alpha = \sum_{\mathbf{x}} \Pr(\mathbf{r} \mid \mathbf{x}) \Pr(\mathbf{x}) \qquad (2.144)$$

does not depend on k or the value of x_k. Within the iterative decoding procedure, the prior probabilities are actually replaced by the extrinsic probabilities from the outer decoders. Although not quite technically correct, we will follow usual practice and refer to these extrinsics as prior probabilities.

Noting that:

$$\Pr\left(x_j = 1\right) = E_j \, e^{\hat{\Lambda}_j/2} \qquad (2.145)$$

$$\Pr\left(x_j = -1\right) = E_j \, e^{-\hat{\Lambda}_j/2} \qquad (2.146)$$

where $E_j = e^{\hat{\Lambda}_j/2}/(1 + e^{\hat{\Lambda}_j})$, the extrinsic LLR for the AWGN multiple-access channel (2.2) is:

$$\tilde{\lambda}_k^{(n)} = \log \left[\frac{\displaystyle\sum_{\mathbf{x}:x_k=1} \exp\left(-\frac{1}{2\sigma^2}|\mathbf{r} - \mathbf{SAx}|^2 + \frac{1}{2}\sum_{j\neq k} \hat{\Lambda}_j^{(n-1)} x_j\right)}{\displaystyle\sum_{\mathbf{x}:x_k=-1} \exp\left(-\frac{1}{2\sigma^2}|\mathbf{r} - \mathbf{SAx}|^2 + \frac{1}{2}\sum_{j\neq k} \hat{\Lambda}_j^{(n-1)} x_j\right)} \right] \qquad (2.147)$$

for $k = 1, 2, \ldots, K$, and all symbol intervals. The constant α (2.144) cancels in the LLR and (2.145), (2.146) are used to get the prior LLR in the exponent, with the constant $\prod_j E_j$ terms also canceling. This sequence of extrinsic LLR values (2.147) are deinterleaved, and forwarded to a bank of K single-user APP error control decoders, where they are used as *a priori* LLRs.

The APP decoders for each code $\mathcal{C}_k$ are typically implemented using standard algorithms, e.g., the forward-backward algorithm [93] or the belief propagation algorithm [94]. The two output sequences from each of the decoders in the APP decoder bank shown in Figure 2.27 are the extrinsic LLR sequence for the code symbols, and the APP LLR sequence for the information symbols, $\mathcal{L}_k^{(n)}$, respectively, both of which are derived from the code constraints ignoring the constraints imposed by the multiple-access channel.

The extrinsic LLR sequences are interleaved, and forwarded to the multiuser APP detector as *a priori* LLRs, $\hat{\Lambda}_k^{(n)}$, $k = 1, 2, \ldots, K$, completing a full iteration. When the iterative decoder has converged, or a stopping condition has been met, the decoded information symbols are determined based on the *a posteriori* output sequences $\mathcal{L}_k$, $k = 1, 2, \ldots, K$, and forwarded to the user sinks.

As illustrated in Figure 2.27, the iterative joint multiuser decoder is closely related to the case of separate detection and decoding depicted in Figure 2.26. The iterative joint multiuser decoder is obtained from the separated structure by allowing the detection and decoding components to iteratively exchange information.

The complexity of the multiuser APP is, like the ML multiuser detector $\mathcal{O}(2^K)$, which is still infeasible for large K. Lower complexity approximate APP multiuser detectors are thus required. Direct approximations to the APP multiuser detector have been suggested based on list detection [87,95–97], and the sphere detector [98]. For the remainder of this chapter, we focus on linear detectors, which have in particular been shown to perform well in an iterative joint multiuser decoder [88].

2.3.3 Linear Detection in Iterative Joint Multiuser Decoding

The application of linear multiuser detection can significantly reduce the complexity of the iterative joint multiuser decoder. Linear multiuser detectors commonly consist of a linear estimation step, followed by a hard decision function, mapping the filter output to the closest valid signal points. For binary transmission, a sign function is used to map signal estimates to signal points. In contrast, the multiuser APP works in the LLR domain, operating on and updating LLR values for the code bits based on prior LLR inputs. To use linear multiuser detectors as a lower complexity replacement for the multiuser APP in an iterative joint multiuser decoder, conversion between the "signal" and "LLR" domains is required. A generic linear detector with such domain conversion blocks is shown in Figure 2.28, where $U : \mathbb{R} \mapsto \mathbb{R}$ maps an LLR into a signal estimate, while $T : \mathbb{R} \mapsto \mathbb{R}$ maps a signal estimate to an LLR. The linear multiuser detector takes the prior signal estimates $\hat{x}_k^{(n-1)}$ and the received signal $\mathbf{r}$, and produces (hopefully improved) signal estimates $\tilde{x}_k^{(n)}$. The linear detector, together with the LLR-signal conversion block (contents of dashed box) is a drop-in replacement for the multiuser APP, trading reduced performance for reduced complexity.

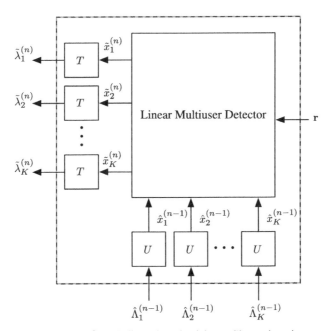

Figure 2.28. Generic linear iterative joint multiuser decoder.

The prevailing approach for converting an LLR into a signal estimate is the per-symbol conditional mean (using the input extrinsic LLRs):

$$\hat{x}_k = U(\hat{\Lambda}_k) = \tanh(\hat{\Lambda}_k/2) \tag{2.148}$$

Depending on the linear detector in question and the system operating point, the use of extrinsic LLRs in this way does not always result in the best performance. In fact, it is still an open problem to determine the optimal prior information. In [23] it is argued that even for suboptimal detectors, extrinsic probabilities should be exchanged, rather than posterior probabilities. The use of posterior probabilities leads to biased linear estimators. In contrast, it was observed in [99] that for moderately loaded systems the use of a posterior probabilities can actually improve performance significantly. Similar observations were reported for large systems over a wide range of system loads in [100], supported by semi-analytical bounds. Further numerical studies on this problem are reported in [101], and in Section 2.3.7 below. Roughly speaking, when sub-optimal approximations to the multiuser APP are used, such as a linear detector, there appears to be a trade-off between accurate estimation of the interfering signals, and bias, or correlation over iteration.

The linear multiuser detector in Figure 2.28 takes these conditional mean signal estimates and produces (hopefully improved) output signal estimates. For a linear detector, we require the output to be a linear function of the inputs $\mathbf{r}$ and $\hat{\mathbf{x}}$, i.e.:

$$\tilde{\mathbf{x}}^{(n)} = \mathbf{F}^{(n)}\mathbf{r} + \mathbf{B}^{(n)}\hat{\mathbf{x}}^{(n-1)} \tag{2.149}$$

for some choice of feed-forward matrix $\mathbf{F}^{(n)}$ and a feedback matrix $\mathbf{B}^{(n)}$, both of which could depend on the iteration index.

One obvious approach is to use interference cancellation, as described in Section 2.2.1. This is developed in Section 2.3.4. This may not, however, make full use of the statistical structure of the residual signal. A more sophisticated approach would be to use the per-user LMMSE filter introduced in Section 2.1.2. This approach is explored in more depth in Section 2.3.5.

The conversion of the output signal estimate into an LLR is commonly based on a Gaussian assumption for $\tilde{x}_k$; namely it is assumed that:

$$p(\tilde{x}_k|x_k) = \frac{1}{\sqrt{2\pi\tilde{\sigma}_k^2}}\exp\left[-\frac{(\tilde{x}_k - \tilde{\mu}_k x_k)^2}{2\tilde{\sigma}_k^2}\right] \tag{2.150}$$

where the parameters $\tilde{\mu}_k$ and $\tilde{\sigma}_k^2$ depend on the underlying statistics of the signal, which in turn depend on the particular linear multiuser detector under consideration.

This Gaussian assumption leads to:

$$\tilde{\lambda}_k = T(\tilde{x}_k) = \frac{2\tilde{\mu}_k}{\tilde{\sigma}_k^2}\tilde{x}_k \tag{2.151}$$

In the large-system limit,[6] the Gaussian assumption has been shown to hold for the output of a broad class of linear detectors, including linear parallel cancellation and

---

[6]For systems of size $K \geq 16$, the Gaussian assumption is quite accurate [58].

LMMSE filtering [58]. For LMMSE filtered outputs in particular, the Gaussian assumption has been shown to hold for most cases of interest [102], including systems with only a few users.

The statistical parameters $\tilde{\mu}_k$, $\tilde{\sigma}_k^2$, can be determined analytically for each symbol interval as shown in the following sections for particular cancellation schemes, or estimated from all the filter outputs $\tilde{x}_k$ over the complete codeword block. In this case, under an assumption of ergodicity:

$$\tilde{\mu}_k = \mathsf{E}\{\tilde{x}_k\, x_k\} \tag{2.152}$$

$$= \left(\frac{6\mathsf{E}^2\left[\tilde{x}_k^2\right] - 2\mathsf{E}\left[\tilde{x}_k^4\right]}{4}\right)^{1/4}, \tag{2.153}$$

$$\tilde{\sigma}_k^2 = \mathsf{E}\{(\tilde{x}_k - \tilde{\mu}_k x_k)^2\} \tag{2.154}$$

$$= \mathsf{E}\left[\tilde{x}_k^2\right] - \tilde{\mu}_k^2, \tag{2.155}$$

where the required moments can be estimated according to:

$$\mathsf{E}\left[\tilde{x}_k^2\right] \simeq \frac{1}{L}\sum_{i=1}^{L} \tilde{x}_k^2[i] \tag{2.156}$$

$$\mathsf{E}\left[\tilde{x}_k^4\right] \simeq \frac{1}{L}\sum_{i=1}^{L} \tilde{x}_k^4[i] \tag{2.157}$$

2.3.4 Parallel Interference Cancellation

Let the filters in (2.149) be independent of iteration, $\mathbf{F} = \mathbf{S}^t$ and $\mathbf{B} = (\mathbf{R} - \mathbf{I})\mathbf{A}$, corresponding to parallel cancellation of $\hat{\mathbf{x}}^{(n)}$.

The cancellation output estimate for user k at iteration n is determined as:

$$\tilde{x}_k^{(n)} = \mathbf{s}_k^t\left(\mathbf{r} - \sum_{j\neq k} \mathbf{s}_j a_j \hat{x}_j^{(n-1)}\right) \tag{2.158}$$

$$= a_k x_k + \underbrace{\mathbf{s}_k^t \sum_{j\neq k} \mathbf{s}_j a_j\left(x_j - \hat{x}_j^{(n-1)}\right) + \mathbf{s}_k^t \mathbf{n}}_{\tilde{n}_k^{(n)}} \tag{2.159}$$

$$= a_k x_k + \tilde{n}_k^{(n)} \tag{2.160}$$

where we assume that $\tilde{n}_k^{(n)}$ is a zero-mean Gaussian random variable. To conform to accepted terminology, we refer to this case as the single-user matched filter (SUMF) parallel interference canceller. The statistics $\tilde{\mu}_k^{(n)}$, $\left(\tilde{\sigma}_k^{(n)}\right)^2$ for this approach are:

$$\tilde{\mu}_{\text{SUMF},k}^{(n)} = a_k \tag{2.161}$$

$$\left(\tilde{\sigma}_{\text{SUMF},k}^{(n)}\right)^2 = \sum_{j \neq k} \left(\mathbf{s}_k^t \mathbf{s}_j\right)^2 a_j^2 \left(1 - \left(\hat{x}_j^{(n-1)}\right)^2\right) + \sigma^2 \tag{2.162}$$

$$= \mathbf{s}_k^t \mathbf{S} \mathbf{A} \mathbf{W}_k^{(n)} \mathbf{A} \mathbf{S}^t \mathbf{s}_k + \sigma^2, \tag{2.163}$$

where $\mathbf{W}_k^{(n)}$ is determined from (2.31). It follows that:

$$\tilde{\lambda}_k^{(n)} = \frac{2a_k}{\mathbf{s}_k^t \mathbf{S} \mathbf{A} \mathbf{W}_k \mathbf{A} \mathbf{S}^t \mathbf{s}_k + \sigma^2} \tilde{x}_{\text{SUMF},k}^{(n)} \tag{2.164}$$

$$= \frac{2a_k \mathbf{s}_k^t \left(\mathbf{r} - \sum_{j \neq k} \mathbf{s}_j a_j x_j^{(n-1)}\right)}{\mathbf{s}_k^t \mathbf{S} \mathbf{A} \mathbf{W}_k \mathbf{A} \mathbf{S}^t \mathbf{s}_k + \sigma^2} \tag{2.165}$$

This approach was first suggested in [36].

2.3.5 Per-User LMMSE Filters with Priors

The linear filter in Figure 2.28 takes as inputs the received signal $\mathbf{r}$ and conditional mean estimates, $\hat{\mathbf{x}}$ (2.148) from all the users. The conditional mean estimates for different users and symbol intervals are approximately independent due to the interleavers/deinterleavers, inspiring application of the LMMSE filter with priors in (2.19). This filter outputs the sum of the vector of priors and the vector of the LMMSE filtered noise hypothesis following from cancellation with all the priors. However, iterative decoding is based on processing extrinsic information where the prior of the data symbol in question is removed from the output. Imposing this principle on the LMMSE filter with priors leads us to the per-user LMMSE filters in (2.25) and (2.39), respectively. The per-user LMMSE-based detectors with priors are well suited as approximate linear APP detectors, which has previously been demonstrated in [21–23,25]. The filter in (2.25) has the structure of interference cancellation based on the conditional mean estimates of the interfering users, followed by LMMSE filtering of the cancellation output.

The per-user LMMSE estimate at iteration n is determined by (2.25) with $\bar{\mathbf{x}} = \hat{\mathbf{x}}^{(n-1)}$ and the filter $\mathbf{V}_k^{(n-1)}$ computed according to (2.24):

$$\tilde{x}_k^{(n)} = a_k \mathbf{s}_k^t \left(\mathbf{S} \mathbf{A} \mathbf{V}_k^{(n-1)} \mathbf{A} \mathbf{S}^t + \sigma^2 \mathbf{I}\right)^{-1} \left(\mathbf{r} - \sum_{j \neq k} \mathbf{s}_j a_j \hat{x}_j^{(n-1)}\right) \tag{2.166}$$

The corresponding statistical parameters for the conversion are determined as:

$$\tilde{\mu}_k^{(n)} = \mathsf{E}\{\tilde{x}_k^{(n)} x_k\} \tag{2.167}$$

$$= a_k \mathbf{s}_k^t \left(\mathbf{SAV}_k^{(n-1)}\mathbf{AS}^t + \sigma^2\mathbf{I}\right)^{-1}\mathbf{s}_k a_k \tag{2.168}$$

$$\left(\tilde{\sigma}_k^{(n)}\right)^2 = \mathsf{E}\left\{\left(\tilde{x}_k^{(n)}\right)^2\right\} - \left(\tilde{\mu}_k^{(n)}\right)^2 \tag{2.169}$$

$$= \tilde{\mu}_k^{(n)} - \left(\tilde{\mu}_k^{(n)}\right)^2 \tag{2.170}$$

and thus:

$$\tilde{\lambda}_k^{(n)} = \frac{2}{1 - a_k \mathbf{s}_k^t \left(\mathbf{SAV}_k^{(n-1)}\mathbf{AS}^t + \sigma^2\mathbf{I}\right)^{-1}\mathbf{s}_k a_k}\,\tilde{x}_k^{(n)} \tag{2.171}$$

$$= \frac{2a_k \mathbf{s}_k^t \left(\mathbf{SAV}_k^{(n-1)}\mathbf{AS}^t + \sigma^2\mathbf{I}\right)^{-1}\left(\mathbf{r} - \sum_{j\neq k}\mathbf{s}_j a_j \hat{x}_j^{(n-1)}\right)}{1 - a_k \mathbf{s}_k^t \left(\mathbf{SAV}_k^{(n-1)}\mathbf{AS}^t + \sigma^2\mathbf{I}\right)^{-1}\mathbf{s}_k a_k} \tag{2.172}$$

This approach was independently proposed in [21,22] for iterative multiuser decoding, and in [25] for iterative joint equalization and decoding.

The nominal LMMSE detector in (2.39) is also derived based on priors. The corresponding estimate is determined by:

$$\tilde{x}_k^{(n)} = a_k \mathbf{s}_k^t \left(\mathbf{SAW}_k^{(n-1)}\mathbf{AS}^t + \sigma^2\mathbf{I}\right)^{-1}\left(\mathbf{r} - \sum_{j\neq k}\mathbf{s}_j a_j \hat{x}_j^{(n-1)}\right) \tag{2.173}$$

with the filter $\mathbf{W}_k^{(n-1)}$ computed according to (2.31). The corresponding statistical parameters for the conversion to LLR are determined as:

$$\tilde{\mu}_k^{(n)} = a_k \mathbf{s}_k^t \left(\mathbf{SAW}_k^{(n-1)}\mathbf{AS}^t + \sigma^2\mathbf{I}\right)^{-1}\mathbf{s}_k a_k \tag{2.174}$$

$$\left(\tilde{\sigma}_k^{(n)}\right)^2 = a_k \mathbf{s}_k^t \left(\mathbf{SAW}_k^{(n-1)}\mathbf{AS}^t + \sigma^2\mathbf{I}\right)^{-1}\mathbf{s}_k a_k \tag{2.175}$$

and thus, the extrinsic LLR is simply determined as:

$$\tilde{\lambda}_k^{(n)} = 2\tilde{x}_k^{(n)} \tag{2.176}$$

$$= 2a_k \mathbf{s}_k^t \left(\mathbf{SAW}_k^{(n-1)}\mathbf{AS}^t + \sigma^2\mathbf{I}\right)^{-1}\left(\mathbf{r} - \sum_{j\neq k}\mathbf{s}_j a_j \hat{x}_j^{(n-1)}\right) \tag{2.177}$$

This approach was first proposed in [23] and further developed in [24]. It was independently derived based on nonlinear MMSE arguments in [27].

2.3.6 Transfer Function Convergence Analysis

The variance evolution technique in [36] for analysis of the iterative decoding with linear parallel interference cancellation provides insight into the character of the component detector and decoders. Variance transfer analysis works by scalar parametrization of the iterative process using the average error variance on $\tilde{x}$ (output of canceller), and $\hat{x}$ (output of decoder). This technique was a precursor for the development of SNR tracking [103] and extrinsic information transfer (EXIT) charts [104,105]. In this section, we focus on variance transfer analysis for the parallel interference cancellation of Section 2.3.4, where it is possible to obtain closed-form detector characteristics for large systems with random spreading.

This semi-analytical approach can be used to evaluate the evolution of the system performance over iteration, including the identification of bottlenecks which prevent successful convergence. The technique is also a useful design tool for maximizing the system load for a fixed power allocation [106]. The EXIT charts approach has been applied in [107] for evaluating iterative multiuser decoders, while EXIT charts and multiuser efficiency tracking have been combined for convergence analysis and optimal power allocation in [24].

Suppose that $\tilde{x} = x + \tilde{v}$ and $\hat{x} = x + \hat{v}$, with error variances:

$$\tilde{\sigma}^2 = \mathsf{E}\{(\tilde{x} - x)^2\} = \mathsf{E}\{\tilde{v}^2\} \tag{2.178}$$

$$\sigma^2 = \mathsf{E}\{(\hat{x} - x)^2\} = \mathsf{E}\{\hat{v}^2\} \tag{2.179}$$

Let the code for each user $\mathcal{C}_k$ have a unit input signal power ($a_k = 1$) variance transfer curve:

$$\sigma^2 = f_k(\tilde{\sigma}_k^2)$$

Typically, such curves are found via monte-carlo simulation of the component code. From this unit power curve, the characteristic for an input with power a_k^2 is easily found as $f_k(\tilde{\sigma}_k^2/a_k^2)$.

Suppose now that the spreading matrix $\mathbf{S}$ has *i.i.d.* zero mean entries with variance $1/N$. Assume that K is sufficiently large so that the initial noise variance $(\tilde{\sigma}^{(1)})^2$ is the same for each user. Further let user k be a specific user of interest. At the output of APP decoder k we have $\hat{x}_k^{(1)} = x_k + \hat{v}_k^{(1)}$ where:

$$\left(\hat{\sigma}_k^{(1)}\right)^2 = \mathsf{E}\left\{\left(\hat{v}_k^{(1)}\right)^2\right\} = f_k\left(\left(\tilde{\sigma}_k^{(1)}\right)^2/a_k^2\right)$$

The noise samples $\hat{v}_k$, for $k = 1, 2, \ldots, K$, are assumed to be mutually independent and also independent of $\mathbf{S}$. These independence assumptions are not strictly true, but yield accurate results in the limit of large block length and independent interleaving of each user.

With reference to (2.162), the cancelled signal in the second iteration[7] for user k is:

$$s_k^t \left(\mathbf{r} - \sum_{j \neq k} s_j a_j \hat{x}_j^{(1)} \right) = \underbrace{a_k x_k}_{\text{signal}} + \underbrace{s_k^t \left(\sum_{j \neq k} s_j a_j \hat{v}_j^{(1)} + \mathbf{n} \right)}_{\text{noise}} \qquad (2.180)$$

$$= a_k \left(x_k + \tilde{v}_k^{(2)} \right) = a_k \tilde{x}_k^{(2)} \qquad (2.181)$$

which has signal power a_k^2 and noise power:

$$\left(\tilde{\sigma}_k^{(2)} \right)^2 = \sigma^2 + \sum_{j \neq k} \mathsf{E}\left\{ \left(s_k^t s_j \right)^2 \right\} a_j^2 \mathsf{E}\left\{ \left(\hat{v}_k^{(1)} \right)^2 \right\} \qquad (2.182)$$

$$= \sigma^2 + \frac{1}{N} \sum_{j \neq k} a_j^2 f_j \left(\left(\tilde{\sigma}_k^{(1)} \right)^2 / a_j^2 \right) \qquad (2.183)$$

For sufficiently large K we therefore have at iteration n:

$$\left(\tilde{\sigma}^{(n+1)} \right)^2 = \sigma^2 + \alpha \sum_{k=1}^{K} \frac{a_k^2}{K} f_k \left(\left(\tilde{\sigma}^{(n)} \right)^2 / a_k^2 \right) \qquad (2.184)$$

$$= \sigma^2 + \alpha \bar{f} \left(\left(\tilde{\sigma}^{(n)} \right)^2 \right) \qquad (2.185)$$

where $\alpha = K/N$, and:

$$\bar{f}(\tilde{\sigma}^2) \triangleq \sum_{k=1}^{K} \frac{a_k^2}{K} f_k \left(\tilde{\sigma}^2 / a_k^2 \right) \qquad (2.186)$$

is defined as the system *average* variance transfer curve.

For equal received powers, $\mathbf{A} = \mathbf{I}$ and each user with the same code with characteristic f, we obtain the following linear input-output characteristic for the canceller:

$$\left(\tilde{\sigma}^{(n+1)} \right)^2 = \sigma^2 + \alpha f \left(\left(\tilde{\sigma}^{(n)} \right)^2 \right)$$

$$= \sigma^2 + \alpha \left(\hat{\sigma}^{(n)} \right)^2$$

Other special case of interest is where there is some number of users classes, with each user in the same class using the same code. For example, suppose there are λK users in class one, using a code with transfer function codes f_1, and $(1 - \lambda)K$ users

---

[7]This is the first iteration where cancellation takes place. In iteration one the signal to be cancelled is zero.

in class two using a code with transfer function f_2. With equal unit powers for all users:

$$\bar{f}(\tilde{\sigma}^2) = \lambda f_1(\tilde{\sigma}^2) + (1 - \lambda) f_2(\tilde{\sigma}^2) \tag{2.187}$$

which is a linear combination of the component code characteristics.

Alternatively, we could consider only one code, but two or more power classes. For example, with two power classes:

$$\bar{f}(\tilde{\sigma}^2) = \lambda a_1^2 f(\tilde{\sigma}^2/a_1^2) + (1 - \lambda) a_2^2 f(\tilde{\sigma}^2/a_2^2). \tag{2.188}$$

The additional factor of a_k inside the argument of the f_k gives an extra degree of freedom in shaping the average code curve. This is explored further in [24,108].

2.3.7 Numerical Examples

In this section, the bit error rate performance and convergence behavior of the iterative interference cancellation joint multiuser decoders are investigated and contrasted against each other. A symbol- and chip-synchronous CDMA system, transmitting over a common AWGN channel, is simulated for the evaluation. BPSK modulation is used for both code symbols and spreading chips. Random spreading sequences are assumed, changing for every bit interval. All users apply the same rate 1/2 (5,7) convolutional code, transmitting frames of 1000 information bits. Each user is assigned a random interleaver of length $L = 2000$ code bits for each frame transmitted. Perfect power control is assumed such that all users are received at the same power. The length of the spreading sequences is set to $N = 8$.

2.3.7.1 *Separate Multiuser Detection and Single-User Decoding* We first consider the case of separate multiuser detection, followed by single-user decoding. In Figure 2.29, the BER performance of simple matched filtering (MF) and LMMSE filtering, respectively, is evaluated as a function of E_b/N_0 and compared for a different number of active users. The single-user (SU) performance curve is included for reference. As predicted by information theory [19], the LMMSE filtered case provides a higher spectral efficiency than the MF filtered case, accommodating a higher load $\alpha = K/N$ for specific target BER and E_b/N_0. Assuming a BER target of 10^{-3} at $10 \log(E_b/N_0) = 8$ dB, the MF case can accommodate at most three users at a loss of 3 dB to the single-user case, while the LMMSE case can accommodate at most seven users at a loss of 4 dB.

2.3.7.2 *Single-User Matched Filter Parallel Interference Cancellation* The single-user matched-filter parallel interference cancellation scheme in Section (2.3.4) provides a low-complexity approach for approximating the iterative joint multiuser decoder. As expected, allowing for an iterative exchange of information significantly improves performance as compared to separate detection and decoding in terms of BER and spectral efficiency (load). Following standard iterative principles,

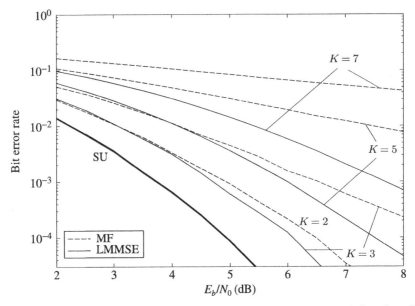

Figure 2.29. The BER as a function of E_b/N_0 for separate detection and decoding. The matched-filter (MF) detector (dashed lines) and the LMMSE-filter detector (solid lines) are considered for $K = 1, 2, 3, 5, 7$.

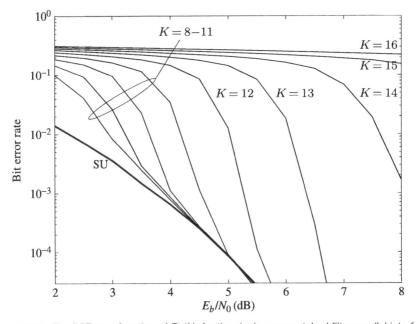

Figure 2.30. The BER as a function of E_b/N_0 for the single-user matched filter parallel interference cancellation with $K = 8\text{--}16$. Extrinsic information is forwarded from the single-user decoders. The single-user performance is included for reference.

extrinsic information is exchanged between all decoder components. However, some performance improvement have been reported in the literature for exchanging *a posteriori* information between the single-user decoders and the multiuser interference canceller. Results for both cases are included in this section.

In Figure 2.30, the BER is presented as a function of E_b/N_0 after ten iterations. Performance curves for $K = 8-16$ are included together with the single-user performance for reference. The expected thresholding effect of iterative decoders are clearly observed. As the load increases, the E_b/N_0 required for convergence towards single-user performance also increases. Single-user performance is obtained within an E_b/N_0 of 6 dB for loads up to $\alpha = 1.5$. For additional iterations, it is expected that a BER of 10^{-4} or less is achieved at $E_b/N_0 = 8$ dB for a load of $\alpha = 1.75$ ($K = 14$).

The BER improvements over iterations are shown in Figure 2.31 for a load of $\alpha = 1.25$ ($K = 10$). The performance for the first ten iterations are included, indicating that for $K = 10$, most of the gain is obtained within the first six iterations. The improvements of the BER as a function of the number of iterations are shown in Figure 2.32 for $n = 1-10$, iterations, $E_b/N_0 = 8$ dB and $K = 8-16$. Single-user performance is below 10^{-7} and is therefore not included in the plot. The single-user matched-filter PIC converges to single-user performance for $K = 8-13$ users within 10 iterations. For additional iterations, the plot is indicating that convergence to a low BER may also occur for $K = 14$, corresponding to a load of $\alpha = 1.75$.

The exchange of *a posteriori* information between the single-user decoders and the multiuser interference canceller have been reported to improve performance. In

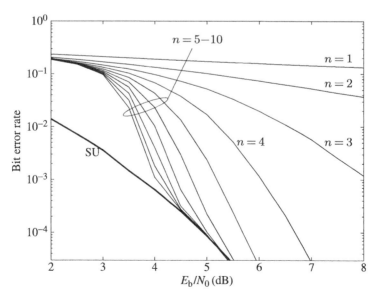

Figure 2.31. The BER as a function of E_b/N_0 for the single-user matched filter parallel interference cancellation from $n = 1-10$ iterations and with $K = 10$. Extrinsic information is forwarded from the single-user decoders. The single-user performance is included for reference.

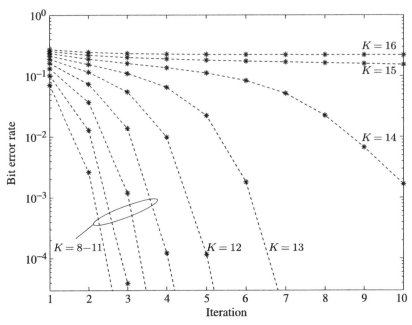

Figure 2.32. The BER as a function of the number of iterations for the single-user matched filter parallel interference cancellation with $K = 8-16$ and $E_b/N_0 = 8$ dB. Extrinsic information is forwarded from the single-user decoders.

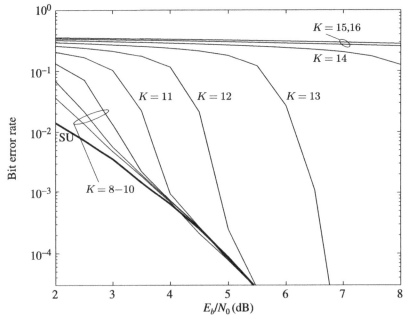

Figure 2.33. The BER as a function of E_b/N_0 for the single-user matched filter parallel interference cancellation with $K = 8-16$. *A posteriori* information is forwarded from the single-user decoders. The single-user performance is included for reference.

Figure 2.33, the BER is presented for this case as a function of E_b/N_0 after 10 iterations. Performance curves for $K = 8-16$ are included together with the single-user performance for reference. Compared to Figure 2.30, a 0.5 dB performance gain is observed for $K = 8-12$ after ten iterations, exchanging posterior information. For $K = 13$ virtually identical BER performance is obtained, while the exchange of extrinsic information for $K = 14$ leads to significantly better BER performance as compared to exchanging posterior information.

In Figure 2.34, a comparison of the BER as a function of the number of iterations at $E_b/N_0 = 8$ dB for the single-user matched-filter PIC is presented, where either extrinsic or posterior information is forwarded from the single-user decoders and $K = 6, 10, 13, 14, 15$. It is observed that the exchange of posterior information leads to faster convergence even for $K = 13$, however for $K = 14$ the scheme is not converging within 8 dB. This is in contrast to the exchange of extrinsic information for $K = 14$, which allows for convergence within $8-9$ dB. It is an open problem to determine the optimal information to exchange within iterative multiuser decoders, applying suboptimal multiuser detectors.

2.3.7.3 Per-User LMMSE Filtering
The performance of parallel interference cancellation is significantly improved by including instantaneous LMMSE filtering of the canceller output. This approach is in fact the per-user LMMSE filter detector, given non-uniform prior distribution of the data symbols, which leads to the per-user LMMSE decoder detailed in Section (2.3.5). The per-user nominal LMMSE decoder

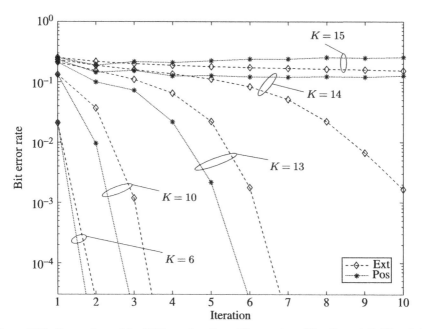

Figure 2.34. Comparison of the BER as a function of the number of iterations at $E_b/N_0 = 8$ dB for the single-user matched filter parallel interference cancellation, where either extrinsic or posterior information is forwarded from the single-user decoders and $K = 6, 10, 13, 14, 15$.

also detailed in Section (2.3.5) provides similar performance improvements. In this section, the BER performance of these two decoders[8] are investigated and discussed. In fact, it turns out that there is no noticeable performance difference between the two decoders. The presented performance curves are therefore valid for both decoders. As for the single-user matched-filter PIC case, we compare the performance of the decoders, exchanging extrinsic information and posterior information, respectively.

In Figure 2.35, the BER is presented as a function of E_b/N_0 after ten iterations. Performance curves for $K = 8-20$ are included together with the single-user perform-ance for reference. Single-user performance is obtained within an E_b/N_0 of 6 dB for loads up to $\alpha = 2.125$ ($K = 17$). For additional iterations, it is expected that a BER below 10^{-4} is achieved at $E_b/N_0 = 8-9\,\text{dB}$ for a load of $\alpha = 2.5$ ($K = 20$). Compared to Figure 2.30, significant performance improvements are obtained when a per-user LMMSE filter is introduced. The improvements come at an increased computational complexity in terms of a matrix inverse required for each user at each symbol interval and each iteration.

The BER improvements over iterations are shown in Figure 2.36 for a load of $\alpha = 1.75$ ($K = 14$). The performance for the first ten iterations are included, indicating that for $K = 14$, most of the gain is again obtained within the first six iterations. Single-user performance is achieved after five iterations at 5.5 dB.

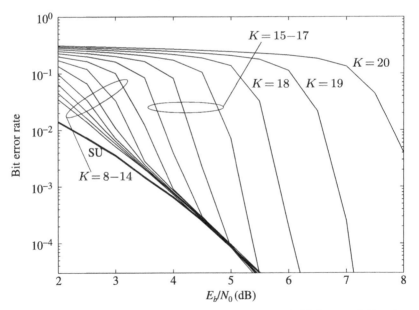

Figure 2.35. The BER as a function of E_b/N_0 for the per-user LMMSE filter parallel interference cancellation with $K = 8-20$. Extrinsic information is forwarded from the single-user decoders. The single-user performance is included for reference.

---

[8]These decoders are referred to as the LMMSE decoders.

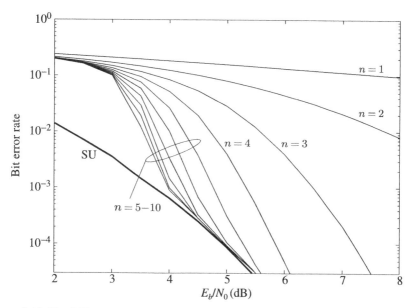

Figure 2.36. The BER as a function of E_b/N_0 for the per-user LMMSE filter parallel interference cancellation from $n = 1-10$ iterations and with $K = 14$. Extrinsic information is forwarded from the single-user decoders. The single-user performance is included for reference.

The improvements of the BER as a function of the number of iterations are shown in Figure 2.37 for $n = 1-10$ iterations, $E_b/N_0 = 8$ dB, and $K = 8-20$. The LMMSE decoders converge to single-user performance for $K = 8-19$ users within ten iterations. For additional iterations, the plot is indicating that convergence to a low BER may also occur for $K = 20$, corresponding to a load of $\alpha = 2.5$.

In Figure 2.38, the BER is presented as a function of E_b/N_0 after ten iterations for the case where posterior information is exchange. Performance curves for $K = 8-20$ are included together with the single-user performance for reference. In this case, the LMMSE decoders converge for up to $K = 20$ within 8 dB. Compared to Figure 2.35, no significant performance improvements are observed for exchanging posterior information for $K = 8-15$ after ten iterations. In contrast, for $K = 16-18$, a 0.4 dB performance gain is observed exchanging posterior information, increasing to a 0.8 dB gain for $K = 19$ and several dBs gain for $K = 20$.

In Figure 2.39, a comparison of the BER as a function of the number of iterations at $E_b/N_0 = 8$ dB for the LMMSE decoders is presented, where either extrinsic or posterior information is forwarded from the single-user decoders and $K = 13, 17, 19,$ 20. For $K = 13$, both cases practically converge after the same number of iterations. However, for $K = 17$ the exchange of posterior information converges one iteration faster. As the load increases, the benefit in convergence rate also increases. The exchange of posterior information for the cases investigated here clearly provide a performance improvement.

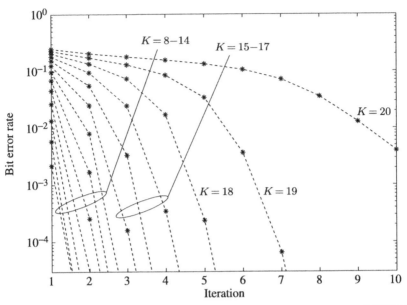

Figure 2.37. The BER as a function of the number of iterations for the per-user LMMSE filter parallel interference cancellation with $K = 8$–20 and $E_b/N_0 = 8$ dB. Extrinsic information is forwarded from the single-user decoders.

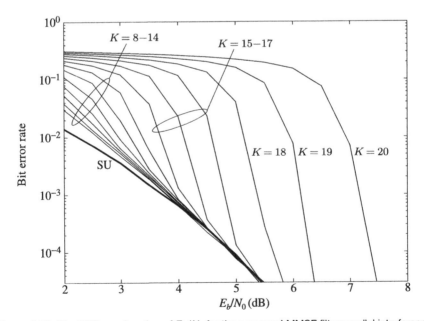

Figure 2.38. The BER as a function of E_b/N_0 for the per-user LMMSE filter parallel interference cancellation with $K = 8$–20. *A posteriori* information is forwarded from the single-user decoders. The single-user performance is included for reference.

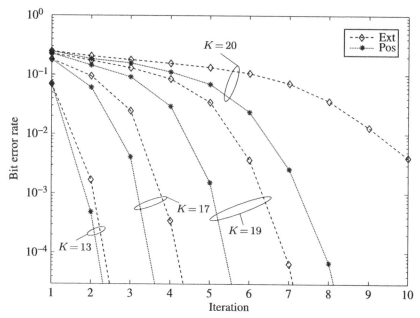

Figure 2.39. Comparison of the BER as a function of the number of iterations at $E_b/N_0 = 8$ dB for the per-user LMMSE filter parallel interference cancellation, where either extrinsic or posterior information is forwarded from the single-user decoders and $K = 13, 17, 19, 20$.

2.3.7.4 *Comparison of Single-User Matched-Filter PIC and LMMSE Decoders*
In this section, we compare the BER performance of the single-user matched-filter PIC and LMMSE decoders. In Figure 2.40, the BER is presented as a function of E_b/N_0 after ten iterations. Results for exchanging both extrinsic and posterior information are included for both the single-user matched-filter PIC and the LMMSE decoders. As observed in previous sections, the exchange of posterior information provides significant improvements. Also, the additional investment in computational complexity by using a LMMSE decoder provides capacity gains of the order of 40%–50% in terms of load, i.e., similar performance is obtained for $K = 18$ with the LMMSE decoder and for $K = 12$ with the single-user matched-filter PIC decoder.

In Figure 2.41, the improvements of the BER as a function of the number of iterations are shown for $n = 1$–10 iterations and $E_b/N_0 = 8$ dB for a similar set of examples, leading to similar conclusions as for Figure 2.40.

To summarize the numerical results, in Figure 2.42, we compare the spectral efficiency achieved at a bit error rate of 10^{-4} by the six different schemes presented here, namely separate detection/decoding with the matched filter (Sep MF) and the LMMSE filter (Sep LMMSE), respectively, the iterative joint decoding parallel interference canceller (Joint SUMF), and the iterative joint decoding per-user LMMSE filtering interference canceller (Joint LMMSE). Ten iterations are performed for the iterative joint decoding schemes, and the use of both extrinsic and posterior information at the input to the multiuser detector is considered.

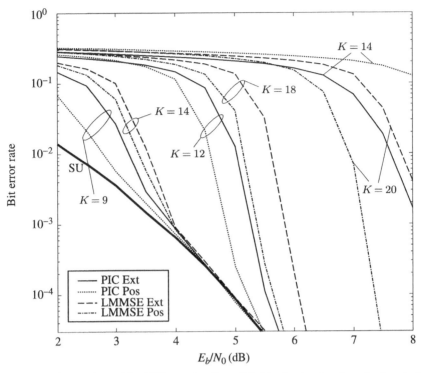

Figure 2.40. Comparison of the BER as a function of E_b/N_0 after ten iterations for the single-user matched filter and the per-user LMMSE filter parallel interference cancellation, where either extrinsic or posterior information is forwarded from the single-user decoders.

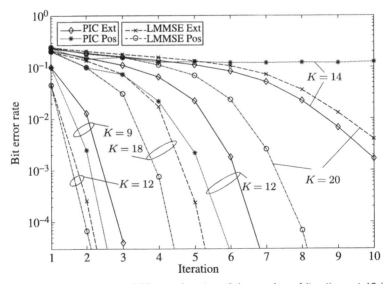

Figure 2.41. Comparison of the BER as a function of the number of iterations at $10 \log(E_b/N_0) = 8$ dB for the single-user matched filter and the per-user LMMSE filter parallel interference cancellation, where either extrinsic or posterior information is forwarded from the single-user decoders.

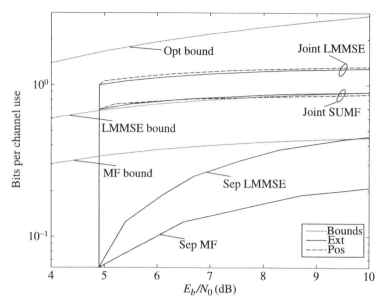

Figure 2.42. A comparison of the spectral efficiency achieved for a bit error rate of 10^{-4}. Six schemes are shown here, namely separate detection/decoding with the matched filter (Sep MF) and the LMMSE filter (Sep LMMSE), respectively, the iterative joint decoding parallel interference canceller (Joint SUMF), and the iterative joint decoding per-user LMMSE filtering interference canceller (Joint LMMSE). Also included are the fundamental limits on spectral efficiency for optimal joint detection (Opt bound), separate detection/decoding with the matched filter (MF bound), and the LMMSE filter (LMMSE bound), respectively.

In Figure 2.42, the spectral efficiency in terms of bits per channel use is plotted for the six schemes together with the corresponding fundamental bounds[9] for rate $R = 1/2$ coding. The spectral efficiency for each of the schemes is determined as βR, where β is the maximum load supportable at a specific E_b/N_0 and with a bit error rate of 10^{-4} or less. Based on Figure 2.42, we confirm that joint iterative decoding provides significantly better performance than separate detection/decoding. It is also confirmed that per-user LMMSE filtering offers a significant improvement in efficiency over simple parallel interference cancellation. The performance of all the schemes suffer a considerable penalty as compared to the maximum spectral efficiency for rate $R = 1/2$ coding. The simple convolutional code used is thus not an appropriate choice for approaching the fundamental limits. As shown in [12], simpler codes such as repetition codes may in fact provide a higher efficiency at moderate to large E_b/N_0.

It is also interesting to observe that the difference in spectral efficiency from using either extrinsic information or posterior information is only of minor significance for the examples considered here. For the parallel interference canceller, posterior information provides a slight advantage for low E_b/N_0, while for higher E_b/N_0 extrinsic information becomes the better choice. The same trend is observed for the per-user LMMSE case. However, up to $E_b/N_0 = 10$ dB, posterior information maintains an

---

[9]The bounds are based on asymptotic spectral efficiency, detailed in [19], as $K \to \infty$ with K/N remaining constant.

advantage over extrinsics. Although these advantages in terms of spectral efficiency are minor, for specific implementations there could still be gains of several dBs.

2.4 CONCLUDING REMARKS

In this chapter we have presented a comprehensive overview of iterative techniques for multiuser detection and multiuser decoding. The presentation demonstrates that interference cancellation is fundamental to most relevant iterative algorithms, providing efficient methods for implementing multiuser detection and decoding schemes.

Interference cancellation can conveniently be constructed as a modular structure based on repetitive application of a basic *interference cancellation unit*. The modular structure allows for realizing a wide range of cancellation modes, including common parallel and serial cancellation schedules. The modular structure is also well suited for hardware implementation, forwarding a common residual error signal to individual single-user cancellation units.

For uncoded communications systems, we considered both linear and non-linear methods. Linear constraints lead to matrix filtering schemes such as the decorrelator and the LMMSE detectors, both involving matrix inversion. Linear iterative detectors are therefore closely related to iterative techniques for solving linear equation systems, some of which date back to the early nineteenth century. The linear detection problem can also be cast as an unconstrained optimization problem, allowing for the application of well-known decent algorithms. A convenient feature of inversion and decent algorithms, such as series expansion, Gauss-Seidel iteration, and conjugate gradient, is that they can directly be interpreted and implemented as interference cancellation structures.

The same level of understanding is not yet available for non-linear iterative techniques. Here, we have presented a new general framework for non-linear cancellation based on constrained optimization. Enforcing particular non-linear constraints through appropriately selected barrier functions, the solution to the constrained optimization problem leads to existing non-linear interference cancellation structures with well-known tentative decision functions such as hard decision, clipped soft decision, and hyperbolic-tangent soft decision. The framework allows for the application of arbitrary barrier functions, which can potentially lead to new and better non-linear tentative decision functions.

Among the linear and non-linear iterative schemes considered here, two schemes stand out in terms of convergence and BER performance. The conjugate gradient cancellation structure is the preferred linear alternative, while a serial cancellation scheme based on the clipped soft decision function is superior among the considered iterative algorithms for uncoded systems.

For coded multiuser systems, a thorough overview of the canonical joint iterative decoder was presented, emphasizing the close relation to turbo decoding techniques. Using an optimal joint multiuser detector within the canonical decoder is not feasible due to complexity. As a viable low-complexity alternative, we defined a generic linear iterative joint detector with suitable interfaces for converting LLR values to corresponding signal estimates, and signal estimates to corresponding LLR values. Linear cancellation is particularly well-suited for iterative multiuser decoding as it naturally incorporates prior LLR information into the cancellation structure in the form of estimates of the current multiple access interference. To demonstrate the principles, parallel cancellation schemes based on matched-filter and LMMSE filtering were derived and discussed.

In a joint iterative decoder with optimal components, the principles of belief propagation dictates the exchange of extrinsic information between components. For suboptimal components, these principles do not necessarily apply. Here, we considered the exchange of posterior information as an alternative, leading to some performance improvements. For most of our numerical examples, complexity savings in terms of iterations required for convergence, or system capacity improvements for a fixed number of iterations were observed. Despite these advantages, only marginal improvements of the overall spectral efficiency were noted. A formal framework and proper justification for the exchange of reliability information between sub-optimal components in an iterative decoder are still open problems.

In summary, we conclude that the area of iterative multiuser detection and decoding is now well-developed and mature. In the process of preparing this chapter we have, however, recognized some remaining challenges. A rigorous framework for the design and analysis of non-linear interference cancellation is still lacking. We have proposed a framework based on constrained optimization that may provide additional insight. Still, more efforts are required in this direction.

Finally, the most obvious shortcoming of multiuser detection and decoding is the lack of practical implementation and commercial acceptance. To realize the information-theoretic advantages of multiuser processing, future wireless communications systems must be specifically designed to support practical implementation of multiuser detection and decoding. Only then will we enjoy the obvious benefits of more than thirty years of research efforts in this area.

REFERENCES

1. C. Sankaran and A. Ephremides. Solving a class of optimum multiuser detection problems with polynomial complexity. *IEEE Trans. Inform. Theory*, 44:1958–1961, September 1998.

2. S. Ulukus and R. Yates. Optimum multiuser detection is tractable for synchronous CDMA using *M*-sequences. *IEEE Commun. Lett.*, 2:89–91, April 1998.

3. C. Schlegel and A. Grant. Polynomial complexity optimal detection of certain multiple access systems. *IEEE Trans. Inform. Theory*, 46(6):2246–2248, September 1998.

4. S. Verdú. Computational complexity of optimum multiuser detection. *Algorithmica*, pages 303–312, May 1989.

5. T. R. Giallorenzi and S. G. Wilson. Multiuser ML sequence estimator for convolutionally coded asynchronous DS-CDMA systems. *IEEE Trans. Commun.*, 44(8):997–1008, August 1996.

6. C. Berrou and A. Glavieux. Near optimum error correcting coding and decoding: Turbo-codes. *IEEE Trans. Commun.*, 44:1261–1271, October 1996.

7. S. Benedetto, D. Divsalar, G. Montorsi, and F. Pollara. Serial concatenation of interleaved codes: Performance analysis, design and iterative decoding. *IEEE Trans. Inform. Theory*, 44(3):909–926, May 1998.

8. R. Gallager. *Low-density parity check codes*. MIT Press, 1963.

9. D. Divsalar, H. Jin, and R. McEliece. Coding theorems for 'turbo-like' codes. In *36th Allerton Conference on Communication, Control and Computing*, pages 201–210, 1998.

10. S. Verdú. *Multiuser Detection*. Cambridge University Press, Cambridge, 1998.

11. L. K. Rasmussen, P. D. Alexander, and T. J. Lim. A linear model for CDMA signals received with multiple antennas over multipath fading channels. Chapt. 2 in: *CDMA Techniques for 3rd Generation Mobile Systems*, edited by F. Swarts, P. van Rooyen, I. Oppermann and M. Lötter, Kluwer Academic Publisher, September 1998.

12. C. Schlegel and A. Grant. *Coordinated Multiuser Communications*. Springer, 2006.

13. S. Verdú. Minimum probability of error for asynchronous Gaussian multiple–access channels. *IEEE Trans. Inform. Theory*, 32(1):85–96, January 1986.

14. S. Moshavi. Multi-user detection for DS-CDMA communications. *IEEE Commun. Mag.*, 34(10):124–136, October 1996.

15. R. Lupas and S. Verdú. Linear multiuser detectors for synchronous code–division multiple–access channels. *IEEE Trans. Inform. Theory*, 35(1):123–136, January 1989.

16. Z. Xie, R. T. Short, and C. K. Rushforth. A family of suboptimum detectors for coherent multiuser communications. *IEEE J. Select. Areas Commun.*, 8(4):683–690, May 1990.

17. P. B. Rapajic and B. S. Vucetic. Adaptive receiver structures for asynchronous CDMA systems. *IEEE J. Select. Areas Commun.*, 12(4):685–697, May 1994.

18. U. Madhow and M. L. Honig. MMSE interference suppression for direct–sequence spread–spectrum CDMA. *IEEE Trans. Commun.*, 42(12):3178–3188, December 1994.

19. S. Verdú and S. Shamai. Spectral efficiency of CDMA with random spreading. *IEEE Trans. Inform. Theory*, 45(2):622–640, March 1999.

20. D. Tse and S. Hanly. Linear multiuser receivers: Effective interference, effective bandwidth and user capacity. *IEEE Trans. Inform. Theory*, 45(2):641–657, March 1999.

21. X. Wang and V. Poor. Iterative (turbo) soft interference cancellation and decoding for coded CDMA. *IEEE Trans. Commun.*, 47(7):1046–1061, July 1999.

22. H. El Gamal and E. Geraniotis. Iterative multiuser detection for coded CDMA signals in AWGN and fading channels. *IEEE J. Select. Areas Commun.*, 18(1):30–41, January 2000.

23. J. Boutros and G. Caire. Iterative multiuser decoding: Unified framework and asymptotic performance analysis. *IEEE Trans. Inform. Theory*, 48:1772–1793, July 2002.

24. G. Caire, R. Müller, and T. Tanaka. Iterative multiuser joint decoding: Optimal power allocation and low-complexity implementation. *IEEE Trans. Inform. Theory*, 50:1950–1972, September 2004.

25. M. Tüchler, R. Koetter, and A. C. Singer. Turbo equalization: Principles and new results. *IEEE Trans. Commun.*, 50(5):754–767, May 2002.

26. J. Luo, K. Pattipati, P. Whillett, and F. Hasegawa. Near optimal multiuser detection in synchronous CDMA. *IEEE Commun. Lett.*, 5:361–363, September 2001.

27. P. H. Tan and L. K. Rasmussen. Asymptotically optimal nonlinear MMSE multiuser detection based on multivariate Gaussian approximation. *IEEE Trans. Commun.*, 54(8): 1427–1438, August 2006.

28. R. R. Müller and J. B. Huber. *Broadband Wireless Communications*, chapter Iterative soft-decision interference cancellation for CDMA, pages 110–115. Springer-Verlag, London, United Kingdom, 1998.

29. W. van Etten. Maximum likelihood receiver for multiple channel transmission systems. *IEEE Trans. Commun.*, 24(2):276–283, February 1976.

30. R. Kohno, M. Hatori, and H. Imai. Cancellation techniques of co-channel interference in asynchronous spread spectrum multiple access systems. *Electronics and Commun.*, 66-A:20–29, 1983.

31. K. S. Schneider. Optimum detection of code division multiplexed signals. *IEEE Trans. Aerosp. Electron. Systems*, 15:181–185, January 1979.

32. R. W. Lucky, J. Salz, and E. J. Weldon Jr. *Principles of Data Communication*. McGraw-Hill, New York, 1968.

33. H. V. Poor. *An Introduction to Signal Detection and Estimation*. Springer-Verlag, 1994.

34. D. P. Taylor. The estimate feedback equalizer: A suboptimum nonlinear receiver. *IEEE Trans. Commun.*, 21:979–990, February 1973.

35. S. Gollamudi and Y.-F. Huang. Iterative nonlinear MMSE multiuser detection. *In Proc. Int. Conf. Acoustics, Speech & Sig.*, pages 2595–2598, Phoenix, USA, March 1999.

36. P. D. Alexander, A. J. Grant, and M. C. Reed. Iterative detection on code-division multiple-access with error control coding. *European Trans. Telecommun.*, 9(5):419–426, Sept.–Oct. 1998.

37. M. K. Varanasi and B. Aazhang. Multistage detection in asynchronous code–division multiple–access communications. *IEEE Trans. Commun.*, 38(4):509–519, April 1990.

38. M. K. Varanasi and B. Aazhang. Near–optimum detection in synchronous code–division multiple–access systems. *IEEE Trans. Commun.*, 39(5):725–736, May 1991.

39. P. Dent, B. Gudmundson, and M. Ewerbring. CDMA-IC: A novel code division multiple access scheme based on interference cancellation. In *IEEE Int. Symp. Person. Indoor Mobile Radio Commun.*, pages 98–102, October 1992.

40. K. Jamal and E. Dahlman. Multi-stage interference cancellation for DS-CDMA. In *IEEE Veh. Technol., Conf.*, pages 671–675, Atlanta, USA, Apr. 1996.

41. P. Patel and J. Holtzman. Analysis of a simple successive interference cancellation scheme in a DS/CDMA system. *IEEE J. Select. Areas Commun.*, 12(5):796–807, June 1994.

42. H. Elders-Boll, H.-D. Schotten, and A. Busboom. Efficient implementation of linear multiuser detectors for asynchronous CDMA systems by linear interference cancellation. *European Trans. Telecommun.*, 9(5):427–438, Sept–Oct 1998.

43. A. J. Grant and C. B. Schlegel. Convergence of linear interference cancellation multiuser receivers. *IEEE Trans. Commun.*, 49:1824–1834, October 2001.

44. L. B. Nelson and H. V. Poor. Iterative multiuser receivers for CDMA channels: An EM-based approach. *IEEE Trans. Commun.*, 44(12):1700–1710, December 196.

45. T. Tanaka and M. Okada. Approximate belief propagation, density evolution, and statistical neurodynamics for CDMA multiuser detection. *IEEE Trans. Inform. Theory*, 51:700–706, February 2005.

46. A. Yener, R. D. Yates, and S. Ulukus. A nonlinear programming approach to CDMA multiuser detection. In *Asilomar Conf. Sig., Sys., Comp.*, pages 1579–1583, Pacific Grove, USA, October 1999.

47. P. H. Tan, L. K. Rasmussen, and T. J. Lim. Iterative interference cancellation as maximum-likelihood detection in CDMA. In *Int. Conf. Info., Commun., Sig. Proc.*, Singapore, December 1999.

48. P. H. Tan, L. K. Rasmussen, and T. J. Lim. Constrained maximum-likelihood detection in CDMA. *IEEE Trans. Commun.*, 49:142–153, January 2001.

49. A. Yener, R. D. Yates, and S. Ulukus. CDMA multiuser detection: a nonlinear programming approach. *IEEE Trans. Commun.*, 50(6):1016–1024, June 2002.

50. L. K. Rasmussen. *Iterative detection methods for multi-user direct sequence CDMA systems*, chapter online in subject category Multiuser Communications, online April 2003. The Wiley Encyclopedia of Telecommunications. Wiley and Son, April 2004.

51. W. H. Press, S. A. Teukolsky, B. P. Flannery, and W. T. Vettering. *Numerical Recipes in C*. Cambridge University Press, 2nd edition, 1992.

52. G. Golub and C. Van Loan. *Matrix Computations*. The John Hopkins University Press, 3rd edition, 1996.

53. R. A. Horn and C. R. Johnson. *Matrix Analysis*. Oxford University Press, Cambridge, 1985.

54. O. Axelsson. *Iterative Solution Methods*. Cambridge University Press, 1994.

55. S. P. Boyd and L. Vandenberghe. *Convex Optimization*. Cambridge University Press, 2004.

56. S. Moshavi, E. G. Kanterakis, and D. L. Schilling. Multistage linear receivers for DS-CDMA systems. *Int. J. Wireless Inform. Networks*, 3:1–17, January 1996.

57. D. Guo, L. K. Rasmussen, S. Sun, and T. J. Lim. A matrix-algebraic approach to linear parallel interference cancellation in CDMA. *IEEE Trans. Commun.*, 41(1):152–161, Jan. 2000.

58. D. Guo, S. Verdú, and L. K. Rasmussen. Asymptotic normality of linear multiuser receiver outputs. *IEEE Trans. Inform. Theory*, 48(12):3080–3095, December 2002.

59. R. R. Müller and S. Verdú. Design and analysis of low-complexity interference mitigation on vector channels. *IEEE J. Select. Areas Commun.*, 19:1429–1441, August 2001.

60. M. J. Juntti, B. Aazhang, and J. O. Lilleberg. Iterative implementations of linear multiuser detection for asynchronous CDMA systems. *IEEE Trans. Commun.*, 46(4):503–508, April 1998.

61. R. M. Buehrer, S. P. Nicoloso, and S. Gollamudi. Linear versus nonlinear interference cancellation. *IEEE J. Commun. Networks*, 1:118–133, June 1999.

62. P. H. Tan and L. K. Rasmussen. Linear interference cancellation in CDMA based on iterative solution techniques for linear equation systems. *IEEE Trans. Commun.*, 48:2099–2108, December 2000.

63. X. Wang and H. V. Poor. Space-time multiuser detection in multipath CDMA channels. *IEEE Trans. Sig. Proc.*, 47:2356–2374, September 1999.

64. L. K. Rasmussen, T. J. Lim, and A.-L. Johansson. A matrix-algebraic approach to successive interference cancellation in CDMA. *IEEE Trans. Commun.*, 48(1):145–151, Jan. 2000.

65. L. G. F. Trichard, J. S. Evans, and I. B. Collings. Large system analysis of linear multistage parallel interference cancellation. *IEEE Trans. Commun.*, 50:1778–1786, November 2002.

66. L. G. F. Trichard, J. S. Evans, and I. B. Collings. Optimal multistage linear multiuser receivers. *IEEE Trans. Wireless Commun.*, 4:1092–1101, May 2005.

67. D. Guo, L. K. Rasmussen, and T. J. Lim. Linear parallel interference cancellation in random-code CDMA. *IEEE J. Selected Areas Commun.*, 17:2074–1081, December 1999.

68. J. S. Goldstein, I. S. Reed, and L. L. Scharf. A multistage representation of the Wiener filter based on orthogonal projections. *IEEE Trans. Inform. Theory*, 44(7):2943–2959, November 1998.

69. M. L. Honig and W. Xiao. Performance of reduced-rank linear interference suppression. *IEEE Trans. Inform. Theory*, 47:1928–1946, July 2001.

70. M. Joham, Y. Sun, M. D. Zoltowski, M. Honig, and J. S. Goldstein. A new backward recursion for the multi-stage nested Wiener filter employing Krylov subspace methods. In *IEEE Military Commun. Conf.*, volume 2, pages 28–31, Washington, DC, USA, October 2001.

71. D. Divsalar, M. K. Simon, and D. Raphaeli. Improved parallel interference cancellation for CDMA. *IEEE Trans. Commun.*, 46(2):258–268, Feb. 1998.

72. L. L. Scharf, L.T. McWhorter, E. K. P. Chong, J. S. Goldstein, and M. D. Zoltowski. Algebraic equivalence of conjugate gradient direction and multistage Wiener filters. In *Workshop Adap. Sensor Array Proc.*, pages 388–392, Hong Kong, China, April 2003.

73. N. B. Mandayam and S. Verdú. Analysis of an approximate decorrelating detector. *Wireless Personal Commun.*, 6:97–111, June 1998.

74. D. Johnsson. Some limit theorems for the eigenvalues of a sample covariance matrix. *J. Mult. Anal.*, 12:1–38, 1982.

75. Z. D. Bai and Y. Q. Yin. Limit of the smallest eigenvalue of a large dimensional sample covariance matrix. *Ann. Prob.*, 21:1275–1294, 1993.

76. Z. D. Bai, J. W. Silverstein, and Y. Q. Yin. A note on the largest eigenvalue of a large dimensional sample covariance matrix. *J. Mult. Anal.*, 26:166–168, 1988.

77. A. J. Grant and C. B. Schlegel. Convergence of linear interference cancellation multiuser receivers. *IEEE Trans. Commun.*, 49:1824–1834, October 2001.

78. D. R. Brown. Multistage parallel interference cancellation: Convergence behavior and improved performance through limit cycle mitigation. *IEEE Trans. Signal Processing*, 53:283–294, January 2005.

79. L. K. Rasmussen and I. J. Oppermann. Ping-pong effects in linear parallel interference cancellation for CDMA. *IEEE Trans. Wireless Commun.*, 2(2):357–363, Mar. 2003.

80. C. Schlegel, S. Roy, P. D. Alexander, and Z. Xiang. Multi-user projection receivers. *IEEE J. Select. Areas Commun.*, 14(8):1610–1618, October 1996.

81. C. Schlegel, P. D. Alexander, and S. Roy. Coded asynchronous CDMA and its efficient detection. *IEEE Trans. Inform. Theory*, 44(7):2837–2847, November 1998.

82. P. D. Alexander, L. K. Rasmussen, and C. B. Schlegel. A linear receiver for coded multiuser CDMA. *IEEE Trans. Commun.*, 45(5):605–610, May 1997.

83. L. Brunel and J. Boutros. Code division multiple access based on independent codes and turbo decoding. *Annales des Telecommunications*, 54(7–8):401–410, Jul.–Aug. 1999.

84. M. Moher. An iterative multiuser decoder for near-capacity communications. *IEEE Trans. Commun.*, 46(7):870–880, July 1998.

85. M. Moher and T. A. Gulliver. Cross-entropy and iterative decoding. *IEEE Trans. Inform. Theory*, 44(7):3097–3104, November 1998.

86. F. N. Brännström, T. M. Aulin, and L. K. Rasmussen. Iterative detectors for trellis code multiple access. *IEEE Trans. Commun.*, 50:1478–1485, September 2002.

87. M. C. Reed, C. B. Schlegel, P. D. Alexander, and J. Asenstorfer. Iterative multiuser detection for CDMA with FEC: Near-single-user performance. *IEEE Trans. Commun.*, 46:1693–1699, December 1998.

88. H. V. Poor. Iterative multiuser detection. *IEEE Signal Processing Mag.*, 21(1):81–88, January 2004.

89. J. Hagenauer. Forward error correcting for CDMA systems. In *IEEE Int. Symp. Spread Spectrum Techn. App.*, pages 566–569, Mainz, Germany, September 1996.

90. M. L. Honig, G. K. Woodward, and Y. Sun. Adaptive iterative multiuser decision feedback detection. *IEEE Trans. Wireless Commun.*, 3(2):477–485, March 2004.

91. R. R. Müller and W. H. Gerstacker. On the capacity loss due to separation of detection and decoding. *IEEE Trans. Inform. Theory*, 50(8):1769–1778, August 2004.

92. D. Guo and S. Verdú. Randomly spread CDMA: Asymptotics via statistical physics. *IEEE Trans. Inform. Theory*, 51:1983–2010, June 2005.

93. L. R. Bahl, J. Cocke, F. Jelinek, and J. Raviv. Optimal decoding of linear codes for minimizing symbol error rate. *IEEE Trans. Inform. Theory*, 20:284–287, March 1974.

94. F. R. Kschischang, B. J. Frey, and H.-A. Loeliger. Factor graphs and the sum-product algorithm. *IEEE Trans. Inform. Theory*, 47:498–519, February 2001.

95. P. D. Alexander, M. C. Reed, J. Asenstorfer, and C. B. Schlegel. Iterative multiuser interference reduction: Turbo CDMA. *IEEE Trans. Commun.*, 47:1008–1014, July 1999.

96. A. B. Reid, A. J. Grant, and P. D. Alexander. List detection for multi-access channels. In *Global Telecommun. Conf.*, pages 1083–1087, Taipei, Taiwan, November 2002.

97. C. Kuhn and J. Hagenauer. Iterative list-sequential (LISS) detector for fading multiple-access channels. In *IEEE Global Commun. Conf.*, pages 330–335, Dallas, USA, December 2004.

98. H. Vikalo, B. Hassibi, and T. Kailath. Iterative decoding for MIMO channels via modified sphere decoding. *IEEE Trans. Wireless Commun.*, 3:2299–2311, November 2004.

99. L. K. Rasmussen, A. J. Grant, and P. D. Alexander. An extrinsic Kalman filter for iterative multiuser decoding. *IEEE Trans. Inform. Theory*, 50(4):642–647, April 2004.

100. M. L. Honig and R. Ratasuk. Large-system performance of iterative multiuser decision-feedback detection. *IEEE Trans. Commun.*, 51(8):1368–1377, August 2003.

101. R. Milner and L. K. Rasmussen. Weighted extrinsic feedback in the iterative multiuser decoding of coded CDMA. In *Aus. Commun. Theory Workshop*, pages 60–65, Brisbane, Australia, February 2006.

102. H. V. Poor and S. Verdú. Probability of error in MMSE multiuser detection. *IEEE Trans. Inform. Theory*, 43:858–871, May 1997.

103. H. El Gamal and A. R. Hammons Jr. Analyzing the turbo decoder using the Gaussian approximation. *IEEE Trans. Inform. Theory*, 47:671–686, February 2001.

104. S. Ten Brink. Convergence of iterative decoding. *IEEE Electron. Lett.*, 35(10):806–808, May 1999.

105. S. Ten Brink. Convergence behavior of iteratively decoded parallel concatenated codes. *IEEE Trans. Commun.*, 49:1727–1737, October 2001.

106. Z. Shi and C. Schlegel. Joint iterative decoding of serially concatenated error control coded CDMA. *IEEE J. Select. Areas Commun.*, 19:1646–1653, August 2001.

107. K. Li and X. Wang. EXIT chart analysis of turbo multiuser detection. *IEEE Trans. Wireless Commun.*, 4(1):300–311, January 2005.

108. A. Grant. Co-channel interference reduction in rayleigh fading channels. *Digital Signal Processing*, 15(5), September 2006.

3

BLIND MULTIUSER DETECTION IN FADING CHANNELS

Daryl Reynolds, H. Vincent Poor, and Xiaodong Wang

3.1 INTRODUCTION

The three primary characteristics that distinguish wireless communications from wireline communications are dynamism, fading, and interference. In this chapter, we address all three of these characteristics by considering adaptive methods for multiuser detection in fading channels.

Fading refers to variations in the gain of a communications channel. The primary source of fading is multipath, which leads to constructive and destructive self-interference of signals at a communications receiver. Depending on various properties of the communication link (bandwidth, mobility, etc.) fading can vary with time (i.e., time-selective fading) and frequency (frequency-selective fading). Of course, given the fact that the multipath profile depends on geometry, fading is also dependent on position and angle of arrival.

A key issue in the mitigation of fading is that the fading is caused by the channel, which is not known *a priori* to the receiver. There are two approaches to dealing with this issue. One is to treat the channel gains as random quantities with known distributions (e.g., Rayleigh, Rician, etc.) and to devise optimal receiver algorithms based on this statistical model. Examples of this approach can be found in [27,42,59]. An alternative approach, which is considered here,

Advances in Multiuser Detection. Edited by Michael L. Honig
Copyright © 2009 John Wiley & Sons, Inc.

is for the receiver to use the outputs of the channel to adapt to the fadings, i.e., adaptive systems essentially estimate the channel and then use the channel estimates in a multiuser receiver corresponding to a known channel. This approach has the advantage of also allowing adaptation to other unknown quantities, such as interference, and also has performance advantages over systems that simply treat the fading as being random and unknowable. A general overview of adaptive receivers for multiuser systems can be found in [51]. Note that it is impossible to give just treatment to the full spectrum of multiuser detection strategies for fading channels in this chapter, so we have chosen to present a sampling of modern *blind* approaches. Other modern approaches to multiuser detection in fading include turbo multiuser detection (discussed in a previous chapter of this book), multiuser detection with multiple-antennas (also discussed in a previous chapter of this book), non-coherent multiuser detection [23,33,44], robust multiuser detection [3,38,50,57], cooperative multiuser detection [41], and decision-feedback multiuser detection [37].

In this chapter, we specifically address the problem of blind adaptation in which the system does not make use of training symbols, but rather learns the optimal receiver directly from data-modulated signals. This approach has the obvious advantage of not requiring additional overhead to transmit training data. We consider several aspects of this problem, including the quasi-static case, in which the fading can be considered to be constant over a processing window, and the fast-fading (or time-selective) case, in which the fading can change within the processing window. We also consider both frequency-flat and frequency-selective cases, and we consider the use of multiple receive antennas. Performance issues are also addressed. We begin in Section 2 with a general approach to this problem based on subspace methods. Here, we see that such techniques provide a powerful framework for addressing a variety of issues arising in fading multiuser channels. One limitation of many adaptive techniques, including these subspace methods, is that they are primarily applicable to systems in which the signaling multiplex (e.g., the spreading codes in a code-division multiple-access (CDMA) system) is time-invariant. In many systems, however, this condition is violated (e.g., in so-called "long code" CDMA systems). Thus, alternative methods are often needed, and in Section 3, we consider methods for such problems based on Markov chain Monte Carlo (MCMC) techniques. This provides a further powerful technique for the treatment of fading in multiuser settings. In Section 4, we turn to the problem of fast fading, to which a sequential expectation-maximization (EM) approach can be successfully applied. Finally, in Section 3.6, we consider situations in which it is advantageous to move computational complexity from the receiver to the transmitter, e.g., a cellular downlink. In particular, we develop transmitter precoding strategies that are designed to jointly combat multiple access interference and exploit any available transmit antenna or multipath diversity. We include analyses of achievable diversity and signal-to-interference-plus-noise ratio (SINR) and bit-error-rate (BER) performance for the adaptive case.

3.2 SIGNAL MODELS AND BLIND MULTIUSER DETECTORS FOR FADING CHANNELS

In this section, we present signal models as well as their corresponding basic blind multiuser detection algorithms for several systems. We begin with the most generally applicable model: asynchronous CDMA with multiple antennas operating over fading multipath channels. We then describe two special cases, namely, synchronous multi-path CDMA system with no intersymbol interference (ISI), and synchronous multi-antenna CDMA system. We will see that all these system share essentially the same signal model, which allows for some unification in system design and analysis.

3.2.1 Asynchronous Multi-Antenna Multipath CDMA

We consider a general asynchronous CDMA uplink with K users signaling through their respective multipath channels and employing P receive antennas. We start with the continuous-time signal model. Let the channel impulse response between the k-th user's transmitter and the p-th receive antenna be:

$$g_k^{(p)}(t) = \sum_{l=1}^{L} \alpha_{l,k}^{(p)} \delta\left(t - \tau_{l,k}^{(p)}\right) \tag{3.1}$$

where L is the total number of paths in the channel; $\alpha_{l,k}^{(p)}$ and $\tau_{l,k}^{(p)}$ are respectively the complex path gain and the delay of the k-th user's l-th path corresponding to the p-th receive antenna, $\tau_{1,k}^{(p)} < \tau_{2,k}^{(p)} < \cdots < \tau_{L,k}^{(p)}$. The continuous-time signal transmitted from the k-th user is:

$$x_k(t) = \sum_{i=0}^{M-1} b_k[i] s_k(t - iT) \tag{3.2}$$

where $s_k(t)$ is the spreading waveform of the k-user, given by:

$$s_k(t) = \frac{1}{\sqrt{N}} \sum_{j=0}^{N-1} s_{j,k} \psi(t - jT_c), \quad 0 \le t < T \tag{3.3}$$

and N is the processing gain; $\{s_{j,k}\}_{j=0}^{N-1}$ is a signature sequence of ± 1's assigned to the k-th user; and $\psi(\cdot)$ is a chip waveform of duration $T_c = \frac{T}{N}$ and with unit energy, i.e., $\int_0^{T_c} \psi^2(t)\, dt = 1$.

The received continuous-time signal at the p-th receive antenna is given by:

$$r^{(p)}(t) = \sum_{k=1}^{K} \left(x_k(t) \star g_k^{(p)}(t)\right) + n^{(p)}(t) \tag{3.4}$$

$$
= \sum_{k=1}^{K} \sum_{i=0}^{M-1} b_k[i] \left\{ s_k(t-iT) \star g_k^{(p)}(t) \right\} + n^{(p)}(t)
$$

$$
= \sum_{k=1}^{K} \sum_{i=0}^{M-1} b_k[i] \sum_{l=1}^{L} \alpha_{l,k}^{(p)} s_k\left(t - iT - \tau_{l,k}^{(p)}\right) + n^{(p)}(t) \tag{3.5}
$$

where $\star$ denotes convolution and $n^{(p)}(t)$ is complex additive white Gaussian noise at the p-th receive antenna with power $E\left\{\left|n^{(p)}(t)\right|^2\right\} = \eta$.

At the receiver, the received signal $r^{(p)}(t)$ is match-filtered with respect to $\psi(t)$ and sampled at the chip-rate. Let:

$$
\iota \triangleq \max_{1 \le k \le K, \, 1 \le p \le P} \left\{ \left\lceil \frac{\tau_{L,k}^{(p)} + T_c}{T} \right\rceil \right\} \tag{3.6}
$$

be the maximum delay spread in terms of symbol intervals. Substituting (3.3) into (3.5), the q-th signal sample during the i-th symbol is given by:

$$
r_q^{(p)}[i] = \int_{iT+qT_c}^{iT+(q+1)T_c} r^{(p)}(t)\psi(t-iT-qT_c)\,dt
$$

$$
= \int_{iT+qT_c}^{iT+(q+1)T_c} \psi(t-iT-qT_c) \sum_{k=1}^{K} \sum_{m=0}^{M-1} b_k[m] \sum_{l=1}^{L} \alpha_{l,k}^{(p)} \frac{1}{\sqrt{N}}
$$

$$
\times \sum_{j=0}^{N-1} s_{j,k}\psi\left(t-mT-\tau_{l,k}^{(p)}-jT_c\right)dt + n_q^{(p)}[i]
$$

$$
= \sum_{k=1}^{K} \sum_{m=i-\iota}^{i} b_k[m] \sum_{l=1}^{L} \alpha_{l,k}^{(p)} \frac{1}{\sqrt{N}} \sum_{j=0}^{N-1} s_{j,k}
$$

$$
\times \int_{iT+qT_c}^{iT+(q+1)T_c} \psi(t-iT-qT_c)\psi\left(t-mT-\tau_{l,k}^{(p)}-jT_c\right)dt + n_q^{(p)}[i]
$$

$$
= \sum_{k=1}^{K} \sum_{m=0}^{\iota} b_k[i-m]
$$

$$
\times \underbrace{\sum_{j=0}^{N-1} s_{j,k} \frac{1}{\sqrt{N}} \sum_{l=1}^{L} \alpha_{l,k}^{(p)} \overbrace{\int_{0}^{T_c} \psi(t)\psi\left(t-\tau_{l,k}^{(p)}+mT-jT_c+qT_c\right)dt}^{h_k^{(p)}[mN+q-j]} + n_q^{(p)}[i],}_{\tilde{s}_k^{(p)}[mN+q]}
$$

$$
q = 0, \ldots, N-1; \quad i = 0, \ldots, M-1 \tag{3.7}
$$

where $n_q^{(p)}[i] = \int_{iT+qT_c}^{iT+(q+1)T_c} n^{(p)}(t)\psi(t - iT - qT_c)\,dt$. Denote:

$$
\underbrace{\underline{r}^{(p)}[i]}_{N \times 1} \triangleq \begin{bmatrix} r_0^{(p)}[i] \\ \vdots \\ r_{N-1}^{(p)}[i] \end{bmatrix}, \qquad
\underbrace{\underline{b}[i]}_{K \times 1} \triangleq \begin{bmatrix} b_1[i] \\ \vdots \\ b_K[i] \end{bmatrix}, \qquad
\underbrace{\underline{n}^{(p)}[i]}_{N \times 1} \triangleq \begin{bmatrix} n_0^{(p)}[i] \\ \vdots \\ n_{N-1}^{(p)}[i] \end{bmatrix},
$$

$$
\underbrace{\underline{\tilde{S}}^{(p)}[j]}_{N \times K} \triangleq \begin{bmatrix} \tilde{s}_1^{(p)}[jN] & \cdots & \tilde{s}_K^{(p)}[jN] \\ \vdots & \vdots & \vdots \\ \tilde{s}_1^{(p)}[jN+N-1] & \cdots & \tilde{s}_1^{(p)}[jN+N-1] \end{bmatrix}, \quad j = 0, \ldots, \iota
$$

Then (3.7) can be written in terms of vector convolution as:

$$
\underline{r}^{(p)}[i] = \underline{\tilde{S}}^{(p)}[i] \star \underline{b}[i] + \underline{n}^{(p)}[i], \quad p = 1, 2, \ldots, P \tag{3.8}
$$

By stacking m successive sample vectors, we define the following quantities:

$$
\underbrace{\mathbf{r}^{(p)}[i]}_{Nm \times 1} \triangleq \begin{bmatrix} \underline{r}^{(p)}[i] \\ \vdots \\ \underline{r}^{(p)}[i+m-1] \end{bmatrix}, \qquad
\underbrace{\mathbf{n}^{(p)}[i]}_{Nm \times 1} \triangleq \begin{bmatrix} \underline{n}^{(p)}[i] \\ \vdots \\ \underline{n}^{(p)}[i+m-1] \end{bmatrix},
$$

$$
\underbrace{\mathbf{b}[i]}_{K(m+\iota) \times 1} \triangleq \begin{bmatrix} \underline{b}[i-\iota] \\ \vdots \\ \underline{b}[i+m-1] \end{bmatrix}, \qquad
\underbrace{\tilde{\mathbf{S}}^{(p)}}_{Nm \times K(m+\iota)} \triangleq \begin{bmatrix} \underline{\tilde{S}}^{(p)}[\iota] & \cdots & \underline{\tilde{S}}^{(p)}[0] & \cdots & \mathbf{0} \\ \vdots & \ddots & \ddots & \ddots & \vdots \\ \mathbf{0} & \cdots & \underline{\tilde{S}}^{(p)}[\iota] & \cdots & \underline{\tilde{S}}^{(p)}[0] \end{bmatrix}
$$

We can then write (3.8) in a matrix forms as:

$$
\mathbf{r}^{(p)}[i] = \tilde{\mathbf{S}}^{(p)}\mathbf{b}[i] + \mathbf{n}^{(p)}[i], \quad p = 1, \ldots, P \tag{3.9}
$$

Finally denote:

$$
\underbrace{\mathbf{r}[i]}_{PNm \times 1} \triangleq \begin{bmatrix} \mathbf{r}^{(1)}[i] \\ \vdots \\ \mathbf{r}^{(P)}[i] \end{bmatrix}, \qquad
\underbrace{\tilde{\mathbf{S}}[i]}_{PNm \times K} \triangleq \begin{bmatrix} \tilde{\mathbf{S}}^{(1)}[i] \\ \vdots \\ \tilde{\mathbf{S}}^{(P)}[i] \end{bmatrix}, \qquad
\underbrace{\mathbf{n}[i]}_{PNm \times 1} \triangleq \begin{bmatrix} \mathbf{n}^{(1)}[i] \\ \vdots \\ \mathbf{n}^{(P)}[i] \end{bmatrix}
$$

Then (3.9) can be written as:

$$
\mathbf{r}[i] = \tilde{\mathbf{S}}\mathbf{b}[i] + \mathbf{n}[i] \tag{3.10}
$$

The smoothing factor m is chosen according to $m = \left\lceil \frac{NP+K}{NP-K} \right\rceil \iota$ so that the matrix $\tilde{\mathbf{S}}$ is a "tall" matrix, i.e., $PNm \geq K(m+\iota)$. Note that because we are match filtering with respect to the transmitted waveforms and not the received waveforms, $\mathbf{r}[i]$ does not, strictly speaking, represent a sufficient statistic for detecting $\mathbf{b}[i]$. However, the loss in inferential information is usually minimal and can be reduced in any case via oversampling.

Let the autocorrelation matrix of the augmented received signal and its eigen-decomposition be:

$$\mathbf{C}_r \triangleq E\left[\mathbf{r}[i]\mathbf{r}[i]^H\right] = \tilde{\mathbf{S}}\tilde{\mathbf{S}}^H + \eta \mathbf{I}_{NPm} \tag{3.11}$$

$$= \mathbf{U}_s\mathbf{\Lambda}_s\mathbf{U}_s^H + \eta \mathbf{U}_n\mathbf{U}_n^H \tag{3.12}$$

where $\mathbf{\Lambda}_s = \mathrm{diag}(\lambda_1, \ldots, \lambda_{K(m+\iota)})$ contains the largest $K(m+\iota)$ eigen-values of $\mathbf{C}_r$; $\mathbf{U}_s = [\mathbf{u}_1, \ldots, \mathbf{u}_{K(m+\iota)}]$ contains the eigen-vectors corresponding to the eigen-values in $\mathbf{\Lambda}_s$; and $\mathbf{U}_n = [\mathbf{u}_{K(m+\iota)+1}, \ldots, \mathbf{u}_{PNm}]$ contains the $[PNm - K(m+\iota)]$ eigen-vectors corresponding to the smallest eigen-value η of $\mathbf{C}_r$. It is not difficult to see that the columns $\mathbf{U}_s$, which are orthogonal to the columns of $\mathbf{U}_n$, form an orthonormal basis for the signal space spanned by $\tilde{\mathbf{S}}$. For this reason, $\mathbf{U}_s$ is often called the signal subspace and $\mathbf{U}_n$ is called the noise subspace.

Suppose that user 1 is the user of interest. The column of $\tilde{\mathbf{S}}$ that corresponds to the symbol $b_1[i]$ in $\mathbf{b}[i]$ is the $(K\iota + 1)$-th column, which we denote here as $\tilde{\mathbf{s}}_1$. The (exact) linear minimum mean-square error (MMSE) detector for user 1 given by [46–48]:

$$\mathbf{w}_1 = \mathbf{C}_r^{-1}\tilde{\mathbf{s}}_1 \tag{3.13}$$

$$= \mathbf{U}_s\mathbf{\Lambda}_s^{-1}\mathbf{U}_s^H\tilde{\mathbf{s}}_1 \tag{3.14}$$

In blind multiuser detection, the receiver is assumed to have knowledge only of the spreading waveform of the user of interest, e.g., $\{s_{j,1}\}_{j=0}^{N-1}$. The so-called composite signature waveform, $\tilde{\mathbf{s}}_1$, is a function of both the spreading waveform and the channel. We next address a blind method for estimating $\tilde{\mathbf{s}}_1$. First, the sample autocorrelation of the received signals is formed, and its eigen-decomposition is computed:

$$\hat{\mathbf{C}}_r \triangleq \frac{1}{M}\sum_{i=0}^{M-1}\mathbf{r}[i]\mathbf{r}[i]^H \tag{3.15}$$

$$= \hat{\mathbf{U}}_s\hat{\mathbf{\Lambda}}_s\hat{\mathbf{U}}_s^H + \hat{\mathbf{U}}_n\mathbf{\Lambda}_n\hat{\mathbf{U}}_n^H \tag{3.16}$$

These (blindly) estimated parameters replace the exact noise and subspace parameters in (3.12). From (3.7):

$$\tilde{s}_k^{(p)}[n] = \sum_{j=0}^{N-1} s_{j,k} h_k^{(p)}[n-j], \quad n = 0, 1, \ldots, (\iota+1)N - 1 \tag{3.17}$$

with:

$$h_k^{(p)}[n] \triangleq \frac{1}{\sqrt{N}} \sum_{l=1}^{L} \alpha_{l,k}^{(p)} \int_0^{T_c} \psi(t)\psi\left(t - \tau_{l,k}^{(p)} + nT_c\right), \quad n = 0, 1, \ldots, \mu - 1 \qquad (3.18)$$

where the length of the channel response $\left\{ h_k^{(p)}[n] \right\}_{n=0}^{\mu-1}$ satisfies:

$$\mu \triangleq \left\lceil \frac{\tau_{L,k}^{(p)}}{T_c} \right\rceil = \left\lceil \frac{\tau_{L,k}^{(p)}}{T} \cdot \frac{T}{T_c} \right\rceil \leq \iota N \qquad (3.19)$$

Denote:

$$\tilde{\mathbf{s}}_k^{(p)} \triangleq \begin{bmatrix} \tilde{s}_k^{(p)}[0] \\ \vdots \\ \tilde{s}_k^{(p)}[(\iota + 1)N - 1] \end{bmatrix}_{(\iota+1)N \times 1}, \qquad \mathbf{h}_k^{(p)} \triangleq \begin{bmatrix} h_k^{(p)}[0] \\ \vdots \\ h_k^{(p)}[\iota N - 1] \end{bmatrix}_{\mu \times 1}$$

and:

$$\boldsymbol{\Xi}_k \triangleq \begin{bmatrix} s_{0,k} & & & \\ s_{1,k} & s_{0,k} & & \\ \vdots & s_{1,k} & \ddots & \\ \vdots & \vdots & \ddots & s_{0,k} \\ s_{N-1,k} & \vdots & & s_{1,k} \\ & s_{N-1,k} & & \vdots \\ & & \ddots & \vdots \\ & & & s_{N-1,k} \end{bmatrix}_{(\iota+1)N \times \mu}$$

Then (3.17) can be written in matrix form as:

$$\tilde{\mathbf{s}}_k^{(p)} = \boldsymbol{\Xi}_k \mathbf{h}_k^{(p)} \qquad (3.20)$$

Finally, let $\tilde{\mathbf{s}}_k$ be the $(K\iota + k)$-th column of the matrix $\tilde{\mathbf{S}}$. Then we have:

$$\tilde{\mathbf{s}}_k \triangleq \underbrace{\begin{bmatrix} \tilde{\mathbf{s}}_k^{(1)} \\ \vdots \\ \tilde{\mathbf{s}}_k^{(P)} \\ \mathbf{0}_{(m-\iota-1)PN \times 1} \end{bmatrix}}_{PNm \times 1} = \underbrace{\overbrace{\begin{bmatrix} \boldsymbol{\Xi}_k & & \\ & \ddots & \\ & & \boldsymbol{\Xi}_k \\ \mathbf{0}_{(m-\iota-1)PN \times P\mu} \end{bmatrix}}^{\tilde{\boldsymbol{\Xi}}_k}}_{PNm \times P\mu} \underbrace{\overbrace{\begin{bmatrix} \mathbf{h}_k^{(1)} \\ \vdots \\ \mathbf{h}_k^{(P)} \end{bmatrix}}^{\mathbf{h}_k}}_{P\mu \times 1} \qquad (3.21)$$

Notice that in (3.21), the composite signature waveform $\tilde{s}_k$ has be decomposed into a known part, $\widetilde{\Xi}_k$, and a part (the channel) which must be estimated blindly. In order to estimate the desired user's channel $\mathbf{h}_1$, we make use of the fact that the noise subspace is orthogonal to the composite signature waveforms, i.e., $\mathbf{U}_n^H \tilde{s}_1 = \mathbf{U}_n^H \widetilde{\Xi}_1 \mathbf{h}_1 = 0$, which suggests the following estimation approach:

$$\hat{\mathbf{h}}_1 = \arg \min_{\|\mathbf{h}\|=1} \left\| \hat{\mathbf{U}}_n^H \widetilde{\Xi}_1 \mathbf{h} \right\|^2$$

$$= \arg \min_{\|\mathbf{h}\|=1} \mathbf{h}^H \underbrace{\left(\widetilde{\Xi}_1^H \hat{\mathbf{U}}_n \hat{\mathbf{U}}_n^H \widetilde{\Xi}_1 \right)}_{\hat{\mathbf{Q}}} \mathbf{h}$$

$$= \text{minimum eigen-vector of } \hat{\mathbf{Q}} \qquad (3.22)$$

Note, however, that even if the noise subspace estimation is perfect, (3.22) specifies $\mathbf{h}_1$ only up to a phase ambiguity, i.e., if $\hat{\mathbf{h}}_1$ is the solution to (3.22), so is $e^{j\phi}\hat{\mathbf{h}}_1$ for any ϕ. Moreover, the constraint $\| \mathbf{h} \| = 1$ also implies a scale ambiguity.

Once $\hat{\mathbf{h}}_1$ is available, the detector for user 1 can be estimated blindly as:

$$\hat{\mathbf{w}}_1 = \hat{\mathbf{U}}_s \hat{\Lambda}_s^{-1} \hat{\mathbf{U}}_s^H \widetilde{\Xi}_1 \hat{\mathbf{h}}_1 \qquad (3.23)$$

3.2.2 Synchronous Multipath CDMA

Here we consider a special case of the general signal model presented above. In particular, we assume a K-user *synchronous* multipath CDMA system with no ISI. Such a system is realized by either neglecting the ISI when the multipath delay spread is small compared with the symbol interval, or by inserting guard intervals between symbols when the delay spread is large. The received N-dimensional signal during the i-th symbol interval, a special case of (3.10), can be written as [7,46]:

$$\mathbf{r}[i] = \sum_{k=1}^{K} b_k[i] \sum_{l=1}^{L} \mathbf{s}_{l,k} h_{l,k} + \mathbf{n}[i] \qquad (3.24)$$

where L is the number of resolvable paths, $\mathbf{s}_{l,k}$ and $h_{l,k}$ are respectively the delayed version of the spreading waveform (with zero-padding when a guard interval is inserted) and the complex channel fading gain corresponding to the l-th path of the k-th user; $\mathbf{n}[i] \sim \mathcal{N}_c(\mathbf{0}, \eta \mathbf{I}_N)$ is the circularly symmetric complex white Gaussian noise vector. Denote:

$$\mathbf{S}_k \triangleq [\mathbf{s}_{1,k} \; \mathbf{s}_{2,k} \cdots \mathbf{s}_{L,k}] \qquad (3.25)$$

$$\mathbf{h}_k \triangleq [h_{1,k} \; h_{2,k} \cdots h_{L,k}]^T \qquad (3.26)$$

Then (3.24) can be rewritten as:

$$\mathbf{r}[i] = \sum_{k=1}^{K} \underbrace{\mathbf{S}_k \mathbf{h}_k}_{\tilde{\mathbf{s}}_k} b_k[i] + \mathbf{n}[i] \tag{3.27}$$

$$= \tilde{\mathbf{S}} \mathbf{b}[i] + \mathbf{n}[i] \tag{3.28}$$

where:

$$\tilde{\mathbf{S}} \triangleq [\tilde{\mathbf{s}}_1 \ \tilde{\mathbf{s}}_2 \ \cdots \ \tilde{\mathbf{s}}_K] \tag{3.29}$$

$$\mathbf{b}[i] \triangleq [b_1[i] \ b_2[i] \ \cdots \ b_K[i]]^T \tag{3.30}$$

Let the autocorrelation matrix of the received signal $\mathbf{r}[i]$ be:

$$\mathbf{C}_r \triangleq E\left[\mathbf{r}[i]\mathbf{r}[i]^H\right] = \tilde{\mathbf{S}}\tilde{\mathbf{S}}^H + \eta \mathbf{I}_N \tag{3.31}$$

$$= \mathbf{U}_s \mathbf{\Lambda}_s \mathbf{U}_s^H + \eta \mathbf{U}_n \mathbf{U}_n^H \tag{3.32}$$

where in (3.32), $\mathbf{\Lambda}_s = \mathrm{diag}(\lambda_1, \ldots, \lambda_K)$ contains the largest K eigen-values of $\mathbf{C}_r$; $\mathbf{U}_s = [\mathbf{u}_1, \ldots, \mathbf{u}_K]$ contains the eigen-vectors corresponding to the eigen-values in $\mathbf{\Lambda}_s$; and $\mathbf{U}_n = [\mathbf{u}_{K+1}, \ldots, \mathbf{u}_N]$ contains the $(N - K)$ eigen-vectors corresponding to the smallest eigen-value η of $\mathbf{C}_r$. Suppose that user 1 is the user of interest. Since $\tilde{\mathbf{s}}_1 \triangleq \mathbf{S}_1 \mathbf{h}_1$, and $\mathbf{U}_n^H \tilde{\mathbf{s}}_1 = \mathbf{0}$, it then follows that $\mathbf{h}_1$ can be obtained from the following relationship:

$$\mathbf{h}_1 = \arg \min_{\|\mathbf{h}\|=1} \left\| \mathbf{U}_n^H (\mathbf{S}_1 \mathbf{h}) \right\|^2$$

$$= \arg \min_{\|\mathbf{h}\|=1} \mathbf{h}^H \underbrace{\left(\mathbf{S}_1^H \mathbf{U}_n \mathbf{U}_n^H \mathbf{S}_1 \right)}_{\mathbf{Q}} \mathbf{h}$$

$$= \text{minimum eigen-vector of } \mathbf{Q} \tag{3.33}$$

The sample autocorrelation of the received signals is formed as:

$$\hat{\mathbf{C}}_r \triangleq \frac{1}{M} \sum_{i=0}^{M-1} \mathbf{r}[i]\mathbf{r}[i]^H \tag{3.34}$$

$$= \hat{\mathbf{U}}_s \hat{\mathbf{\Lambda}}_s \hat{\mathbf{U}}_s^H + \hat{\mathbf{U}}_n \hat{\mathbf{\Lambda}}_n \hat{\mathbf{U}}_n^H \tag{3.35}$$

and blind channel estimation is performed, as above, by replacing the quantities in (3.33) by their estimates, i.e.:

$$\hat{\mathbf{Q}} = \mathbf{S}_1^H \hat{\mathbf{U}}_n \hat{\mathbf{U}}_n^H \mathbf{S}_1 \tag{3.36}$$

$$\hat{\mathbf{h}}_1 = \text{minimum eigen-vector of } \hat{\mathbf{Q}} \tag{3.37}$$

Finally, the estimated blind linear MMSE detector is given by:

$$\hat{\mathbf{w}}_1 = \hat{\mathbf{U}}_s \hat{\boldsymbol{\Lambda}}_s^{-1} \hat{\mathbf{U}}_s^H \mathbf{S}_1 \hat{\mathbf{h}}_1 \tag{3.38}$$

3.2.3 Synchronous Multi-Antenna CDMA

We next consider a K-user synchronous CDMA system employing P receive antennas. Let $h_{p,k}$ be the complex fading gain between the transmit antenna and the p-th receive antenna for the k-th user. At the p-th receive antenna, the received discrete-time signal during the i-th symbol interval is given by:

$$\mathbf{r}^{(p)}[i] = \sum_{k=1}^{K} h_{p,k} b_k[i] \mathbf{s}_k + \mathbf{n}^{(p)}[i], \quad p = 1, 2, \ldots, P \tag{3.39}$$

where $\mathbf{s}_k$ is the spreading waveform of the k-th user; $\mathbf{n}^{(p)}[i] \sim \mathcal{N}_c(\mathbf{0}, \eta \mathbf{I}_N)$ is the circularly symmetric complex white Gaussian noise vector at antenna p. It is assumed that the noise vectors at different antennas are independent.

Denote:

$$\mathbf{h}_k \triangleq [h_{1,k}\ h_{2,k} \cdots h_{P,k}]^T$$

$$\tilde{\mathbf{s}}_k \triangleq \mathbf{h}_k \otimes \mathbf{s}_k$$

$$\mathbf{r}[i] \triangleq [\mathbf{r}^{(1)}[i]^T\ \mathbf{r}^{(2)}[i]^T \cdots \mathbf{r}^{(P)}[i]^T]^T$$

$$\mathbf{n}[i] \triangleq [\mathbf{n}^{(1)}[i]^T\ \mathbf{n}^{(2)}[i]^T \cdots \mathbf{n}^{(P)}[i]^T]^T$$

$$\mathbf{b}[i] \triangleq [b_1[i]\ b_2[i] \cdots b_K[i]]^T$$

$$\tilde{\mathbf{S}} \triangleq [\tilde{\mathbf{s}}_1\ \tilde{\mathbf{s}}_2 \cdots \tilde{\mathbf{s}}_K]$$

Then (3.39) can be written as:

$$\mathbf{r}[i] = \sum_{k=1}^{K} b_k[i] \tilde{\mathbf{s}}_k + \mathbf{n}[i] \tag{3.40}$$

$$= \tilde{\mathbf{S}} \mathbf{b}[i] + \mathbf{n}[i] \tag{3.41}$$

with $\mathbf{n} \sim \mathcal{N}_c(\mathbf{0}, \eta \mathbf{I}_{PN})$.

Let the autocorrelation matrix of the received signal and its eigen-decomposition be:

$$\mathbf{C}_r \triangleq E[\mathbf{r}[i]\mathbf{r}[i]^H] = \tilde{\mathbf{S}}\tilde{\mathbf{S}}^H + \eta \mathbf{I}_{PN} \tag{3.42}$$

$$= \mathbf{U}_s \boldsymbol{\Lambda}_s \mathbf{U}_s^H + \eta \mathbf{U}_n \mathbf{U}_n^H \tag{3.43}$$

where $\boldsymbol{\Lambda}_s = \mathrm{diag}(\lambda_1, \ldots, \lambda_K)$ contains the largest K eigen-values of $\mathbf{C}_r$; $\mathbf{U}_s = [\mathbf{u}_1, \ldots, \mathbf{u}_K]$ contains the eigen-vectors corresponding to the eigen-values in $\boldsymbol{\Lambda}_s$; and $\mathbf{U}_n = [\mathbf{u}_{K+1}, \ldots, \mathbf{u}_{PN}]$ contains the $(PN - K)$ eigen-vectors corresponding to the smallest eigen-value η of $\mathbf{C}_r$.

The linear MMSE detector for user 1 is then given by:

$$\mathbf{w}_1 = \mathbf{C}_r^{-1}\tilde{\mathbf{s}}_1 \tag{3.44}$$

$$= \mathbf{U}_s\mathbf{\Lambda}_s^{-1}\mathbf{U}_s^H\tilde{\mathbf{s}}_1 \tag{3.45}$$

Since $\tilde{\mathbf{s}}_1 \triangleq \mathbf{h}_1 \otimes \mathbf{s}_1$, and $\mathbf{U}_n^H\tilde{\mathbf{s}}_1 = \mathbf{0}$, it then follows that $\mathbf{h}_1$ satisfies:

$$\mathbf{h}_1 = \arg \min_{\|\mathbf{h}\|=1} \left\| \mathbf{U}_n^H(\mathbf{h} \otimes \mathbf{s}_1) \right\|^2$$

$$= \arg \min_{\|\mathbf{h}\|=1} \mathbf{h}^H\mathbf{Q}\mathbf{h}$$

$$= \text{minimum eigen-vector of } \mathbf{Q} \tag{3.46}$$

with:

$$\mathbf{Q} \triangleq \left(\mathbf{I}_P \otimes \mathbf{s}_1^H\right)\mathbf{U}_n\mathbf{U}_n^H(\mathbf{I}_P \otimes \mathbf{s}_1) \tag{3.47}$$

Note that $\mathbf{h}_1$ is determined uniquely by (3.46) up to a scale and phase ambiguity if and only if $\text{rank}(\mathbf{Q}) = P - 1$.

A blind estimate of the linear MMSE detector in this case is then given by the following procedure:

$$\hat{\mathbf{C}}_r \triangleq \frac{1}{M}\sum_{i=0}^{M-1}\mathbf{r}[i]\mathbf{r}[i]^H \tag{3.48}$$

$$= \hat{\mathbf{U}}_s\hat{\mathbf{\Lambda}}_s\hat{\mathbf{U}}_s^H + \hat{\mathbf{U}}_n\hat{\mathbf{\Lambda}}_n\hat{\mathbf{U}}_n^H \tag{3.49}$$

$$\hat{\mathbf{Q}} = \left(\mathbf{I}_P \otimes \mathbf{s}_1^H\right)\hat{\mathbf{U}}_n\hat{\mathbf{U}}_n^H(\mathbf{I}_P \otimes \mathbf{s}_1) \tag{3.50}$$

$$\hat{\mathbf{h}}_1 = \text{minimum eigen-vector of } \hat{\mathbf{Q}} \tag{3.51}$$

$$\hat{\mathbf{w}}_1 = \hat{\mathbf{U}}_s\hat{\mathbf{\Lambda}}_s^{-1}\hat{\mathbf{U}}_s^H(\hat{\mathbf{h}}_1 \otimes \mathbf{s}_1) \tag{3.52}$$

3.2.4 Remarks

From the above the discussion, it is seen that the three systems considered here share the same signal model of the form:

$$\mathbf{r}[i] = \tilde{\mathbf{S}}\mathbf{b}[i] + \mathbf{n}[i] \tag{3.53}$$

Furthermore, for these three systems, a linear MMSE decision on the i-th symbol of user 1 is made based on the output of the linear MMSE detector $\mathbf{w}_1^H\mathbf{r}[i]$, where $\mathbf{w}_1$ has the form:

$$\mathbf{w}_1 = \mathbf{C}_r^{-1}\tilde{\mathbf{s}}_1 = \mathbf{U}_s\mathbf{\Lambda}_s^{-1}\mathbf{U}_s^H\tilde{\mathbf{s}}_1 \tag{3.54}$$

Moreover, the composite signature waveform $\tilde{\mathbf{s}}_1$ of the desired user is determined by the original signature waveform of this user, and a channel vector $\mathbf{h}_1$, which is given by the minimum eigen-vector of a matrix $\mathbf{Q}$. The matrix $\mathbf{Q}$ is in turn determined by the noise subspace $\mathbf{U}_n$ and the original signature waveform of the desired user.

3.3 PERFORMANCE OF BLIND MULTIUSER DETECTORS

In this section, we analyze the performance of blind multiuser detection. We focus on the synchronous multipath CDMA system described in Section 3.2.2. The results, however, are directly applicable to the synchronous multi-antenna CDMA system described in Section 3.2.3, and the asynchronous multipath multi-antenna CDMA system described in Section 3.2.1, with appropriate interpretation of the quantities such as $\boldsymbol{\Lambda}_s$, $\mathbf{U}_s$, $\mathbf{U}_n$, and $\mathbf{Q}$ in the corresponding system. The mathematical proofs of the results presented below can be found in [18,19].

3.3.1 Complex Gaussian Distribution

Let $\mathbf{x} \in \mathbb{C}^n$ be a complex random vector. We say that $\mathbf{x}$ is complex Gaussian distributed if the vector $\begin{bmatrix} \Re\mathbf{x} \\ \Im\mathbf{x} \end{bmatrix} \in \mathbb{R}^{2n}$ has a (real-valued) Gaussian distribution. Hence, a complex Gaussian vector $\mathbf{x}$ is completely specified by its mean $\boldsymbol{\mu} \triangleq E[\mathbf{x}]$, and the following covariance matrix:

$$\mathrm{Cov}\left\{ \begin{bmatrix} \Re\mathbf{x} \\ \Im\mathbf{x} \end{bmatrix} \right\} = \begin{bmatrix} \mathrm{Cov}\{\Re\mathbf{x}, \Re\mathbf{x}\} & \mathrm{Cov}\{\Re\mathbf{x}, \Im\mathbf{x}\} \\ \mathrm{Cov}\{\Im\mathbf{x}, \Re\mathbf{x}\} & \mathrm{Cov}\{\Im\mathbf{x}, \Im\mathbf{x}\} \end{bmatrix} \tag{3.55}$$

An equivalent characterization of the complex Gaussian vector $\mathbf{x}$ is through the following two complex-valued covariance matrices:

$$\mathbf{C} \triangleq E[(\mathbf{x} - \boldsymbol{\mu})(\mathbf{x} - \boldsymbol{\mu})^H]$$

$$\bar{\mathbf{C}} \triangleq E[(\mathbf{x} - \boldsymbol{\mu})(\mathbf{x} - \boldsymbol{\mu})^T]$$

We call $\mathbf{C}$ the *Hermitian* covariance matrix and $\bar{\mathbf{C}}$ the *symmetric* covariance matrix. The real-valued covariance matrix (3.55) can be expressed by $\mathbf{C}$ and $\bar{\mathbf{C}}$, i.e.:

$$\mathrm{Cov}\left\{ \begin{bmatrix} \Re\mathbf{x} \\ \Im\mathbf{x} \end{bmatrix} \right\} = \frac{1}{2} \begin{bmatrix} \Re\mathbf{C} + \Re\bar{\mathbf{C}} & \Im\bar{\mathbf{C}} - \Im\mathbf{C} \\ \Im\bar{\mathbf{C}} + \Im\mathbf{C} & \Re\mathbf{C} - \Re\bar{\mathbf{C}} \end{bmatrix} \tag{3.56}$$

Hence, in what follows, we use the following notation to represent a complex Gaussian vector:

$$\mathbf{x} \sim \mathcal{N}_c(\boldsymbol{\mu}, \mathbf{C}, \bar{\mathbf{C}}) \tag{3.57}$$

When $\bar{\mathbf{C}} = \mathbf{0}$, $\mathbf{x}$ is said to have a *circularly symmetric* complex Gaussian distribution. In this case, $\Re\mathbf{x}$ and $\Im\mathbf{x}$ are independent and have the same (real-valued) Gaussian distribution.

3.3.2 Performance of Blind Multiuser Detectors with Known Channels

First we assume that the receiver has the knowledge of the original signature waveform and the channel of the *desired user*, i.e., $\mathbf{S}_1$ and $\mathbf{h}_1$ in the synchronous multipath CDMA case, $\mathbf{s}_1$ and $\mathbf{h}_1$ in the synchronous multi-antenna CDMA case, and $\tilde{\Xi}_1$ and $\mathbf{h}_1$ in the asynchronous multipath multi-antenna CDMA case. Equivalently, the composite signature waveform $\tilde{\mathbf{s}}_1$ is assumed known. This corresponds to systems where the desired user's channel is obtained through non-blind methods, such as pilot channels or pilot symbols.

As mentioned earlier, we focus on the synchronous multipath CDMA system described in Section 3.2.2. Based on the sample autocorrelation matrix $\mathbf{C}_r$ in (3.34) and its eigen-decomposition (3.35), we can obtain two forms of the estimated linear MMSE detector, namely, the direct-matrix-inversion (DMI) detector, and the subspace detector, given respectively by:

$$\hat{\mathbf{w}}_1 = \hat{\mathbf{C}}_r^{-1}\tilde{\mathbf{s}}_1, \quad \text{[DMI blind detector]} \tag{3.58}$$

and:

$$\hat{\mathbf{w}}_1 = \hat{\mathbf{U}}_s\hat{\mathbf{\Lambda}}_s^{-1}\hat{\mathbf{U}}_s^H\tilde{\mathbf{s}}_1. \quad \text{[subspace blind detector]} \tag{3.59}$$

In what follows, we first present results on the asymptotic distributions of the DMI blind detector (3.58) and the subspace blind detector (3.59), when the number of received signals M is large. We then give expressions of the output SINR as well as the BER for the two blind detectors.

The following result gives the asymptotic distribution of the DMI blind detector and that of the subspace blind detector.

Theorem 1 *Let*:

$$\mathbf{w}_1 = \mathbf{C}_r^{-1}\tilde{\mathbf{s}}_1 = \mathbf{U}_s\mathbf{\Lambda}_s^{-1}\mathbf{U}_s^H\tilde{\mathbf{s}}_1 \tag{3.60}$$

be the exact linear MMSE detector, and let $\tilde{\mathbf{w}}_1$ be the DMI blind detector given by (3.58) or the subspace blind detector given by (3.59). Then:

$$\sqrt{M}(\hat{\mathbf{w}}_1 - \mathbf{w}_1) \longrightarrow \mathcal{N}_c\left(\mathbf{0}, \mathbf{C}_w^0, \bar{\mathbf{C}}_w^0\right), \quad \text{in distribution, as } M \to \infty$$

with:

$$\mathbf{C}_w^0 = \left(\mathbf{w}_1^H\tilde{\mathbf{s}}_1\right)\mathbf{U}_s\mathbf{\Lambda}_s^{-1}\mathbf{U}_s^H + \tau\mathbf{U}_n\mathbf{U}_n^H + \mathbf{U}_s\mathbf{\Lambda}_s^{-1}\mathbf{U}_s^H\tilde{\mathbf{S}}\left[\mu\left(\tilde{\mathbf{S}}^T\mathbf{w}_1^*\mathbf{w}_1^T\tilde{\mathbf{S}}^*\right) + \nu\mathbf{D}\right]$$
$$\times \tilde{\mathbf{S}}^H\mathbf{U}_s\mathbf{\Lambda}_s^{-1}\mathbf{U}_s^H \tag{3.61}$$

$$\bar{\mathbf{C}}_w^0 = \mathbf{w}_1\mathbf{w}_1^T + \mathbf{U}_s\mathbf{\Lambda}_s^{-1}\mathbf{U}_s^H\tilde{\mathbf{S}}\left[\mu\left(\mathbf{w}_1^T\tilde{\mathbf{S}}^*\tilde{\mathbf{S}}^H\mathbf{w}_1\right)\mathbf{I}_K + \nu\bar{\mathbf{D}}\right]\tilde{\mathbf{S}}^T\mathbf{U}_s^*\mathbf{\Lambda}_s^{-1}\mathbf{U}_s^T \tag{3.62}$$

where:

$$\mathbf{D} \triangleq \mathrm{diag}\left\{|\tilde{\mathbf{s}}_1^H \mathbf{w}_1|^2, |\tilde{\mathbf{s}}_2^H \mathbf{w}_1|^2, \ldots, |\tilde{\mathbf{s}}_K^H \mathbf{w}_1|^2\right\} \tag{3.63}$$

$$\bar{\mathbf{D}} \triangleq \mathrm{diag}\left\{(\tilde{\mathbf{s}}_1^H \mathbf{w}_1)^2, (\tilde{\mathbf{s}}_2^H \mathbf{w}_1)^2, \ldots, (\tilde{\mathbf{s}}_K^H \mathbf{w}_1)^2\right\} \tag{3.64}$$

$$\mu \triangleq \left|E[b^2]\right|^2 \tag{3.65}$$

$$\nu \triangleq E\left[|b|^4\right] - 2E\left[|b|^2\right]^2 - \left|E[b^2]\right|^2 \tag{3.66}$$

$$\tau \triangleq \begin{cases} \frac{1}{\eta}\tilde{\mathbf{s}}_1^H \mathbf{U}_s \mathbf{\Lambda}_s^{-1} \mathbf{U}_s^H \tilde{\mathbf{s}}_1 & \textit{DMI blind detector} \\ \eta\tilde{\mathbf{s}}_1^H \mathbf{U}_s \mathbf{\Lambda}_s^{-1}(\mathbf{\Lambda}_s - \eta\mathbf{I}_K)^{-2}\mathbf{U}_s^H \tilde{\mathbf{s}}_1 & \textit{subspace blind detector} \end{cases} \tag{3.67}$$

and where b denotes the transmitted modulation symbols.

Denote $\Delta\mathbf{w}_1 \triangleq \hat{\mathbf{w}}_1 - \mathbf{w}_1$ as the error of the estimated detector. The blind detector $\hat{\mathbf{w}}_1$ given by (3.58) or (3.59) is correlated with the received signal $\mathbf{r}[i]$ in (3.24) to yield the decision statistic:

$$\hat{\mathbf{w}}_1^H \mathbf{r}[i] = \mathbf{w}_1^H \left(\sum_{k=1}^K \tilde{\mathbf{s}}_k b_k[i] + \mathbf{n}[i]\right) + \Delta\mathbf{w}_1^H \mathbf{r}[i] \tag{3.68}$$

The SINR at the output of this estimated detector is then given by:

$$\begin{aligned} \mathrm{SINR} &\triangleq \frac{\left|E\left[\hat{\mathbf{w}}_1^H \mathbf{r}[i] \,|\, b_1[i]\right]\right|^2}{E\left[\mathrm{Var}\left(\hat{\mathbf{w}}_1^H \mathbf{r}[i] \,|\, b_1[i]\right)\right]} \\ &= \frac{\left(\mathbf{w}_1^H \tilde{\mathbf{s}}_1\right)^2 E\left[|b|^2\right]}{E\left[|b|^2\right]\sum_{k=2}^K |\mathbf{w}_1^H \tilde{\mathbf{s}}_k|^2 + \eta\|\mathbf{w}_1\|^2 + \frac{1}{M}\mathrm{tr}(\mathbf{C}_w^0 \mathbf{C}_r)} \end{aligned} \tag{3.69}$$

Denote:

$$\tilde{\mathbf{R}} \triangleq \tilde{\mathbf{S}}^H \tilde{\mathbf{S}} \tag{3.70}$$

Then after some manipulations, the quantities in (3.69) are given in terms of $\tilde{\mathbf{R}}$ and $(\eta, \mu, \nu, N, K, M)$ as follows:

$$\mathbf{w}_i^H \tilde{\mathbf{s}}_j = \tilde{\mathbf{s}}_i^H \mathbf{w}_j = \left[\left(\mathbf{I}_K + \eta\tilde{\mathbf{R}}^{-1}\right)^{-1}\right]_{i,j} \tag{3.71}$$

$$\|\mathbf{w}_1\|^2 = \left[\left(\tilde{\mathbf{R}} + \eta\mathbf{I}_K\right)^{-1}\tilde{\mathbf{R}}\left(\tilde{\mathbf{R}} + \eta\mathbf{I}_K\right)^{-1}\right]_{1,1} \tag{3.72}$$

$$\tau\eta = \begin{cases} \mathbf{s}_1^H \mathbf{w}_1, & \text{DMI} \\ \eta^2 \left[\left(\tilde{\mathbf{R}} + \eta \mathbf{I}_K \right)^{-1} \tilde{\mathbf{R}}^{-1} \right]_{1,1}, & \text{subspace} \end{cases} \tag{3.73}$$

and:

$$\mathrm{tr}(\mathbf{C}_w^0 \mathbf{C}_r) = K \mathbf{w}_1^H \tilde{\mathbf{s}}_1 + (N - K)\tau\eta + \mu \sum_{i=1}^{K} \sum_{j=1}^{K} \left(\tilde{\mathbf{s}}_i^H \mathbf{w}_i \right) \left(\tilde{\mathbf{s}}_i^H \mathbf{w}_j \right) \left(\mathbf{w}_j^H \tilde{\mathbf{s}}_j \right)$$

$$+ \nu \sum_{k=1}^{K} \left| \tilde{\mathbf{s}}_k^H \mathbf{w}_1 \right|^2 \left(\tilde{\mathbf{s}}_k^H \mathbf{w}_k \right) \tag{3.74}$$

We next give the expressions of $\tilde{\mathbf{R}}$ in terms of the spreading waveforms and the channels of all users, for the following four systems:

1. *Synchronous Flat-Fading CDMA:* This corresponds to $L = 1$ in (3.24). Let $\mathbf{H} \triangleq \mathrm{diag}(h_{1,1}, h_{1,2}, \ldots, h_{1,K})$ and $\mathbf{S} \triangleq [\mathbf{s}_1, \mathbf{s}_2, \ldots, \mathbf{s}_K]$. Then we have $\tilde{\mathbf{S}} = \mathbf{SH}$. Denote $\mathbf{R} \triangleq \mathbf{S}^H \mathbf{S}$. Hence we have:

$$\tilde{\mathbf{R}} = \mathbf{H}^H \mathbf{R} \mathbf{H} \tag{3.75}$$

2. *Synchronous Multipath CDMA:* Consider the signal model (3.27). Let $\mathcal{S} \triangleq [\mathbf{S}_1, \mathbf{S}_2, \ldots, \mathbf{S}_K]$ and $\mathbf{H} \triangleq \mathrm{diag}(\mathbf{h}_1, \mathbf{h}_2, \ldots, \mathbf{h}_K)$. Then we have $\tilde{\mathbf{S}} = \mathcal{S}\mathbf{H}$. Hence we have:

$$\tilde{\mathbf{R}} = \mathbf{H}^H \mathcal{S}^H \mathcal{S} \mathbf{H} \tag{3.76}$$

3. *Synchronous Multi-Antenna CDMA:* Consider the signal model (3.41). Note that:

$$[\tilde{\mathbf{R}}]_{ij} = \tilde{\mathbf{s}}_i^H \tilde{\mathbf{s}}_j = (\mathbf{h}_i \otimes \mathbf{s}_i)^H (\mathbf{h}_j \otimes \mathbf{s}_j)$$

$$= \left(\mathbf{h}_i^H \mathbf{h}_j \right) \left(\mathbf{s}_i^H \mathbf{s}_j \right) \tag{3.77}$$

Denote $\mathbf{H} \triangleq [\mathbf{h}_1, \mathbf{h}_2, \ldots, \mathbf{h}_K]$. It then follows that:

$$\tilde{\mathbf{R}} = \mathbf{R} \circ \left(\mathbf{H}^H \mathbf{H} \right) \tag{3.78}$$

where $\circ$ denotes element-wise matrix product.

4. *Asynchronous Multipath Multi-Antenna CDMA:* Consider the general signal model (3.21). Let $\widetilde{\Xi} \triangleq \left[\widetilde{\Xi}_1, \widetilde{\Xi}_2, \ldots, \widetilde{\Xi}_K \right]$, and $\mathbf{H} \triangleq \mathrm{diag}(\mathbf{h}_1, \mathbf{h}_2, \ldots, \mathbf{h}_K)$. Then we have $\tilde{\mathbf{S}} = \widetilde{\Xi}\mathbf{H}$. Hence:

$$\tilde{\mathbf{R}} = \mathbf{H}^H \widetilde{\Xi}^H \widetilde{\Xi} \mathbf{H} \tag{3.79}$$

3.3.3 Performance of Blind Multiuser Detector with Blind Channel Estimation

Next we present results on the asymptotic distributions of the blind detector with blind channel estimation described in Section 3.2.2 and the corresponding detector output SINR. The following result gives the asymptotic distribution of the blind detector with blind channel estimation, given by (3.34)–(3.38).

Theorem 2 *Let:*

$$\mathbf{w}_1 = \mathbf{U}_s \mathbf{\Lambda}_s^{-1} \mathbf{U}_s^H \tilde{\mathbf{s}}_1 \tag{3.80}$$

be the exact linear MMSE detector, $\mathbf{h}_1$ be the true channel of user 1, and $\hat{\mathbf{w}}_1$ be the estimated blind detector given by (3.34)–(3.38). Let $\mathbf{C}_w^0$ and $\bar{\mathbf{C}}_w^0$ be the quantities given by (3.61) and (3.62) respectively. Then there exists a phase factor $e^{\jmath\phi}$ such that:

$$\sqrt{M}\left(\hat{\mathbf{w}}_1 - \|\mathbf{h}_1\|^{-1} e^{\jmath\phi}\, \mathbf{w}_1\right) \;\longrightarrow\; \mathcal{N}_c(\mathbf{0},\, \|\mathbf{h}_1\|^{-2}\, \mathbf{C}_w,\, \|\mathbf{h}_1\|^{-2}\, \bar{\mathbf{C}}_w),$$

in distribution, as $M \to \infty$

with:

$$\mathbf{C}_w = \mathbf{C}_w^0 + \beta_1 \mathbf{\Psi} \mathbf{Q}^\dagger \mathbf{\Psi}^H + \beta_2\left(\mathbf{\Psi} \mathbf{Q}^\dagger \mathbf{S}_1^H \mathbf{U}_n \mathbf{U}_n^H + \mathbf{U}_n \mathbf{U}_n^H \mathbf{S}_1 \mathbf{Q}^\dagger \mathbf{\Psi}^H\right) \tag{3.81}$$

$$\bar{\mathbf{C}}_w = \bar{\mathbf{C}}_w^0 \tag{3.82}$$

where:

$$\mathbf{\Psi} \triangleq \mathbf{U}_s \mathbf{\Lambda}_s^{-1} \mathbf{U}_s^H \mathbf{S}_1 \tag{3.83}$$

$$\beta_1 \triangleq \eta \tilde{\mathbf{s}}_1^H \mathbf{U}_s \mathbf{\Lambda}_s (\mathbf{\Lambda}_s - \eta \mathbf{I}_K)^{-2} \mathbf{U}_s^H \tilde{\mathbf{s}}_1 \tag{3.84}$$

$$\beta_2 \triangleq \eta \tilde{\mathbf{s}}_1^H \mathbf{U}_s (\mathbf{\Lambda}_s - \eta \mathbf{I}_K)^{-2} \mathbf{U}_s^H \tilde{\mathbf{s}}_1 \tag{3.85}$$

and $\mathbf{Q}^\dagger$ is the pseudo-inverse of the matrix $\mathbf{Q} \triangleq \mathbf{S}_1^H \mathbf{U}_n \mathbf{U}_n^H \mathbf{S}_1$.

We can also obtain the asymptotic distribution of the channel estimate, as given by the following result.

Corollary 1 *Let $\mathbf{h}_1$ be the true channel of user 1, and let $\hat{\mathbf{h}}_1$ be the channel estimate given by (3.37). Then there exists a phase factor $e^{\jmath\phi}$, such that:*

$$\sqrt{M}\left(\hat{\mathbf{h}}_1 - \|\mathbf{h}_1\|^{-1} e^{\jmath\phi} \mathbf{h}_1\right) \;\longrightarrow\; \mathcal{N}_c\left(\mathbf{0},\, \beta_1 \|\mathbf{h}_1\|^{-2} \mathbf{Q}^\dagger,\, \mathbf{0}\right), \quad \text{in distribution, as } M \to \infty$$

where β_1 is given by (3.84). Thus asymptotically $\hat{\mathbf{h}}_1$ is circularly symmetric complex Gaussian.

Hence the blind detector with blind channel estimation can be written as:

$$\hat{\mathbf{w}}_1 = \|\mathbf{h}_1\|^{-1} e^{j\phi}\, \mathbf{w}_1 + \Delta \mathbf{w}_1 \tag{3.86}$$

It is straightforward to show that the detector output SINR is given by:

$$\text{SINR} = \frac{(\mathbf{w}_1^H \tilde{\mathbf{s}}_1)^2}{\displaystyle\sum_{k=2}^{K} |\mathbf{w}_1^H \tilde{\mathbf{s}}_k|^2 + \eta \|\mathbf{w}_1\|^2 + \frac{1}{M}\text{tr}(\mathbf{C}_w \mathbf{C}_r)} \tag{3.87}$$

where $\mathbf{w}_1^H \tilde{\mathbf{s}}_k$ and $\|\mathbf{w}_1\|^2$ are given respectively by (3.71) and (3.72), and using (3.81) after some manipulations, we obtain:

$$\text{tr}(\mathbf{C}_w \mathbf{C}_r) = \text{tr}(\mathbf{C}_w^0 \mathbf{C}_r) + \beta_1 \, \text{tr}\left[\left(\mathbf{S}_1^T \mathbf{U}_s \boldsymbol{\Lambda}_s^{-1} \mathbf{U}_s^H \mathbf{S}_1\right)\mathbf{Q}^\dagger\right] \tag{3.88}$$

where $\text{tr}(\mathbf{C}_w^0 \mathbf{C}_r)$ is given by (3.74) and:

$$\beta_1 = \eta \left[\mathbf{I}_K + \eta \tilde{\mathbf{R}}^{-1}\right]_{1,1} \tag{3.89}$$

where $\tilde{\mathbf{R}}$ is given by (3.70). Hence it is clear that the effect of channel estimation on the output SINR of the blind detector is to add the second term in (3.88) to the denominator of (3.87).

3.3.4 Numerical Results

The simulated system is a synchronous multipath CDMA system. The number of users is $K = 18$. Each user's original spreading sequence is randomly generated and has length 21. The channel of each user has length 11 and is randomly generated. Both the spreading sequences and the channels of all users are fixed throughout the simulations. Note that the length of the composite signature waveform of each user (i.e., $\tilde{\mathbf{s}}_k$) is 31. Here we insert a guard interval of length 10 chips between two consecutive symbols to avoid intersymbol interference. Note such a setup is merely for the purpose of verifying the theoretical results in this paper.

As noted earlier, when the channel is unknown and is blindly estimated, the blind channel estimator introduces a phase ambiguity $\alpha = e^{j\phi}$ which must be estimated for detection purpose. In the simulations, we employ the following simple phase estimator:

$$\hat{\alpha} = \left[\frac{1}{M}\sum_{i=1}^{M}\left(\hat{\mathbf{w}}_1^H \mathbf{r}[i]\right)^2\right]^{\frac{1}{2}} = \left(\hat{\mathbf{w}}_1^H \hat{\mathbf{C}}_r \hat{\mathbf{w}}_1^\star\right)^{\frac{1}{2}} \tag{3.90}$$

Note that the above phase estimator still contains a phase ambiguity of π for BPSK and $\frac{\pi}{2}$ for QPSK, which is inherent to any blind detector.

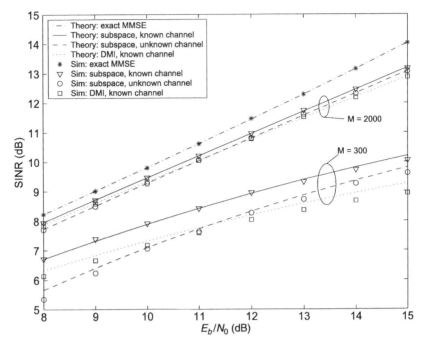

Figure 3.1. Output SINR versus input E_b/N_0 for BPSK.

In Figure 3.1, assuming BPSK modulation the simulated output SINR of three blind detectors, namely, the DMI detector with known channel, the subspace detector with known channel, and the subspace detector with unknown channel, are plotted as a function of input E_b/N_0 of each user, for two block sizes, i.e., $M = 300$ and $M = 2000$. The corresponding theoretical values, as well as the theoretical and simulated SINR of the exact MMSE detector, are also plotted in the same figure. Moreover, the SINR results for QPSK modulation for the same simulated system are plotted in Figure 3.2. It is seen from these figures that the analytical expressions match very well with the simulation results. Furthermore, in unknown channels, the simple phase estimator (3.90) incurs little performance loss compared with the case where the phase ambiguity is perfectly known. Finally, we note that by invoking a Gaussian assumption on the detector output, the SINR can be translated into the bit error rate, as shown in [19].

3.3.5   Adaptive Implementation

Note that the blind multiuser detection and channel estimation strategies presented in previous sections operate in "batch" mode, i.e., a set of M received signals are buffered before the estimation and MMSE filtering steps. In this section we develop a sequential adaptive strategy that provides blind updates of the channel and detector with each received signal. First, we address adaptive implementation of the blind channel

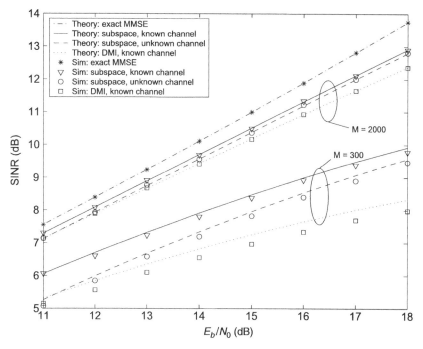

Figure 3.2. Output SINR versus input E_b/N_0 for QPSK.

estimator discussed above. Suppose the signal subspace $\mathbf{U}_s$ is known. Denote by $\mathbf{z}[i]$ the projection of the received signal $\mathbf{r}[i]$ onto the noise subspace, i.e.:

$$\mathbf{z}[i] \triangleq \mathbf{r}[i] - \mathbf{U}_s\mathbf{U}_s^H\mathbf{r}[i] \tag{3.91}$$

$$= \mathbf{U}_n\mathbf{U}_n^H\mathbf{r}[i] \tag{3.92}$$

Since $\mathbf{z}[i]$ lies in the noise subspace, it is orthogonal to any signal in the signal subspace. In particular, it is orthogonal to $\tilde{\mathbf{s}}_1 = \mathbf{\Xi}_1\mathbf{h}_1$. Hence $\mathbf{h}_1$ is the solution to the following constrained optimization problem:

$$\min_{\mathbf{h}_1 \in \mathbb{C}^{P\mu}} E\left[\left\|(\mathbf{\Xi}_1\mathbf{h}_1)^H\mathbf{z}[i]\right\|^2\right], \quad \text{s.t. } \|\mathbf{h}_1\| = 1 \tag{3.93}$$

In order to obtain a sequential algorithm to solve the above optimization problem, we write it in the following (trivial) state space form:

$$\mathbf{h}_1[i] = \mathbf{h}_1[i], \qquad \text{(state equation)}$$

$$0 = \left(\mathbf{\Xi}_1^H\mathbf{z}[i]\right)^H\mathbf{h}_1[i], \quad \text{(observation equation)}$$

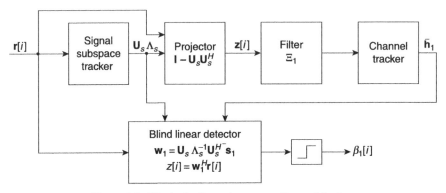

Figure 3.3. Blind adaptive subspace multiuser detector.

The standard Kalman filter can then be applied to the above system, as follows. (We define $\mathbf{x}[i] \triangleq \boldsymbol{\Xi}_1^H \mathbf{z}[i]$):

$$\mathbf{k}[i] = \boldsymbol{\Sigma}[i-1]\mathbf{x}[i]\left(\mathbf{x}[i]^H \boldsymbol{\Sigma}[i-1]\mathbf{x}[i]\right)^{-1} \tag{3.94}$$

$$\mathbf{h}_1[i] = \mathbf{h}_1[i-1] - \mathbf{k}[i]\left(\mathbf{x}[i]^H \mathbf{h}_1[i]\right) / \left\|\mathbf{h}_1[i-1] - \mathbf{k}[i]\left(\mathbf{x}[i]^H \mathbf{h}_1[i]\right)\right\| \tag{3.95}$$

$$\boldsymbol{\Sigma}[i] = \boldsymbol{\Sigma}[i-1] - \mathbf{k}[i]\mathbf{x}[i]^H \boldsymbol{\Sigma}[i-1] \tag{3.96}$$

Note that (3.95) contains a normalization step to satisfy the constraint $\|\mathbf{h}_1[i]\| = 1$.

Since the subspace blind detector may be written in closed-form as a function of the signal subspace components, one may use a suitable subspace tracking algorithm in conjuction with this detector and a channel estimator to form an *adaptive* detector that is able to track changes in the number of users and their composite signature waveforms. Figure 3.3 contains a block diagram of such a receiver. The received signal $\mathbf{r}[i]$ is fed into a subspace tracker that sequentially estimates the signal subspace components $(\mathbf{U}_s, \boldsymbol{\Lambda}_s)$. The signal $\mathbf{r}[i]$ is then projected onto the noise subspace to obtain $\mathbf{z}[i]$, which is in turn passed through a linear filter that is determined by the signature sequence of the desired user. The output of this filter is fed into a channel tracker that estimates the channel state of the desired user. Finally, the linear MMSE detector is constructed in closed-form based on the estimated signal subspace components and the channel state.

3.3.6 Algorithm Summary

The adaptive receiver algorithm is summarized as follows:

ALGORITHM 3.1

[Blind Adaptive Subspace Multiuser Detection]
1. Using a suitable signal subspace tracking algorithm, update the signal subspace components $\mathbf{U}_s[i]$, $\boldsymbol{\Lambda}_s[i]$, $\eta^2[i]$, and the subspace rank $r[i]$ at each time slot i.

2. Update the channel estimate $\mathbf{h}[i]$ using (3.94)–(3.96).
3. Form detector and perform differential detection:

$$\mathbf{w}[i] = \mathbf{U}_s[i]\mathbf{\Lambda}_s[i]^{-1}\mathbf{U}_s[i]^H\tilde{\mathbf{s}}_1[i],$$
$$z_1[i] = \mathbf{w}[i]^H\mathbf{r}[i],$$
$$\hat{\beta}_1[i] = \text{sign}\{\Re(z_1[i]z_1[i-1]^\star)\}$$

We next give a simulation example illustrating the performance of the blind adaptive receiver in an asynchronous CDMA system with multipath channels. The processing gain $N = 15$ and the spreading codes are Gold codes of length 15. Each user's channel has $L = 3$ paths. The delay of each path $\tau_{l,k}$ is uniformly distributed on $[0, 10T_c]$. Hence, as in the preceding example, the maximum delay spread is one symbol interval. The fading gain of each path in each user's channel is generated from a complex Gaussian distribution and is fixed for all simulations. The path gains in each user's channel are normalized so that all users' signals arrive at the receiver with the same power. The received signal is sampled at twice the chip-rate. Figure 3.4 shows the performance of subspace blind adaptive receiver using the NAHJ subspace tracking algorithm [31], in terms of output SINR. During the first 1000 iterations, there are eight total users. At iteration 1000, four new users are added to the system. At iteration 2000, one additional known user is added and three existing users vanish. We see that this blind adaptive receiver can closely track the dynamics of the channel.

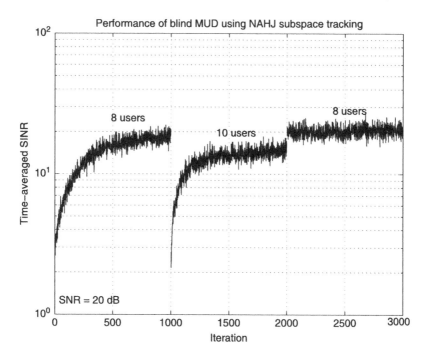

Figure 3.4. Performance of the blind adaptive subspace multiuser detector in an asynchronous CDMA system with multipath.

3.4 BAYESIAN MULTIUSER DETECTION FOR LONG-CODE CDMA

In the previous sections, we discussed subspace blind multiuser detection techniques. The key underlying assumption for this approach is that the users employ periodic spreading codes, i.e., the sequence $\{s_{j,k}\}_{j=0}^{N-1}$, $\forall k$, in (3.3) is independent of the symbol time index i. On the other hand, existing CDMA standards (such as IS-95, WCDMA, CDMA2000) employ long spreading codes on the reverse link, i.e., PN sequences with very long periods. The theme of this section is the design of a blind multiuser receiver for an uplink asynchronous coded CDMA system employing long spreading sequences. It is assumed that the blind receiver has only the knowledge of the spreading sequences and the initial delays of the desired users within the cell.

3.4.1 System Descriptions

3.4.1.1 Signal and Channel Model
Because we are assuming long-code CDMA instead of short-code CDMA, the signal models in Section 3.2 must be modified slightly. Consider a K-user uplink CDMA system, employing normalized long pseudo-random spreading sequences, and signaling through multipath channels with additive white Gaussian noise and other unknown interference. The k-th user's spreading waveform in (3.3) corresponding to the i-th symbol interval is now written as:

$$s_k^{[i]}(t) = \frac{1}{\sqrt{N}} \sum_{j=0}^{N-1} s_{j,k}[i]\psi(t - jT_c), \quad 0 \le t < T \tag{3.97}$$

where $\{s_{j,k}[i], j = 0, \ldots, N-1\}$, $\forall k$, $\forall i$ is the segment of the k-th user's signature sequence corresponding to the i-th symbol. For simplicity, we consider the single receive antenna case. Then the received signal in (3.5) is now modified as:

$$r(t) = \sum_{k=1}^{K} \sum_{i=0}^{M-1} b_k[i] \sum_{l=1}^{L} \alpha_{l,k} s_k^{[i]}(t - iT - \tau_{l,k}) + v(t) \tag{3.98}$$

where $v(t)$ is the ambient noise plus the unknown interfering signals, as will be explained below. At the receiver, the received signal $r(t)$ is filtered by a chip-matched filter and sampled at the chip-rate. Define $\gamma_k \triangleq \lfloor \frac{\tau_{k,1}}{T_c} \rfloor - 1$ as the initial delay in terms of chips for the k-th user's signal, $\mu \triangleq \max_k \lceil \frac{\tau_{k,L} - \tau_{k,1}}{T_c} \rceil$ as the maximum channel delay among all users, $\nu \triangleq \max_{1 \le k \le K} \{ \lceil \frac{\tau_{k,L}}{T} \rceil \}$ as the maximum delay spread among the K users in terms of symbol intervals,[1] and $\mathbf{h}_k \triangleq [h_k[\gamma_k + 1], \ldots, h_k[\gamma_k + \mu]]^T$ as the channel response for the k-th user. We assume that both the maximum initial delay $\max_k\{\gamma_k\}$ and the maximum channel delay μ are less than N. Hence, the maximum symbol delay satisfies $\nu \le 2$.

---

[1]This definition differs slightly from (3.6) in order to simplify the presentation.

The received discrete-time signal vector corresponding to the i-th symbol interval is given by:

$$\mathbf{r}[i] = \sum_{k=1}^{K} \left(b_k[i]\mathbf{C}_{k,i}^{(0)} + b_k[i-1]\mathbf{C}_{k,i-1}^{(1)} + b_k[i-2]\mathbf{C}_{k,i-2}^{(2)} \right)\mathbf{h}_k + \mathbf{v}[i],$$

$$i = 0, 1, \ldots, M-1 \qquad (3.99)$$

where $\mathbf{r}[i] \triangleq [r_0[i], \; r_1[i], \ldots, r_{N-1}[i]]^T$ is defined similar to (3.7), and $\mathbf{v}[i] \triangleq [v_0[i], \ldots, v_{N-1}[i]]^T$ is a vector of noise matched filter outputs, as in (3.7). It is easy to verify that $\mathbf{C}_{k,i}^{(n)}$ can be expressed as:

$$
\begin{bmatrix}
\left[\mathbf{C}_{k,i}^{(0)}\right]_{N\times L} \\[2ex]
\left[\mathbf{C}_{k,i}^{(1)}\right]_{N\times L} \\[2ex]
\left[\mathbf{C}_{k,i}^{(2)}\right]_{N\times L}
\end{bmatrix}
=
\begin{bmatrix}
\mathbf{0}_{\gamma_k \times 1} & \mathbf{0} & \cdots & \mathbf{0} \\
s_{0,k}[i] & & & \\
s_{1,k}[i] & s_{0,k}[i] & & \\
\vdots & s_{1,k}[i] & \ddots & \\
s_{N-1,k}[i] & \vdots & & s_{0,k}[i] \\
& s_{N-1,k}[i] & & s_{1,k}[i] \\
& & \ddots & \vdots \\
& & & s_{N-1,k}[i] \\
\mathbf{0} & \mathbf{0} & \cdots & \mathbf{0}
\end{bmatrix}_{3N\times L}
\qquad (3.100)
$$

3.4.1.2 Noise Model

In the simplest case, it is assumed that $v(t)$ in (3.98) contains the channel ambient noise only, which is a white complex Gaussian process, as in (3.5). It is further assumed that the chip waveform $\psi(t)$ is a rectangle pulse with duration T_c. Hence all the noise samples $\{v_q[i]\}_{q,i}$ are *i.i.d.* zero mean complex Gaussian random variables with variance η^2. Moreover, $\{\mathbf{v}[i]\}_i$ is a sequence of zero-mean *i.i.d.* complex Gaussian vectors, i.e.:

$$\mathbf{v}[i] \sim \mathcal{N}_c(\mathbf{0}, \eta^2\mathbf{I}) \qquad (3.101)$$

In cellular DS-CDMA, the same uplink/downlink pair of frequency bands are reused for each cell. Therefore, a signal transmitted in one cell may cause interference in neighboring cells, resulting in out-cell multiple-access interference (OMAI). In addition, narrowband communication systems sometimes can overlay with CDMA systems, and thus cause narrowband interference (NBI) to the latter [24,26]. Hence, in general, the noise component $\mathbf{v}[i]$ in (3.99) consists of white Gaussian noise (WGN), OMAI and NBI, i.e., $\mathbf{v} = \mathbf{v}_{\mathrm{WGN}} + \mathbf{v}_{\mathrm{OMAI}} + \mathbf{v}_{\mathrm{NBI}}$. The WGN has zero mean and covariance $\Sigma_{\text{WGN}} = \eta^2 \mathbf{I}$. The OMAI has the same structure as the in-cell CDMA signals, i.e.:

$$\mathbf{v}_{\text{OMAI}}[i] = \sum_{k=K+1}^{K+K'} \left[b_k[i]\mathbf{C}_{k,i}^{(0)} + b_k[i-1]\mathbf{C}_{k,i-1}^{(1)} + b_k[i-2]\mathbf{C}_{k,i-2}^{(2)} \right] \mathbf{h}_k \qquad (3.102)$$

where K' denote the total number of out-cell users. When K' is large, by the central limit theorem, $\mathbf{v}_{\text{OMAI}}[i]$ approaches a Gaussian vector with zero mean and a covariance matrix, denoted by Σ_{OMAI}. Note that both the encoded bits $\{b_k[i]\}_{k,i}$ and the elements of the spreading sequences $\{s_{j,k}[i]\}_{k,i,j}$ are zero-mean independent random variables. After some manipulations, the element of Σ_{OMAI} can be written as:

$$\Sigma_{\text{OMAI}}[u,v] = \sum_{k=1}^{K} \sum_{\substack{j,l \\ j-l=u-v}} h_k[j]h_k^*[l], \quad u,v = 0,1,\ldots,N-1 \qquad (3.103)$$

The NBI signal is typically modeled as a correlated Gaussian process. For example, it can be represented as an m_0th-order autoregressive (AR) signal [28], where $m_0 \ll N$, i.e.:

$$v_{\text{NBI}}[n] = -\sum_{j=1}^{m_0} a_j v_{\text{NBI}}[n-j] + e[n] \qquad (3.104)$$

where $v_{\text{NBI}}[n]$ denote the noise sample component due to the NBI signal; $e[n]$ is a white Gaussian process with variance ε^2. Hence $\mathbf{v}_{\text{NBI}}[i] \triangleq [v_{\text{NBI}}[iN], v_{\text{NBI}}[iN+1], \ldots, v_{\text{NBI}}[iN+N-T]]^T$ is Gaussian with zero mean and a covariance matrix, denoted by Σ_{NBI}.

Combining these three components, the noise vectors $\{\mathbf{v}[i]\}_i$ can be modeled as colored Gaussian vectors with zero mean and a covariance matrix $\Sigma = \Sigma_{\text{WGN}} + \Sigma_{\text{OMAI}} + \Sigma_{\text{NBI}}$, i.e.:

$$\mathbf{v}[i] \sim \mathcal{N}_c(\mathbf{0}, \Sigma) \qquad (3.105)$$

Note that when the OMAI and the NBI are present, the noise vectors $\{\mathbf{v}[i]\}_i$ are correlated. Nevertheless, we ignore such temporal correlations to simplify the algorithms.

3.4.1.3 Blind Bayesian Multiuser Detection

The binary information bits $\{d_k[n]\}_n$ for user k are encoded using some channel code (e.g., block code, convolutional code, or turbo code). A code-bit interleaver is used to reduce the influence of the error bursts at the input of the channel decoder. The interleaved code bits are then mapped to BPSK symbols $\{a_k[i]\}_{i=1}^{M-1}$. These BPSK symbols are differentially encoded to yield the symbol stream $\{b_k[i]\}_{i=0}^{M-1}$. Differential encoding is used to resolve the phase ambiguity inherent in any blind receiver, and is given by:

$$\begin{cases} b_k[0] = 1, \\ b_k[i] = b_k[i-1]\,a_k[i], \quad i = 1,2,\ldots,M-1 \end{cases} \qquad (3.106)$$

Each symbol $b_k[i]$ is then modulated by a spreading waveform and transmitted through a multipath channel. The received signal is given by (3.98).

The Bayesian multiuser detector assumes the knowledge of the spreading sequence and the initial delay information for each in-cell user, i.e., $\{\mathbf{C}_{k,i}^{(0)}, \mathbf{C}_{k,i}^{(1)}, \mathbf{C}_{k,i}^{(2)}\}_{k=1;i=0}^{K;M-1}$ in (3.98) are known to the receiver. Note that the initial delay is the first non-zero channel coefficient, which may be obtained through CDMA timing acquisition techniques.

For convenience, define the *a priori* log-likelihood ratios (LLR) of the interleaved code bits as:

$$\rho_k[i] \triangleq \log \frac{P(a_k[i] = +1)}{P(a_k[i] = -1)} \tag{3.107}$$

Further define $\mathbf{Y} \triangleq \{\mathbf{r}[0], \mathbf{r}[1], \ldots, \mathbf{r}[M-1]\}$. The Bayesian multiuser detector estimates *a posteriori* probabilities of the code bits:

$$P(a_k[i] = +1 \mid \mathbf{Y}), \quad i = 1, \ldots, M-1; \quad k = 1, \ldots, K \tag{3.108}$$

based on the received signals $\mathbf{Y}$, the signal model (3.98) and the prior information $\{\rho_k[i]\}_{k=1;i=1}^{K;M-1}$, without knowing the channel response $\{\mathbf{h}_k\}_{k=1}^{K}$ and the noise parameters (i.e., η^2 for white Gaussian noise; $\boldsymbol{\Sigma}$ for colored Gaussian noise). Note that although $\mathbf{Y}$ is directly determined by the differentially encoded symbols $\{b_k[i]\}_{k,i}$ as seen in the signal model (3.98), the channel decoders require the posterior distributions of the code bits $\{a_k[i]\}_{k,i}$.

Note that such a Bayesian multiuser detector can be employed as the front-end soft-input soft-output demodulator in a turbo receiver for coded CDMA systems [49,54,55].

3.4.1.4 *The Gibbs Sampler*

The blind Bayesian multiuser detectors discussed below are based on the Gibbs sampler [11], a Markov chain Monte Carlo (MCMC) procedure for numerical Bayesian computation. Let $\boldsymbol{\theta} = [\theta_1, \theta_2, \ldots, \theta_d]^T$ be a vector of unknown parameters. Let $\mathbf{Y}$ be the observed data. Suppose that we are interested in finding the *a posteriori* marginal distribution of some parameter, say θ_j, conditioned on the observation $\mathbf{Y}$, i.e., $p(\theta_j | \mathbf{Y})$. Direct evaluation involves integrating out the rest of the parameters from the joint posterior density $p(\boldsymbol{\theta} | \mathbf{Y})$, which in most cases is computationally infeasible. The basic idea behind the Gibbs sampler is to generate random samples from the joint posterior distribution $p(\boldsymbol{\theta} | \mathbf{Y})$, and then to estimate any marginal distribution using these samples. Given the samples at time $(n-1)$, $\boldsymbol{\theta}^{(n-1)} = [\theta_1^{(n-1)} \cdots \theta_d^{(n-1)}]^T$, at the nth iteration, this algorithm performs the following operation to obtain samples at time n, $\boldsymbol{\theta}^{(n)} = [\theta_1^{(n)} \cdots \theta_d^{(n)}]^T$:

- For $i = 1, \ldots, d$, draw $\theta_i^{(n)}$ from the conditional distribution:

$$p\left(\theta_i \mid \theta_1^{(n)}, \ldots, \theta_{i-1}^{(n)}, \theta_{i+1}^{(n-1)}, \ldots, \theta_d^{(n-1)}, \mathbf{Y}\right)$$

It is known that under regularity conditions [2,12,13]:

- The distribution of $\boldsymbol{\theta}^n$ converges geometrically to $p(\boldsymbol{\theta} | \mathbf{Y})$, as $n \to \infty$.
- $\frac{1}{N} \sum_{n=1}^{N} f(\boldsymbol{\theta}^{(n)}) \xrightarrow{a.s.} \int f(\boldsymbol{\theta}) p(\boldsymbol{\theta} | \mathbf{Y}) \, d\boldsymbol{\theta}$, as $N \to \infty$, for any integrable function f.

3.4.2 Bayesian MCMC Multiuser Detectors

3.4.2.1 White Gaussian Noise   We first consider the problem of computing the *a posteriori* bit probabilities in (3.108) under the assumption that the ambient noise distribution is white and Gaussian. That is, the pdf of $\mathbf{v}[i]$ in (3.99) is given by:

$$p(\mathbf{v}[i]) = \left(\frac{1}{\pi\eta^2}\right)^P \exp\left(-\frac{\|\mathbf{v}[i]\|^2}{\eta^2}\right) \tag{3.109}$$

Denote:

$$\mathbf{b}_k \triangleq \{b_k[i]\}_{i=0}^{M-1}, \qquad \mathbf{B} \triangleq \{\mathbf{b}_k\}_{k=1}^K$$

$$\mathbf{A} \triangleq \{a_k[i]\}_{k=1;i=1}^{K;M-1}, \qquad \mathbf{H} \triangleq \{\mathbf{h}_k\}_{k=1}^K$$

$$\mathbf{S}_{k,i}(\mathbf{b}_k) \triangleq b_k[i]\mathbf{C}_{k,i}^{(0)} + b_k[i-1]\mathbf{C}_{k,i-1}^{(1)} + b_k[i-2]\mathbf{C}_{k,i-2}^{(2)}$$

Then (3.99) can be written as:

$$\mathbf{r}[i] = \sum_{k=1}^K \mathbf{S}_{k,i}(\mathbf{b}_k)\mathbf{h}_k + \mathbf{v}[i], \quad i = 0, 1, \dots, M-1 \tag{3.110}$$

The problem is solved under a Bayesian framework, by treating the unknown quantities $\mathbf{H}$, η^2, and $\mathbf{B}$ as realizations of random variables with some prior distributions. The Gibbs sampler is then employed to calculate the marginal distribution of those unknown parameters. Note that although the code bits $\mathbf{A}$ are of interest, it is more convenient to sample the differentially encoded bits $\mathbf{B}$ in the Gibbs sampler.

Prior Distributions   In principle, prior distributions are used to incorporate the prior knowledge about the unknown parameters, and less restrictive (i.e., non-informative) priors should be employed when such knowledge is limited. The priors should also be chosen such that the conditional posterior distributions are easy to compute and simulate. To that end, we choose the following prior distributions $p(\mathbf{H})$, $p(\eta^2)$, and $p(\mathbf{B})$, for the unknown parameters:

1. For the unknown channel $\mathbf{h}_k$, a complex Gaussian prior distribution is assumed:

$$p(\mathbf{h}_k) \sim \mathcal{N}_c(\mathbf{h}_{k0}, \mathbf{\Sigma}_{k0}) \tag{3.111}$$

   Note that large value of $\mathbf{\Sigma}_{k0}$ corresponds to less informative prior.

2. For the noise variance η^2, an inverse chi-square prior distribution is assumed:

$$p(\eta^2) = \frac{(\nu_0\lambda_0)^{\nu_0}}{\Gamma(\nu_0)}\left(\frac{1}{\eta^2}\right)^{\nu_0+1} \exp\left(-\frac{\nu_0\lambda_0}{\eta^2}\right) \sim \chi^{-2}(2\nu_0, \lambda_0) \tag{3.112}$$

   Small value of $\nu_0\lambda_0$ corresponds to the less informative priors.

3. The data bit sequence $\mathbf{b}_k$ is a Markov chain, encoded from $\{a_k[i]\}_{i=1}^{M-1}$. Its prior distribution can be expressed as:

$$p(\mathbf{b}_k) = p(b_k[0])p(b_k[1] \mid b_k[0]) \cdots p(b_k[M-1] \mid b_k[M-2])$$

$$= p(b_k[0])p(a_k[1] = b_k[1]b_k[0]) \cdots p(a_k[M-1] = b_k[M-1]b_k[M-2])$$

$$= \frac{1}{2} \prod_{i=1}^{M-1} \frac{\exp(\rho_k[i] \, b_k[i-1] \, b_k[i])}{1 + \exp(\rho_k[i] \, b_k[i-1] \, b_k[i])} \tag{3.113}$$

Notice that we set $p(b_k[0]) = \frac{1}{2}$ to count for the phase ambiguity in $b_k[0]$.

Conditional Posterior Distributions   The following conditional posterior distributions are required by the Bayesian MCMC multiuser detector.

1. The conditional distribution of the k-th user's channel response $\mathbf{h}_k$ given η^2, $\mathbf{B}$, $\mathbf{H}_k$, and $\mathbf{Y}$ is (where $\mathbf{H}_k \triangleq \mathbf{H} \setminus \mathbf{h}_k$):

$$p(\mathbf{h}_k \mid \mathbf{B}, \eta^2, \mathbf{H}_k, \mathbf{Y}) \sim \mathcal{N}_c(\mathbf{h}_{k*}, \mathbf{\Sigma}_{k*}) \tag{3.114}$$

with:

$$\mathbf{\Sigma}_{k*}^{-1} \triangleq \mathbf{\Sigma}_{k0}^{-1} + \frac{1}{\eta^2} \sum_{i=0}^{M-1} \mathbf{S}_{k,i}^{\mathrm{H}}(\mathbf{b}_k) \mathbf{S}_{k,i}(\mathbf{b}_k) \tag{3.115}$$

and:

$$\mathbf{h}_{k*} \triangleq \mathbf{\Sigma}_{k*} \left[\mathbf{\Sigma}_{k0}^{-1} \mathbf{h}_{k0} + \frac{1}{\eta^2} \sum_{i=0}^{M-1} \mathbf{S}_{k,i}^{\mathrm{H}}(\mathbf{b}_k) \left(\mathbf{r}[i] - \sum_{j \neq k} \mathbf{S}_{j,i}(\mathbf{b}_j) \mathbf{h}_j \right) \right] \tag{3.116}$$

2. The conditional distribution of the noise variance η^2 given $\mathbf{H}$, $\mathbf{B}$, and $\mathbf{Y}$ is given by:

$$p(\eta^2 \mid \mathbf{H}, \mathbf{B}, \mathbf{Y}) \sim \chi^{-2}\left(2[\nu_0 + MP], \frac{\nu_0 \lambda_0 + s^2}{\nu_0 + MP} \right) \tag{3.117}$$

with:

$$s^2 \triangleq \sum_{i=0}^{M-1} \left\| \mathbf{r}[i] - \sum_{k=1}^{K} \mathbf{S}_{k,i}(\mathbf{b}_k) \mathbf{h}_k \right\|^2 \tag{3.118}$$

3. The conditional distribution of the data bit $b_k[i]$ given $\mathbf{H}$, η^2, $\mathbf{B}_{ki}$, and $\mathbf{Y}$ can be obtained from (where $\mathbf{B}_{ki} \triangleq \mathbf{B} \setminus b_k[i]$):

$$\frac{P(b_k[i] = +1 \mid \mathbf{H}, \eta^2, \mathbf{B}_{ki}, \mathbf{Y})}{P(b_k[i] = -1 \mid \mathbf{H}, \eta^2, \mathbf{B}_{ki}, \mathbf{Y})}$$

$$= \exp\left(b_k[i+1]\rho_k[i+1] + b_k[i-1]\rho_k[i] - \frac{\Delta s^2}{\eta^2} \right) \tag{3.119}$$

with:

$$\Delta s^2 \triangleq -4\mathcal{R}\left\{\mathbf{h}_k^H\left(\mathbf{C}_{k,i}^{(0)T}\mathbf{r}_k[i] + \mathbf{C}_{k,i}^{(1)T}\mathbf{r}_k[i+1] + \mathbf{C}_{k,i}^{(2)T}\mathbf{r}_k[i+2]\right)\right\} \quad (3.120)$$

$$\mathbf{r}_k[l] \triangleq \mathbf{r}[l] - \sum_{j \neq k} \mathbf{S}_{j,l}(\mathbf{b}_j)\mathbf{h}_j - \mathbf{S}_{k,l}(\mathbf{b}_{kl}^0)\mathbf{h}_k \quad (3.121)$$

where $\mathbf{b}_{kl}^0 \triangleq \{b_k[0], \ldots, b_k[l-1], 0, b_k[l+1], \ldots, b_k[M-1]\}$.

Using the above conditional posterior distributions, the Gibbs sampling implementation of the blind Bayesian multiuser detector in white Gaussian noise proceeds iteratively as follows. Note that the samples of code bits **A** are computed based on the samples of differentially encoded bits **B** in ($\star$).

ALGORITHM 3.2

---

1. Draw the initial values $\theta^{(0)} = \{\mathbf{H}^{(0)}, \eta^{2(0)}, \mathbf{B}^{(0)}\}$ from their prior distributions.
2. For $n = 1, 2, \ldots$
   (a) For $k = 1, 2, \ldots, K$
      i. Draw $\mathbf{h}_k^{(n)}$ from $p(\mathbf{h}_k \mid \mathbf{H}_k^{(n-1)}, \eta^{2(n-1)}, \mathbf{B}^{(n-1)}, \mathbf{Y})$ given by (3.114).
   (b) Draw $\eta^{2(n)}$ from $p(\eta^2 \mid \mathbf{H}^{(n)}, \mathbf{B}^{(n-1)}, \mathbf{Y})$ given by (3.117).
   (c) For $i = 0, 1, \ldots, M-1$
      i. For $k = 1, 2, \ldots, K$
         A. Draw $b_k^{(n)}[i]$ from $P(b_k[i] \mid \mathbf{h}^{(n)}, \eta^{2(n)}, \mathbf{B}_{ki}^{(n-1)}, \mathbf{Y})$ given by (3.119).
         B. Compute $a_k^{(n)}[i] = b_k^{(n)}[i]\, b_k^{(n)}[i-1]$ ($\star$).
      where
      $$\mathbf{H}_k^{(n-1)} \triangleq \left\{\mathbf{h}_1^{(n)}, \ldots, \mathbf{h}_{k-1}^{(n)}, \mathbf{h}_{k+1}^{(n-1)}, \ldots, \mathbf{h}_K^{(n-1)}\right\} \text{ and}$$
      $$\mathbf{B}_{ki}^{(n-1)} \triangleq \{b_1^{(n)}[0], \ldots, b_1^{(n)}[M-1], \ldots, b_k^{(n)}[i-1],$$
      $$b_k^{(n-1)}[i+1], \ldots, b_K^{(n-1)}[M-1]\}.$$

To ensure convergence, the above procedure is usually carried out for $(n_0 + P)$ iterations and samples from the last P iterations are used to calculate the Bayesian estimates of the unknown quantities. The posterior symbol probabilities (3.108) can be approximated as:

$$P(a_k[i] = +1 \mid \mathbf{Y}) \cong \frac{1}{P}\sum_{n=n_0+1}^{n_0+P} \delta_{ki}^{(n)}, \quad k = 1, \ldots, K; i = 1, \ldots, M-1 \quad (3.122)$$

where $\delta_{ki}^{(n)}$ is an indicator such that $\delta_{ki}^{(n)} = 1$, if $a_k^{(n)}[i] = +1$ and $\delta_{ki}^{(n)} = 0$, if $a_k^{(n)}[i] = -1$.

3.4.2.2 *Colored Gaussian Noise* We next discuss the blind Bayesian multiuser detector for colored Gaussian noise. It is assumed that $\mathbf{v}[i]$ in (3.99) have a complex joint Gaussian distribution, i.e.:

$$p(\mathbf{v}[i]) = (\pi|\boldsymbol{\Sigma}|)^{-N} \exp\left(-\mathbf{v}^H[i]\,\boldsymbol{\Sigma}^{-1}\,\mathbf{v}[i]\right) \tag{3.123}$$

As mentioned before, the noise vectors $\{\mathbf{v}[i]\}_i$ are temporally correlated. However, in what follows we ignore this correlation to reduce the receiver complexity.

Prior Distributions The unknown quantities in this case are $(\mathbf{H}, \boldsymbol{\Sigma}^{-1}, \mathbf{B})$, which are assumed to be independent with each other. As in the case of Gaussian noise, the prior distribution of $\mathbf{H}$ and $\mathbf{B}$ are given respectively by (3.111) and (3.113). For the noise covariance matrix $\boldsymbol{\Sigma}$, an inverse complex Wishart prior is assumed, i.e.:

$$p(\boldsymbol{\Sigma}) = \frac{|\boldsymbol{\Psi}|^m \cdot |\boldsymbol{\Sigma}|^{-(m+P+1)} \cdot \exp\left[-\mathrm{tr}\left(\boldsymbol{\Psi}\boldsymbol{\Sigma}^{-1}\right)\right]}{2^{mP}\Gamma_P(m)} \sim \mathcal{W}_c^{-1}(\boldsymbol{\Psi}, m) \tag{3.124}$$

where $\mathrm{tr}(\cdot)$ denotes the trace of a matrix; $\Gamma_P(m) \triangleq \prod_{i=1}^{P}\Gamma(m+1-i)$. Small values of m and $\boldsymbol{\Psi}$ correspond to less informative prior. The inverse of the covariance matrix $\boldsymbol{\Sigma}$ has a complex Wishart distribution, i.e.:

$$p(\boldsymbol{\Sigma}^{-1}) \sim \mathcal{W}_c(\boldsymbol{\Psi}^{-1}, m) \tag{3.125}$$

A random matrix with a Wishart distribution with m degrees of freedom (3.125) can be generated by $\sum_{i=1}^{m+1}\mathbf{u}_i\mathbf{u}_i^H$, where $\{\mathbf{u}_i\}$ are *i.i.d.* Gaussian random vectors with zero mean and covariance $\boldsymbol{\Psi}^{-1}$.

Conditional Posterior Distributions The following conditional posterior distributions are required by the blind Bayesian multiuser detector.

1. The conditional distribution of the k-th user's channel response $\mathbf{h}_k$ given $\boldsymbol{\Sigma}^{-1}$, $\mathbf{B}$, $\mathbf{H}_k$, and $\mathbf{Y}$ is (where $\mathbf{H}_k \triangleq \mathbf{H} \setminus \mathbf{h}_k$):

$$p(\mathbf{h}_k\,|\,\mathbf{B}, \boldsymbol{\Sigma}^{-1}, \mathbf{H}_k, \mathbf{Y}) \sim \mathcal{N}_c(\mathbf{h}_{k\star}, \Sigma_{k\star}) \tag{3.126}$$

with:

$$\Sigma_{k\star}^{-1} \triangleq \Sigma_{k0}^{-1} + \sum_{i=0}^{M-1}\mathbf{S}_{k,i}^H(\mathbf{b}_k)\,\boldsymbol{\Sigma}^{-1}\,\mathbf{S}_{k,i}(\mathbf{b}_k) \tag{3.127}$$

and:

$$\mathbf{h}_{k\star} \triangleq \Sigma_{k\star}\left[\Sigma_{k0}^{-1}\mathbf{h}_{k0} + \sum_{i=0}^{M-1}\mathbf{S}_{k,i}(\mathbf{b}_k)\boldsymbol{\Sigma}^{-1}\left(\mathbf{r}[i] - \sum_{j\neq k}\mathbf{S}_{j,i}(\mathbf{b}_j)\mathbf{h}_j\right)\right] \tag{3.128}$$

2. The conditional distribution of the noise covariance matrix $\boldsymbol{\Sigma}$ given $\mathbf{H}$, $\mathbf{B}$, and $\mathbf{Y}$ is:

$$p(\boldsymbol{\Sigma} \mid \mathbf{H}, \mathbf{B}, \mathbf{Y}) \sim \mathcal{W}_c^{-1}(\boldsymbol{\Psi} + \mathbf{Q}, m + M) \tag{3.129}$$

with:

$$\mathbf{Q} \triangleq \sum_{i=0}^{M-1} \left(\mathbf{r}[i] - \sum_{k=1}^{K} \mathbf{S}_{k,i}(\mathbf{b}_k)\mathbf{h}_k\right)\left(\mathbf{r}[i] - \sum_{k=1}^{K} \mathbf{S}_{k,i}(\mathbf{b}_k)\mathbf{h}_k\right)^H \tag{3.130}$$

Therefore, the conditional distribution of the inverse covariance matrix $\boldsymbol{\Sigma}^{-1}$ given $\mathbf{H}$, $\mathbf{B}$, and $\mathbf{Y}$ is:

$$p(\boldsymbol{\Sigma}^{-1} \mid \mathbf{H}, \mathbf{B}, \mathbf{Y}) \sim \mathcal{W}_c\big((\boldsymbol{\Psi} + \mathbf{Q})^{-1}, m + M\big) \tag{3.131}$$

3. The conditional distribution of the data bit $b_k[i]$ given $\mathbf{H}$, $\boldsymbol{\Sigma}^{-1}$, $\mathbf{B}_{ki}$, and $\mathbf{Y}$ can be obtained from (where $\mathbf{B}_{ki} \triangleq \mathbf{B} \setminus b_k[i]$):

$$\frac{P(b_k[i] = +1 \mid \mathbf{H}, \boldsymbol{\Sigma}^{-1}, \mathbf{B}_{ki}, \mathbf{Y})}{P(b_k[i] = -1 \mid \mathbf{H}, \boldsymbol{\Sigma}^{-1}, \mathbf{B}_{ki}, \mathbf{Y})} = \exp\Big(b_k[i+1]\,\rho_k[i+1].$$
$$+\, b_k[i-1]\,\rho_k[i] - \mathrm{tr}\Big(\Delta \mathbf{Q} \boldsymbol{\Sigma}^{-1}\Big)\Big) \tag{3.132}$$

with:

$$\Delta \mathbf{Q} \triangleq -4\mathcal{R}\Big\{\mathbf{r}_k[i]\mathbf{h}_k^H \mathbf{C}_{k,i}^{(0)T} + \mathbf{r}_k[i+1]\mathbf{h}_k^H \mathbf{C}_{k,i+1}^{(1)T} + \mathbf{r}_k[i+2]\mathbf{h}_k^H \mathbf{C}_{k,i+2}^{(2)T}\Big\} \tag{3.133}$$

$$\mathbf{r}_k[l] \triangleq \mathbf{r}[l] - \sum_{j \neq k} \mathbf{S}_{j,l}(\mathbf{b}_j)\mathbf{h}_j - \mathbf{S}_{k,l}(\mathbf{b}_{kl}^0)\mathbf{h}_k \tag{3.134}$$

where $\mathbf{b}_{kl}^0 \triangleq \{b_k[0], \ldots, b_k[l-1], 0, b_k[l+1], \ldots, b_k[M-1]\}$.

Using the above conditional posterior distribution, the Gibbs sampling implementation of the blind Bayesian multiuser detector in colored Gaussian noise proceeds iteratively as follows. As in the case of white Gaussian noise, the *a posteriori* symbol probability $P[a_k[i] = +1 \mid \mathbf{Y}]$ can also be computed by (3.122).

ALGORITHM 3.3

[Blind Bayesian Multiuser Detection in Colored Gaussian Noise]
1. Draw the initial values $\theta^{(0)} = \{\mathbf{H}^{(0)}, \boldsymbol{\Sigma}^{-1(0)}, \mathbf{B}^{(0)}\}$ from their prior distributions.
2. For $n = 1, 2, \ldots$
    (a) For $k = 1, 2, \ldots, K$
        i. Draw $\mathbf{h}_k^{(n)}$ from $p[\mathbf{h}_k \mid \mathbf{H}_k^{(n-1)}, \boldsymbol{\Sigma}^{-1(n-1)}, \mathbf{B}^{(n-1)}, \mathbf{Y}]$ given by (3.126).
    (b) Draw $\boldsymbol{\Sigma}^{-1(n)}$ from $p[\boldsymbol{\Sigma}^{-1} \mid \mathbf{H}^{(n)}, \mathbf{B}^{(n-1)}, \mathbf{Y}]$ given by (3.131).

(c) For $i = 0, 1, \ldots, M - 1$

    i. For $k = 1, 2, \ldots, K$

        A. Draw $b_k^{(n)}[i]$ from $P(b_k[i] \mid \mathbf{h}^{(n)}, \boldsymbol{\Sigma}^{-1(n)}, \mathbf{B}_{ki}^{(n-1)}, \mathbf{Y})$ given by (3.132).

        B. Compute $a_k^{(n)}[i] = b_k^{(n)}[i]\, b_k^{(n)}[i-1]$ ($\bigstar$).

where

$$\mathbf{H}_k^{(n-1)} \triangleq \{ \mathbf{h}_1^{(n)}, \ldots, \mathbf{h}_{k-1}^{(n)}, \mathbf{h}_{k+1}^{(n-1)}, \ldots, \mathbf{h}_K^{(n-1)} \} \text{ and}$$

$$\mathbf{B}_{ki}^{(n-1)} \triangleq \{ b_1^{(n)}[0], \ldots, b_1^{(n)}[M-1], \ldots, b_k^{(n)}[i-1],$$
$$b_k^{(n-1)}[i+1], \ldots, b_K^{(n-1)}[M-1] \}$$

3.4.3 Simulation Examples

In this section, we provide a number of simulation examples to illustrate the performance of the blind Bayesian multiuser detectors. We consider a CDMA system with processing gain $N = 10$. The long spreading sequences of all users are generated randomly. In all the simulations described in this section, the following *non-informative conjugate* prior distributions are used in the Gibbs sampler. For the case of white Gaussian noise:

$$p\big(\mathbf{h}_k^{(0)}\big) \sim \mathcal{N}(\mathbf{h}_{k0}, \boldsymbol{\Sigma}_{k0}) \quad : \quad \mathbf{h}_{k0} = \mathbf{0}, \quad \boldsymbol{\Sigma}_{k0} = 1000\,\mathbf{I}$$
$$p\big(\eta^{2(0)}\big) \sim \mathcal{X}^{-2}(\nu_0, \lambda_0) \quad : \quad \nu_0 = 2, \quad \lambda_0 = 0.3$$

and for the case of colored Gaussian noise:

$$p\big(\mathbf{h}_k^{(0)}\big) \sim \mathcal{N}(\mathbf{h}_{k0}, \boldsymbol{\Sigma}_{k0}) \quad : \quad \mathbf{h}_{k0} = \mathbf{0}, \quad \boldsymbol{\Sigma}_{k0} = 1000\,\mathbf{I}$$

$$p(\boldsymbol{\Sigma}^{-1(0)}) \sim \mathcal{W}_c(\boldsymbol{\Psi}^{-1}, m) \quad : \quad \boldsymbol{\Psi} = 0.01\,\mathbf{I}, \quad m = 2$$

In all the simulations related to colored Gaussian noise, signal-to-noise ratio (SNR) is used to denote the in-cell user signal to WGN ratio, SIR is used to denote the in-cell user signal to NBI ratio. The NBI is modeled as a 2nd order AR model with coefficients $a_1 = 1.8$, $a_2 = -0.81$ in (3.104). The OMAI is generated according to (3.102) with energy 12 dB below the in-cell user and the number of out-cell users is set as $K' = 6K$. The number of path for each user is $L = 3$; the transmitter delay ι_k is generated randomly with the restriction $\iota_k < P$. For each data block, the Gibbs sampling is performed for 100 iterations, with the first 50 iterations as the "burning-in" period, i.e., $n_0 = P = 50$ in (3.122).

Convergence Behavior of Bayesian Multiuser Detectors   We first illustrate the convergence behavior of proposed blind Bayesian multiuser detector in white Gaussian noise. In Figure 3.5, we plot the first 100 samples drawn by the Gibbs sampler of $\mathbf{h}_1$ and η^2. The corresponding true values of these quantities are also

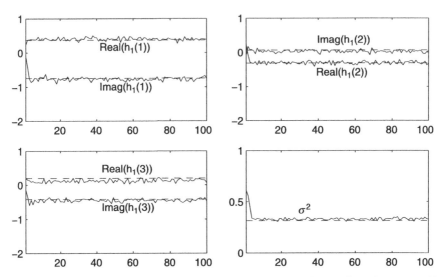

Figure 3.5. Samples drawn by the Gibbs sampler for the case of white Gaussian noise with $K = 3$, $E_b/N_0 = 8\,dB$ for all the users.

shown in the same figure with dashed lines. It is seen that the Gibbs sampler reaches convergence within the first several iterations.

Next, we illustrate the convergence behavior of the blind Bayesian multiuser detector in colored Gaussian noise. The channel responses of in-cell users are generated randomly with normalized energy, and the channel response of the out-cell users are generated randomly with energy 12 dB below. In Figure 3.6, we plot the first 100 samples drawn by the Gibbs sampler of $\mathbf{h}_2$ and $\boldsymbol{\Sigma}^{-1}(1, 2)$. The corresponding true values of $-\mathbf{h}_2$ and $\boldsymbol{\Sigma}^{-1}(1, 2)$ are also shown in the same figure with dashed lines. Again, it is seen that the Gibbs sampler reaches convergence within the first several iterations. The channel response samples converges to $-\mathbf{h}_k$ or $\mathbf{h}_k$ randomly due to the phase ambiguity. It is seen that $\boldsymbol{\Sigma}^{-1}(1, 2)$ is far from zero, which indicates that the noise covariance matrix is not diagonal any more with the existence of OMAI and NBI.

Performance of Bayesian Multiuser Detectors   Figure 3.7 illustrates the performance of the Gibbs multiuser detector in white Gaussian noise with different number of in-cell users. The bit error rate for $\{a_k[i]\}$ is averaged among all the users, and then plotted. The RAKE receiver and the nonlinear parallel interference concellation (PIC) receiver [40] are also implemented assuming *perfect channel knowledge*. The performance of the RAKE receiver and the performance of PIC after five iterations are also shown in the same figure for the purpose of comparison. It is seen that at reasonable SNR, the performance of the Gibbs multiuser detector is better than that of the other two methods where perfect channel knowledge is assumed. The performance gain over the other two methods increases as the number of users increases.

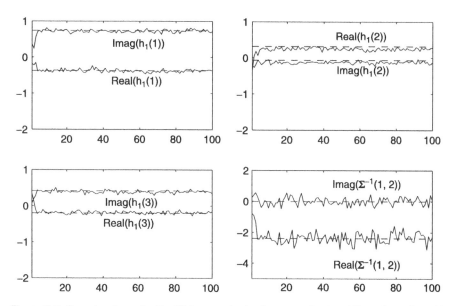

Figure 3.6. Samples drawn by the Gibbs sampler for the case of colored Gaussian noise with $K = 2$, $K' = 18$; for each in-cell user, SNR = 20 dB and SIR = -15 dB.

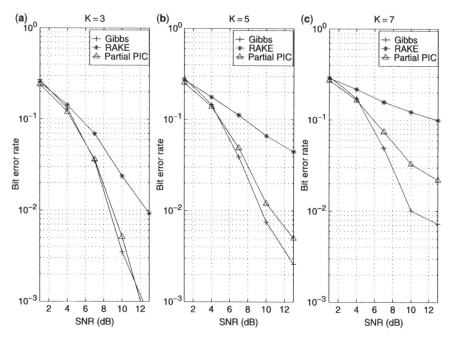

Figure 3.7. Performance of blind Bayesian multiuser detector in white Gaussian noise, assuming all the in-cell users have same energy, (a) $K = 3$; (b) $K = 5$; (c) $K = 7$.

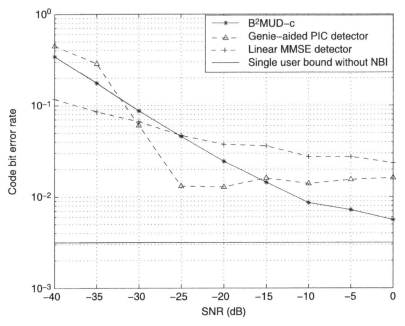

Figure 3.8. Performance of blind Bayesian multiuser detector in colored Gaussian noise with $K = 3$, $K' = 18$ and fixed SNR $= 15$ dB for all in-cell users.

In order to demonstrate the performance of the Gibbs multiuser detector in colored Gaussian noise, in Figure 3.8, we compare its performance with that of the following receiver schemes: (1) Linear MMSE multiuser detector, where we assume that the multipath channels for all in-cell users are known to the receiver. (2) Genie-aided PIC detector, where we assume that a genie provides the receiver with an observation of signal-free NBI corrupted by additive ambient noise and OMAI with the same statistics. We can then use a Kalman filter to obtain an estimate of the NBI signal [43]. After subtracting the estimated NBI signal from the observation $\{\mathbf{r}_i\}_i$, a PIC receiver is implemented assuming perfect channel knowledge. (3) Single-user bound without NBI, where we assume that there is no NBI. Rake receiver is implemented for single user CDMA system with the same component of white ambient noise and OMAI. It is clear that this detector provides a lower bound to the system we discussed here.

Note that the three approaches given above assume perfect channel knowledge as well as other side information about the channel. For example, in the linear MMSE detector, the covariance matrix of the combined NBI, OMAI and noise is assumed known; in genie-aided PIC detector, a genie observation is assumed to be available for estimating NBI signal; in single user bound, both NBI and other in-cell users are assumed perfectly known. Hence, such performance comparisons are unfavorable to the Gibbs multiuser detector. Nevertheless, it is seen in Figure 3.8 that at reasonable SIR, the Gibbs detector outperforms the other two receivers (linear MMSE detector and genie-aided PIC detector) and approaches a near single-user bound performance, which demonstrates that the blind Bayesian multiuser detector under the colored Gaussian noise assumption is effective for combating unknown NBI and OMAI.

Finally, as mentioned earlier, when employed in the context of iterative demodulation and decoding, the Bayesian multiuser detector becomes the key component of a blind turbo multiuser detector, which performs joint channel estimation, demodulation, and decoding. See [55] for details.

3.5 MULTIUSER DETECTION FOR LONG-CODE CDMA IN FAST-FADING CHANNELS

In the preceding two sections, we have assumed that fading channels remain fixed for the duration of an entire data burst, i.e., the slow-fading case. In this section, we treat the fast-fading scenario where the fading channels vary from symbol to symbol. A sequential multiuser detection algorithms is outlined for asynchronous long-code CDMA uplink over unknown multipath fast fading channels. With the prior knowledge of only the signature waveforms, the delays and the second-order statistics of the fading channel, the receivers sequentially estimate the channel using the sequential EM algorithm. The snapshot estimates for each path are tracked by linear MMSE filters and the user data are detected by maximum likelihood sequence estimation conditioned on the channel estimates.

3.5.1 Channel Model and Sequential EM Algorithm

The received continuous-time signal model is of similar form as (3.98), except now the channel responses depend on the symbol index i, i.e.:

$$r(t) = \sum_{k=1}^{K} \sum_{i=0}^{M-1} b_k[i] \sum_{l=1}^{L} \alpha_{l,k}[i] s_k^{[i]}(t - iT - \tau_{l,k}) + v(t) \qquad (3.135)$$

Accordingly the received discrete-time signal vector corresponding to the i-th symbol interval is given by:

$$\mathbf{r}[i] = \sum_{k=1}^{K} \left(b_k[i]\mathbf{C}_{k,i}^{(0)} + b_k[i-1]\mathbf{C}_{k,i-1}^{(1)} + b_k[i-2]\mathbf{C}_{k,i-2}^{(2)} \right) \mathbf{h}_k[i] + \mathbf{v}[i],$$
$$i = 0, 1, \ldots \qquad (3.136)$$

Our goal is to sequentially track the fading channel responses $\{\mathbf{h}_k[i], i = 0, 1, \ldots\}, \forall k$, and at the same time to estimate the transmitted data symbols $\{b_k[i], i = 0, 1, \ldots\}, \forall k$. To that end, we make use of the sequential EM algorithm [39,52]. Next we briefly introduce this technique.

Suppose $\mathbf{y}_1, \mathbf{y}_2, \ldots$ are a sequence of observations with probability density function (pdf) $f(\mathbf{y} \mid \mathbf{t})$, where $\mathbf{t} \in \mathbb{C}^m$ is a static parameter vector, for some m.

A class of sequential estimators derived from the maximum-likelihood principle is given by:

$$\mathbf{t}^{[i+1]} = \mathbf{t}^{[i]} + \mathbf{\Pi}\left(\mathbf{y}_{i+1}, \mathbf{t}^{[i]}\right) \mathbf{s}\left(\mathbf{y}_{i+1}, \mathbf{t}^{[i]}\right) \tag{3.137}$$

where $\mathbf{t}^{[i]}$ is the estimate of $\mathbf{t}$ at the i-th step; $\mathbf{\Pi}(\mathbf{y}_{i+1}, \mathbf{t}^{[i]})$ is an $m \times m$ matrix defined below; and:

$$\mathbf{s}\left(\mathbf{y}_{i+1}, \mathbf{t}^{[i]}\right) \triangleq \left[\frac{\partial}{\partial \theta_1^{\star}} \log f(\mathbf{y}_{i+1} \mid \mathbf{t}), \ldots, \frac{\partial}{\partial \theta_m^{\star}} \log f(\mathbf{y}_{i+1} \mid \mathbf{t})\right]^{T} \Bigg|_{\mathbf{t}=\mathbf{t}^{[i]}} \tag{3.138}$$

is the score (i.e., the gradient of the log-likelihood function). Let $\mathbf{H}(\mathbf{y}_i, \mathbf{t}^{[i]})$ denote the Hessian matrix of $\log f(\mathbf{y}_i \mid \mathbf{t}^{[i]})$, where:

$$H_{j,k}\left(\mathbf{y}_i, \mathbf{t}^{[i]}\right) = \frac{\partial^2}{\partial \theta_j^{\star} \partial \theta_k} \log f(\mathbf{y}_i \mid \mathbf{t}) \Bigg|_{\mathbf{t}=\mathbf{t}^{[i]}}, \quad j = 1, \ldots, m, \, k = 1, \ldots, m \tag{3.139}$$

Let $\mathbf{x}_i$ denote a "complete" data related to $\mathbf{y}_i$, for $i = 1, 2, \ldots$. The complete data $\mathbf{x}_i$ is selected in the (sequential) EM algorithms such that $\mathbf{y}_i$ can be obtained through a many-to-one mapping $\mathbf{x}_i \rightarrow \mathbf{y}_i$, and their knowledge makes the estimation problem easy (e.g., the conditional density $f(\mathbf{x}_i \mid \mathbf{t})$ can be easily obtained). Denote the Fisher information matrix of the data $\mathbf{y}_i$ and $\mathbf{x}_i$, respectively, as:

$$\mathbf{I}\left(\mathbf{t}^{[i]}\right) = -E\left[\mathbf{H}\left(\mathbf{y}_i, \mathbf{t}^{[i]}\right)\right] \quad \text{and} \quad \mathbf{I}_c\left(\mathbf{t}^{[i]}\right) = -E\left[\mathbf{H}\left(\mathbf{x}_i, \mathbf{t}^{[i]}\right)\right]$$

Different algorithms are characterized by different choices of the function $\mathbf{\Pi}\left(\mathbf{y}_{i+1}, \mathbf{t}^{[i]}\right)$ in (3.137):

- The sequential EM algorithm:

$$\mathbf{\Pi}\left(\mathbf{y}_{i+1}, \mathbf{t}^{[i]}\right) = \frac{1}{i} \mathbf{I}_c^{-1}\left(\mathbf{t}^{[i]}\right) \tag{3.140}$$

  The consistency and asymptotic normality of the algorithm is reported in [39].
- The Newton-Raphson algorithm:

$$\mathbf{\Pi}\left(\mathbf{y}_{i+1}, \mathbf{t}^{[i]}\right) = -\mathbf{H}^{-1}\left(\mathbf{y}_{i+1}, \mathbf{t}^{[i]}\right) \tag{3.141}$$

- A stochastic approximation procedure:

$$\mathbf{\Pi}\left(\mathbf{y}_{i+1}, \mathbf{t}^{[i]}\right) = \frac{1}{i} \mathbf{I}^{-1}\left(\mathbf{t}^{[i]}\right) \tag{3.142}$$

Note that, for independent and identically distributed (i.i.d.) observations $\{\mathbf{y}_i\}_i$, if i in (3.142) is substituted by $[i + 1]$, we obtain the maximum-likelihood estimator (MLE) of $\mathbf{t}$ for exponential families [39]. The asymptotic distribution of this procedure can be found in [8,34].

- If $\Pi\left(\mathbf{y}_{i+1}, \mathbf{t}^{[i]}\right)$ is a constant diagonal matrix with small elements, then (3.137) is the conventional steepest-descent algorithm. Some other choices of $\Pi\left(\mathbf{y}_{i+1}, \mathbf{t}^{[i]}\right)$ are suggested in [39].

- For *time-variant* parameters $\{\mathbf{t}[i]\}_i$, a conventional approach suggested in [10,22] is to substitute the converging series $1/i$ in (3.140) with a small positive constant λ_0. The new estimator is given by:

$$\hat{\boldsymbol{\theta}}[i+1] = \hat{\boldsymbol{\theta}}[i] + \lambda_0 \mathbf{I}_c^{-1}\left(\hat{\boldsymbol{\theta}}[i]\right)\mathbf{s}\left(\mathbf{y}_{i+1}, \hat{\boldsymbol{\theta}}[i]\right) \tag{3.143}$$

where $\hat{\boldsymbol{\theta}}[i]$ is the estimate of $\mathbf{t}[i]$.

3.5.2 Sequential Blind Multiuser Detector

In the sequential EM-based multiuser detector, at time i, we have the following steps:

1. Localization:

$$\hat{\mathbf{h}}[i] = \hat{\mathbf{h}}[i] + \lambda_0 \mathbf{I}_c^{-1}\left(\tilde{\mathbf{h}}[i]\right)\mathbf{s}\left(\mathbf{r}[i], \tilde{\mathbf{h}}[i]\right) \tag{3.144}$$

   where $\tilde{\mathbf{h}}[i]$ is a one-step prediction of $\mathbf{g}[i]$ provided by Step 3.

2. Multiuser detection:

$$\hat{\mathbf{b}}[i] = \arg\max_{\mathbf{b}[i]} f\left(\mathbf{r}[i] \mid \tilde{\mathbf{h}}[i], \mathbf{b}[i]\right) \tag{3.145}$$

3. Tracking:

$$\tilde{h}_{kl}[i+1] = \boldsymbol{\alpha}_{kl}^T[i][\hat{h}_{kl}(i-q)\cdots\hat{h}_{kl}[i]]^T, \quad k = 1,\ldots,K, l = 1,\ldots,L \tag{3.146}$$

   where $\boldsymbol{\alpha}_{kl}[i]$ is a $(q+1)$-order linear MMSE filter for $\hat{h}_{kl}[i]$.

Based on the signal model (3.136), the quantities in the algorithm above [i.e., $\mathbf{s}\left(\mathbf{r}(i), \tilde{\mathbf{h}}(i)\right)$, $\mathbf{I}_c\left(\tilde{\mathbf{h}}(i)\right)$, $\hat{\mathbf{b}}(i)$, and $\boldsymbol{\alpha}_{kl}(i)$]. In particular, a direct evaluation of the score and the information matrix of the incomplete data is prohibitive. By introducing appropriate complete data, an approximation to $\mathbf{s}\left(\mathbf{z}(i), \tilde{\mathbf{h}}(i)\right)$ based on the EM algorithm can be obtained. As a byproduct, a low-complexity detector of the data $\mathbf{b}(i)$ can also be obtained. The instantaneous MMSE filter $\boldsymbol{\alpha}_{kl}(i)$ can also be derived. Due to space limitations, we do not provide the detailed derivations here. They can be found in [21].

3.5.3 Simulation Results

Next we illustrate the performance of sequential EM multiuser detector in an unknown multipath fading CDMA channel by a simulation example. The delays of users' paths were randomly generated and then fixed for the simulation. Each user has two ($L = 2$) equal-energy paths. The time-variant fading coefficients were randomly generated from Clarke's model to simulate a Rayleigh fading channel with carrier frequency

1850 MHz, data rate 144 kb/s, vehicle speed = 52 miles/hour, i.e., normalized bandwidth-time product $BT = 0.001$ for all paths. We consider a reverse link of an asynchronous CDMA system with five users ($K = 5$). The spreading sequences of each user and the data bits of each user are independently and randomly generated. The processing gain is $N = 12$. All users have equal signal amplitudes. To remove the sign ambiguity of the tracked fading process, the transmitted bits are differentially encoded.

From simulation results, after 200 symbols, the estimator enters a steady state. A transition state is plotted in Figure 3.9. The average bit error rates (BER) of five users versus $\frac{E_b}{N_0}$ are plotted in Figure 3.10, where E_b is the energy per bit. The figure shows the performance of the pilot-assisted multiuser receiver with pilot insertion rates 1/500 and 1/100 reaches an error floor due to ISI and MAI. The figure also shows the performance of the pilot-assisted single-user RAKE receiver in a single-user environment with pilot insertion rate 1/500 and the performance of the proposed sequential-EM and approximate sequential-EM receivers, including analytical bounds for each. It is seen that significant performance gains are achieved by the proposed receivers compared to pilot-assisted receivers. Both the performance of the approximate sequential-EM receiver with three iterations and the performance of the sequential-EM receiver are very close to that for known channels. This can be attributed to the quality of the channel estimation for both receivers, as seen in Figures 3.9 and 3.10. Moreover, the performance difference between the sequential-EM receiver

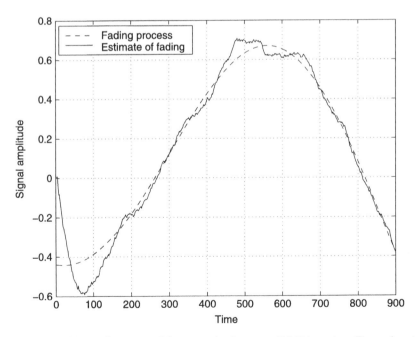

Figure 3.9. Tracking performance of the approximate sequential-EM receiver. The real part of User 1's first path in a multiuser multipath fading channel with $K = 5$, $L = 2$, $BT = 0.001$, $E_b/N_0 = 8$ dB.

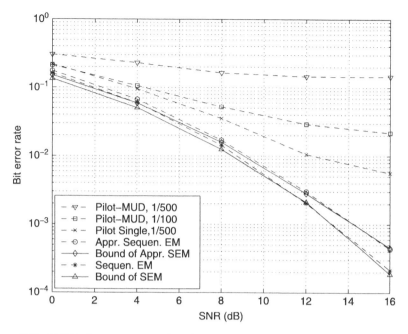

Figure 3.10. BER Performance comparison between the proposed multiuser receivers and the pilot-assisted multiuser receivers in a multipath fading channel with the number of users $K = 5$ and 1, the number of paths $L = 2$, processing gain $N = 12$, $BT = 0.001$.

and the approximate sequential-EM receiver is small in the low and medium SNR regions, and it is about 1 dB at high SNRs. Furthermore, it is seen from the simulation results that the proposed multiuser receivers in a multiuser channel even outperform the RAKE receiver in a single-user channel. This is because the RAKE receiver makes the assumption that the delayed signals from different paths for each user are orthogonal, which effectively neglects the intersymbol interference.

3.6 TRANSMITTER-BASED MULTIUSER PRECODING FOR FADING CHANNELS

The receiver-based multiuser detection techniques presented in this chapter allow system designers to trade receiver complexity for improved multiuser system performance. In many applications, however, it is useful to have the option of moving complexity away from the receiver to the transmitter. Cellular service providers, for example, would prefer to keep mobile unit costs to a minimum so they can continue to entice customers with free phones. Similarly, heterogeneous ad hoc or wireless sensor networks may be composed of nodes with widely varying power constraints and computational capabilities, making the option of moving complexity where it can be managed most efficiently an attractive option.

In this section, we present *transmitter-based* techniques for enhancing multiuser system performance, i.e., transmitter-based multiuser detection. More specifically, we develop linear transmitter precoding strategies for multiple-access interference suppression and diversity exploitation when no receiver channel state information (CSI) is available and receivers are restricted to low-complexity matched-filter detection. In the context of cellular systems, transmitter precoding would be appropriate for the downlink, where the complexity of mobile units should be kept to a minimum. There has been a significant volume of work in the area of joint transmitter/receiver design when receiver CSI or feedback *is* available [9,35,36,53] and on diversity transmission without receiver CSI, but with maximum-likelihood reception [15,16,20]. Our focus, however, is on precoding jointly for diversity and interference suppression in systems that require ultra-low complexity receivers, that is, matched filtering without receiver-based channel estimation. Precoding for multiple access systems, as previously developed, focuses on transmitter-based MAI suppression. The authors in [45], for example, developed minimum mean square error precoders for synchronous CDMA in additive white Gaussian noise channels. They also presented an extension to multipath channels, but a RAKE receiver is required and the channel is assumed perfectly known at the receiver. These initial results were promising, showing that precoding outperformed decorrelating receiver-based multiuser detection in some cases. In [1], the authors considered transmitter precoding for multipath fading channels but, in contrast to the present work, their prefilter is applied to the output of the spread spectrum encoder, rather than applying the filter first, followed by spreading. It was shown that this approach has inferior average performance unless the spreading codes themselves are allowed to be adaptive. In [5], the authors developed a simple but remarkable precoding technique for exploiting multipath diversity that requires no receiver CSI. This technique, called pre-rake diversity combining, will be used in the present work.

Our approach differs from that of existing information-theoretic precoding work [4] in that we are precoding to minimize mean square error instead of maximizing capacity, mutual information or some other information-theoretic criterion. That is, we are optimizing performance, in terms of interference suppression and diversity gain, while keeping the rate fixed. In contrast to these works, we are also interested in very low complexity joint decoding and detection (via the matched filter). The QR-decomposition, "writing-on-dirty-paper" based pre-subtraction approach [56] and related non-linear TH precoding schemes are also not immediately applicable because it can require more receiver complexity than we are willing to tolerate here. Portions of this material were first published in [30,32].

3.6.1 Basic Approach and Adaptation

3.6.1.1 Uplink Signal Model and Blind Channel Estimation For the uplink channel we assume, as in earlier sections, a synchronous multipath CDMA signal model, shown in Figure 3.11. Extensions to other models are relatively straightforward.

3.6.1.2 Downlink Signal Model and Matched Filter Detection The downlink signal transmitted from the base station through the synchronous multipath

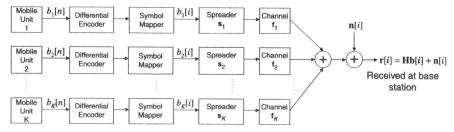

Figure 3.11. The uplink of a K-user CDMA system.

channel during the i-th symbol interval can be written:

$$\mathbf{x}[i] = \mathbf{SMb}[i] \tag{3.147}$$

where:

$$\mathbf{S} \triangleq [\mathbf{s}_1 \ \mathbf{s}_2 \ \cdots \ \mathbf{s}_K] \tag{3.148}$$

is the matrix of spreading waveforms and $\mathbf{M} \in \mathbb{C}^{K \times K}$ is a complex precoding filter that we will optimize in the following section. Throughout this section, we assume that the CDMA system is operating in the time division duplex mode, so that the downlink and uplink operate using the same carrier frequency in different time slots. We also assume that the time elapsing between uplink and downlink transmissions is sufficiently small compared to the coherence time of the channel that the channel impulse response is the same for the uplink and downlink. Then from (3.25) and (3.26), the received signal at User 1's mobile unit can be written as:

$$\mathbf{r}_1[i] = \underbrace{\left[\bar{\mathbf{S}}_1 \mathbf{h}_1 \ \bar{\mathbf{S}}_2 \mathbf{h}_1 \ \cdots \ \bar{\mathbf{S}}_K \mathbf{h}_1\right]}_{\mathbf{H}_1} \mathbf{Mb}[i] + \mathbf{n}_1[i] \tag{3.149}$$

where $\bar{\mathbf{S}}_1, \bar{\mathbf{S}}_2, \ldots, \bar{\mathbf{S}}_K$ contain shifted versions of their respective signature waveforms as in (3.25), except that the L shifts are the same for each user's waveform since all spreading codes have been transmitted over User 1's downlink channel. The matrix $\mathbf{H}_1$ is similarly defined as an extension to (3.29). Detection of the downlink information bits is accomplished via matched filtering of the received signal $\mathbf{r}_1[i]$ with User 1's signature waveform, $\mathbf{s}_1$. Figure 3.12 contains a block diagram of the signal processing that takes place at the base station when an adaptive implementation is employed. We will see more details in Section 3.6.1.4.

3.6.1.3 Transmitter Precoding for a Synchronous Multipath Downlink

We seek to choose the precoding matrix $\mathbf{M}$ so as to provide the best downlink performance possible when the mobile units are constrained to the use of a matched filter receiver. We choose the minimum mean-square error criterion, so

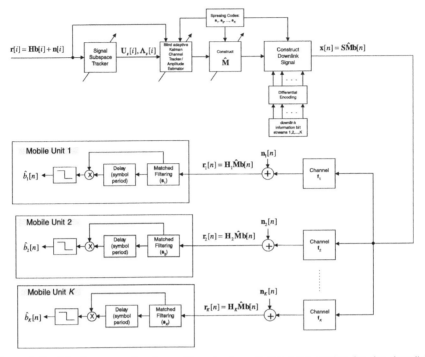

Figure 3.12. Adaptive precoding transmitter structure at the base station for the downlink signal.

$\mathbf{M}$ is chosen to minimize:

$$J = E\left[\left\|\begin{bmatrix} b_1 \\ b_2 \\ \vdots \\ b_K \end{bmatrix} - \begin{bmatrix} \mathbf{s}_1^H \mathbf{r}_1 \\ \mathbf{s}_2^H \mathbf{r}_2 \\ \vdots \\ \mathbf{s}_K^H \mathbf{r}_K \end{bmatrix}\right\|^2\right] \tag{3.150}$$

where we have dropped the time index for clarity. It is easy to see that:

$$\begin{bmatrix} \mathbf{s}_1^H \mathbf{r}_1 \\ \mathbf{s}_2^H \mathbf{r}_2 \\ \vdots \\ \mathbf{s}_K^H \mathbf{r}_K \end{bmatrix} = \underbrace{\begin{bmatrix} \mathbf{s}_1^H \mathbf{H}_1 \\ \mathbf{s}_2^H \mathbf{H}_2 \\ \vdots \\ \mathbf{s}_K^H \mathbf{H}_K \end{bmatrix}}_{\mathcal{H}} \mathbf{Mb} + \underbrace{\begin{bmatrix} \mathbf{s}_1^H \mathbf{n}_1 \\ \mathbf{s}_2^H \mathbf{n}_2 \\ \vdots \\ \mathbf{s}_K^H \mathbf{n}_K \end{bmatrix}}_{\mathbf{v}} \tag{3.151}$$

Then:

$$J = E\left[\|\mathbf{b} - \mathcal{H}\mathbf{Mb} - \mathbf{v}\|^2\right] \tag{3.152}$$

The following proposition gives the optimal precoding matrix.

Proposition 1   *The choice of* **M** *that minimizes* J *is* $\mathbf{M} = \mathcal{H}^{-1}$.

The proof appears in [30].

Denote by $\hat{\mathbf{H}}_i \, (1 \leq i \leq K)$ the matrix $\mathbf{H}_i$ where the channel $\mathbf{h}_i$ has been replaced with the blind estimate $\hat{\mathbf{h}}_i$ obtained from (3.33). Then we may form an initial blind estimate of **M** at the base station as:

$$\hat{\mathbf{M}} = \begin{bmatrix} \mathbf{s}_1^H \hat{\mathbf{H}}_1 \\ \mathbf{s}_2^H \hat{\mathbf{H}}_2 \\ \vdots \\ \mathbf{s}_K^H \hat{\mathbf{H}}_K \end{bmatrix}^{-1} \tag{3.153}$$

There remain amplitude and phase ambiguities in $\hat{\mathbf{M}}$ that are addressed in the following sections.

Notice that choosing the optimal precoding matrix by minimizing J places no explicit constraint on average transmit power. In fact, it was found in a related work on MMSE precoding [45] that unconstrained optimization with simple power scaling provides superior performance at high SNR to constrained optimization. As a result, we shall focus on the former. We will also suppress the power scale factor for simplicity.

3.6.1.4 Adaptive Implementation

In this section, we apply the adaptive implementation discussed in Section 3.3.5 to the transmitter precoding strategy in Section 3.6.1.3 in order to update the precoding matrix **M** at each time slot. This sequential updating allows the implementation to adapt as the channel changes and as users enter and leave the system. A block diagram of the signal processing at the base station appears in Figure 3.12. Note that we have suppressed the signal processing necessary for detection of the uplink bits. The uplink signal received at the base station is used in a signal subspace tracker, along with the known spreading codes of all users, to construct channel estimates as discussed in the following section. Recall that since we are assuming TDD mode, the uplink channel estimates also serve as downlink channel estimates that can be used to construct **M**. As previously mentioned, these channel estimates have amplitude and phase ambiguities. Since nearly all cellular CDMA systems employ power control, it is likely that the base station has some knowledge of each users' transmit power. This information, coupled with estimates of the received power, can be used to estimate the channel amplitude for each user. More specifically, let the diagonal matrices $\mathbf{A} = \mathrm{diag}(\alpha_1, \alpha_2, \ldots, \alpha_K)$ and $\mathbf{P} = \mathrm{diag}(\sqrt{p_1}, \sqrt{p_2}, \ldots, \sqrt{p_K})$ contain the unknown channel amplitudes and the known uplink transmit powers, respectively. Also define the uplink signal mixing matrix $\bar{\bar{\mathbf{S}}}$ such that $\tilde{\mathbf{S}} = \bar{\bar{\mathbf{S}}} \mathbf{A} \mathbf{P}$, so that the columns of $\bar{\bar{\mathbf{S}}}$ have unit norm.

We propose an estimator based on the following fact.

Proposition 2

$$\mathbf{A} = \left[\bar{\bar{\mathbf{S}}}^H \mathbf{U}_s (\mathbf{\Lambda}_s - \eta^2 \mathbf{I}_K)^{-1} \mathbf{U}_s^H \bar{\bar{\mathbf{S}}} \mathbf{P}^2 \right]^{-\frac{1}{2}} \tag{3.154}$$

where $\mathbf{U}_s$, $\mathbf{\Lambda}_s$ *are signal subspace components derived from an eigendecomposition of the autocorrelation matrix,* $\mathbf{C}_r$, *of the received signal, as in* (3.31) *and* (3.32).

The proof appears in [30].

We may obtain an estimate, $\hat{\mathbf{A}}$, of $\mathbf{A}$ by replacing $\bar{\mathbf{S}}$, $\mathbf{U}_s$, $\mathbf{\Lambda}_s$ and η^2 of (3.154) with their respective estimates obtained from subspace tracking and blind Kalman channel estimation as discussed in Section 3.3.5. The channel phase ambiguity can be circumvented by the use of differential encoding and decoding of the data. After channel and amplitude estimation, the amplitude corrected channel information is then used, along with the known spreading codes, to construct the precoding matrix $\hat{\mathbf{M}}$. Finally, the downlink information bits are differentially encoded and filtered with $\hat{\mathbf{M}}$ before spreading and transmission. At the mobile unit, matched filtering and differential detection are performed to obtain estimates of the downlink information bits.

3.6.1.5 Algorithm Summary   Making use of the sequential adaptive Kalman channel estimator discussed in Section 3.3.5, we can produce an adaptive implementation of transmitter precoding at the base station as follows.

ALGORITHM 3.4

[Sequential adaptive transmitter precoding for synchronous multipath CDMA]
1. Using a suitable signal subspace tracking algorithm, e.g., NAHJ-FST, update the signal subspace components $\mathbf{U}_s[i]$, $\mathbf{\Lambda}_s[i]$, and $\eta^2[i]$ at each time slot i using the uplink signals.
2. Track the channels $\{\mathbf{h}_k\}_{k=1}^K$ as follows:

$$\mathbf{z}[i] = \mathbf{r}[i] - \mathbf{U}_s[i]\mathbf{U}_s[i]^H\mathbf{r}[i]$$

$$\mathbf{x}[i] = \mathbf{S}_k^H\mathbf{z}[i]$$

$$\mathbf{k}[i] = \mathbf{\Sigma}[i-1]\,\mathbf{x}[i]\left(\mathbf{x}[i]^H\mathbf{\Sigma}[i-1]\mathbf{x}[i]\right)^{-1}$$

$$\mathbf{h}_k[i] = \left[\mathbf{h}_k[i-1] - \mathbf{k}[i]\big(\mathbf{x}[i]^H\mathbf{h}_k[i-1]\big)\right]/\|\mathbf{h}_k[i-1]$$

$$\qquad - \mathbf{k}[i]\big(\mathbf{x}[i]^H\mathbf{h}_k[i-1]\big)\|$$

$$\mathbf{\Sigma}[i] = \mathbf{\Sigma}[i-1] - \mathbf{k}[i]\,\mathbf{x}[i]^H\mathbf{\Sigma}[i-1].$$

3. Calculate the channel amplitudes via (3.154) using the channel estimates, the signal subspace parameters, the known spreading codes, and the known transmit powers.
4. Using (3.151) and the information from steps 1–3, calculate $\mathcal{H}$ and set $\mathbf{M} = \mathcal{H}^{-1}$.
5. Differentially encode the downlink bit streams for each user to form $\mathbf{b}[i]$.
6. Transmit the precoded downlink signal $\mathbf{x}[i] = \mathbf{SMb}[i]$.
7. Perform matched filtering and differential detection at the mobile units.

3.6.2 Precoding with Multiple Transmit Antennas

3.6.2.1 Downlink Signal Model Without receiver channel state information (CSI), it is difficult to fully exploit receive antenna diversity because the only diversity combining available at the receiver is (non-coherent) addition of the antenna outputs. This provides no diversity in fading environments [14]. In a block fading environment with transmitter CSI, however, we can employ selection diversity with multiple receive antennas simply by adding a few bits to each frame to instruct the receiver to use the "best" antenna. This possibility notwithstanding, we consider a K-user downlink CDMA system in flat block fading with two transmit antennas and a single receive antenna for each user. Extensions to more than two transmit antennas are straightforward. The discrete-time BPSK modulated signal transmitted from antenna $a \in \{1, 2\}$ is:

$$\mathbf{x}^{(a)} = \alpha \mathbf{S}\mathbf{M}^{(a)}\mathbf{b} \tag{3.155}$$

where, similar to (3.147), the columns of $\mathbf{S} \in \mathbb{C}^{N \times K}$ are the normalized spreading codes of the K users, $\mathbf{b} \in \{\pm 1\}^K$ contains the downlink bits corresponding to the K users, N is the processing gain, and $\mathbf{M}^{(a)} \in \mathbb{C}^{K \times K}$ is a complex precoding matrix used for multiple-access interference (MAI) suppression and transmitter antenna diversity exploitation and is optimized in later sections. The scalar α is a transmit power factor that will be addressed in a later section. For now, we assume $\alpha = 1$.

The goal is to choose $\mathbf{M}^{(1)}$ and $\mathbf{M}^{(2)}$ to optimize downlink performance when no receiver CSI is available and the receiver is constrained to matched filter detection. The precoders must not only suppress interference, but they must also exploit available diversity. We are interested in situations in which we have either perfect or partial CSI available at the transmitter.

3.6.2.2 Precoder Design for Orthogonal Spreading Codes After chip-matched filtering, the noise free received signal at user 1's mobile unit is:

$$\mathbf{r}_1 = h_1^{(1)}\mathbf{x}^{(1)} + h_1^{(2)}\mathbf{x}^{(2)} \tag{3.156}$$

$$= h_1^{(1)}\mathbf{S}\mathbf{M}^{(1)}\mathbf{b} + h_1^{(2)}\mathbf{S}\mathbf{M}^{(2)}\mathbf{b} \tag{3.157}$$

where $h_b^{(a)}$ is the complex channel gain between transmit antenna a and user b's mobile unit and where we have set $\alpha = 1$. The channel gains are assumed mutually independent. The mobile units are restricted to (spreading-code) matched filter detection. If $\mathbf{s}_1 \triangleq [\mathbf{S}]_{:,1}$ and we have orthogonal spreading codes, the decision statistic for user 1 is:

$$d_1 = \mathbf{s}_1^H\mathbf{r}_1 + \eta n_1 \tag{3.158}$$

$$= h_1^{(1)}\left[\mathbf{M}^{(1)}\mathbf{b}\right]_1 + h_1^{(2)}\left[\mathbf{M}^{(2)}\mathbf{b}\right]_1 + \eta n_1 \tag{3.159}$$

where $n_1 \sim \mathcal{N}_c(0, 1)$ and is independent of $\mathbf{b}$ and the channel and η^2 is the noise power. For now, define $\mathbf{M}^{(a)}$, $a \in \{1, 2\}$ to be diagonal matrices whose elements are given by:

$$\left[\mathbf{M}^{(a)}\right]_{i,i} = \frac{h_i^{(a)*}}{\sqrt{\left|h_i^{(1)}\right|^2 + \left|h_i^{(2)}\right|^2}} \tag{3.160}$$

Then we have:

$$d_1 = \sqrt{\left|h_1^{(1)}\right|^2 + \left|h_1^{(2)}\right|^2} b_1 + \eta n_1 \tag{3.161}$$

and the corresponding bit estimate is:

$$\hat{b}_1 = \text{sign}\{\text{Re}[d_1]\} \tag{3.162}$$

This achieves an instantaneous SNR of:

$$\text{SNR}_1 = \frac{\left|h_1^{(1)}\right|^2 + \left|h_1^{(2)}\right|^2}{\eta^2} \tag{3.163}$$

which provides full two branch diversity for every user and has a $\mathcal{X}_4^2$ distribution when the channel gains are independent complex Gaussian random variables. Precoding for this scenario reduces to maximal ratio weighting [14], which has the same performance as beamforming to a single receive antenna. Note that $\mathbf{M}^{(1)}$, $\mathbf{M}^{(2)}$ in (3.160) are normalized in the sense that the sum of the average (with respect to $\mathbf{b}$) transmit power from both antennas is K. That is:

$$E_{\mathbf{b}}\left\{\left\|\mathbf{SM}^{(1)}\mathbf{b}\right\|^2\right\} + E_{\mathbf{b}}\left\{\left\|\mathbf{SM}^{(2)}\mathbf{b}\right\|^2\right\} = \text{tr}\left(\mathbf{M}^{(1)H}\mathbf{M}^{(1)}\right)$$

$$+ \text{tr}\left(\mathbf{M}^{(2)H}\mathbf{M}^{(2)}\right) = K \tag{3.164}$$

for every channel realization. Therefore, we can set $\alpha = 1$ in (3.155).

3.6.2.3 Precoder Design for Non-Orthogonal Spreading Codes

Let $\boldsymbol{\rho}_1^T \triangleq \mathbf{s}_1^H \mathbf{S}$. The decision statistic for user 1, assuming for the moment that $\alpha = 1$, is:

$$d_1 = \underbrace{h_1^{(1)} \boldsymbol{\rho}_1^T \mathbf{M}^{(1)} \mathbf{b}}_{d_1^{(1)}} + \underbrace{h_1^{(2)} \boldsymbol{\rho}_1^T \mathbf{M}^{(2)} \mathbf{b}}_{d_1^{(2)}} + \eta n_1 \tag{3.165}$$

Our goal is to choose $\mathbf{M}^{(1)}$, $\mathbf{M}^{(2)}$ to maximize the collective performance of all users in some sense, assuming no receiver CSI and matched filter detection. We form MMSE cost functions for the optimization of $\mathbf{M}^{(1)}$, $\mathbf{M}^{(2)}$ by stacking $d_k^{(1)}$ ($1 \leq k \leq K$) and $d_k^{(2)}$ ($1 \leq k \leq K$), respectively. The result for transmit antenna $a \in \{1, 2\}$ is:

$$
J^{(a)} = E \left[\left\| \begin{bmatrix} |h_1^{(a)}|^2 \left(|h_1^{(1)}|^2 + |h_1^{(2)}|^2 \right)^{-\frac{1}{2}} b_1 \\ |h_2^{(a)}|^2 \left(|h_2^{(1)}|^2 + |h_2^{(2)}|^2 \right)^{-\frac{1}{2}} b_2 \\ \vdots \\ |h_K^{(a)}|^2 \left(|h_K^{(1)}|^2 + |h_K^{(2)}|^2 \right)^{-\frac{1}{2}} b_K \end{bmatrix} - \begin{bmatrix} h_1^{(a)} \boldsymbol{\rho}_1^T \\ h_2^{(a)} \boldsymbol{\rho}_2^T \\ \vdots \\ h_K^{(a)} \boldsymbol{\rho}_K^T \end{bmatrix} \mathbf{M}^{(a)} \mathbf{b} - \begin{bmatrix} \eta n_1 \\ \eta n_2 \\ \vdots \\ \eta n_K \end{bmatrix} \right\|^2 \right]
\tag{3.166}
$$

$$
= E \left[\| \mathbf{D}^{(a)} \mathbf{b} - \mathbf{H}^{(a)} \mathbf{R} \mathbf{M}^{(a)} \mathbf{b} - \mathbf{n} \|^2 \right]
\tag{3.167}
$$

where:

$$
\mathbf{D}^{(a)} \triangleq \mathrm{diag}\left(|h_1^{(a)}|^2 \left[|h_1^{(1)}|^2 + |h_1^{(2)}|^2 \right]^{-\frac{1}{2}}, |h_2^{(a)}|^2 \left[|h_2^{(1)}|^2 + |h_2^{(2)}|^2 \right]^{-\frac{1}{2}}, \dots \right.
$$

$$
\left. |h_K^{(a)}|^2 \left[|h_K^{(1)}|^2 + |h_K^{(2)}|^2 \right]^{-\frac{1}{2}} \right)
\tag{3.168}
$$

$$
\mathbf{H}^{(a)} \triangleq \mathrm{diag}\left(h_1^{(a)}, h_2^{(a)}, \dots, h_K^{(a)} \right) \quad a \in \{1, 2\}, \quad \mathbf{R} \triangleq \mathbf{S}^H \mathbf{S}
\tag{3.169}
$$

$$
\mathbf{n} \triangleq \begin{bmatrix} \eta n_1 \\ \eta n_2 \\ \vdots \\ \eta n_K \end{bmatrix}
\tag{3.170}
$$

and where the expectations are with respect to $\mathbf{n}$ and $\mathbf{b}$. At this stage, the cost functions implicitly assume that the channel gains are deterministic and known at the transmitter.

The motivation behind the construction of the cost functions is self evident except, perhaps, for the presence of $\mathbf{D}^{(1)}$ and $\mathbf{D}^{(2)}$. This is related to the transmit power constraint and power loading. If we allow for an infinite peak-to-average power ratio at the transmitter, we can replace $\mathbf{D}^{(1)}$ and $\mathbf{D}^{(2)}$ in (3.167) with $\mathbf{I}_K$ and the resulting optimal precoding matrix will completely eliminate the detrimental effects of fading.[2] Because real transmitters cannot operate with an infinite dynamic range, this is not a reasonable assumption. Therefore, we will insist that the sum of the average (with respect to $\mathbf{b}$) transmit power from all antennas be equal to the number of users. For diversity

[2]The precoding matrix for this situation can be found using (3.171)–(3.172) and by solving for α using the procedure in Section 3.6.4.2. Essentially, the transmitter will increase power (perhaps without bound) during fades and decrease power during channel peaks, resulting in infinite peak-to-average transmit power.

transmission (instead of multiplexing [58]) with this power constraint and orthogonal codes, the best precoding scheme is maximum ratio weighting, as in 3.6.2.2. It is therefore important that the precoding matrices that minimize the cost functions $J^{(1)}$, $J^{(2)}$ reduce to (3.160) when spreading codes are orthogonal. It is easy to verify that this is true when $\mathbf{D}^{(a)}$ satisfies (3.168).

Proposition 3   *The choice of* $\mathbf{M}^{(1)}$ *that minimizes* $J^{(1)}$ *and the choice of* $\mathbf{M}^{(2)}$ *that minimizes* $J^{(2)}$ *are given by*:

$$\mathbf{M}^{(1)} = \mathbf{R}^{-1}\left[\mathbf{H}^{(1)}\right]^{-1}\mathbf{D}^{(1)} \tag{3.171}$$

$$\mathbf{M}^{(2)} = \mathbf{R}^{-1}\left[\mathbf{H}^{(2)}\right]^{-1}\mathbf{D}^{(2)} \tag{3.172}$$

The proof appears in [32].

These results show that optimal precoding with perfect transmitter CSI and non-orthogonal codes is maximum ratio weighting followed by transmitter-based decorrelation. Precoder design when partial channel knowledge (defined as quantities statistically dependent upon the true channel) is available at the transmitter is discussed in [32].

3.6.3   Precoding for Multipath ISI Channels

The conventional technique for diversity exploitation in multipath is RAKE reception, i.e., maximum ratio combining of each path at the receiver. Because we are considering applications that do not allow for receiver channel information, we will, instead, use a form of pre-rake diversity combining. We will see that Propositions 1 and 2 can be applied with minor modifications to fully exploit multipath and transmit antenna diversity while completely eliminating multiple-access interference.

3.6.3.1   *Prerake-Diversity Combining*   We will assume a synchronous, block fading, L-path multipath channel [6] with no intersymbol interference,[3] where the impulse response between transmit antenna a and user k's receive antenna can be modelled as:

$$h_k^{(a)}(t) = \sum_{l=0}^{L-1} h_k^{(a)}[l]\delta(t - lT_c) \tag{3.173}$$

where $\left\{h_k^{(a)}[l]\right\}_{l=0}^{L-1}$ is a set of *i.i.d.* complex Gaussian random variables and T_c is the chip duration, i.e., the symbol duration divided by the processing gain. Synchronism is a reasonable assumption for downlink transmissions (where precoding is most practical) and intersymbol interference can be eliminated using guardbands or it can simply be neglected if the channel delay spread is small relative to the symbol interval.

---

[3]Inter-chip interference constitutes the multipath interference in this model and in Figure 3.13.

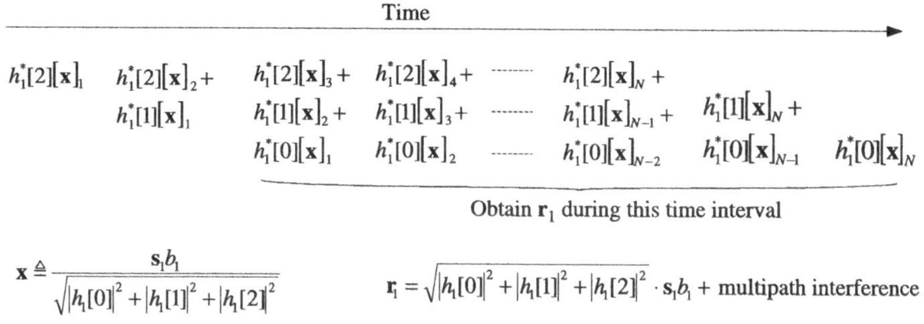

Time

$$h_1^*[2][\mathbf{x}]_1 \quad h_1^*[2][\mathbf{x}]_2 + \quad h_1^*[2][\mathbf{x}]_3 + \quad h_1^*[2][\mathbf{x}]_4 + \quad \cdots \cdots \quad h_1^*[2][\mathbf{x}]_N +$$
$$h_1^*[1][\mathbf{x}]_1 \quad h_1^*[1][\mathbf{x}]_2 + \quad h_1^*[1][\mathbf{x}]_3 + \quad \cdots \cdots \quad h_1^*[1][\mathbf{x}]_{N-1} + \quad h_1^*[1][\mathbf{x}]_N +$$
$$h_1^*[0][\mathbf{x}]_1 \quad h_1^*[0][\mathbf{x}]_2 \quad \cdots \cdots \quad h_1^*[0][\mathbf{x}]_{N-2} \quad h_1^*[0][\mathbf{x}]_{N-1} \quad h_1^*[0][\mathbf{x}]_N$$

Obtain $\mathbf{r}_1$ during this time interval

$$\mathbf{x} \triangleq \frac{\mathbf{s}_1 b_1}{\sqrt{|h_1[0]|^2 + |h_1[1]|^2 + |h_1[2]|^2}} \qquad \mathbf{r}_1 = \sqrt{|h_1[0]|^2 + |h_1[1]|^2 + |h_1[2]|^2} \cdot \mathbf{s}_1 b_1 + \text{multipath interference}$$

Figure 3.13. Pre-rake diversity combining (precoding) for a single-antenna, single user CDMA signal in a 3-path multipath channel with processing gain N. The N elements of $\mathbf{x}$ are scaled and transmitted over $N + 2$ chip intervals. The desired portion of the received signal $\mathbf{r}_1$ indicates full diversity is achievable if the multipath/interchip interference can be suppressed via additional precoding.

The general idea behind pre-rake diversity combining [5] is to transmit precoded versions of the chip stream $\mathbf{Sb} \in \mathbb{C}^{N \times 1}$ during L consecutive chip intervals so that after the L-th transmission, all paths add up coherently at the receiver. Figure 3.13 illustrates the approach for the single antenna, single user case in a 3-path channel. The discrete-time transmitted signal for this scenario is:

$$[\tilde{\mathbf{x}}]_i = \left(\sum_{l=0}^{L-1} |h_1[l]|^2 \right)^{-\frac{1}{2}} \sum_{l=0}^{L-1} h_1^*[L-1-l] \cdot [\mathbf{s}_1 b_1]_{i-l}, \quad i = 1, 2, \ldots, N+L-1 \quad (3.174)$$

The desired portion of the noise-free received signal $\mathbf{r}_1$ is available between relative chip intervals L and $N + L - 1$ and is given by:

$$\mathbf{r}_1 = \left(\sum_{l=0}^{L-1} |h_1[l]|^2 \right)^{\frac{1}{2}} \mathbf{s}_1 b_1 + \text{multipath/interchip interference} \quad (3.175)$$

In the next section we will show that MMSE precoding for a K-user system can fully exploit multipath and transmit antenna diversity for every user while completely eliminating multipath and multiuser interference.

3.6.3.2 Precoder Design

For a K-user multi-antenna CDMA system using pre-rake diversity combining, the discrete-time signal transmitted from antenna a is:

$$\mathbf{x}^{(a)} = \sum_{l=0}^{L-1} \tilde{\mathbf{S}}[l] \mathbf{M}^{(a)}[l] \mathbf{b} \quad (3.176)$$

where $\mathbf{x} \in \mathbb{C}^{(N+L-1) \times 1}$, $\mathbf{M}^{(a)}[l]$ is the precoding matrix for transmit antenna a and path l, and $\tilde{\mathbf{S}}[l]$ is defined by:

$$\tilde{\mathbf{S}}[l] \triangleq \begin{bmatrix} \mathbf{0}_{L-1-l,K} \\ \mathbf{S} \\ \mathbf{0}_{l,K} \end{bmatrix} \tag{3.177}$$

Then we have:

$$\mathbf{x}^{(a)} = \breve{\mathbf{S}} \mathcal{M}^{(a)} \mathbf{b} \tag{3.178}$$

where:

$$\mathcal{M}^{(a)} \triangleq \begin{bmatrix} \mathbf{M}^{(a)}[0] \\ \mathbf{M}^{(a)}[1] \\ \vdots \\ \mathbf{M}^{(a)}[L-1] \end{bmatrix}_{KL \times K} , \quad \breve{\mathbf{S}} \triangleq \left[\tilde{\mathbf{S}}[0] \; \tilde{\mathbf{S}}[1] \cdots \tilde{\mathbf{S}}[L-1] \right]_{(N+L-1) \times KL} \tag{3.179}$$

so that the total average transmit power from antenna a required to send a single symbol vector is:

$$P^{(a)} = E_{\mathbf{b}} \left[\left\| \mathbf{x}^{(a)} \right\|^2 \right] \tag{3.180}$$

$$= \text{tr} \left(\mathcal{M}^{(a)H} \breve{\mathbf{S}}^H \breve{\mathbf{S}} \mathcal{M}^{(a)} \right) \tag{3.181}$$

The noise free received signal at user 1's mobile unit due to the signal transmitted from antenna a is given by:

$$\mathbf{r}_1^{(a)} = \sum_{l=0}^{L-1} \sum_{i=0}^{L-1} h_1^{(a)}[i] \mathbf{S}[i-l] \mathbf{M}^{(a)}[l] \, \mathbf{b} \tag{3.182}$$

$$= \sum_{l=0}^{L-1} h_1^{(a)}[l] \mathbf{S} \mathbf{M}^{(a)}[l] \, \mathbf{b} + \underbrace{\sum_{l=0}^{L-1} \sum_{\substack{i=0 \\ i \neq l}}^{L-1} h_1^{(a)}[i] \mathbf{S}[i-l] \mathbf{M}^{(a)}[l] \, \mathbf{b}}_{\text{inter-chip interference}} \tag{3.183}$$

where $\mathbf{S}[p] \in \mathbb{C}^{N \times K}$ is a matrix of p-shifted spreading codes with zero padding. If $p = 1$, for example, the k-th column of $\mathbf{S}[p]$ is $[0[\mathbf{s}_k]_1[\mathbf{s}_k]_2 \cdots [\mathbf{s}_k]_{N-1}]^T$. For negative p, the spreading codes are shifted up and zeros are inserted at the bottom of the matrix.

As before, we assume matched filter detection so that the decision statistic for user 1 due to the signal transmitted from antenna a is:

$$d_1^{(a)} = \mathbf{s}_1^H \mathbf{r}_1^{(a)} \tag{3.184}$$

Stacking decision statistics from each user, we have:

$$\mathbf{d}^{(a)} \triangleq \left[d_1^{(a)} \ d_2^{(a)} \cdots d_K^{(a)} \right]^T \tag{3.185}$$

$$= \sum_{i=0}^{L-1} \sum_{l=0}^{L-1} \mathbf{H}^{(a)}[i]\mathbf{R}[i-l]\mathbf{M}^{(a)}[l]\mathbf{b} \tag{3.186}$$

$$= \mathcal{H}^{(a)}\mathcal{R}\mathcal{M}^{(a)}\mathbf{b} \tag{3.187}$$

where:

$$\mathbf{H}^{(a)}[i] \triangleq \mathrm{diag}\left(h_1^{(a)}[i], h_2^{(a)}[i], \ldots, h_k^{(a)}[i] \right) \tag{3.188}$$

$$\mathbf{R}[i-l] \triangleq \mathbf{S}^H\mathbf{S}[i-l] \tag{3.189}$$

$$\mathcal{H}^{(a)} \triangleq \left[\mathbf{H}^{(a)}[0] \ \mathbf{H}^{(a)}[1] \ \cdots \ \mathbf{H}^{(a)}[L-1] \right]_{K \times KL} \tag{3.190}$$

and:

$$\mathcal{R} \triangleq \begin{bmatrix} \mathbf{R}[0] & \mathbf{R}[-1] & \cdots & \mathbf{R}[-(L-1)] \\ \mathbf{R}[1] & \ddots & & \ddots \\ & & \ddots & & \mathbf{R}[-1] \\ \mathbf{R}[L-1] & & \mathbf{R}[1] & \mathbf{R}[0] \end{bmatrix}_{KL \times KL} \tag{3.191}$$

The cost function for determining the optimal precoder supermatrix $\mathcal{M}^{(a)}$ is formed as:

$$J_{\mathrm{mp}}^{(a)} = E\left[\left\| \mathcal{D}^{(a)}\mathbf{b} - \mathcal{H}^{(a)}\mathcal{R}\mathcal{M}^{(a)}\mathbf{b} - \mathbf{n} \right\|^2 \right] \tag{3.192}$$

where $\mathcal{D}^{(a)}$ is a diagonal power loading matrix whose elements are given by:

$$\left[\mathcal{D}^{(a)} \right]_{i,i} = \frac{\displaystyle\sum_{l=0}^{L-1} \left| h_i^{(a)}[l] \right|^2}{\left[\displaystyle\sum_{a=1}^{2} \sum_{j=0}^{L-1} \left| h_i^{(a)}[j] \right|^2 \right]^{\frac{1}{2}}} \tag{3.193}$$

The optimal precoding supermatrix for antenna a, assuming perfect or partial channel information, can be found using straightforward modifications of Propositions 1 and 2.

Proposition 4 *The precoding supermatrix* $\mathcal{M}^{(a)}$ *that minimizes* $J_{\text{mp}}^{(a)}$ *satisfies*:

$$\mathcal{M}^{(a)} = \left[\mathcal{H}^{(a)}\mathcal{R}\right]^{\dagger}\mathcal{D}^{(a)} \tag{3.194}$$

for $a \in \{1, 2\}$.

The proof appears in [32]. The individual precoding matrices $\{\mathbf{M}^{(a)}[l]\}_{l=0}^{L-1}$ can be found from the optimal precoding supermatrix via:

$$\mathbf{M}^{(a)}[l] \overset{\text{set}}{=} \mathcal{M}_{Kl+1:Kl+K,:}^{(a)} \tag{3.195}$$

3.6.4 Performance Analyses

3.6.4.1 Performance of Transmitter Precoding with Blind Channel Estimation
Previously in this chapter, we developed analytical tools to investigate the performance of blind and group-blind linear MMSE multiuser detection. In this section, we adapt these tools to the analysis of single-antenna transmitter precoding with blind channel estimation. In particular, we will derive SINR and BER expressions that take residual multiple-access interference and channel estimation error into account.

Notice that the estimate $\hat{\mathbf{M}}$ given by (3.153) is not a consistent estimate of $\mathbf{M}$ because of the unknown phase and scaling factors. However, there is a diagonal matrix $\mathbf{\Phi}$ so that $\hat{\mathbf{M}}\mathbf{\Phi}^{-1}$ is a consistent estimate. The matrix $\mathbf{\Phi}$ is of the form:

$$\mathbf{\Phi} \triangleq \text{diag}\left(\|\mathbf{h}_1\|e^{\jmath\phi_1}, \|\mathbf{h}_2\|e^{\jmath\phi_2}, \ldots, \|\mathbf{h}_K\|e^{\jmath\phi_K}\right) \tag{3.196}$$

where ϕ_k, $k = 1, \ldots, K$ are phase factors that depend on how the estimation is implemented. With this in mind, we state the following result, which is proved in [32].

Theorem 3 *Let* $\hat{\mathbf{M}}$ *be given by (3.153), and let* $\mathbf{b}$ *be i.i.d. QPSK symbols independent of* $\hat{\mathbf{M}}$. *Then*:

$$\sqrt{M}\left([\hat{\mathbf{M}}\mathbf{\Phi}^{-1} - \mathbf{M}]\mathbf{b}\right) \longrightarrow \mathcal{N}_c(\mathbf{0}, \ \mathbf{C}_m) \text{ in distribution as } M \to \infty$$

with:

$$\mathbf{C}_m = \mathcal{H}^{-1}\mathbf{D}\mathcal{H}^{-H} \tag{3.197}$$

where the diagonal elements of $\mathbf{D}$ *are given by*:

$$[\mathbf{D}]_{i,i} = \beta_i \sum_{k=1}^{K} \sum_{l=1}^{K} [\mathcal{H}^{-1}\mathcal{H}^{-H}]_{k,l}\mathbf{s}_i^H\mathbf{S}_k\mathbf{Q}_i^{\dagger}\mathbf{S}_l^H\mathbf{s}_i \tag{3.198}$$

and:

$$\beta_i \triangleq \eta^2\mathbf{h}_i^H\mathbf{U}_s\mathbf{\Lambda}_s(\mathbf{\Lambda}_s - \eta\mathbf{I}_K)^{-2}\mathbf{U}_s^H\mathbf{h}_i \tag{3.199}$$

while the off-diagonal elements can be ignored with good accuracy. Here $\mathbf{Q}_i^\dagger$ *denotes the Moore-Penrose generalized inverse* [17] *of the matrix* $\mathbf{Q}_i \triangleq \mathbf{S}_i^H \mathbf{U}_n \mathbf{U}_n^H \mathbf{S}_i$.

The proof appears in [30].

The SINR at the output of the matched filter for User 1 is given by [18,19]:

$$\text{SINR} \triangleq \frac{|E\{\mathbf{s}_1^H \mathbf{r}_1[i] \mid b_1[i]\}|^2}{E\{\text{Var}(\mathbf{s}_1^H \mathbf{r}_1[i] \mid b_1[i])\}} \tag{3.200}$$

Now suppose that the phase and amplitude factors in $\mathbf{\Phi}$ have been determined. Write the estimated matrix, $\hat{\mathbf{M}}$, as $\hat{\mathbf{M}}\mathbf{\Phi}^{-1} = \mathbf{M} + \Delta\mathbf{M}$, where $\Delta\mathbf{M}$ is the estimation error. Dropping the time index for clarity, the received signal can then be written as:

$$r_1 = \mathbf{s}_1^H \mathbf{r}_1 \tag{3.201}$$

$$= (\mathbf{s}_1^H \mathbf{H}_1)\hat{\mathbf{M}}\mathbf{\Phi}^{-1}\mathbf{b} + \mathbf{s}_1^H \mathbf{n}_1 \tag{3.202}$$

$$= (\mathbf{s}_1^H \mathbf{H}_1)\mathbf{M}\mathbf{b} + (\mathbf{s}_1^H \mathbf{H}_1)\Delta\mathbf{M}\mathbf{b} + \mathbf{s}_1^H \mathbf{n}_1 \tag{3.203}$$

$$= (\mathbf{s}_1^H \mathbf{H}_1)[\mathbf{M}]_{:,1}b_1 + (\mathbf{s}_1^H \mathbf{H}_1)[\mathbf{M}]_{:,2:K}[\mathbf{b}]_{2:K} + (\mathbf{s}_1^H \mathbf{H}_1)\Delta\mathbf{M}\mathbf{b} + \mathbf{s}_1^H \mathbf{n}_1 \tag{3.204}$$

where the notation $[\mathbf{M}]_{:,2:K}$ indicates the matrix composed of columns two through K of the matrix $\mathbf{M}$. According to Theorem 3, for large M the third term in (3.204) is also Gaussian distributed (independent of the other terms) with variance:

$$v_1^2 = \frac{1}{M}(\mathbf{s}_1^H \mathbf{H}_1)\mathbf{C}_m \mathbf{H}_1^H \mathbf{s}_1 \tag{3.205}$$

Since $\mathbf{M}$ represents an MMSE detector, we can also make the approximate assumption that the multiple-access interference is Gaussian distributed [25]. We can therefore calculate the BER via a single Q-function as:

$$P_b(e) \cong Q(\sqrt{\text{SINR}}) \tag{3.206}$$

with:

$$\text{SINR} = \frac{[(\mathbf{s}_1^H \mathbf{H}_1)[\mathbf{M}]_{:,1}]^2}{\sum_{k=2}^{K} |(\mathbf{s}_1^H \mathbf{H}_1)[\mathbf{M}]_{:,k}|^2 + \eta^2 \|\mathbf{s}_1\|^2 + \frac{1}{M}(\mathbf{s}_1^H \mathbf{H}_1)\mathbf{C}_m \mathbf{H}_1^H \mathbf{s}_1} \tag{3.207}$$

Notice that the first term in the denominator of the SINR expression is due to residual multiple-access interference. The second term is the ambient noise, and the third term is due to the channel estimation error.

A comparison of these results with simulation appears in [30], where good agreement is found.

3.6.4.2 Performance and Achievable Diversity for Multi-Antenna Precoding
We have seen that with perfect channel knowledge at the transmitter and orthogonal spreading codes, we can achieve full transmit diversity with precoding. We will see here that non-zero spreading code crosscorrelations lead to an SNR loss, but full diversity is still achievable.

The General Case   Stacking decision statistics from all users obtained using the optimal $\mathbf{M}^{(1)}, \mathbf{M}^{(2)}$, we define the composite received signal as:

$$\mathbf{r} \triangleq \begin{bmatrix} d_1 \\ d_2 \\ \vdots \\ d_K \end{bmatrix} \tag{3.208}$$

$$= \alpha \left[\mathbf{H}^{(1)} \mathbf{R} \mathbf{M}^{(1)} + \mathbf{H}^{(2)} \mathbf{R} \mathbf{M}^{(2)} \right] \mathbf{b} + \eta \mathbf{n} \tag{3.209}$$

$$= \alpha \mathbf{E} \mathbf{b} + \eta \mathbf{n} \tag{3.210}$$

where:

$$\mathbf{E} \triangleq \operatorname{diag}\left(\sqrt{|h_1^{(1)}|^2 + |h_1^{(2)}|^2}, \sqrt{|h_2^{(1)}|^2 + |h_2^{(2)}|^2}, \ldots, \sqrt{|h_K^{(1)}|^2 + |h_K^{(2)}|^2} \right) \tag{3.211}$$

Because $\mathbf{E}$ is diagonal, multiple access interference is completely eliminated. Furthermore, we have seen in Section 3.6.2.2 that with orthogonal codes, we can set $\alpha = 1$ (i.e., no transmit power adjustment is necessary) and achieve full transmit diversity. In general, however, we must set $\alpha \leq 1$ to constrain average transmit power. For our purposes, average transmit power normalization requires:

$$\alpha^2 E_{\mathbf{b}}\left[\|\mathbf{S}\mathbf{M}^{(1)}\mathbf{b}\|^2 \right] + \alpha^2 E_{\mathbf{b}}\left[\|\mathbf{S}\mathbf{M}^{(2)}\mathbf{b}\|^2 \right] = K \tag{3.212}$$

for every channel realization. That is, the sum of the average transmit power from the two antennas is equal to the number of users. Dropping the antenna superscripts for notational convenience, we have:

$$E_{\mathbf{b}}\left[\|\mathbf{S}\mathbf{M}\mathbf{b}\|^2 \right] = \operatorname{tr}\left(\mathbf{M}^H \mathbf{R} \mathbf{M} \right) \tag{3.213}$$

$$= \operatorname{tr}\left[(\mathbf{H}^{-1}\mathbf{D})^H \mathbf{R}^{-1} (\mathbf{H}^{-1}\mathbf{D}) \right] \tag{3.214}$$

For transmit antenna $a \in \{1, 2\}$, the diagonal structures of $\mathbf{D}$ and $\mathbf{H}$ yield:

$$\operatorname{tr}\left(\mathbf{M}^{(a)H} \mathbf{R} \mathbf{M}^{(a)} \right) = \sum_{i=1}^{K} \left[\mathbf{R}^{-1} \right]_{i,i} \frac{|h_i^{(a)}|^2}{|h_i^{(1)}|^2 + |h_i^{(2)}|^2} \tag{3.215}$$

Summing the average transmit power contributions from each antenna, we have:

$$\mathrm{tr}\!\left(\mathbf{M}^{(1)H}\mathbf{R}\mathbf{M}^{(1)}\right) + \mathrm{tr}\!\left(\mathbf{M}^{(2)H}\mathbf{R}\mathbf{M}^{(2)}\right) = \sum_{i=1}^{K} \left[\mathbf{R}^{-1}\right]_{i,i} \tag{3.216}$$

$$= \sum_{i=1}^{K} \frac{1}{\lambda_i} \tag{3.217}$$

where $\{\lambda_i\}_{i=1}^{K}$ are the eigen-values of $\mathbf{R}$. Therefore, by (3.212), the power scaling factor α must satisfy:

$$\alpha = \sqrt{\frac{1}{\dfrac{1}{K}\displaystyle\sum_{i=1}^{K}\dfrac{1}{\lambda_i}}} \tag{3.218}$$

Notice that α^2 is simply the inverse of the average of the diagonal elements of $\mathbf{R}^{-1}$. It is interesting to relate this result to the performance of receiver-based decorrelating multiuser detection (MUD), in which the performance of user k is dependent upon the inverse of $[\mathbf{R}^{-1}]_{k,k}$, but is not dependent upon the other diagonal elements of $\mathbf{R}^{-1}$. In this sense, we can think of the performance of precoding, which is the same for every user, as the performance of decorrelating MUD "averaged" over every user. This interpretation is supported by the simulation results reported in [45].

Assuming all channel gains are independent and have the same statistics, the average bit-error-probability (BEP) of every user will be the same and is given by [29, p. 825]:

$$\overline{\mathrm{Pr}}(\varepsilon) = E\!\left[Q\!\left(\frac{\alpha}{\eta}\sqrt{\left|h_1^{(1)}\right|^2 + \left|h_1^{(2)}\right|^2}\right)\right] \tag{3.219}$$

$$= \frac{1}{4}\left(\mu^3 - 3\mu + 2\right) \tag{3.220}$$

where:

$$\mu \triangleq \sqrt{\frac{\gamma}{1+\gamma}}, \quad \gamma \triangleq \frac{\xi_h \alpha^2}{2\eta^2}, \quad \xi_h \triangleq E\!\left[\left|h_i^{(a)}\right|^2\right], \quad a = 1, 2;\ i = 1, 2, \ldots, K \tag{3.221}$$

This performance is the same as two-branch maximum ratio combining with an SNR penalty of $10\log_{10}\alpha^2$ dB. Hence diversity, defined here as the slope of the BEP curve, is unaffected by signature waveform crosscorrelations, but we do suffer SNR loss.

Equicorrelated Spreading Codes   As an important special case, we consider the scenario in which the normalized spreading code crosscorrelations satisfy:

$$
\mathbf{s}_k^H \mathbf{s}_l = \begin{cases} 1 & k = l \\ \rho & k \neq l \end{cases} \tag{3.222}
$$

for some $\rho \in [0, 1)$. It is easy to show using the matrix inversion lemma [17, p. 19] that:

$$
\mathbf{R}^{-1} = \frac{1}{1-\rho}\mathbf{I}_K - \underbrace{\frac{\rho}{(1-K)\rho^2 + (K-2)\rho + 1}}_{\check{\rho}}\mathbf{1}_{K,K} \tag{3.223}
$$

which yields:

$$
\operatorname{tr}\left(\mathbf{M}^{(1)H}\mathbf{R}\mathbf{M}^{(1)}\right) + \operatorname{tr}\left(\mathbf{M}^{(2)H}\mathbf{R}\mathbf{M}^{(2)}\right) = K\left(\frac{1}{1-\rho} - \check{\rho}\right) \tag{3.224}
$$

and:

$$
\alpha(\rho, K) = \left[\frac{1}{1-\rho} - \frac{\rho}{(1-K)\rho^2 + (K-2)\rho + 1}\right]^{-\frac{1}{2}} \tag{3.225}
$$

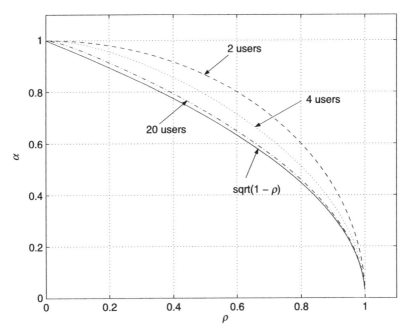

Figure 3.14. The power scaling factor, α as function of the signature crosscorrelations, ρ, for different numbers of users.

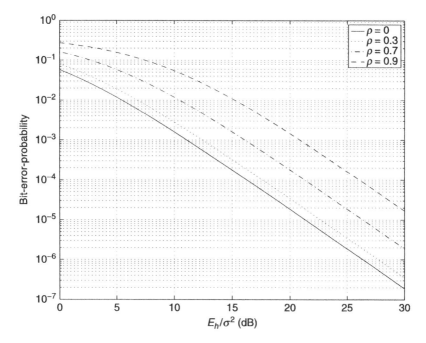

Figure 3.15. The bit-error-probability for the equicorrelated spreading code case, averaged over all 20 users and their channel gains, versus E_h/η^2 for precoding with two transmit antennas and one receiver antenna and for crosscorrelation values of $\rho = 0$, 0.3, 0.7, 0.9. The transmit energy per user per bit is one.

Clearly:

$$\lim_{K \to \infty} \alpha(\rho, K) = \sqrt{1 - \rho} \qquad (3.226)$$

In fact, $\alpha(\rho, K)$ tends to its limit rather quickly, as we see from Figure 3.14, which plots $\alpha(\rho, K)$ as a function of ρ, for several values of K. Note that ρ is, implicitly, a function of the number of users, K, and the processing gain, N. In order to increase K while maintaining a constant ρ, the processing gain (and, hence, the bandwidth) will, in general, have to be increased as well.

For a moderate or large number of users, the performance is nearly equivalent to two-branch maximum ratio combining with a SNR penalty of $10 \log_{10}(1 - \rho)$ dB. Figure 3.15 illustrates the BEP performance, calculated using (3.220), for 20 users and for various values of the crosscorrelation parameter ρ. The $10 \log_{10}(1 - \rho)$ dB SNR penalty is clearly visible.

3.7 CONCLUSION

In this chapter, we have treated the three main problems arising in wireless systems—dynamism, fading and interference—in a unified setting. We have examined several power approaches to these problems, including subspace methods, Markov chain Monte Carlo, the sequential EM algorithm, and transmitter-based techniques. We have seen that these methods can perform quite well, thereby allowing for effective signal reception in multiuser fading environments.

REFERENCES

1. M. Brandt-Pearce and A. Dharap. Transmitter-based multiuser interference rejection for the down-link of a wireless CDMA system in a multipath environment. *IEEE J. Select. Areas Commun.*, 18(3):407–417, Mar. 2000.

2. K. S. Chan. Asymptotic behavior of the Gibbs sampler. *J. Amer. Stat. Assoc.*, 88:320–326, 1993.

3. B. Chen, C. Tsai, and C. Hsu. Robust adaptive MMSE/DFE multiuser detection in multi-path fading channel with impulse noise. *IEEE Trans. Sig. Proc.*, 53(1): 306, Jan. 2005.

4. U. Erez, S. Shamai (Shitz), and R. Zamir. Capacity and lattice-strategies for cancelling known interference. Submitted to *IEEE Trans. Inform. Theory*, 2002.

5. R. Esmailzadeh, E. Sourour, and M. Nakagawa. Pre-rake diversity combining in time-division duplex CDMA mobile communications. *IEEE Trans. Vehicular Tech.*, 48(3):795–801, May 1999.

6. J. Evans and D. N. C. Tse. Large system performance of linear multiuser receivers for multi-path fading channels. *IEEE Trans. Inform. Theory*, 46(6):2059–2078, Sept. 2000.

7. J. Evans and D. N. C. Tse. Linear multiuser receivers for multipath fading channels. *IEEE Trans. Inform. Theory*, 46(6):2059–2078, Sep. 2000.

8. V. Fabian. On asymptotic normality in stochastic approximation. *Ann. Math. Stat.*, 39(4):1327–1332, 1968.

9. R. F. H. Fischer, C. Windpassinger, A. Lampe, and J. B. Huber. Tomlinson-Harashima precoding in space-time transmission for low-rate backward channel. In *Proc. International Zurich Seminar on Broadband Communications. Accessing, Transmission, Networking*, pages 7-1–7-6, Zurich, Switzerland, Feb. 2002.

10. L. Frenkel and M. Feder. Recursive expectation-maximization (EM) algorithms for time-varying parameters with applications to multiple target tracking. *IEEE Trans. Sig. Proc.*, 47(2):306–320, Feb. 1999.

11. A. E. Gelfand and A. F. W. Smith. Sampling-based approaches to calculating marginal densities. *J. Amer. Stat. Assoc.*, 85:398–409, 1990.

12. S. Geman and D. Geman. Stochastic relaxation, Gibbs distribution, and the Bayesian restoration of images. *IEEE Trans. Pattern Anal. Machine Intell.*, PAMI-6(11):721–741, Nov. 1984.

13. W. R. Gilks, S. Richardson, and D. J. Spiegelhalter. *Markov Chain Monte Carlo in Practice*. Chapman & Hall, 1995.

14. B. Hochwald, T. L. Marzetta, and C. B. Papadias. A transmitter diversity scheme for wide-band CDMA systems based on space-time spreading. *IEEE J. Select. Areas Commun.*, 19(1):48–60, Jan. 2001.

15. B. M. Hochwald and T. L. Marzetta. Unitary space-time modulation for multiple-antenna communications in Rayleigh flat fading. *IEEE Trans. Inform. Theory*, 46(2):543–564, March 2000.

16. B. M. Hochwald and W. Sweldens. Differential unitary space-time modulation. *IEEE Trans. Commun.*, 48(12):2041–2052, Dec. 2000.

17. R. A. Horn and C. R. Johnson. *Matrix Analysis*. Cambridge, UK: Cambridge University Press, 1985.

18. A. Høst-Madsen and X. Wang. Performance of blind and group-blind multiuser detection. *IEEE Trans. Inform. Theory*, 48(7):1849–1872, July 2002.

19. A. Høst-Madsen, X. Wang, and S. Bahng. Asymptotic analysis of blind multiuser detection with blind channel estimation. *IEEE Trans. Sig. Proc.*, 52(6):1722–1738, June 2004.

20. B. L. Hughes. Differential space-time modulation. *IEEE Trans. Inform. Theory*, 46(7):2567–2578, Nov. 2000.

21. Q. Li, C. N. Georghiades, and X. Wang. Blind multiuser detection in uplink CDMA with multipath fading: A sequential EM approach. *IEEE Trans. Commun.*, 52(1):0090, Jan. 2004.

22. L. Ljung and T. Söderström. *Theory and Practice of Recursive Identification*. MIT Press: Cambridge, MA, 1987.

23. M. L. McCloud and L. L. Scharf. MMSE multiuser detection for noncoherent non-orthogonal multipulse modulation. In *Proc. 2000 IEEE International Symposium on Information Theory (ISIT'00)*, page 356, Sorrento, Italy, June 2000.

24. H. V. Poor and L. A. Rusch. Narrowband interference suppression in spread spectrum CDMA. *IEEE Personal Communications Magazine*, 1(3):14–27, Aug. 1994.

25. H. V. Poor and S. Verdú. Probability of error in MMSE multiuser detection. *IEEE Trans. Inform. Theory*, IT-43(3):858–871, May 1997.

26. H. V. Poor. Active interference suppression in CDMA overlay systems. *IEEE J. Select. Areas Commun.*, 19(1), Jan. 2001.

27. H. V. Poor and M. Tanda. Multiuser detection in flat fading non-gaussian channels. *IEEE Trans. Commun.*, 50(11): 1769, Nov. 2002.

28. H. V. Poor and X. Wang. Code-aided interference suppression in DS/CDMA communications—Part I: Interference suppression capability. *IEEE Trans. Commun.*, COM-45(9):1101–1111, Sep. 1997.

29. J. G. Proakis. *Digital Communications*. McGraw-Hill, 1995.

30. D. Reynolds, A. Høst Madsen, and X. Wang. Adaptive transmitter precoding for time division duplex CDMA in fading multipath channels: Strategy and analysis. *EURASIP Journal of App. Sig. Proc.*, 2002(12):1325–1334, Dec. 2002.

31. D. Reynolds and X. Wang. Adaptive group-blind multiuser detection based on a new subspace tracking algorithm. *IEEE Trans. Commun.*, 49(7):1135–1141, July 2001.

32. D. Reynolds, X. Wang, and K. N. Modi. Interference suppression and diversity exploitation for multi-antenna CDMA with ultra-low complexity receivers. *IEEE Trans. Sig. Proc.*, 53(8):3226–3237, Aug. 2005.

33. A. Russ and M. K. Varanasi. Noncoherent multiuser detection for nonlinear modulation over the Rayleigh-fading channel. *IEEE Trans. Inform. Theory*, 47(1):295, Jan. 2001.

34. J. Sacks. Asymptotic distribution of stochastic approximation procedures. *Ann. Math. Stat.*, 29(2):373–405, 1958.

35. H. Sampath, P. Stoica, and A. Paulraj. Generalized linear precoder and decoder design for MIMO channels using the weighted MMSE criterion. *IEEE Trans. Commun.*, 49(12):2198–2206, Dec. 2001.

36. A. Scaglione, P. Stoica, S. Barbarossa, G. Giannakis, and H. Sampath. Optimal designs for space-time linear precoders and decoders. *IEEE Trans. Sig. Proc.*, 50(5):1051–1064, May 2002.

37. B. C. Skelton and D. P. Taylor. Multiuser detectors with decision feedback for asynchronous spread-spectrum multiple access in multipath fading channels. *IEEE Trans. Commun.*, 49(4):594, April 2001.

38. P. Spasojevic and X. Wang. Improved robust multiuser detection in non-Gaussian channels. *Sig. Proc. Letters*, 8(3):83, Mar. 2001.

39. D. M. Titterington. Recursive parameter estimation using incomplete date. *J. Roy. Statist. Soc.*, 46(2):257–267, 1984.

40. M. K. Varanasi and B. Aazhang. Multistage detection in asynchronous code division multiple-access communications. *IEEE Trans. Commun.*, COM-38(4):509–519, April 1990.

41. L. Venturino, X. Wang, and M. Lops. Multiuser detection for cooperative networks and performance analysis. *IEEE Trans. Sig. Proc.*, 54(9): 3315, Sept. 2006.

42. S. Verdú. *Multiuser Detection.* Cambridge, UK: Cambridge University Press, 1998.

43. R. Vijayan and H. V. Poor. Nonlinear techniques for interference suppression in spread spectrum systems. *IEEE Trans. Commun.*, COM-38(7):1060–1065, July 1990.

44. E. Visotsky and U. Madhow. Noncoherent multiuser detection for CDMA systems with nonlinear modulation: A non-Bayesian approach. *IEEE Trans. Inform. Theory*, 47(4):1352, May 2001.

45. B. R. Vojčić and W. M. Jang. Transmitter precoding in synchronous multiuser communications. *IEEE Trans. Commun.*, 46(10):1346–1355, October 1998.

46. X. Wang and H. V. Poor. Blind adaptive multiuser detection in multipath CDMA channels based on subspace tracking. *IEEE Trans. Sig. Proc.*, 46(11):3030–3044, Nov. 1998.

47. X. Wang and H. V. Poor. Blind equalization and multiuser detection for CDMA communications in dispersive channels. *IEEE Trans. Commun.*, COM-46(1):91–103, Jan. 1998.

48. X. Wang and H. V. Poor. Blind multiuser detection: A subspace approach. *IEEE Trans. Inform. Theory*, 44(2):677–691, March 1998.

49. X. Wang and H. V. Poor. Iterative (Turbo) soft interference cancellation and decoding for coded CDMA. *IEEE Trans. Commun.*, 47(7):1046–1061, July 1999.

50. X. Wang and H. V. Poor. Robust multiuser detection in non-Gaussian channels. *IEEE Trans. Sig. Proc.*, 47(2):289–305, Feb. 1999.

51. X. Wang and H. V. Poor. *Wireless Communication Systems: Advanced Techniques for Signal Reception.* Prentice-Hall, Upper Saddle River, NJ, 2004.

52. E. Weinstein, A. V. Oppenheim, M. Feder, and J. R. Buck. Iterative and sequential algorithms for multisensor signal enhancement. *IEEE Trans. Sig. Proc.*, 42(4):846–859, April 1994.

53. K. Wong, R. D. Murch, and K. B. Letaief. Optimizing time and space MIMO antenna system for frequency selective fading channels. *IEEE Trans. Commun.*, 19(7):1395–2206, July 2001.

54. Z. Yang, B. Lu, and X. Wang. Bayesian Monte Carlo multiuser receiver for space-time coded multi-carrier CDMA systems. *IEEE J. Select. Areas Commun.*, 19(8):1625–1637, Aug. 2001.

55. Z. Yang and X. Wang. Blind turbo multiuser detection for long-code multipath CDMA. *IEEE Trans. Commun.*, 50(1):112–125, Jan. 2002.

56. W. Yu and J. M. Cioffi. Trellis precoding for the broadcast channel. In *Proc. IEEE Globecom 2001*, pages 1344–1348, San Antonio, TX, USA, Nov. 2001.

57. K. Z. Arifi, S. Shahbazpanahi, B. B. Gershman, and L. Zhi-Quan. Robust blind multiuser detection based on the worst-case performance optimization of the MMSE receiver. *IEEE Trans. Sig. Proc.*, 53(1): 295, Jan. 2005.

58. L. Zheng and D. N. C. Tse. Diversity and multiplexing: A fundamental tradeoff in multiple-antenna channels. *IEEE Trans. Inform. Theory*, 49(5):1073–1096, May 2003.

59. Z. Zvonar and D. Brady. Multiuser detection in single-path fading channels. *IEEE Trans. Commun.*, COM-42(2/3/4):1729–1739, Feb./Mar./Apr. 1994.

4

PERFORMANCE WITH RANDOM SIGNATURES

Matthew J. M. Peacock, Iain B. Collings, and Michael L. Honig

4.1 RANDOM SIGNATURES AND LARGE SYSTEM ANALYSIS

In this chapter, we study the performance of multiuser detectors where the users are assigned *random* signatures. The signatures may result from spreading in time and frequency, as in variants of Code-Division Multiple Access (CDMA), and spreading in space when multiple antennas are available at the transmitter and/or receiver. Random signatures are commonly used in cellular CDMA systems, and are often used to model channel gains across multiple receive antennas.

In general, there are three main objectives of this performance analysis: (1) It should provide an efficient method for estimating performance metrics such as Signal-to-Interference Plus Noise Ratio (SINR) and bit error rate as a function of system parameters; (2) It should provide a means for optimizing system parameters in different scenarios; and (3) It should provide insight into the relative performance of different receivers. Meeting these objectives becomes challenging when the system model has many parameters (e.g., number of users, bandwidth, spreading factor, power and signature assignments, channel gains, modulation format, etc.), and the performance metrics take on complicated forms. Furthermore, what is often desired are *averages* of the performance metrics over realizations of random parameters, such as channels and signatures, which can be quite difficult without resorting to Monte Carlo simulation.

Advances in Multiuser Detection. Edited by Michael L. Honig
Copyright © 2009 John Wiley & Sons, Inc.

An analytical approach, which circumvents the preceding difficulties, and which has become popular in recent years, is to evaluate the performance under a *large-system limit* (also referred to as *asymptotic analysis*). The essential idea is straightforward:

Scale the system size to infinity, letting size parameters grow *proportionately*.

This is depicted in Figure 4.1, where each block denotes a matrix or vector in the system model (e.g., as defined in the next section). The rows and columns of each matrix are scaled up proportionately, i.e., keeping the ratio of rows to columns fixed. The length of each vector is then scaled accordingly.

The benefit of asymptotic analysis is that, under certain conditions, the performance metrics converge to *deterministic quantities*, which may be determined in a straightforward manner by solving a set of nonlinear equations. These limiting values generally do not depend on the distribution of the underlying random variables. (This can be interpreted as a consequence of a law of large numbers.) Hence, this eliminates the need for computing an explicit average. Moreover, the nonlinear equations are specified in terms of a few key parameters, such as the system load (users per dimension), the noise power, and transmitted power distribution(s).

Interestingly, the results obtained in the large system limit not only give a concise characterization of performance, but also typically give an excellent estimate of the performance of practical, finite-size systems. In fact, the large system results often apply to systems that are surprisingly small.

Much of the large system analysis to date applies to models in which the transmitted data is spread over the transmit and/or receive dimensions, and corresponds to a

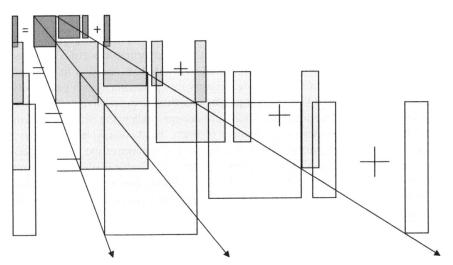

Figure 4.1. Illustration of large-system limit. Each rectangle represents a matrix or a vector in the system model. The system size is scaled so that the rows and columns of the matrices increase *proportionally*.

spreading, or signature matrix, which is random with independent, identically distributed (*i.i.d.*) elements, or a random orthogonal matrix. In some situations, this randomness exists naturally, e.g., for a multi-antenna link with rich scattering, where the fading across different pairs of transmit/receive antennas is independent. For CDMA systems it is the matrix of signatures, which contains the required randomness. Although not strictly adhering to any current CDMA standard, this random signature model does characterize the long pseudo-noise sequences in many CDMA cellular systems and provides a lower bound on performance achievable with a deterministic choice of signatures.

During the past few years, large system analysis has been applied to numerous variations of CDMA and Multi-Input–Multi-Output (MIMO) channel models (e.g., corresponding to multiple antennas at the transmitter and/or receiver). This analysis relies upon results from the mathematical literature on large random matrices. A comprehensive review of this work, along with the relevant random matrix literature, is given in [60]. We remark that for most readers with engineering backgrounds, the derivations of many of the random matrix results presented in [60], which have appeared in the mathematical literature, are likely to be challenging to follow. Our intent is not to give a similar overview of these large system results. Rather, our objective is to provide insight into the nature of large system analysis, and the relevant results in the random matrix literature, by presenting some relatively simple derivations from first principles.

The approach we take for deriving large system performance is based on different incremental expansions of the received covariance matrix. These expansions lead to a set of nonlinear scalar equations for performance metrics such as SINR and spectral efficiency, and are straightforward to solve numerically. In addition, the form of the results often give insight into the effect of system parameters and channel characteristics on performance. This approach can be applied to a broad class of problems in a systematic, unified manner. For example, it can be applied to a wider class of signature and channel matrices than previously considered, including circulant and Toeplitz, which model linear time-invariant channels. The key assumption is that the distribution of the vector of transmitted symbols must be invariant to rotation.[1]

We start by presenting a general linear system model with additive noise. The model can be applied to a range of CDMA and MIMO systems by changing the random "mixing matrix," which relates inputs (e.g., transmitted symbols) to outputs in the absence of noise. Of course, the performance measures considered depend on the properties (e.g., distribution of eigen-values) of the random mixing matrix. We illustrate our approach to large system analysis by starting with the case in which the mixing matrix is *i.i.d.*, corresponding to the CDMA uplink with ideal channels (no channel-induced distortion). Applying the matrix-expansion approach with an optimal (Minimum Mean Squared Error (MMSE)) linear receiver leads to the Tse-Hanly fixed-point equation for output SINR, originally presented in [58]. We then extend this analysis to account for the effect of frequency-selective channels. In

---

[1] Although this assumption is needed for the analysis, numerical examples indicate that the results apply to a larger class of distributions, e.g., corresponding to symbols from a finite-alphabet.

addition to computing the output SINR for the linear MMSE receiver, we also compute the spectral efficiency with the optimal (e.g., maximum-likelihood) receiver.

We then consider the large system performance of *adaptive* linear multi-user detection, motivated by the scenario in which the system parameters, including channels and signatures, are unknown *a priori*. The filter is estimated via a training sequence, and the performance (output SINR) depends on the length of the training sequence, in addition to all other system parameters. Our focus is on least squares estimation, which is widely used in practice. Large system analysis leads to a relatively simple relation between the respective output SINRs of the least squares and MMSE filters, which applies to a general class of channel models. Application of these results to the optimization of training overhead in packet data transmissions is also discussed.

Finally, in an appendix we compute the eigen-value distributions of sums and products of unitarily invariant, positive-semidefinite random matrices. These results have been previously used to derive large system performance metrics, and originate from the theory of *free probability*, which pertains to non-commutative random variables (random matrices being a canonical example). Here we simply show how these results (so-called *R*- and *S*-transforms) can be derived from the matrix expansion approach.

We emphasize that our intent in this chapter is to focus on the main ideas behind the large system derivations. The discussion and derivations are therefore not meant to be rigorous, and intentionally sidestep technical issues concerning the convergence of the random quantities to deterministic limits. A short section on bibliographic notes, at the end of the chapter, gives pointers to some of the relevant literature.

4.2 SYSTEM MODELS

We start with the general linear model[2]

$$\mathbf{y}(i) = \mathbf{M}\mathbf{b}(i) + \mathbf{n}(i) \tag{4.1}$$

where $\mathbf{b}(i)$ is the $K \times 1$ vector of transmitted symbols at time i, $\mathbf{y}(i)$ is the $N \times 1$ vector of received symbols, $\mathbf{n}(i)$ is a vector of noise samples, and $\mathbf{M}$ is the $N \times K$ *mixing matrix*, which maps the inputs to outputs in the absence of noise. We assume complex baseband signaling, so that the elements of $\mathbf{b}$, $\mathbf{n}$, $\mathbf{M}$, and $\mathbf{y}$ are complex in general.[3] Throughout this chapter the noise $\mathbf{n}$ is white, circularly symmetric Gaussian with zero mean and covariance $\sigma^2 \mathbf{I}_N$. Also, unless specified otherwise (i.e., in Section 4.8), the transmitted symbols are *i.i.d.* with zero mean and unit variance.

---

[2]**Notation:** Throughout this chapter all vectors are defined as column vectors and designated with bold lower case; all matrices are given in bold upper case; $(\cdot)^\dagger$ denotes Hermitian (i.e., complex conjugate) transpose; $(\cdot)^\ddagger$ denotes the operation $\mathbf{X}^\ddagger = \mathbf{X}\mathbf{X}^\dagger$; tr$[\cdot]$ denotes the matrix trace; $|\cdot|$ and $\|\cdot\|$ denote the Euclidian and induced spectral norms, respectively; $\mathbf{I}_N$ denotes the $N \times N$ identity matrix; and, expectation is denoted $\mathbf{E}[\cdot]$.

[3]We will write $\mathbf{M} \in \mathbb{C}^{N \times K}$ or $\mathbf{M} \in \mathbb{R}^{N \times K}$ to denote an $N \times K$ matrix with complex or real elements, respectively.

The large system analysis, which follows, can be applied to a broad class of mixing matrices $\mathbf{M}$, which appear in (4.1). Namely, the following analysis can be applied to any $\mathbf{M}$ that has a well-defined asymptotic eigen-value distribution, to be defined in the next section, provided that the symbol vector $\mathbf{b}$ is unitarily invariant.[4] Of course, the final results depend on the eigen-values of $\mathbf{M}$, which in turn depend on the particular system assumptions. We also remark that numerical examples indicate that the large system results apply to a larger class of non-unitarily invariant symbol vectors, e.g., those selected from finite alphabets.

We now specify three particular cases, motivated by the CDMA uplink and down-link, which we will use to illustrate large system analysis. We also highlight other relevant applications, namely, dispersive channels with multiple antennas and/or multiple users.

4.2.1 Uplink CDMA Without Multipath

The simplest model we consider corresponds to the synchronous CDMA uplink with K users in the absence of multipath. In this case (4.1) becomes:

$$\mathbf{y}(i) = \mathbf{SAb}(i) + \mathbf{n}(i) \tag{4.2}$$

where $\mathbf{y}$ is the vector of received chips corresponding to a transmitted symbol, and $\mathbf{M} = \mathbf{SA}$, where $\mathbf{S} \in \mathbb{C}^{N \times K}$ is the signature matrix, and $\mathbf{A} \in \mathbb{R}^{K \times K}$ is a diagonal matrix containing the symbol amplitudes. The signature for user k is the kth column of $\mathbf{S}$. The amplitudes are assumed to be positive and real-valued. (Phase offsets can be absorbed in the signatures or channel matrix, to be defined.) Given amplitude A_k for user k, the transmitted *power* is therefore $P_k = A_k^2$. Note that the matrix A_k can also include the effect of random channel attenuation, corresponding to flat, or non-frequency-selective fading.

A key assumption in what follows is that the elements of $\mathbf{S}$ are *random*. This corresponds to existing CDMA cellular systems in which the signatures are generated via a pseudo-random number generator (e.g., see the overview of CDMA cellular systems in [47, Ch. 10]). Because of this assumption, performance measures of interest, such as SINR, achievable rate (capacity), and even probability of error, are random, unless they are conditioned on a particular realization of $\mathbf{S}$. We will consider the following two special random signature models.

- *i.i.d. signatures*: The elements of $\mathbf{S}$ are assumed to be *i.i.d.* complex random variables with zero mean and variance $1/\sqrt{N}$. For technical reasons, we will also assume that the elements have finite positive moments.
- *Isometric signatures*:[5] The random signatures are *orthogonal*, so that $\mathbf{S}^\dagger \mathbf{S} = \mathbf{I}_K$. (Note that this is possible only for $K \leq N$.) Equivalently, we assume that $\mathbf{S}$ is

---

[4]That is, the distribution does not change under rotation. This is true, for example, when the transmitted symbols are *i.i.d.* Gaussian.

[5]We will sometimes use the abbreviation "iso." for isometric.

obtained by extracting $K \le N$ columns from an $N \times N$ *Haar-distributed* unitary random matrix. That is, a unitary random matrix Ω is Haar-distributed if its probability distribution is invariant to left or right multiplication by any constant unitary matrix. For example, if $\mathbf{X}$ is a square random matrix with *i.i.d.*, zero-mean, complex Gaussian entries, then the unitary matrix $\mathbf{X}(\mathbf{X}^{\dagger}\mathbf{X})^{-1/2}$ is Haar.

For both models the signatures, symbols, noise, and amplitudes are all mutually independent. The matrix $\mathbf{A}$ then determines how the transmitted energy varies across symbols (or users).

For uplink CDMA, isometric signatures require coordinated assignments across users, e.g., by the base station. Also, in that scenario synchronization becomes an important issue, since timing offsets on the order of a chip could potentially cause a significant increase in correlation between pairs of signatures. This is less of an issue for downlink CDMA, where all signals are generated at the base station.

The random signature model (4.2) also applies to a single-user narrowband communications link with multiple antennas at both the transmitter and receiver. In that case, the matrix $\mathbf{S}$ represents the *channel*, and the amplitude matrix determines the allocation of powers across transmit antennas. That is, the *(i, j)*th element of $\mathbf{S}$ is the channel gain from transmit antenna j to receive antenna i. The "signature" corresponding to a particular transmit antenna consists of the channel gains across receive antennas. The *i.i.d.* model then corresponds to the assumption of a rich scattering environment [56, Ch. 7]. Here the isometric model is not relevant, since the channel is provided by nature.

4.2.2 Downlink CDMA

A shortcoming of the model (4.2) is that it does not account for multipath introduced by the channel. After analyzing (4.2), we will therefore proceed to the more complicated model:

$$\mathbf{y}(i) = \mathbf{HSAb}(i) + \mathbf{n}(i) \tag{4.3}$$

where the mixing matrix $\mathbf{M} = \mathbf{HSA}$, and $\mathbf{H} \in \mathbb{C}^{M \times N}$ represents the effect of the channel. This model applies to the CDMA *downlink*, since the channel matrix $\mathbf{H}$ is the same for each user. (The corresponding uplink model is presented next.) That is, $\mathbf{y}$ is the vector of chips received by a particular user (mobile) with channel matrix $\mathbf{H}$. For CDMA, the received chips are received sequentially in time, the rows of the channel matrix $\mathbf{H}$ are shifted versions of the channel impulse response (e.g., see [17, Sec. 12.4]), and typically $M = N$, so that $\mathbf{H}$ is square.

We remark that the preceding model (4.3) also applies to multi-carrier (MC)-CDMA, in which the signatures are defined in the *frequency*-domain, as opposed to the time-domain [10,33]. In that case, by adding a cyclic prefix to the transmitted symbol vector $\mathbf{b}$, the channel matrix $\mathbf{H}$ becomes diagonal, where the diagonal elements are the complex channel gains across frequencies. Specifically, with the cyclic prefix the corresponding channel matrix $\mathbf{H}$ is circulant instead of Toeplitz

[17, Sec. 12.4].[6] We then factor the circulant matrix $\mathbf{H} = \mathbf{WDW}^\dagger$, where $\mathbf{W}^\dagger$ is the Discrete Fourier Transform (DFT) matrix, $\mathbf{W}$ is the inverse DFT matrix, $\mathbf{W}^\dagger\mathbf{W} = \mathbf{WW}^\dagger = \mathbf{I}$, and $\mathbf{D}$ is diagonal with diagonal elements equal to the DFT of the first row of $\mathbf{H}$. The received vector is then:

$$\mathbf{y} = \mathbf{WDW}^\dagger\mathbf{SAb} + \mathbf{n}. \tag{4.4}$$

Writing the signature matrix as $\mathbf{S} = \mathbf{W}\tilde{\mathbf{S}}$ and premultiplying $\mathbf{y}$ by $\mathbf{W}^\dagger$ gives:

$$\tilde{\mathbf{y}} = \mathbf{D}\tilde{\mathbf{S}}\mathbf{Ab} + \tilde{\mathbf{n}} \tag{4.5}$$

where $\tilde{\mathbf{n}} = \mathbf{W}^\dagger\mathbf{n}$ has the same statistics as $\mathbf{n}$, since $\mathbf{W}^\dagger$ is unitary. That is, for MC-CDMA $\tilde{\mathbf{S}}$ is the signature matrix in the frequency domain, which the transmitter transforms to $\mathbf{S}$ in the time-domain, and at the receiver we transform the received signal back to the frequency-domain. Of course, the models (4.4) and (4.5) are equivalent in the sense that MC-CDMA has the same performance (e.g., achievable rate) as CDMA with spreading in the time-domain.

For the large system analysis, which follows, $\mathbf{H}$ is allowed to be an *arbitrary* matrix, and the performance depends only on the eigen-values of $\mathbf{HH}^\dagger$ (e.g., the channel gains across frequency when $\mathbf{H}$ is circulant). That is, $\mathbf{H}$ can be deterministic, corresponding to a particular multipath profile, or random with given first-order distribution for the sub-channel gains. (Higher-order statistics across frequency do not affect the large system results, which follow.)

The model (4.3) also applies to a single-user multi-antenna link, where $\mathbf{S}$ represents a *precoding* matrix, which precedes the MIMO channel $\mathbf{H}$. Assuming a rich scattering environment, the channel $\mathbf{H}$ has *i.i.d.* elements. (In practice, it is desirable to select a precoding matrix $\mathbf{S}$ to diagonalize $\mathbf{H}$, rather than choose a random $\mathbf{S}$, independent of $\mathbf{H}$, as assumed in the subsequent analysis.) Another interpretation of (4.3) corresponds to the correlated MIMO channel model proposed in [39]. Namely, in some situations it may be possible to represent the MIMO channel as the product of matrices $\mathbf{\Phi D\Psi}$, where $\mathbf{\Phi}$ and $\mathbf{\Psi}$ are *i.i.d.*, and $\mathbf{D}$ is diagonal. In (4.3) this corresponds to taking $\mathbf{H} = \mathbf{\Phi D}$, and $\mathbf{S}$ to be *i.i.d.* This can be interpreted as modeling the presence of a few scatterers interspersed between the transmitter and receiver. The rank of the scattering array matrix $\mathbf{D}$ determines the *richness* of the MIMO channel. Namely, as the rank increases, the overall channel matrix becomes *i.i.d.* The subsequent large system analysis can be directly applied to this model.

Finally, we remark that choosing $\mathbf{S} = \mathbf{A} = \mathbf{I}_K$ leads to the classical single-user channel model with inter-symbol interference. Namely, in that case $\mathbf{y}$ consists of samples at the output of a dispersive channel. If the samples are in time, as in a single-carrier system, then the channel matrix $\mathbf{H}$ is Toeplitz, so that $\mathbf{y}$ is the convolution of the transmitted symbols with the channel impulse response. If the samples are

---

[6]The addition of any finite cyclic prefix does not affect the large system performance, since the associated loss in information rate from adding the prefix is normalized by the degrees of freedom (e.g., processing gain), which tend to infinity.

in frequency, as in a multi-carrier system, then the channel matrix $\mathbf{H}$ is diagonal. In that case, the analogous large system analysis to that considered here would let the block sizes of the transmitted and received symbols grow proportionally. This leads to the classical results on MMSE filtering with an infinitely long filter, and capacity results with infinitely long code words [17, Chs. 4, 11].

4.2.3 Multi-Cell Downlink or Multi-Signature Uplink

The downlink model (4.3) applies to a single, isolated cell. Extending this model so that it accounts for interference from other cells gives:

$$\mathbf{y}(i) = \sum_{j=1}^{J} \mathbf{H}_j \mathbf{S}_j \mathbf{A}_j \mathbf{b}_j(i) + \mathbf{n}(i) \tag{4.6}$$

where $\mathbf{y}(i)$ is the received signal at a particular mobile, and the sum is over signals from nearby base stations. That is, the columns of $\mathbf{S}_j$ are signatures assigned to users in cell j, and the matrix $\mathbf{H}_j$ represents the channel from the corresponding base station to the user (mobile) of interest. If each user is assigned a single signature, then $\mathbf{S}_j$ is $N \times K_j$, where K_j is the number of users in cell j. Comparing with (4.1), in this case $\mathbf{b}$ in (4.1) is obtained by stacking $\mathbf{b}_1, \ldots, \mathbf{b}_K$, and taking $\mathbf{M} = [\mathbf{S}_1\mathbf{A}_1\mathbf{H}_1 \ \mathbf{S}_2\mathbf{A}_2\mathbf{H}_2 \ldots \mathbf{S}_J\mathbf{A}_J\mathbf{H}_J]$.

The model (4.6) also applies to the CDMA uplink, where each user can be assigned multiple signatures. This gives a convenient way to vary the transmitted data rate, and is used in current (third generation) CDMA cellular standards [47, Ch. 10]. According to this interpretation, the sum in (4.6) is over users, i.e., the columns of $\mathbf{S}_j$ are the signatures assigned to user j, $\mathbf{b}_j$ is the corresponding vector of transmitted symbols from user j, and $\mathbf{H}_j$ is the channel for user j. In this context, the model (4.3) applies to a single-user link, where the columns of $\mathbf{S}$ are the signatures assigned to the user, and $\mathbf{A}$ contains the amplitudes across signatures.

Henceforth, we will primarily refer to the downlink CDMA interpretation of (4.6). We let K_j denote the number of users in cell j, so that the signature matrix $\mathbf{S}_j$ is $N \times K_j$. The total number of assigned signatures is $K = \sum_{j=1}^{J} K_j$. Each random signature matrix $\mathbf{S}_j$ can be either *i.i.d.* or isometric, and can be chosen differently across users (i.e., $\mathbf{S}_j$ may be *i.i.d.*, and $\mathbf{S}_l$, $l \neq j$, may be isometric). The signature matrices and channels are all assumed to be mutually independent. Also, we will assume that all channel matrices can be diagonalized by the same similarity transformation. This applies to transmission through linear, time-invariant channels, in which case the channel matrices can be assumed to be circulant.

The model (4.6) also applies to the narrowband uplink with multiple antennas at the base station and the mobiles. In that scenario, $\mathbf{H}_j$ is *i.i.d.*, assuming a rich scattering environment, and $\mathbf{S}_j$ is a precoding matrix. This includes the special case $\mathbf{S}_j = \mathbf{I}_K$, and as discussed earlier, this model encompasses the MIMO channel model in [39] with $\mathbf{H}_j = \boldsymbol{\Phi}_j\mathbf{D}_j$, where $\boldsymbol{\Phi}_j$ is *i.i.d.*, $\mathbf{D}_j$ is diagonal, $\mathbf{S}_j$ is *i.i.d.*, and all of these are chosen independently across users.

Finally, we remark that the large system analysis, which follows, also applies to the system model:

$$\mathbf{y}(i) = \sum_{j=1}^{J} \mathbf{M}_j \mathbf{b}_j(i) + \mathbf{n}(i) \tag{4.7}$$

where the conditions previously stated for (4.1) apply to each $\mathbf{M}_j$ and $\mathbf{b}_j$, and in addition, the matrices $\mathbf{M}_j$ are diagonalized by the same similarity transformation. For example, this is true when the $\mathbf{M}_j$'s are circulant, in which case the eigen-vectors are columns of the DFT matrix. This model is a generalization of (4.6), where $\mathbf{S}_j \mathbf{A}_j \mathbf{b}_j$ in (4.6) is replaced by $\mathbf{b}_j$ in (4.7). (That is, in (4.7) we do not require $\mathbf{E}[\mathbf{b}_j^{\ddagger}]$ to be the scaled identity matrix.)

4.2.4 Model Limitations

Here we briefly discuss some limitations of the preceding models. Namely, although these models can be used to obtain insight into scenarios of practical interest, they do not account for significant features of some wireless systems. For example, the CDMA uplink is typically designed to be asynchronous, whereas these models are synchronous. Asynchronous transmissions may alter the performance somewhat, depending on the transmitted pulse shape and the span of the filter (e.g., see [10,29,30]). With random signatures and channels, and bandwidth efficient pulse shaping, we expect that the general observations and trends for the synchronous models considered here will carry over to asynchronous models. (This is shown to be true for the MMSE linear receiver in [29].)

These models also assume that all channels are stationary. Specifically, for the MC-CDMA model with diagonal $\mathbf{H}$, the diagonal entries of $\mathbf{H}$ are chosen from the same first-order distribution. The performance results for SINR assume that each user spreads power across all sub-channels (entries of $\mathbf{H}$), and the channel gains are constant over a symbol. The results for spectral efficiency assume that each user codes across all sub-channels, and the channel gains are stationary over a codeword. The discussion of least squares estimation of the receiver filters in Section 4.8 assumes that the channel is constant for the duration of training. Of course, these results can be applied to time-varying channels provided that the channels remain approximately stationary over the duration of interest.[7]

The analysis of fixed (non-adaptive) receivers assumes that the channels are known perfectly at the receiver. Hence these models do not account for random phase and frequency offsets, which may be encountered with multi-carrier transmission.

The random signature models do not account for correlated signature elements. This is of little importance for CDMA, since signatures are accurately modeled as *i.i.d.* However, it can become an issue for the multi-antenna interpretations of the models considered, since the signatures elements correspond to spatial channel

---

[7]Extensions to certain classes of time-varying channel models, such as those described in [14], may also be possible.

gains, which can be highly correlated in many situations of interest (e.g., see [56, Ch. 7]). Also, as mentioned in Section 4.2.2, we assume that the precoding matrix $\mathbf{S}$ in (4.3) is independent of the channel $\mathbf{H}$. This is likely to be true for downlink CDMA. For the multi-antenna interpretation this corresponds to random spreading across antennas, and does not account for any selection of $\mathbf{S}$ that exploits channel state information at the transmitter.

4.3 LARGE SYSTEM LIMIT

The large system limit lets the number of rows and columns of the signature matrices tend to infinity with fixed ratio. For the simplest model (4.2), the large system limit takes K and N to infinity with fixed ratio $\beta = K/N$. This also applies to the model (4.3), in which case the size of $\mathbf{H}$, $\mathbf{S}$, $\mathbf{A}$, $\mathbf{b}$, and $\mathbf{n}$ all grow, as illustrated in Figure 4.1. For the model (4.6), the large system limit is defined by *fixing J*, and letting $K_1, \ldots, K_J$ and N all tend to infinity with fixed ratios $\beta_j = K_j/N$, $j = 1, \ldots, J$. Furthermore, for the general case where the channel matrices are $M \times N$, we also let $M \to \infty$ with fixed $\alpha = M/N$.

In the large system limit, many performance measures of interest, which depend on the random signatures and channel matrices, become *deterministic*. Furthermore, in this limit the performance does not depend on the distribution of signature elements. We first illustrate this for linear receivers, and then provide some mathematical background that will be useful in subsequent sections.

4.3.1 SINR of Linear Filters

For the simplest model (4.2), a linear receiver for user k can be denoted as $\mathbf{c}_k \in \mathbb{C}^{N \times 1}$, and the estimated symbol for user k is then:

$$\hat{b}_k = \mathbf{c}_k^\dagger \mathbf{y} = A_k(\mathbf{c}_k^\dagger \mathbf{s}_k)b_k + \mathbf{c}_k^\dagger(\mathbf{S}_{b_k}\mathbf{A}_{b_k}\mathbf{b}_{b_k} + \mathbf{n}) \tag{4.8}$$

where the subscript b_k indicates that the contribution from the symbol b_k has been removed. That is, $\mathbf{S}_{b_k}$, $\mathbf{A}_{b_k}$, and $\mathbf{b}_{b_k}$ are $\mathbf{S}$, $\mathbf{A}$, and $\mathbf{b}$ with the kth column, row and column, and element removed, respectively.[8] Also, here we drop the dependence on i, since the SINR does not depend on i. The two terms after the right equality are the signal and interference plus noise, respectively. Defining the received covariance matrix as:

$$\mathbf{R} = \mathbf{E}[\mathbf{y}\mathbf{y}^\dagger] = (\mathbf{SA})^{\ddagger} + \sigma^2 \mathbf{I}_N \tag{4.9}$$

the received interference plus noise covariance matrix is therefore:

$$\mathbf{R}_{b_k} = (\mathbf{S}_{b_k}\mathbf{A}_{b_k})^{\ddagger} + \sigma^2 \mathbf{I}_N \tag{4.10}$$

$$= \mathbf{R} - P_k \mathbf{s}_k \mathbf{s}_k^\dagger \tag{4.11}$$

---

[8]We use the subscript b_k, instead of k, to distinguish this rank-reduction from another related rank-reduction to be introduced later.

The output SINR of the filter, conditioned on **S** and **A**, is given by:

$$\rho_k^N = \frac{\mathbf{E}[|\mathbf{c}_k^\dagger(A_k b_k \mathbf{s}_k)|^2]}{\mathbf{E}[|\mathbf{c}_k^\dagger(\mathbf{y} - A_k b_k \mathbf{s}_k)|^2]} = \frac{P_k |\mathbf{c}_k^\dagger \mathbf{s}_k|^2}{\mathbf{c}_k^\dagger \mathbf{R}_{b_k} \mathbf{c}_k} \tag{4.12}$$

where the expectation is over **b** and **n**, and the superscript N denotes the size of the system. For finite N, **S** and **A** are random, and hence the SINR (4.12) is random. However, assuming the *i.i.d.* signature model with particular choices of $\mathbf{c}_k$, in the large system limit the SINR converges to a deterministic quantity. Specifically, for the matched filter receiver $\mathbf{c}_k = \mathbf{s}_k$, taking $(K, N) \to \infty$ with fixed ratio $\beta = K/N$, we have:

$$\mathbf{s}_k^\dagger \mathbf{s}_k \to 1 \tag{4.13}$$

$$\mathbf{s}_k^\dagger \mathbf{R}_{b_k} \mathbf{s}_k = \sum_{m \neq k} P_m |\mathbf{s}_m^\dagger \mathbf{s}_k|^2 + \sigma^2(\mathbf{s}_k^\dagger \mathbf{s}_k) \to \beta \mathbf{E}[P] + \sigma^2 \tag{4.14}$$

where the expectation is over the distribution of interfering powers. Both limits are a consequence of the strong law of large numbers, i.e., the random quantities shown converge to their respective means. (Convergence is in the almost sure sense, provided that the set of powers $\{P_m\}$ converges in distribution almost surely to a deterministic distribution with compact support.) Hence in the large system limit, the SINR for the matched filter receiver converges to:

$$\rho_k^{\text{MF}} = \frac{P_k}{\sigma^2 + \beta \mathbf{E}[P]} \tag{4.15}$$

For the linear MMSE receiver, we choose the filter $\mathbf{c}_k$ to minimize the mean-squared error cost function:

$$\mathcal{C} = \mathbf{E}\left[|b_k - \hat{b}_k|^2\right] \tag{4.16}$$

where the expectation is over **b** and **n**. The solution is $\mathbf{c}_k = A_k \mathbf{R}^{-1} \mathbf{s}_k$, and the corresponding SINR is given by $P_k \bar{\rho}_k^N$, where:

$$\bar{\rho}_k^N = \mathbf{s}_k^\dagger \mathbf{R}_{b_k}^{-1} \mathbf{s}_k. \tag{4.17}$$

Figure 4.2 shows a scatter plot of ρ_k^N versus N with fixed $\beta = K/N = 1/2$, noise variance $1/\sigma^2 = 10\,\text{dB}$, and equal amplitudes $\mathbf{A} = \mathbf{I}$. For each N, the points shown correspond to different realizations of random signatures, again assuming the *i.i.d.* signature model. The figure indicates that as N increases, the variance of the SINR tends to zero, so that the SINR converges to a deterministic limit. For small N, the relatively large variance shown can potentially lead to outages, i.e., the SNR corresponding to a particular realization of signatures can fall significantly below the average value. Here we discuss performance averaged over the signatures, and do not consider performance metrics such as outage probability.

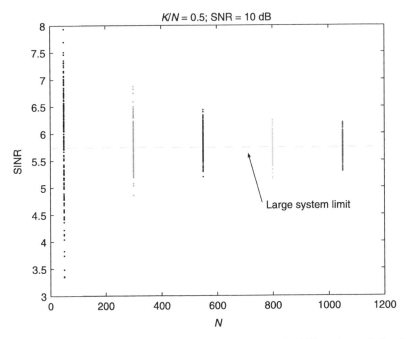

Figure 4.2. Scatter plot of SINR versus system size for the linear MMSE receiver with the signal model (4.2). Each point corresponds to a particular realization of assigned signatures.

Throughout this chapter, we omit proofs of convergence in the large system limit in order to focus on the computation of the actual limits. Namely, in Section 4.5, we will *assume* that the SINR in (4.17) converges, and derive the associated limit. In Section 4.6 we then extend this analysis to the models (4.3) and (4.6). Although the following derivations are not rigorous, they can be made rigorous by showing (uniform) convergence of the appropriate random quantities (e.g., see [43]).

Before proceeding, we first observe that the SINR for the linear MMSE receiver can be expressed in terms of the eigen-values of the received interference plus noise covariance matrix $\mathbf{R}_{b_k}$. Namely, we can rewrite (4.17) as:

$$\bar{\rho}_k^N = \text{tr}[\mathbf{R}_{b_k}^{-1} \mathbf{s}_k \mathbf{s}_k^\dagger] \tag{4.18}$$

Now if $\bar{\rho}_k^N$ converges in the large system limit to $\bar{\rho}_k$, then this limit will not change if we average $\bar{\rho}_k^N$ over $\mathbf{s}_k$. Since $\mathbf{s}_k$ and $\mathbf{R}_{b_k}$ are independent and $\mathbf{E}[\mathbf{s}_k \mathbf{s}_k^\dagger] = \frac{1}{N}\mathbf{I}_N$, we have:

$$\mathbf{E}_{\mathbf{s}_k}(\bar{\rho}_k^N) = \frac{1}{N}\text{tr}[\mathbf{R}_{b_k}^{-1}] \tag{4.19}$$

$$= \frac{1}{N}\sum_{n=1}^{N}\frac{1}{\lambda_n^N + \sigma^2} \tag{4.20}$$

where $\{\lambda_n^N\}_{n=1,\ldots,N}$ are the eigen-values of $\tilde{\mathbf{R}} = (\mathbf{S}_{b_k}\mathbf{A}_{b_k})^{\ddagger}$. We are therefore interested in characterizing the eigen-values of $\tilde{\mathbf{R}}$ in the large system limit.

4.4 RANDOM MATRIX TERMINOLOGY

In this section, we introduce some background terminology and definitions pertaining to random matrices.

4.4.1 Eigen-Value Distributions

For an arbitrary $N \times N$ complex Hermitian matrix $\mathbf{X}_N$ with real-valued eigen-values $\{x_n^N\}_{n=1,\ldots,N}$, its **empirical eigen-value distribution** (e.e.d.) is defined as:

$$F_X^N(x) = \frac{1}{N} \cdot \left| \{x_n^N : x_n^N \le x, n = 1,\ldots,N\} \right| \tag{4.21}$$

The **asymptotic eigen-value distribution** (a.e.d.) of a sequence of $N \times N$ complex Hermitian matrices is the (deterministic) distribution function given by the limit of the e.e.d.'s of the matrices as $N \to \infty$, if it exists.

Using the definition (4.21), we can rewrite (4.20) as:

$$\mathbf{E}[\bar{\rho}_k^N] = \int \frac{1}{\lambda + \sigma^2} dF_{\tilde{\mathbf{R}}}^N(\lambda) \tag{4.22}$$

Taking the large system limit[9] gives the corresponding expression for $\bar{\rho}_k = \lim_{N\to\infty} \bar{\rho}_k^N$, where the e.e.d. $F_{\tilde{\mathbf{R}}}^N$ is replaced by the associated a.e.d. $F_{\tilde{\mathbf{R}}}$.

4.4.2 Stieltjes Transform

The a.e.d. is often given implicitly by its **Stieltjés (or Cauchy) transform**. Namely, the Stieltjés transform of the distribution of a real-valued random variable X is defined as:

$$G_X(z) = \mathbf{E}\left[\frac{1}{X - z}\right] \quad \text{for } z \in \mathbb{C}^+ \tag{4.23}$$

where $\mathbb{C}^+ = \{x \in \mathbb{C} \mid \mathrm{Im}(x) > 0\}$. Similarly, the Stieltjés transform of the e.e.d. of a Hermitian matrix $\mathbf{X}_N \in \mathbb{C}^{N \times N}$, is given by:

$$G_X^N(z) = \int \frac{1}{x - z} dF_X^N(x) = \frac{1}{N} \sum_{n=1}^{N} \frac{1}{x_n^N - z} = \frac{1}{N} \mathrm{tr}[(\mathbf{X}_N - z\mathbf{I}_N)^{-1}]. \tag{4.24}$$

---

[9]The large system limit of a variable X^N will be denoted as $\lim_{N\to\infty} X^N$. It is implicitly assumed that other system parameters scale accordingly.

Comparing with (4.22) and (4.20), we can write the SINR for the model (4.2) with *i.i.d.* **S** and a linear MMSE receiver as:

$$\bar{\rho}_k^N = \lim_{z \to -\sigma^2} G_{\mathbf{R}}^N(z) \tag{4.25}$$

The limit is needed since, strictly speaking, $G_{\mathbf{R}}^N$ is not defined on the real line.

The moments of a random variable X are easily computed from the associated Stieltjés transform. Namely, since $1/(x - z) = -z^{-1} \sum_m^\infty (x/z)^m$ for $|x/z| < 1$, we have:

$$\mathbf{E}[X^m] = \frac{1}{(m - 1)!} \lim_{z \to 0} \frac{\mathrm{d}}{\mathrm{d}z^m}[z^{-1} G_X(z^{-1})] \tag{4.26}$$

Also, the distribution $F_X(x)$ can be obtained from its Stieltjés transform by the inversion formula [60]:

$$\mathrm{d}F_X(x) = \mathrm{d}x \, \frac{1}{\pi} \lim_{z \to x} \mathrm{Im}(G_X(z)) \tag{4.27}$$

4.4.3 Examples

Consider the matrix:

$$\mathbf{Y} = \mathbf{X} + \mathbf{SPS}^\dagger \tag{4.28}$$

where $\mathbf{S} \in \mathbb{C}^{N \times K}$ contains *i.i.d.* elements with zero mean and variance $1/N$, $\mathbf{X} \in \mathbb{C}^{N \times N}$ is Hermitian with an e.e.d., which converges in distribution to the a.e.d. F_X almost surely as $N \to \infty$, and $\mathbf{P} \in \mathbb{R}^{K \times K}$ is diagonal with diagonal entries $\{P_k^K\}_{k=1, \dots, K}$, which converge almost surely in distribution to F_P as $K \to \infty$. It is shown in [50] that as $K, N \to \infty$ with $K/N \to \beta$, the e.e.d. of $\mathbf{Y}$ converges in distribution almost surely to a deterministic distribution function with Stieltjés transform, which satisfies:

$$G_Y(z) = G_X\left(z - \beta \int \frac{p \, \mathrm{d}F_P(p)}{1 + p \, G_Y(z)}\right) \tag{4.29}$$

where G_X is the Stieltjés transform of F_X. For the special case $\mathbf{X} = \mathbf{0}$ and $\mathbf{P} = P\mathbf{I}_K$, so that $G_X(z) = -1/z$, and $F_P(x) = \delta(x - P)$, where $\delta(x)$ is the Dirac delta function, (4.29) becomes:

$$G_Y(z) = \frac{1}{-z + \beta \frac{P}{1 + PG_Y(z)}} \tag{4.30}$$

The corresponding a.e.d., obtained by inverting the Stieltjés transform, is the Marčenko-Pastur distribution with corresponding density:

$$f_Y(x) = (1 - \beta)^+ \delta(x) + \frac{\sqrt{(x - a)^+(b - x)^+}}{2\pi x} \tag{4.31}$$

where $(x)^+ = \max(0, x)$, $a = (1 - \sqrt{\beta})^2$, and $b = (1 + \sqrt{\beta})^2$.

From (4.29) with $\mathbf{X} = 0$ and (4.25), we conclude that as $(K, N) \to \infty$, $\bar{\rho}_k^N$ converges almost surely to the deterministic limit $\bar{\rho}$, which satisfies the fixed point equation:

$$\bar{\rho} = \left(\sigma^2 + \beta \int \frac{p \, dF_P(p)}{1 + p \, \bar{\rho}} \right)^{-1} \tag{4.32}$$

where F_P is the asymptotic distribution of $\mathbf{P} = \mathbf{A}^2$. This relation for the large system SINR was presented in [58], and is known as the Tse-Hanly formula. It was derived by applying (4.29), as shown here. In Section 4.5 we give a simple, direct derivation of (4.32). Note, in particular, that $\bar{\rho}$ depends only on σ^2, β, and $F_P(\cdot)$, and is independent of k and the distribution of the elements of $\mathbf{S}$.

4.4.4 Asymptotic Equivalence

For the large system results, which follow, we need the following definition.

DEFINITION 1 *Let $\{a_N\}_{N=1}$ and $\{b_N\}_{N=1, \ldots}$ denote a pair of infinite sequences of complex-valued random variables indexed by N. These sequences are defined to be* asymptotically equivalent, *denoted $a_N \asymp b_N$, iff $|a_N - b_N| \xrightarrow{a.s.} 0$ as $N \to \infty$, where $\xrightarrow{a.s.}$ denotes almost-sure convergence in the limit considered.*

The relation $\asymp$ is an equivalence relation, and is clearly transitive. Asymptotic equivalence for sequences of $N \times 1$ vectors and $N \times N$ matrices is similarly defined, where the absolute value is replaced by the Euclidean vector norm and the associated induced spectral norm [27], respectively.

Note that $a_N \asymp b_N$ can hold even if neither sequence converges. Another possibility is that a_N and b_N converge, but not to a unique limit, so that $a_N \asymp b_N$ even though $\lim a_N$ and $\lim b_N$ are undefined. We will often state large system results in terms of asymptotic equivalence, which allows for the possibility that the sequences in the relation do not converge to unique limits.

As an example, if $\bar{\rho}_k^N$ in (4.18) converges to a large system limit, then we must have $\bar{\rho}_k \asymp \frac{1}{N} \text{tr}[\mathbf{R}_{b_k}^{-1}]$. Also, we observe that adding a user signature (corresponding to appending a column of $\mathbf{S}$) will not change this large system relation. In other words, the incremental degradation in SINR due to adding a single interferer with fixed power goes to zero in the large system limit. (More generally, we can add or subtract any finite number of users without changing the large system limit.) Hence we can write:

$$\bar{\rho}_k \asymp \frac{1}{N} \text{tr}[\mathbf{R}_{b_k}^{-1}] \asymp \frac{1}{N} \text{tr}[\mathbf{R}^{-1}] \tag{4.33}$$

To prove convergence of the performance metrics considered in the large system limit, the following definition is needed.

DEFINITION 2 *Let $\{\{a_{N,n}\}_{n=1, \ldots, N}\}_{N=1, \ldots}$ and $\{\{b_{N,n}\}_{n=1, \ldots, N}\}_{N=1, \ldots}$ denote a pair of infinite sequences, indexed by N. The Nth element is a complex-valued* *sequence of length N, indexed by n. These sequences are defined to be* uniformly asymptotically equivalent, *denoted* $a_{N,n} \overset{n}{\asymp} b_{N,n}$, iff $\max_{n \leq N} |a_{N,n} - b_{N,n}| \overset{a.s.}{\longrightarrow} 0$ as $N \to \infty$.

The relation $\overset{n}{\asymp}$ is again an equivalence relation, and is clearly transitive. Uniform asymptotic equivalence is, of course, stronger than asymptotic equivalence, and is needed to prove almost sure convergence of the performance measures considered. We will sometimes write $\overset{n}{\asymp}$, instead of $\asymp$, when this is needed to prove almost sure convergence, although the associated proofs are omitted.

4.5 INCREMENTAL MATRIX EXPANSION

In this section, we evaluate the SINR expression for the MMSE receiver (4.17) in the large system limit using an approach based on an incremental expansion of the correlation matrix. We start with the identity:

$$1 = \frac{1}{N}\text{trace }(\mathbf{R}^{-1}\mathbf{R}) \tag{4.34}$$

Substituting for $\mathbf{R}$ from (4.9) gives:

$$1 = \frac{1}{N}\text{trace }[\mathbf{R}^{-1}(\sigma^2\mathbf{I} + \mathbf{SPS}^{\dagger})]$$

$$= \sigma^2 \frac{1}{N}\text{trace }\mathbf{R}^{-1} + \frac{1}{N}\sum_{k=1}^{K} P_k(\mathbf{s}_k^{\dagger}\mathbf{R}^{-1}\mathbf{s}_k) \tag{4.35}$$

where we have represented $\mathbf{SPS}^{\dagger} = \sum_{k=1}^{K} P_k\mathbf{s}_k\mathbf{s}_k^{\dagger}$ as the sum of outer products, and we have used the fact that $\text{tr}[\mathbf{AB}] = \text{tr}[\mathbf{BA}]$.

Next, the Matrix Inversion Lemma (MIL) is used to *remove* the column $\mathbf{s}_k$ from $\mathbf{R}^{-1}$ (equivalently, expanding $\mathbf{R}^{-1}$ as the sum of two terms), leaving the matrix $\mathbf{R}_{b_k}^{-1}$, which is independent of $\mathbf{s}_k$. That is:

$$\mathbf{R}^{-1} = \mathbf{R}_{b_k}^{-1} - P_k\frac{\mathbf{R}_{b_k}^{-1}\mathbf{s}_k\mathbf{s}_k^{\dagger}\mathbf{R}_{b_k}^{-1}}{1 + P_k\mathbf{s}_k^{\dagger}\mathbf{R}_{b_k}^{-1}\mathbf{s}_k} \tag{4.36}$$

and substituting into the summand in (4.35) gives:

$$1 = \sigma^2 \frac{1}{N}\text{trace }\mathbf{R}^{-1} + \frac{1}{N}\sum_{k=1}^{K} P_k\left(\bar{\rho}_k^N - P_k\frac{(\bar{\rho}_k^N)^2}{1 + P_k\bar{\rho}_k^N}\right) \tag{4.37}$$

$$= \sigma^2 \frac{1}{N}\text{trace }\mathbf{R}^{-1} + \frac{1}{N}\sum_{k=1}^{K} \frac{P_k\bar{\rho}_k^N}{1 + P_k\bar{\rho}_k^N} \tag{4.38}$$

where $\bar{\rho}_k$ is the normalized SINR, defined in (4.17).

The removal of the contribution from s_k to the summand in (4.35) is the key step in the derivation of the large system SINR. This approach is therefore called an *incremental matrix expansion*. Extensions of this basic approach will be used later to derive performance measures for the other models presented in Section 4.2.

We now take the limit of (4.38) as $(K, N) \to \infty$ with fixed $\beta = K/N$. As stated in the preceding section, we will assume that the SINR $P_k \bar{\rho}_k^N$ converges to a deterministic limit, and define $\bar{\rho}_k = \lim_{N \to \infty} \bar{\rho}_k^N$. Taking the large system limit of (4.38) and applying (4.33) therefore gives (4.32), where we assume that the set of user powers $\{P_k\}$ converges in distribution almost surely to the continuous power distribution $F_P(\cdot)$.

As previously mentioned, this fixed point equation was first derived in [58] by applying the result (4.29) from [50]. The preceding derivation is presented in [71], although the manipulations are implicit in [49,50]. This derivation is not rigorous since it is based on the assumption that $\bar{\rho}_k^N$ converges to an asymptotic limit. Also, in going from (4.38) to (4.32), we have implicitly assumed that:

$$\lim_{N \to \infty} \sum_{k=1}^{K} Y_k^N = \sum_{k=1}^{K} \lim_{N \to \infty} Y_k^N \tag{4.39}$$

where $Y_k^N = \frac{1}{N} P_k \bar{\rho}_k^N / (1 + P_k \bar{\rho}_k^N)$. To show convergence of $\bar{\rho}_k^N$ to $\bar{\rho}$ in the almost sure sense, it is necessary to show that $\bar{\rho}_k^N \overset{n}{\asymp} \bar{\rho}$. We omit this, and refer the interested reader to [50, Lemma 2.6].

Note that the integral in (4.32) can also be written as $\mathbf{E}_P\left(\frac{P}{1+P\bar{\rho}}\right)$. Comparing (4.32) with the large system SINR for the matched filter given by (4.15) suggests that the term $\beta \mathbf{E}_P\left(\frac{P}{1+P\bar{\rho}_1^N}\right)$, where the expection is over the distribution of P, can be interpreted as the residual, or *effective interference* at the output of the MMSE filter [58].

Of course, an important question is how large does the system have to be for the large system results to be accurate? Figure 4.2 indicates that the average SINR (over signatures) does not vary much with system size and is quite close to the large system limit even for the smallest value of N shown. Because the distribution of the SINR over the signatures for finite N is unknown, there is no analytical description of how fast the SINR converges to the large system limit. However, related results for large N have been presented in [57,74]. Namely, it is shown in [57] that the variance of the SINR decreases as $1/N$, and an explicit expression for the variance is given for large N. Furthermore, it is shown in [74] that for large N the distribution of the residual interference is Gaussian. This enables the computation of error probabilities using the large system results. For example, for uncoded BPSK modulation the error probability for a particular user transmitting with power P with a linear MMSE receiver, averaged over signatures, is accurately approximated by $Q(\sqrt{P\bar{\rho}})$, where $Q(x) = \int_x^\infty \frac{1}{\sqrt{2\pi}} e^{-t^2/2} \, dt$.

Finally, we remark that if we replace σ^2 in the preceding derivation by the complex variable $-z$, then (4.32) becomes the fixed-point equation for the Stieltjés transform of the a.e.d. $G_{\mathbf{R}}(z)$, given by (4.29) with $\mathbf{X} = 0$.

4.6 ANALYSIS OF DOWNLINK MODEL

We now consider the downlink model (4.3). Our objective is again to obtain the large system limit of the SINR with the linear MMSE receiver. Of course, in this case, the answer must depend on properties of the channel matrix $\mathbf{H}$. We focus on the model (4.3) mainly to simplify the notation and discussion. It is straightforward to extend the following analysis to the multi-cell downlink model (4.6) and to the more general linear model (4.1).

4.6.1 MMSE Receiver and SINR

The MMSE filter c_k, which minimizes $\mathbf{E}[|b_k - \mathbf{c}_k^\dagger \mathbf{y}|^2]$, where $\mathbf{y}$ is given by (4.3), is given by:

$$\mathbf{c}_k = A_k \mathbf{R}^{-1} \mathbf{H}\mathbf{s}_k \qquad (4.40)$$

where:

$$\mathbf{R} = \mathbf{E}[\mathbf{y}\mathbf{y}^\dagger] = \sigma^2 \mathbf{I}_M + \widetilde{\mathbf{R}} \qquad (4.41)$$

$$\widetilde{\mathbf{R}} = (\mathbf{H}\mathbf{S}\mathbf{A})^\ddagger \qquad (4.42)$$

and the expectation in (4.41) is with respect to $\mathbf{b}$ and $\mathbf{n}$.

Focusing on the kth transmitted stream, we can write:

$$\mathbf{y} = A_k \mathbf{H}\mathbf{s}_k b_k + \mathbf{y}_k \qquad (4.43)$$

where the first term is the signal and the second term contains the multiuser interference plus noise, i.e.:

$$\mathbf{y}_k = A_{b_k} \mathbf{H}_{b_k} \mathbf{s}_{b_k} \mathbf{b}_{b_k} + \mathbf{n} \qquad (4.44)$$

The subscript b_k again indicates that the contribution corresponding to the symbol b_k is removed. The corresponding SINR at the output of the MMSE receiver is given by:

$$\rho_k^N = \frac{\mathbf{E}[|\mathbf{c}_k^\dagger(\mathbf{y} - \mathbf{y}_k)|^2]}{\mathbf{E}[|\mathbf{c}_k^\dagger \mathbf{y}_k|^2]} \qquad (4.45)$$

$$= \frac{P_k \left| \mathbf{c}_k^\dagger \mathbf{H}\mathbf{s}_k \right|^2}{\mathbf{c}_k^\dagger \mathbf{R}_{b_k} \mathbf{c}_k} \qquad (4.46)$$

where:

$$\mathbf{R} = \mathbf{R}_{b_k} + P_k (\mathbf{H}\mathbf{s}_k)^\ddagger \qquad (4.47)$$

and the expectation in (4.45) is with respect to $\mathbf{n}$ and $\mathbf{b}$. The superscript N again signifies that the associated variable corresponds to a finite size system, as opposed to the asymptotic variables studied in subsequent sections. For finite N, the SINR depends on the realizations of $\mathbf{H}$, $\mathbf{S}$, and $\mathbf{A}$, and is therefore random.

Applying the MIL to (4.47), the SINR (4.46) can be written as:

$$\rho_k^N = P_k\,\bar{\rho}_k^N \tag{4.48}$$

where:

$$\bar{\rho}_k^N = \mathbf{s}_k^\dagger \mathbf{H}^\dagger \mathbf{R}_{b_k}^{-1} \mathbf{H} \mathbf{s}_k \tag{4.49}$$

4.6.2 Large-System SINR

Due to the presence of $\mathbf{H}$ in (4.3), the normalized SINR ρ_k^N does not turn out to be asymptotically equivalent to the Stieltjés transform $G_{\mathbf{R}}^N(z)$, as it was in (4.25) for the basic model (4.2). Rather, we have the following asymptotic equivalences:

$$\bar{\rho}_k^N \overset{k}{\asymp} \bar{\rho}^N = \begin{cases} \frac{1}{N}\,\mathrm{tr}[\mathbf{H}^\ddagger \mathbf{R}^{-1}], & i.i.d.\,\mathbf{S} \\ \frac{1}{N-K}\,\mathrm{tr}[(\mathbf{H}\mathbf{\Pi})^\ddagger \mathbf{R}^{-1}], & iso.\,\mathbf{S} \end{cases} \tag{4.50}$$

where $\mathbf{\Pi}^2 = \mathbf{I}_N - \mathbf{S}\mathbf{S}^\dagger$. The relation for $i.i.d.\,\mathbf{S}$ follows from (4.49), and can be justified by a similar argument as used to show (4.33). The relation for isometric $\mathbf{S}$ follows from [7, Proposition 4], and takes a little more effort to show. That is, consider the general quadratic form $\mathbf{s}_k^\dagger \mathbf{Y}^N \mathbf{s}_k$, where $\mathbf{s}_k$ is selected from $\mathbf{S}$ and $\mathbf{Y}^N$ can be a function of $\mathbf{S}_{b_k}$, but not $\mathbf{s}_k$. Also, we must assume that $\sup_N \|\mathbf{Y}^N\| < \infty$. Now, without lack of generality, we can assume that $\mathbf{S}$ is obtained via a Gram-Schmidt orthogonalization of an $N \times K$ matrix $\mathbf{X} = [\mathbf{x}_1, \ldots, \mathbf{x}_K]$, where the elements of $\mathbf{X}$ are $i.i.d.$, zero mean, unit variance, proper complex Gaussian random variables. That is, $\mathbf{s}_k = \dfrac{\mathbf{\Pi}_k \mathbf{x}_k}{|\mathbf{\Pi}_k \mathbf{x}_k|}$ where $\mathbf{\Pi}_k = \mathbf{I}_N - \mathbf{S}_{b_k}\mathbf{S}_{b_k}^\dagger$. Therefore:

$$\mathbf{s}_k^\dagger \mathbf{Y}^N \mathbf{s}_k = \frac{\mathbf{x}_k^\dagger \mathbf{\Pi}_k^\dagger \mathbf{Y}^N \mathbf{\Pi}_k \mathbf{x}_k}{|\mathbf{\Pi}_k \mathbf{x}_k|^2}. \tag{4.51}$$

Since $\mathbf{x}_k$ is complex Gaussian, and $\mathbf{\Pi}_k$ is an orthogonal projection matrix of rank $N - K + 1$, it follows that $|\mathbf{\Pi}_k \mathbf{x}_k|^2$ is the sum of $N - K + 1$ independent, mean zero, unit variance complex Gaussian random variables. Hence in the large system limit, assuming $K/N \to \beta < 1$, $|\mathbf{\Pi}_k \mathbf{x}_k|^2/(N - K) \to 1$. Therefore:

$$\mathbf{s}_k^\dagger \mathbf{Y}^N \mathbf{s}_k \overset{k}{\asymp} \frac{\mathrm{tr}\left[\mathbf{\Pi}_k \mathbf{Y}^N\right]}{N - K} \overset{k}{\asymp} \frac{\mathrm{tr}\left[\mathbf{\Pi}\mathbf{Y}^N\right]}{N - K} \tag{4.52}$$

where $\mathbf{\Pi} = \mathbf{I}_N - \mathbf{SS}^\dagger$.

4.6.3 Two Important Preliminary Results

We are interested in evaluating the large system limit $\bar{\rho} = \lim_{N \to \infty} \bar{\rho}_k^N$, where $\bar{\rho}_k^N$ is defined in (4.50). Our approach depends on an important observation and two related results, which are presented next.

4.6.3.1 *Rotational Invariance of SINR* We first make the intuitive observation that multiplying the received vector $\mathbf{y}$ in (4.3) by an orthogonal (rotation) matrix should not change the SINR performance metric. This rotation should also not change the eigen-values of the correlation matrix. Hence for purposes of computing the large system SINR, we can replace $\mathbf{H}$ by $\mathbf{VH}$, where $\mathbf{V}$ is a Haar matrix. Furthermore, we can write the singular value decomposition $\mathbf{H} = \mathbf{V}_l \mathbf{D} \mathbf{V}_r^\dagger$, where $\mathbf{V}_l$ and $\mathbf{V}_r$ are unitary matrices, and have as columns the left and right singular vectors of $\mathbf{H}$, and the diagonal matrix $\mathbf{D}$ contains the singular values of $\mathbf{H}$, which we denote $d_1, \ldots, d_N$. (If $\mathbf{H}$ is diagonal, then $\mathbf{V}_l = \mathbf{I}_M$ and $\mathbf{V}_r = \mathbf{I}_N$.) Hence $\mathbf{HS}$ in the model (4.3) can be replaced by $\mathbf{V}\mathbf{V}_l \mathbf{D} \mathbf{V}_r \mathbf{S}$. Now $\mathbf{V}\mathbf{V}_l$ is unitary, and if we further assume that $\mathbf{S}$ is unitarily invariant, then we can replace $\mathbf{V}_r \mathbf{S}$ by $\mathbf{S}$ without affecting the SINR. Hence we conclude that we can replace $\mathbf{HS} = \mathbf{V}_l \mathbf{D} \mathbf{V}_r \mathbf{S}$ by $\mathbf{VDS}$, or equivalently, $\mathbf{H}$ by $\mathbf{VD}$, where $\mathbf{V}$ is Haar, without affecting the SINR. This is crucial, because we can then apply the matrix expansion approach to $\mathbf{V}$ to derive useful asymptotic relations, which cannot be directly obtained without this substitution for $\mathbf{H}$.

Formally, we have the following proposition.

Proposition 1 *For the model (4.3), the distribution of both the Stieltjés transform of the e.e.d. of $\tilde{\mathbf{R}} = (\mathbf{HSA})^\ddagger$ and the SINR are invariant to the substitution of $\mathbf{VD}$ for $\mathbf{H}$, where $\mathbf{V}$ is an $M \times N$ Haar-distributed random unitary matrix, and $\mathbf{D}$ is an $M \times N$ diagonal matrix containing the singular values of $\mathbf{H}$.*

***Proof*:** Let $\mathbf{T}$ be an independent $M \times M$ Haar-distributed random matrix. We first show that the quantities of interest, $G_{\tilde{\mathbf{R}}}^N(z)$ and $\bar{\rho}_k^N$, are unchanged by the substitution of $\mathbf{TH}$ for $\mathbf{H}$. That is:

$$G_{\tilde{\mathbf{R}}}^N(z) = \frac{1}{N}\mathrm{tr}[\mathbf{R}^{-1}] = \frac{1}{N}\mathrm{tr}[(-z\mathbf{I}_N + (\mathbf{HSA})^\ddagger)^{-1}] \tag{4.53}$$

$$= \frac{1}{N}\mathrm{tr}[\mathbf{TT}^\dagger \mathbf{R}^{-1}] = \frac{1}{N}\mathrm{tr}[(-z\mathbf{I}_N + (\mathbf{THSA})^\ddagger)^{-1}] \tag{4.54}$$

and:

$$\bar{\rho}_k^N = \mathbf{s}_k^\dagger \mathbf{H}^\dagger \mathbf{R}^{-1} \mathbf{H} \mathbf{s}_k \tag{4.55}$$

$$= \mathbf{s}_k^\dagger \mathbf{H}^\dagger \mathbf{T}^\dagger \mathbf{T} \mathbf{R}^{-1} \mathbf{T}^\dagger \mathbf{T} \mathbf{H} \mathbf{s}_k \tag{4.56}$$

$$= \mathbf{s}_k^\dagger (\mathbf{T} \mathbf{H})^\dagger (-z\mathbf{I}_N + (\mathbf{T} \mathbf{H} \mathbf{S} \mathbf{A})^\ddagger)^{-1} (\mathbf{T} \mathbf{H}) \mathbf{s}_k \tag{4.57}$$

Writing $\mathbf{T} \mathbf{H} \mathbf{S} = (\mathbf{T} \mathbf{V}_l) \mathbf{D} (\mathbf{V}_r^\dagger \mathbf{S})$, the unitary invariance of $\mathbf{T}$ and $\mathbf{S}$ gives the result.

4.6.3.2 Covariance Matrix Expansion Along Transmit Dimensions
Here we present an alternative expansion to the one given in (4.47) for the covariance matrix $\mathbf{R}$. Namely, Proposition 1 implies that we can replace $\mathbf{H} \mathbf{S}$ with:

$$\mathbf{V} \mathbf{D} \mathbf{S} = \sum_{n=1}^{\alpha^* N} d_n \mathbf{v}_n \tilde{\mathbf{s}}_n^\dagger \tag{4.58}$$

where $\alpha^* = \min(M, N)/N$ and $\tilde{\mathbf{s}}_n$ is the nth row of $\mathbf{S}$. We can therefore expand $\mathbf{R}$ as:

$$\mathbf{R} = [(\mathbf{H}_{t_n} \mathbf{S}_{t_n} + d_n \mathbf{v}_n \tilde{\mathbf{s}}_n^\dagger) \mathbf{A}]^\ddagger - z\mathbf{I}_N \tag{4.59}$$

$$= \mathbf{R}_{t_n} + d_n \mathbf{u}_n \mathbf{v}_n^\dagger + d_n \mathbf{v}_n \mathbf{u}_n^\dagger + d_n^2 c_n \mathbf{v}_n \mathbf{v}_n^\dagger \tag{4.60}$$

for $0 < n \le \alpha^* N$, where:

$$\mathbf{R}_{t_n} = (\mathbf{H}_{t_n} \mathbf{S}_{t_n} \mathbf{A})^\ddagger - z\mathbf{I}_N, \tag{4.61}$$

$$\mathbf{u}_n = \mathbf{H}_{t_n} \mathbf{S}_{t_n} \mathbf{A}^2 \tilde{\mathbf{s}}_n, \tag{4.62}$$

$$c_n = \tilde{\mathbf{s}}_n^\dagger \mathbf{A}^2 \tilde{\mathbf{s}}_n, \tag{4.63}$$

and $\mathbf{H}_{t_n}$ and $\mathbf{S}_{t_n}$ denote $\mathbf{H}$ and $\mathbf{S}$ with their nth column and row removed, respectively. The subscript t_n indicates that the signal component along the nth singular value of $\mathbf{H}$, or *transmit* dimension, has been removed.

In contrast to the use of the MIL in (4.36), in this case the MIL cannot be directly applied to the expansion in (4.60). For this, we require the following asymptotic extension of the MIL.

Lemma 1 (**Asymptotic extension to MIL [43]**) *Let*:

$$\mathbf{Y}_N = \mathbf{X}_N + \mathbf{v}_N \mathbf{u}_N^\dagger + \mathbf{u}_N \mathbf{v}_N^\dagger + c_N \mathbf{u}_N \mathbf{u}_N^\dagger, \tag{4.64}$$

where $\mathbf{v}_N, \mathbf{u}_N \in \mathbb{C}^N$, $c_N \in \mathbb{R}^*$, and $\mathbf{X}_N = \mathbf{Q}_N - z\mathbf{I}_N$, where $\mathbf{Q}_N$ is an $N \times N$ Hermitian matrix and $z \in \mathbb{C}^+$. Assume that as $N \to \infty$:

$$\mathbf{u}_N^\dagger \mathbf{X}_N^{-1} \mathbf{v}_N \asymp 0 \tag{4.65}$$

and:

$$g = \inf_N |\mathbf{u}_N| > 0 \tag{4.66}$$

$$B = \sup_N \max\{\|\mathbf{X}_N\|, |\mathbf{v}_N|, |\mathbf{u}_N|, |c_N|\} < \infty \tag{4.67}$$

Then:

$$\mathbf{Y}_N^{-1}\mathbf{u}_N \asymp \frac{\mathbf{X}_N^{-1}(\mathbf{u}_N - u_N\mathbf{v}_N)}{1 - u_N(v_N - c_N)}, \tag{4.68}$$

$$\mathbf{Y}_N^{-1}\mathbf{v}_N \asymp \frac{\mathbf{X}_N^{-1}(-v_N\mathbf{u}_N + (1 + c_N u_N)\mathbf{v}_N)}{1 - u_N(v_N - c_N)}, \tag{4.69}$$

where:

$$u_N = \mathbf{u}_N^\dagger \mathbf{X}_N^{-1}\mathbf{u}_N, \qquad v_N = \mathbf{v}_N^\dagger \mathbf{X}_N^{-1}\mathbf{v}_N \tag{4.70}$$

and:

$$\epsilon = \inf_N |1 - u_N(v_N - c_N)| > 0 \tag{4.71}$$

where ϵ depends only on B, g, and Im(z).

Proof: This can be shown by repeated applications of the MIL. That is, dropping the subscript N, let:

$$\mathbf{Y} = \mathbf{X}_2 + c\mathbf{u}\mathbf{u}^\dagger$$
$$\mathbf{X}_2 = \mathbf{X}_1 + \mathbf{u}\mathbf{v}^\dagger$$
$$\mathbf{X}_1 = \mathbf{X} + \mathbf{v}\mathbf{u}^\dagger$$

The MIL gives relations for $\mathbf{X}_1^{-1}$, $\mathbf{X}_2^{-1}$, and $\mathbf{Y}^{-1}$, which can be simplified by using (4.65).

The result (4.71) states that the denominator term in (4.68)–(4.69) is uniformly bounded, which is required for manipulations involving asymptotic equivalence relations.

4.6.4 Large System SINR

The large system SINR for the system model (4.3) can be computed by applying the preceding results. Namely, we can expand the received covariance in two different ways, given by (4.47) and (4.60). Applying the MIL to (4.47) and Lemma 1 to (4.60), and taking the large system limit gives a set of asymptotic equivalences, which can be manipulated and solved for the large system SINR $\bar{\rho} = \lim_{N\to\infty} \rho^N$, where $\bar{\rho}^N$ is defined in (4.50). We state the expression for the Stieltjés transform of the a.e.d. of $\tilde{\mathbf{R}}$ as the following theorem. The expression for asymptotic SINR then follows directly. We provide a sketch of the proof of Theorem 1 in Appendix 4.10.

Theorem 1 [43] *As* $(M, N, K) \to \infty$ *with* $M/N \to \alpha > 0$ *and* $K/N \to \beta > 0$, *the e.e.d. of* $\tilde{\mathbf{R}} = (\mathbf{HSA})^{\ddagger}$, $F_{\tilde{\mathbf{R}}}^N$, *almost surely converges in distribution to a deterministic distribution* $F_{\tilde{\mathbf{R}}}$, *whose Stieltjés transform* $G_{\tilde{\mathbf{R}}}(z) = \gamma$, $z \in \mathbb{C}^+$, *satisfies*:

$$\gamma = -z^{-1}\left(1 - \frac{\beta}{\alpha}\bar{\rho}\mathcal{E}\right) \tag{4.72}$$

where:

$$\bar{\rho} = \begin{cases} \alpha^*\mathcal{H}, & i.i.d. \ \mathbf{S}, \\ \dfrac{\alpha^*\mathcal{H}}{1 - \alpha(1 + z\gamma)}, & iso. \ \mathbf{S}, \end{cases} \tag{4.73}$$

$\alpha^* = min(\alpha, \ 1)$:

$$\mathcal{E} = \mathbf{E}\left[\frac{P}{1 + \bar{\rho}P}\right], \tag{4.74}$$

$$\mathcal{H} = \mathbf{E}\left[\frac{H}{-z + \phi H}\right], \tag{4.75}$$

and:

$$\phi = \begin{cases} \beta\mathcal{E}, & i.i.d. \ \mathbf{S}, \\ \dfrac{\beta\mathcal{E}}{1 - \alpha(1 + z\gamma)}, & iso. \ \mathbf{S}. \end{cases} \tag{4.76}$$

The expectations in (4.74) and (4.75) are with respect to the scalar random variables P *and* H, *respectively, where the distributions of* P *and* H *are the a.e.d.'s of* $\mathbf{A}^2$ *and the first* $\alpha^*N = min(M, N)$ *non-zero eigen-values of* $\mathbf{H}^{\ddagger}$, *respectively.*

The large system SINR $\bar{\rho}$ is obtained by letting $z \to -\sigma^2$. For *i.i.d.* **S** (4.73)–(4.76) give the coupled equations:

$$\bar{\rho} = \alpha^*\mathbf{E}\left[\frac{H}{\sigma^2 + H\phi}\right] \qquad \phi = \beta\mathbf{E}\left[\frac{P}{1 + P\bar{\rho}}\right] \tag{4.77}$$

If the channel is flat fading (i.e., all nonzero singular values are the same) with known gain, then H becomes a deterministic constant, and (4.77) essentially reduces to the fixed-point expression (4.32) for the basic model (4.2). For the frequency-selective channel considered, H becomes a random variable, and (4.77) indicates that the SINR is computed by simply averaging the corresponding SINR (i.e., with deterministic H) over the first-order distribution of channel gains.

For isometric $\mathbf{S}$, letting $z \to -\sigma^2$ in Theorem 1 gives:

$$\bar{\rho} = \frac{\alpha^*}{1 - \alpha(1 - \sigma^2 \gamma)} \mathbf{E}\left[\frac{H}{\sigma^2 + H\phi}\right] \qquad \phi = \frac{\beta}{1 - \alpha(1 - \sigma^2 \gamma)} \mathbf{E}\left[\frac{P}{1 + P\bar{\rho}}\right] \qquad (4.78)$$

where the denominator:

$$1 - \alpha(1 - \sigma^2 \gamma) = \beta\left(1 - \mathbf{E}\left[\frac{1}{1 + P\bar{\rho}}\right]\right) = \alpha^*\left(1 - \mathbf{E}\left[\frac{1}{1 + \sigma^2 H\phi}\right]\right) \qquad (4.79)$$

It is easy to verify that if H is a deterministic constant, then $\bar{\rho} = H/\sigma^2$, as expected for $\beta < 1$.

Comparing the expressions for *i.i.d.* and isometric $\mathbf{S}$, isometric $\mathbf{S}$ achieves a higher SINR if and only if $1 - \alpha(1 - \sigma^2 \gamma) < 1$. From (4.79) this is always the case if $\beta \leq 1$, in which case isometric $\mathbf{S}$ corresponds to orthogonal signatures.

Arguments similar to those used to derive the preceding SINR results for the model (4.3) can be used to derive the large system SINR for the more general signal model (4.6). (For details, see [43].) Namely, in this case the large system limit is defined by letting $(M, N, K_j) \to \infty$ with $M/N \to \alpha > 0$ and $K_j/N \to \beta_j > 0$, $j = 1, \ldots, J$. That is, for the downlink interpretation of the model the number of base stations J is fixed, and the number of signatures K_j assigned to a particular base station j tends to infinity in proportion with N. For the uplink interpretation, the number of *users* is fixed, and K_j is the number of signatures assigned to user j. In this case, the Stieltjés transform of the e.e.d. of $\tilde{\mathbf{R}} = \sum_{j=1}^{J} (\mathbf{H}_j \mathbf{S}_j \mathbf{A}_j)^{\ddagger}$ converges to the deterministic function:

$$G_{\tilde{\mathbf{R}}}(z) = -\frac{1}{z}\left(1 - \sum_{j=1}^{J} \beta_j \bar{\rho}_j \mathcal{E}_j\right) \qquad (4.80)$$

where:

$$\bar{\rho}_j = \begin{cases} \mathcal{H}_j, & \text{i.i.d. } \mathbf{S}_j, \\[2mm] \dfrac{\mathcal{H}_j}{1 - \beta_j \bar{\rho}_j \mathcal{E}_j}, & \text{iso. } \mathbf{S}_j. \end{cases} \qquad (4.81)$$

$$\mathcal{E}_j = \mathbf{E}\left[\frac{P_j}{1 + P_j \bar{\rho}_j}\right], \qquad (4.82)$$

$$\mathcal{H}_j = \mathbf{E}\left[\frac{H_j}{-z + \sum_i \phi_i H_i}\right], \qquad (4.83)$$

$$\phi_i = \begin{cases} \mathcal{E}_j, & \text{i.i.d. } \mathbf{S}_j, \\[1mm] \mathcal{E}_j + \phi_j^2 \mathcal{H}_j, & \text{iso. } \mathbf{S}_j. \end{cases} \qquad (4.84)$$

The expectations in (4.82) and (4.83) are with respect to the random variables $\{H_j\}_{j=1,\ldots,J}$ and P_j, where the distributions of H_j and P_j are the a.e.d.'s of $\mathbf{H}_j^{\ddagger}$ and $\mathbf{A}_j^{\ddagger}$, respectively.

Letting $z \rightarrow -\sigma^2$, the asymptotic normalized SINR with *i.i.d.* signatures satisfies the coupled equations:

$$\bar{\rho}_j = \mathbf{E}\left[\frac{H_j}{\sigma^2 + \sum_{i=1}^{J}\phi_i H_i}\right] \qquad \phi_j = \beta_j \mathbf{E}\left[\frac{P_j}{1 + \bar{\rho}_j P_j}\right] \qquad (4.85)$$

for $j = 1, \ldots, J$. With isometric signatures the normalized SINR is given by (4.81) where:

$$\mathcal{H}_j = \mathbf{E}\left[\frac{P_j}{\sigma^2 + \sum_i \phi_i H_i}\right] \qquad \phi_i = \beta_i \frac{\mathcal{E}_i}{1 - \mathcal{H}_i}. \qquad (4.86)$$

These expressions apply even when the H_j's are correlated, provided that they are uniformly bounded, and that the joint e.e.d. converges in distribution almost surely.

Comparing (4.85) with the MMSE expression (4.32) for the model (4.2) with *i.i.d.* signatures, and conditioning on the channels H_j, $j = 1, \ldots, J$, we can identify the effective interference term as $\sum_{i=1}^{J}\beta_i H_i \mathbf{E}[P_i/(1 + \bar{\rho}_i P_i)]$. This is simply the effective interference averaged over the cells. The large system MMSE with random channels is then obtained by averaging over the channel gains. Focusing on a particular user j, isometric signatures achieve a higher SINR than *i.i.d.* signatures when $\beta_j \bar{\rho}_j \mathcal{E}_j < 1$. This is true if $\beta_j < 1$.

4.6.5 Numerical Example

Figure 4.3 shows plots of large system SINR versus system load $\beta = \beta_1 + \beta_2$ corresponding to the model (4.6) with two transmitters. The figure also shows results for $N = 32$ obtained by averaging over 2000 channel and noise realizations with binary transmitted symbols. The distribution of the channel gain H_j, $j = 1, 2$, is assumed to be exponential, corresponding to frequency-selective Rayleigh fading. In this example, Transmitter 1 is assigned 50% more signatures than Transmitter 2 ($\beta_1 = 0.6\beta$, $\beta_2 = 0.4\beta$), the signatures on all transmitters are assigned equal power ($\mathbf{A}_1 = \mathbf{I}_{K_1}$, $\mathbf{A}_2 = \mathbf{I}_{K_2}$), and $\mathbf{E}[H_1] = 1$, $\mathbf{E}[H_2]/\mathbf{E}[H_1] = -3$ dB (i.e., the received signal power from Transmitter 2 is half that of Transmitter 1). Curves are shown for the SNRs 8, 12, 16, and 20 dB. In all cases the asymptotic values essentially match the empirical results, indicating that the asymptotic analysis can be used to predict the performance of finite-size systems of interest. These results show that the SINR improvement with isometric signatures relative to *i.i.d.* signatures ranges between one and two dB.

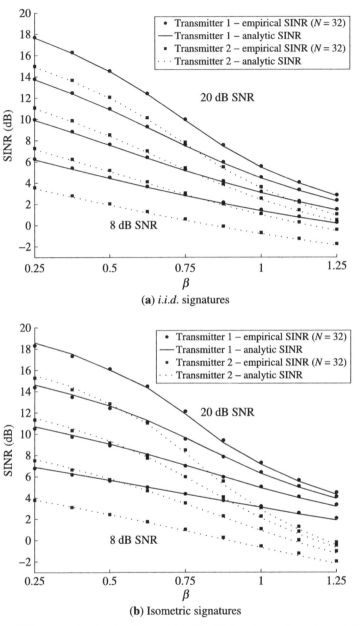

Figure 4.3. SINR versus load β for the model (4.6) with two transmitters. Asymptotic (large system) values are compared with numerically averaged values generated for a finite-size system with $N = 32$. Parameters are $\beta_1 = 0.6\beta$, $\beta_2 = 0.4\beta$, $\mathbf{A}_1 = \mathbf{I}_{K_1}$, $\mathbf{A}_2 = \mathbf{I}_{K_2}$, and $\mathbf{E}[H_1] = 1$, $\mathbf{E}[H_2]/\mathbf{E}[H_1] = -3$ dB.

4.7 SPECTRAL EFFICIENCY

The *spectral efficiency* for transmitter k, denoted as C_k, refers to the maximum achievable rate (i.e., with error control coding) normalized by bandwidth. Taking the bandwidth to be the chip rate, C_k is stated in units of bits per chip. Here we compute the spectral efficiency in the large system limit for linear MMSE and optimal (capacity-achieving) receivers for the models presented in Section 4.2. For the linear receiver, the filter output is passed directly to the decoder. The following results were first derived for the flat fading channel model (4.2) in [18,48,63].

4.7.1 Sum Capacity

With either the matched filter or linear MMSE receiver, the residual interference plus noise at the filter output becomes Gaussian in the large system limit. Hence the large system spectral efficiency for user k is simply given by the Shannon formula:

$$C_k = \log(1 + P_k \bar{\rho}_k) \tag{4.87}$$

where $P_k \bar{\rho}_k$ is the corresponding large system SINR at the filter output.[10] For the system models (4.2) and (4.3), the normalized *sum* spectral efficiency per signature is:

$$\bar{C}_{\text{lin}}^N = \frac{1}{N} \sum_{k=1}^{K} C_k \tag{4.88}$$

and in the large system limit, this becomes an average over the power distribution, i.e.:

$$\bar{C}_{\text{lin}}^N \to \beta \int \log(1 + P\bar{\rho}) \, dF(P) \tag{4.89}$$

since $\bar{\rho}_k$ is independent of k in the large system limit.

With an optimal receiver the normalized sum spectral efficiency per signature is given by [63], [56, Ch. 10]:

$$\bar{C}^N = \frac{1}{N} \log |\sigma^{-2} \mathbf{R}| \tag{4.90}$$

where $|\mathbf{M}|$ is the determinant of the matrix $\mathbf{M}$, and $\mathbf{R} = \mathbf{E}[\mathbf{y}\mathbf{y}^{\dagger}]$ is the input covariance matrix. This is true for all of the models discussed in Section 4.2.[11] For the general model (4.1), we have $\mathbf{R} = \sigma^2 \mathbf{I}_N + \tilde{\mathbf{R}}$ where $\tilde{\mathbf{R}} = \mathbf{M}\mathbf{M}^{\dagger}$, and we can

---

[10]Note that to compute the spectral efficiency in terms of energy per bit, instead of energy per symbol, the power P_k must be replaced by $C_k P_k$ [63].

[11]In the original expression in [63], $\mathbf{R}$ is replaced by $\sigma^2 \mathbf{I}_K + \mathbf{M}^{\dagger}\mathbf{M}$, assuming the general model (4.1). The two expressions are equivalent since $\mathbf{M}^{\dagger}\mathbf{M}$ and $\mathbf{M}\mathbf{M}^{\dagger}$ have the same non-zero eigen-values.

rewrite (4.90) as:

$$\bar{\mathsf{C}}^N = \frac{1}{N} \sum_{n=1}^{N} \log\left(1 + \lambda_n/\sigma^2\right) \tag{4.91}$$

where $\{\lambda_n\}$ is the set of eigen-values of $\tilde{\mathbf{R}}$. Taking the large system limit gives:

$$\bar{\mathsf{C}} = \int \log\left(1 + \lambda/\sigma^2\right) dF_{\tilde{\mathbf{R}}}(\lambda) \tag{4.92}$$

where $F_{\tilde{\mathbf{R}}}(\cdot)$ is the asymptotic eigen-value distribution of $\tilde{\mathbf{R}}$.

The integral in (4.92) can be evaluated in closed-form for the flat fading model (4.2) when $\mathbf{A} = \mathbf{I}$ (equal amplitudes) (see [60]). Another convenient expression for $\bar{\mathsf{C}}$ is obtained by expressing $\bar{\mathsf{C}}^N$ as a function of σ^2 and differentiating, i.e.:

$$\frac{d\bar{\mathsf{C}}^N(\sigma^2)}{d\sigma^2} = -\frac{1}{N} \sum_{n=1}^{N} \frac{\lambda_n}{\sigma^2(\sigma^2 + \lambda_n)} \tag{4.93}$$

$$= \frac{1}{N} \sum_{n=1}^{N} \left(\frac{1}{\sigma^2 + \lambda_n} - \frac{1}{\sigma^2}\right) \tag{4.94}$$

$$= G_{\tilde{\mathbf{R}}}^N(-\sigma^2) - \frac{1}{\sigma^2} \tag{4.95}$$

where $G_{\tilde{\mathbf{R}}}^N(\cdot)$ is the Stieltjés transform of the e.e.d. of $\tilde{\mathbf{R}}$, as defined in (4.24). Since the spectral efficiency goes to zero as the noise level increases to infinity, we have the boundary condition $\lim_{\sigma^2 \to \infty} \bar{\mathsf{C}}^N(\sigma^2) = 0$. Taking the large system limit of (4.95) therefore gives:

$$\bar{\mathsf{C}}(\sigma^2) = \int_{\infty}^{\sigma^2} \left(G_{\tilde{\mathbf{R}}}^N(-z) - \frac{1}{z}\right) dz = -\int_{-\infty}^{-\sigma^2} \left(G_{\tilde{\mathbf{R}}}(z) + \frac{1}{z}\right) dz \tag{4.96}$$

where $G_{\tilde{\mathbf{R}}}(\cdot)$ is the Stieltjés transform of the a.e.d. of $\tilde{\mathbf{R}}$, e.g., given by Theorem 1 for the model (4.3).

Some numerical examples for the downlink multi-cell model (4.6) are shown in Figure 4.4, in which case $G_{\tilde{\mathbf{R}}}(\cdot)$ is given by (4.80)–(4.84). Asymptotic sum spectral efficiencies are plotted versus total system load β with different numbers of cells J as a parameter. Results are shown with both linear MMSE and optimal receivers, and the channel $\mathbf{H}$ is assumed to have exponentially distributed eigen-values with unit variance corresponding to Rayleigh fading. In each case the load (number of signatures) for each cell is $\beta_j = K_j/N = \beta/J$, and $\frac{E_b}{N_0} = 10\,\text{dB}$ for each signature (i.e., the total power assigned to a user is proportional to the number of assigned signatures). Figures 4.4a and 4.4b show results for *i.i.d.* and isometric signatures, respectively. Also shown for comparison are the single-user spectral efficiencies with AWGN

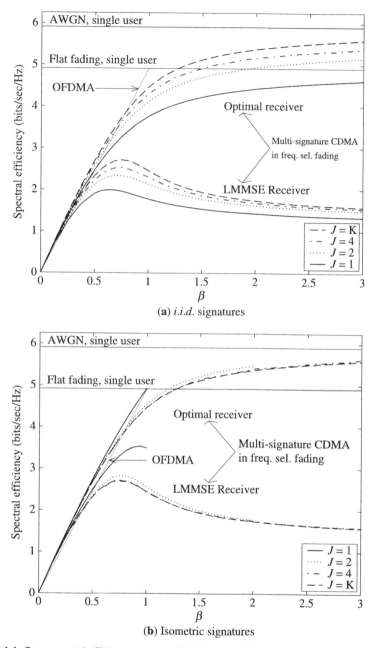

Figure 4.4. Sum spectral efficiency vs β for OFDMA and CDMA with $J = 1, 2, 4$ cells, K signatures, and $\frac{E_b}{N_0} = 10\,\mathrm{dB}$ for each signature. In each case $\beta_j = \beta/J$.

and flat fading, given by:

$$C_{AWGN}(\sigma^2) = \log_2\left(1 + \frac{1}{\sigma^2}\right) \tag{4.97}$$

$$C_{Fading}(\sigma^2) = \int_0^\infty \log_2\left(1 + \frac{x}{\sigma^2}\right)e^{-x}\,dx \tag{4.98}$$

The capacity with fading is the *ergodic* capacity, i.e., assumes that the codeword is spread over fading states. For example, this applies to an Orthogonal Frequency Division Multiplexing (OFDM) system in which the coding is across sub-channels. The straight line labeled "OFDMA (Orthogonal Frequency Division Multiple Access)" applies to the situation in which the signatures occupy nonoverlapping sets of sub-channels, and hence there is no multi-user interference. In that case, the system load β_j designates the fraction of total bandwidth allocated to cell j.

Focusing first on the results for *i.i.d.* signatures, with the optimal receiver the sum spectral efficiency increases monotonically, whereas with the linear MMSE receiver, there is a unique maximum, corresponding to a system load $\beta < 1$. This is due to the fact that a linear receiver is unable to suppress interference effectively when the load becomes large. Hence the achievable rate decreases once the load increases beyond a critical value.

As $\beta \to \infty$ (with fixed J), the optimal spectral efficiency with $K = J$ (single signature per transmitter) appears to approach the AWGN single-user spectral efficiency, whereas with $J = 1$ (single cell with K signatures) the optimal sum spectral efficiency appears to approach the flat fading single-user spectral efficiency. (Of course, the rate per signature tends to zero.) For both the optimal and linear MMSE receivers the spectral efficiency increases with J for fixed β, which implies that "self-interference," corresponding to signatures assigned to the same user, which pass through the same channel, is worse than other-user interference, corresponding to signatures, which pass through a different channel.

Turning to the results with isometric signatures, we observe that for given β and J the spectral efficiency is somewhat higher than with *i.i.d.* signatures (except for $J = K$, when they are the same). (Note that we must have $\beta < J$ with isometric signatures.) For $J = 1$ and $\beta = 1$, $(\mathbf{HS})^\ddagger = \mathbf{H}^\ddagger$, so that each signature in the CDMA scheme corresponds to a single, orthogonal sub-channel, hence the spectral efficiency is the same as for OFDM. However, for $\beta < 1$ the CDMA spectral efficiency is slightly greater than that of OFDM. This is because CDMA spreads over all sub-channels, whereas OFDM is assumed to use the fraction β of available sub-channels. Hence CDMA achieves a higher degree of frequency diversity. We also observe that the CDMA spectral efficiency is insensitive to J, in contrast with the results for *i.i.d.* signatures, and decreases slightly as J increases, due to interference among non-orthogonal signatures assigned to different transmitters. Still, the minimum spectral efficiency with $J = K$ (single signature per transmitter) is greater than the spectral efficiency with *i.i.d.* signatures (for any J).

4.7.2 Capacity Regions

In addition to computing the asymptotic sum capacity for the model (4.6), we can also compute the optimal asymptotic capacity *region*. For example, following the uplink interpretation of (4.6), suppose that there are two users with assigned rates R_1 and R_2. The *achievable rate region* is given by [62], [56, Ch. 6]

$$R_k \leq \bar{C}_k^N(\beta_k, \sigma^2), \quad k = 1, 2; \tag{4.99}$$

$$0 \leq R_1 + R_2 \leq \bar{C}^N(\beta_k, \sigma^2) \tag{4.100}$$

where $\bar{C}_k^N$ is the normalized single-user capacity (summed over signatures) assuming $R_j = 0, j \neq k$. The spectral efficiency region depends on the particular realizations of $\mathbf{H}_k, \mathbf{S}_k$, and is therefore random for finite N. Taking the large system limit gives the asymptotic spectral efficiency region, which is deterministic. The generalization to $K \geq 2$ users is straightforward, and is discussed in [42].

Figure 4.5a shows examples of asymptotic spectral efficiency regions with different values for β_1 and β_2 with the constraint $\beta_1 + \beta_2 = 1$ and per-signature receive SNR $= 8$ dB. Regions are shown for both *i.i.d.* and isometric signatures, and for both the optimal receiver and the linear MMSE receiver with single-signature decoders. Figure 4.5b shows the union of asymptotic spectral efficiency regions over all β_1 such that $\beta_1 + \beta_2 = 1$, where all other parameters are the same as in Figure 4.5a.

As expected, Figure 4.5a shows that the spectral efficiency for a particular transmitter increases with the number of assigned signatures, or load. With *i.i.d.* signatures, Figure 4.5b shows that the boundary of the spectral efficiency region is concave, and the sum capacity $C_1 + C_2$ is maximized with $\beta_1 = \beta_2 = 0.5$, whereas with isometric signatures the boundary is convex, and the sum capacity is *minimized* at $\beta_1 = \beta_2 = 0.5$; the maximum is at either $\beta_1 = 0$ or 1. Still, the minimum sum spectral efficiency with isometric signatures is greater than the maximum spectral efficiency with *i.i.d.* signatures. This again implies that with *i.i.d.* signatures, self-interference is more harmful than interference corresponding to a signature from another transmitter. In contrast, with isometric signatures self-interference is *less* harmful than interference from other transmitters.

Also shown in Figure 4.5b are the regions corresponding to OFDMA with AWGN and Rayleigh fading channels. In that case, the system load β_j designates the fraction of total bandwidth allocated to transmitter j. The capacity region for fully loaded OFDMA ($\beta = 1$) is larger than that for CDMA, since OFDMA eliminates interference. However, OFDMA is limited to $\beta \leq 1$, requires coordinated assignment of sub-channels to users, and is also more susceptible to interference from other cells and co-channel systems. As β increases beyond one, the spectral efficiency region of CDMA becomes larger than that of OFDMA with Rayleigh fading channels, as shown in Figure 4.4.

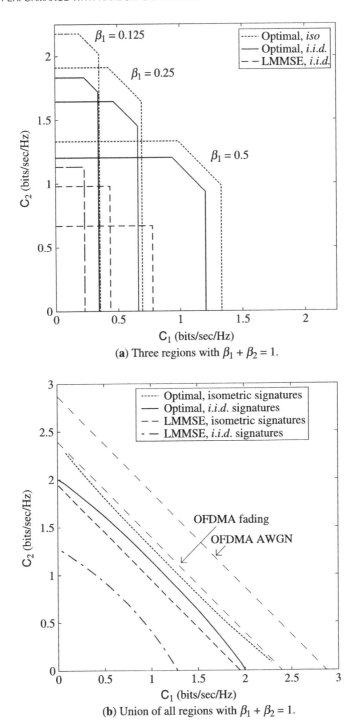

Figure 4.5. Asymptotic two-user capacity regions with *i.i.d.* and isometric signatures. Also shown are the corresponding regions with the linear MMSE receiver, *i.i.d.* signatures, and single-signature decoding. The SNR = 8 dB.

4.8 ADAPTIVE LINEAR RECEIVERS

So far, we have analyzed the performance of linear and optimal receivers, assuming that the receiver has perfect knowledge of the channel and signatures associated with all transmitted data streams. In practice, the optimal receiver must estimate the channel, and acquire knowledge of interfering signatures. In the case of a linear receiver, the filter for each transmitted data stream can be *adapted*, i.e., directly estimated from training symbols [23]. Direct filter estimation has the advantage that the receiver need not have any prior information about the channel or interfering signatures. This helps to mitigate other-cell interference in a cellular network, and interference in ad hoc (e.g., peer-to-peer) networks.

Given a sequence of training symbols, various adaptive filtering algorithms, such as as least squares and stochastic gradient, or Least Mean Square (LMS), have been proposed for both channel estimation and direct estimation of a linear MMSE filter (e.g., see [21,23]). In this section, we consider the performance of an *adaptive least squares (ALS)* algorithm for filter estimation. Least squares estimation is commonly used in practice, and provides reliable estimates given sufficient training, and assuming that the system and estimated parameters are stationary over the estimation interval. Here we indicate how the large system analysis in the preceding sections can be extended to ALS algorithms, and relate the asymptotic SINR of the ALS receiver to that of the MMSE receiver.

We remark that characterizing the transient behavior of ALS filters with random inputs is a long-standing difficult problem. Although a few approximate analyses have been presented for various applications (e.g., see [21, Chs. 9,14], [32], [12], [45]), exact results are generally unavailable.[12]

4.8.1 ALS Receiver

We focus on the signal model (4.3), and let $\hat{\mathbf{c}}_k$ denote the filter used to estimate the transmitted sequence $\{b_k(i)\}$, where $b_k(i)$ is the ith transmitted symbol in symbol stream k (i.e., corresponding to user k). Given the T training symbols $b_k(1), \ldots, b_k(T)$, the ALS receiver chooses $\hat{\mathbf{c}}_k$ to minimize the least squares cost function:

$$\hat{\mathcal{C}}_k(T) = \sum_{i=1}^{T} w_i |b_k(i) - \hat{b}_k(i)|^2 \tag{4.101}$$

where $\hat{b}_k(i) = \hat{\mathbf{c}}_k^\dagger(T)\mathbf{y}(i)$ is the ALS estimate of the data symbol $b_k(m)$. In contrast to the MMSE filter discussed in the preceding section, here the ALS filter depends on the number of training symbols. Also, the sequence $\{w_i\}$ can be used for *data windowing*. For example, taking $w_i = \epsilon^{T-i}$, where $\epsilon \in (0, 1)$ is a constant, corresponds to *exponential windowing*, which discounts the effect of past training symbols. This is especially

---

[12]A notable exception is the statistical characterization of an SINR performance measure with Gaussian inputs presented in [46], which was motivated by radar applications. We say more about this in Section 4.8.2.

useful in time-varying, or non-stationary environments (e.g., with fading or a randomly changing set of interferers), where the filter is updated recursively in time (e.g., see [21, Ch. 9]). We will refer to the sequence $w_i = 1$ for all i as a *rectangular window*.

Minimizing the cost function (4.101) with respect to $\hat{\mathbf{c}}_k$ gives:

$$\hat{\mathbf{c}}_k(T) = \hat{\mathbf{R}}^{-1}(T)\hat{\mathbf{s}}_k(T) \tag{4.102}$$

where the *sample covariance matrix* and estimated *steering vector* are, respectively:

$$\hat{\mathbf{R}}(T) = \frac{1}{T}\sum_{i=1}^{T} w_i \mathbf{y}(i)\mathbf{y}^\dagger(i) + \frac{\mu}{\eta}\mathbf{I}_M \tag{4.103}$$

$$\hat{\mathbf{s}}_k(T) = \frac{1}{T}\sum_{i=1}^{T} w_i b_k^*(i)\mathbf{y}(i) \tag{4.104}$$

where $\mu \in \mathbb{R}^+$ is a *diagonal loading* constant, and $\eta = T/N$. The diagonal loading is needed to ensure that $\hat{\mathbf{R}}(T)$ is non-singular for $T < N$. The particular representation of the diagonal loading shown is convenient for the large system analysis. When implemented recursively in time as a series of rank-one updates with exponential windowing, this adaptive filter is often referred to as the *recursive least-squares (RLS)* filter [21, Ch. 9].

In what follows, we assume constant channels and transmitted powers, so that the sequence of received vectors $\{\mathbf{y}(i)\}$ from (4.3) is wide-sense stationary. For finite T, the ALS receiver then approximates the MMSE receiver. Namely, the covariance matrix $\mathbf{R}$, which appears in the expression for the MMSE receiver (4.40), is approximated by the time-averaged sample covariance matrix $\hat{\mathbf{R}}(T)$, and the signature $\mathbf{s}_k$ is approximated by the time-averaged steering vector $\hat{\mathbf{s}}_k(T)$. With rectangular windowing as $T \to \infty$, conditioned on the signatures, $\hat{\mathbf{R}}(T) \to \mathbf{R}$ and $\hat{\mathbf{s}}_k(T) \to A_k\mathbf{H}\mathbf{s}_k$ almost surely, so that from (4.102) and (4.40), $\hat{\mathbf{c}}_k(T) \to \mathbf{c}_k$ almost surely. That is, the ALS filter converges to the MMSE filter.

With exponential windowing, past observations are discounted, so that the ALS filter is effectively computed with a finite number of observations. Hence as $T \to \infty$, neither $\hat{\mathbf{R}}$ nor $\hat{\mathbf{s}}_k$ converge to deterministic quantities, and $\hat{\mathbf{c}}_k$ is random. Exponential windowing therefore facilitates tracking in a non-stationary environment, but introduces additional estimation error, due to random variations of the filter [21, Ch. 9].

In the absence of a training sequence, an ALS estimate of the MMSE filter can still be computed provided that the steering vector $\mathbf{H}\mathbf{s}_k$ is known. This amounts to knowing the channel and desired user's signature at the receiver. The ALS estimate is then computed as $\hat{\mathbf{R}}^{-1}\mathbf{H}\mathbf{s}_k$, where $\hat{\mathbf{R}}^{-1}$ is given by (4.103). This is sometimes referred to as "blind" or "semi-blind" estimation, since the estimator is not driven by a training sequence (although additional information, i.e., the steering vector, is required) [23,45]. It has also been referred to as the "sample-matrix inversion" method in other applications [46]. Here our emphasis is on the performance of ALS with training, since that is most commonly used, although we will also present results for the semi-blind estimator.

4.8.2 ALS Convergence: Numerical Example

The SINR for the kth stream at the output of the ALS receiver is again given by (4.46), where c_k is replaced by the LS estimate $\hat{c}_k(T)$. The average in (4.45) is over data symbols and noise, conditioned on the channel $\mathbf{H}$, signature matrix $\mathbf{S}$, and c_k. Note, however, that $c_k(T)$ depends on the particular realization of training symbols and noise. We emphasize that the SINR is a function of training samples T. To illustrate this, Figure 4.6 shows a plot of output SINR versus training samples for system model (4.2) with rectangular windowing and a particular realization of signatures and training symbols. Also shown is a plot of SINR averaged over many such realizations for the *i.i.d.* signature model with *i.i.d.* training symbols.

As T increases, the averaged SINR converges to the SINR corresponding to the MMSE receiver. The transient behavior shown indicates the amount of training needed to achieve a target SINR. This becomes important in packet data communications, where training represents overhead, which must be included in each packet. It is desirable to minimize this overhead, while maintaining desired performance objectives (e.g., throughput).

Analytical characterization of the transient behavior of ALS estimators with random inputs is difficult, due to the complicated form of the SINR (4.46) with the

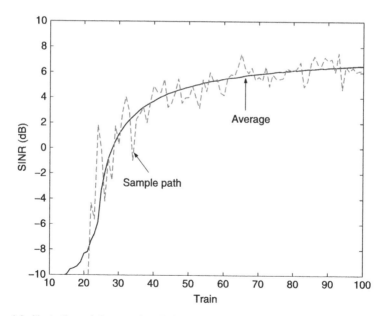

Figure 4.6. Illustration of the transient behavior of the ALS estimator (4.102). Output SINR versus number of input samples is shown for the signal model (4.2) with $K = 10$, $N = 20$, and background SNR of 10 dB.

ALS filter, given by (4.102). We remark that a similar problem arises in radar applications, and has been studied in [46]. There the vectors $\mathbf{y}(i)$, $i = 1, \ldots, T$, which appear in the sample covariance matrix $\hat{\mathbf{R}}$, are assumed to be complex *i.i.d.* Gaussian random vectors with *i.i.d.* elements. In that case, $\hat{\mathbf{R}}$ is a complex *Wishart* matrix (e.g., see [60]). For the semi-blind ALS algorithm, it is then possible to derive the distribution of the weight vector $\hat{\mathbf{c}}_k$.[13] In the next section, we explain how the ALS transient behavior can be characterized by taking a large system limit.

4.8.3 Large System Limit

For the ALS receiver the large-system limit lets K, N, and number of training symbols T all tend to infinity with fixed ratios $\beta = K/N$ and $\eta = T/N > 0$. This limit is therefore the same as that discussed in Section 4.3, but with the additional limit $T \to \infty$ with $T/N \to \eta$. That is, we allow the training interval to grow in proportion with the number of estimated degrees of freedom N. In this limit, it can again be shown that the SINR converges to a *deterministic* limit, independent of the distribution of the transmitted symbols, and the particular channel, signature, and noise realizations. We emphasize that this limit reveals the entire transient behavior of the adaptive filter, not just the steady-state performance. The steady-state performance is obtained by first taking the preceding large system limit, and then letting the normalized training $\eta \to \infty$.

To compute this large system limit, we rewrite the sample covariance matrix in (4.103) and the steering vector in (4.104) as:

$$\hat{\mathbf{R}} = \frac{1}{T}\mathcal{R}\mathbf{W}\mathcal{R}^\dagger + \frac{\mu}{\eta}\mathbf{I}_M \tag{4.105}$$

$$\hat{\mathbf{s}}_k = \begin{cases} \frac{1}{T}\mathcal{R}\mathbf{W}\underline{\mathbf{b}}_k, & \text{with training} \\ \mathbf{H}\mathbf{s}_k, & \text{semi-blind} \end{cases} \tag{4.106}$$

where:

$$\mathcal{R} = \mathbf{H}\mathbf{S}\mathbf{A}\mathbf{B}^\dagger + \mathbf{N} \tag{4.107}$$

and where $\mathbf{B} \in \mathbb{C}^{T \times K}$ contains the transmitted symbols, $\mathbf{N} \in \mathbb{C}^{M \times T}$ contains the noise, and $\mathbf{W} \in \mathbb{R}^{T \times T}$ is a diagonal data windowing matrix, i.e., $\mathbf{W} = \text{diag}(w_1, \ldots, w_T)$. Namely, the ith row of $\mathbf{B}$ is $\mathbf{b}^\dagger(i)$, $i \leq T$, and the ith column of $\mathbf{N}$ is $\mathbf{n}(i)$. The kth column of $\mathbf{B}$ will be denoted as $\mathbf{b}_k$.

The large system analysis can accomodate both *i.i.d.* and orthogonal training sequences, i.e., we will consider the following two cases:

- $\mathbf{B}$ contains *i.i.d.* elements with zero mean, unit variance, and finite positive moments.
- $\mathbf{B}$ contains either random orthogonal rows or columns.

---

[13]The distribution of an SINR type of performance measure, appropriate for radar applications, is also derived in [46]. That performance measure is somewhat different from the SINR considered here.

In the latter case, if $K \leq T$ then $\mathbf{B}^{\dagger}\mathbf{B} = T\mathbf{I}_K$, and we assume that $\frac{1}{\sqrt{T}}\mathbf{B}$ is obtained by extracting K columns from a $T \times T$ Haar random matrix. If $K > T$, then $\mathbf{B}\mathbf{B}^{\dagger} = K\mathbf{I}_T$, and we assume that $\frac{1}{\sqrt{K}}\mathbf{B}^{\dagger}$ is obtained by extracting T columns from a $K \times K$ Haar random matrix. In general, we expect orthogonal training sequences to generate less interference among users, and therefore achieve a higher SINR.

With data windowing we must define $\mathbf{W}$ for each T so that the distribution of its diagonal values converges to something appropriate in the large system limit. Note that if the window length is a constant, then as $(K, N, T) \rightarrow \infty$, a finite number of observations are used to compute an increasing number of filter weights, and hence the SINR tends to zero. It is therefore necessary to scale the window shape with the system size. With exponential windowing we therefore define $L = \frac{1}{1-\epsilon}$ as the 'average' window length, and take $L \rightarrow \infty$ with $L/N \rightarrow \bar{L} > 0$ constant. Of course, other analogous types of scaling rules must be applied to other window shapes. Finally, the normalization of the diagonal loading constant μ in (4.103) accounts for the fact that for fixed K and N, the diagonal loading should decrease as $1/T$.

4.8.4 Analysis and Results

First consider the simplest case in which the received signal is given by (4.2) with constant powers (i.e., $\mathbf{H} = \mathbf{A} = \mathbf{I}$), and where the ALS filter uses a rectangular data window ($\mathbf{W} = \mathbf{I}$) with no diagonal loading. In that case, it can be shown that evaluating the large system SINR for the ALS receiver reduces to evaluating the a.e.d. of the *sample* covariance matrix $\mathcal{R}\mathcal{R}^{\dagger}$, where $\mathcal{R}$ is defined in (4.107). In the absence of noise, we have:

$$\mathcal{R}^{\ddagger} = (\mathbf{S}\mathbf{B}^{\dagger})^{\ddagger} \tag{4.108}$$

$$= \mathbf{S}\mathbf{V}_B\Lambda_B\mathbf{V}_B^{\dagger}\mathbf{S}^{\dagger} \tag{4.109}$$

$$= \mathbf{V}_S\Lambda_B\mathbf{V}_S^{\dagger} \tag{4.110}$$

where we have factored $\mathbf{B}^{\dagger}\mathbf{B} = \mathbf{V}_B\Lambda_B\mathbf{V}_B^{\dagger}$, where Λ_B is the diagonal matrix of eigenvalues and $\mathbf{V}_B$ is orthonormal, and $\mathbf{V}_S = \mathbf{S}\mathbf{V}_B$. Now if $\mathbf{S}$ is Haar (e.g., has *i.i.d.* Gaussian elements), then $\mathbf{V}_S$ is Haar, and the Stieltjés transform of the a.e.d. of $\mathcal{R}^{\ddagger}$ is given by (4.29) with $\mathbf{X} = 0$, where the power distribution F_P is replaced by the a.e.d. of $\mathbf{B}^{\dagger}\mathbf{B}$. With *i.i.d.* training, this a.e.d. is the Marčenko-Pastur distribution with density (4.31).

Hence in this simplified case, we can evaluate the large system SINR for the ALS receiver by applying the random matrix results in Section 4.4. This type of analysis can be extended to account for noise, a general power distribution, diagonal loading, and data windowing (see [71]).

Alternatively, the large system SINR can be evaluated using the incremental matrix expansion technique discussed in Section 4.5. This provides a more general framework, which allows for a frequency-selective channel $\mathbf{H}$, *i.i.d.* or orthogonal signature matrix $\mathbf{S}$, and an *i.i.d.* or orthogonal matrix of training sequences $\mathbf{B}$. In addition, the derivation is more direct and also leads to a convenient relationship, which expresses the SINR for the ALS filter in terms of the SINR for the corresponding MMSE filter.

4.8.4.1 ALS Transient Behavior The following theorem applies for any $\mathbf{H}$, both *i.i.d.* and isometric $\mathbf{S}$, and any data windowing shape. Also, the noise vector $\mathbf{n}(i)$ must be unitarily invariant (not necessarily Gaussian) with elements having zero mean and variance σ^2.

Theorem 2 [43] *For the signal model (4.3) without diagonal loading ($\mu = 0$), the large system SINR for the kth data stream at the output of the ALS filter with i.i.d. training sequences is given by*:

$$\rho_k^{ALS} = \frac{\rho_k^{MMSE}}{\zeta + \frac{\zeta - 1}{\rho_k^{MMSE}}} \tag{4.111}$$

where:

$$\zeta = \frac{\mathcal{W}_{1,1}}{\mathcal{W}_{1,2}}, \tag{4.112}$$

$$\mathcal{W}_{m,n} = \mathbf{E}\left[\frac{W^m}{(1 + Wr)^n}\right], \tag{4.113}$$

the distribution of the scalar random variable W is the a.e.d. of $\mathbf{W}$*, and r is determined by the relation*:

$$\mathcal{W}_{0,1} = 1 - \frac{\alpha}{\eta} \tag{4.114}$$

Furthermore, assuming i.i.d. *transmitted data symbols with* $\mu = 0$*, the SINR for the semi-blind ALS (BALS) filter is given by*:

$$\rho_k^{BALS} = \frac{\rho_k^{MMSE}}{\zeta + (\zeta - 1)\rho_k^{MMSE}}. \tag{4.115}$$

If $W = 1$, corresponding to rectangular windowing, then from (4.113)–(4.114) we have $r = \alpha/(\eta - \alpha)$ and $\zeta = \eta/(\eta - \alpha)$.[14] With exponential weighting it can be shown that the a.e.d. of $\mathbf{W}$ is given by:

$$F_W(w) = 1 + \frac{\bar{L}}{\eta}\ln w, \quad e^{-\eta/L} \le w \le 1 \tag{4.116}$$

---

[14]This corresponding result for $\alpha = 1$ was first presented in [71]. An approximation of the large system SINR for the semi-blind ALS filter, which is accurate for large η, has been presented in [75].

where $\bar{L}$ is the normalized window size defined in Section 4.8.1. From (4.113)–(4.114) we can compute:

$$r = \frac{e^{\alpha/\bar{L}} - 1}{1 - e^{(\alpha - \eta)/\bar{L}}} \tag{4.117}$$

and averaging over W in (4.113) gives:

$$\zeta = \frac{\alpha(1 - e^{-\eta/\bar{L}})}{\bar{L}(1 - e^{(\alpha - \eta)/\bar{L}})(1 - e^{-\alpha/\bar{L}})} \tag{4.118}$$

We observe that ζ depends only on α, η, and the data window shape. *Hence the ALS SINR (with either training or semi-blind) depends on the channel, load, SNR, and power distribution only through the SINR achieved with the linear MMSE filter, ρ_k^{MMSE}. In other words, for a particular α and window shape, the transient response of the ALS filter is the same for all combinations of channels, loads, and SNRs that correspond to the same SINR value ρ_k^{MMSE}, which assumes perfect knowledge of the mixing matrix* **HSA** *and σ^2.*[15]

The following observations are a direct consequence of (4.111) and (4.115). Here we assume $\alpha = 1$.

1. For both ALS filters, as $\eta \to 1$ from above, $\zeta \to \infty$ and the SINR tends to zero. This is true with both rectangular and exponential windowing.

2. If ρ_k^{MMSE} is large, then $\rho_k^{ALS} \approx \rho_k^{MMSE}/\zeta$ and $\rho_k^{BALS} \approx 1/\zeta$. The performance of the semi-blind ALS filter is therefore limited by *estimation error*.[16] That is, increasing the training power does not improve the performance. In contrast, the gap in performance (SINR) associated with the ALS filter with training and the MMSE filter is given by the constant factor $1/\zeta$ when $\eta > 1$. Of course, as $\eta \to \infty$, $\zeta \to 1$ with rectangular windowing, so that $\rho_k^{ALS} \to \rho_k^{MMSE}$.

3. With rectangular windowing, training, and large ρ_k^{MMSE}, the large system $\rho_k^{ALS} \approx \rho_k^{MMSE}(1 - 1/\eta)$, which has been derived by other approximate methods (namely, replacing the received covariance matrix $\hat{\mathbf{R}}$ by its mean, as in [21, Ch. 9]). The large system analysis provides a rigorous justification for this approximation along with a more refined SINR expression, which is valid for all values of ρ_k^{MMSE}. When $\eta = 2$, this approximation states that $\frac{1}{2}\rho_k^{MMSE}\frac{1}{2}\rho_k^{MMSE}$, i.e., the ALS linear filter incurs a three dB loss from the optimal filter.[17]

4. The ALS filter with training performs better than the semi-blind ALS filter if and only if $\rho_k^{MMSE} > 1$ (zero dB), independent of the data windowing, user load, and training interval. In practice, the target SINR at the output of the linear filter is likely to be higher than zero dB, so that training is preferable, when available.

---

[15]This behavior has been observed empirically in [12,45].

[16]This was first observed in [75].

[17]A similar type of result is presented in [46], although the model and performance metric are different from those considered here.

The discussion in Section 4.2 implies that Theorem 2 holds for the general random matrix model (4.1), where $\mathbf{b}$ is unitarily invariant and the eigen-values of $\mathbf{M}^{\ddagger}$ converge to a well-defined a.e.d. in the large system limit. Hence, this result also holds for Toeplitz $\mathbf{M}$, which models a channel with inter-symbol interference. We emphasize, however, that Theorem 2 assumes that successive training vectors $\mathbf{b}(i)$ and $\mathbf{b}(i+1)$ are independent. In the equalization application that is not true in general, since the training vector $\mathbf{b}(i+1)$ is typically a *shifted* version of $\mathbf{b}(i)$, i.e., $\mathbf{b}_{m+1}(i+1) = \mathbf{b}_m(i)$, $m = 1, \ldots, N-1$, where here the subscript denotes the particular element of the vector. Hence Theorem 2 does not apply to that scenario.

Although the large system SINR for the model (4.3) with diagonal loading, data windowing, and orthogonal training can be computed, the expressions are complicated, and are omitted here (see [43]). Generalizations of the relations in Theorem 2, which include diagonal loading and/or orthogonal training sequences, are currently unavailable.

4.8.4.2 Steady-State SINR
The steady-state SINR can be obtained by letting $\eta \to \infty$ in Theorem 2. This leads to the same expressions (4.111) and (4.115) in Theorem 2, where:

$$\zeta = \frac{\widetilde{\mathcal{W}}_{1,1}}{\widetilde{\mathcal{W}}_{1,2}} \tag{4.119}$$

and $\widetilde{\mathcal{W}}_{m,n} = \lim_{\eta \to \infty} \eta \mathcal{W}_{m,n}$.[18] This steady-state relationship also holds for orthogonal training sequences.

With exponential weighting, from (4.118) we have $\zeta = \frac{\alpha}{\bar{L}(1-e^{-\alpha/\bar{L}})}$. As $\bar{L} \to \infty$, i.e., as we increase the window size, $\zeta \to 1$ (using L'Hôpital's rule), and $\rho_k^{\text{ALS}} \to \rho_k^{\text{MMSE}}$, as expected. Expanding $e^{\alpha/\bar{L}}$ in a Taylor series, it is straightforward to show that for large $\bar{L}$, $\zeta \approx [1 + \alpha/(2\bar{L})]/\alpha$. This approximation can also be obtained by approximating the (random) covariance matrix with its average, and has been presented in [45] for $\alpha = 1$.

4.8.5 Numerical Examples

Figure 4.7 shows plots of large system SINR vs. training η for the the signal model (4.3). The channel matrix $\mathbf{H}$ is $N \times N$, so that $\alpha = 1$, and the a.e.d. of $\mathbf{H}^{\ddagger}$ is $f_H(h) = \exp(-h)$ for $h > 0$, corresponding to frequency-selective Rayleigh fading across sub-channel gains. This density is used to compute the expectations over the channel gain H in Theorem 1. The user load $\beta = 1/2$ and the diagonal loading constant $\mu = 0.1$. The transmitted power is the same for each data stream (i.e., $\mathbf{A} = \mathbf{I}_K$), and the background SNR on each data stream, defined as the energy per symbol divided by σ^2, is 10 dB. Curves are shown for both *i.i.d.* and isometric signatures,

---

[18]This is not obvious from the expressions in Theorem 2, since it turns out that $\mathcal{W}_{1,1} \to 0$ and $\mathcal{W}_{1,2} \to 0$ as $\eta \to \infty$. Hence ζ must be computed using L'Hôpital's rule.

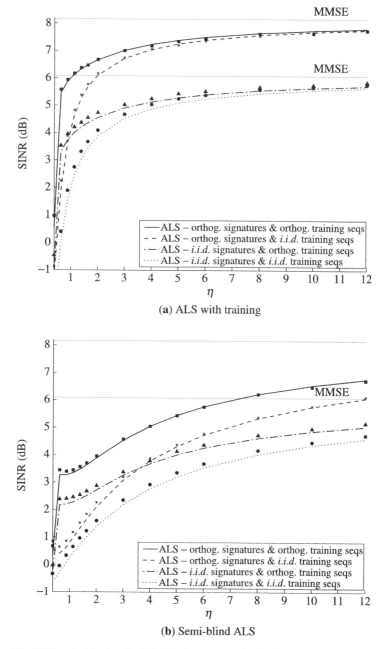

Figure 4.7. SINR vs. training length: CDMA in frequency-selective Rayleigh fading, SNR = 10 dB, $\beta = 0.5$, $\mu = 0.1$, rectangular windowing, equal power per signature, exponential distribution for a.e.d. of $\mathbf{H}^{\ddagger}$. Comparison with empirical values $N = 32$, QPSK modulation.

and *i.i.d.* and orthogonal training sequences. The figure compares the large system value with empirical values obtained by averaging results for a system with $N = 32$ and QPSK modulation.

The large system values for the MMSE receiver are given by Theorem 1. The ALS large system SINR with *i.i.d.* training and no diagonal loading is determined from Theorem 2, and with orthogonal training and/or diagonal loading is determined from [43, Theorem 2, Theorem 3, and Lemma 1]. The results show that large system results accurately predict the empirical values for $N = 32$.

Figure 4.7b shows that for a small number of training symbols (i.e., small η), orthogonal training sequences with *i.i.d.* signatures performs better than *i.i.d.* training with orthogonal signatures, and vice versa for large η. This is because as T increases, the K *i.i.d.* training sequences become "more orthogonal," reducing interference, and because isometric signatures consistently outperform *i.i.d.* signatures.

Figure 4.8 illustrates the performance of the ALS receiver with exponential windowing. Namely, Figure 4.8a shows plots of the capacity difference per-signature:

$$\Delta C_{ALS}^k = \log(1 + \rho_k^{MMSE}) - \log(1 + \rho_k^{ALS}) \tag{4.120}$$

as a function of the exponential window size $\bar{L}$. Different curves are shown corresponding to different training lengths η. These plots were generated with *i.i.d.* signatures, *i.i.d.* training, system load $\beta = 0.75$, and diagonal loading constant $\mu = 0$. Also, the power density across users is $f_P(p) = \frac{3}{4}\delta(p - 1) + \frac{1}{4}\delta(p - \frac{1}{2})$, that is, 3/4 of the users transmit with unit power, and the remaining transmit at half power. The curves shown correspond to the high-power users.

As the exponential window length increases, the SINR approaches that achieved with rectangular windowing (dotted line), as expected for the time-invariant model (4.3). For both types of windows, increasing the number of training symbols η brings diminishing returns. Of course, exponential windowing is included to allow for time-varying channels. The curves shown for exponential windowing are then a valid approximation for a time-varying system in which the system model is stationary during the effective window size.

4.8.6 Optimization of Training Overhead

We now apply the preceding results to optimize the amount of training overhead for packet data transmission. Namely, transmissions consist of packets, each containing an initial training sequence for filter estimation followed by the data. We assume that the signal model is stationary within each packet. Each receiver computes a least squares estimate of the linear MMSE filter from the training sequence, and uses this to demodulate the succeeding data symbols. We also assume that each filter estimate relies upon only the training symbols within the packet, and does not use information from prior packets. In other words, user signatures and channels are independent from packet to packet, corresponding to independent block fading.

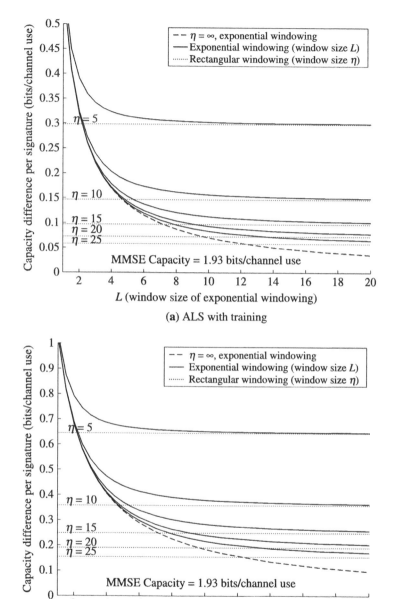

Figure 4.8. Capacity difference between ALS and MMSE receivers vs. exponential window size: CDMA in frequency-selective fading, *i.i.d.* signatures, *i.i.d.* training, SNR = 10 dB, $\beta = 0.75$, $\mu = 0$, and the power distribution $f_P(p) = \frac{3}{4}\delta(p-1) + \frac{1}{4}\delta(p-\frac{1}{2})$. Results for high-power users are shown. Each curve corresponds to a particular value of normalized training η. As the exponential window size $L \to \infty$, the performance with exponential windowing approaches the corresponding value with rectangular windowing (dotted line). Note that the scale of the vertical axis of Figure 4.8b is twice that of Figure 4.8a.

This scenario leads to the following tradeoff: a short training interval means that the SINR associated with the ALS filter is relatively small, whereas a large training interval leaves insufficient room in the packet for transmitting the data symbols. Hence, there is an optimal amount of training that maximizes the achievable rate per packet. Here we apply the preceding results to optimize the training overhead per packet assuming an ALS receiver, which directly estimates the filter. Optimization of training overhead has also been studied for MIMO channels in [20] and for dispersive single-input/single-output channels in [64], assuming an *optimal (maximum-likelihood)* receiver. (See also [37], which considers a time-varying channel model.) In that work, the training symbols are used to estimate the channel directly. Although suboptimal, linear interference suppression has relatively low complexity, and can suppress interference for which the corresponding channel estimates may not be available (e.g., which originates in other cells in a cellular network).

Consider a packet, or coherence block, containing B symbols in which the first T are training symbols, and the rest are data symbols. There are K equal-power data streams, which are coded independently with capacity-achieving codes having rate $R_c = \log_2(1 + \rho^{\mathrm{ALS}})$. (This, of course, assumes that the residual interference plus noise at the receiver output is Gaussian, which is true in the large system limit.) To apply the preceding large system results for the ALS filter, we let (K, N, T) all tend to infinity with fixed ratios, as before, and also let $B \to \infty$ with fixed $\bar{B} = B/N$. In other words, both the packet length and training interval T are normalized by the number of filter coefficients to be estimated. The number of information bits per block summed over users is $KR_c(B - T)$ and the number of transmit dimensions per block is NB, so that the *normalized capacity*, or information bits per transmit dimension, is:

$$\bar{C} = \beta(1 - \eta/\bar{B})R_c \tag{4.121}$$

Our objective is to maximize $\bar{C}$ over η. As η increases, ρ^{ALS} increases, so that R_c (achievable rate per data symbol) increases, but the fraction of data symbols $1 - \eta/\bar{B}$ decreases. The optimal training length, denoted as η^*, balances this tradeoff.

The preceding tradeoff for the model (4.2) is illustrated in Figure 4.9, which is taken from [52]. The figure shows capacity $\bar{C}$ in (4.121) as a function of normalized training for fixed load $\beta = 1$, $SNR = 10$ dB, and normalized block length $\bar{B} = 10$. Different curves are shown for different diagonal loading factors. Without diagonal loading, the normalized optimal training length $\eta^* \approx 2.9$, i.e., devoting 29 percent of the packet to training maximizes the achievable rate. These results show that the achievable rate is insensitive to training lengths η in the range of 2.5 to 4, but decreases rapidly as η decreases below two. Of course, in general the optimal amount of training depends on the load, SNR, and block length. These results also show that a small diagonal loading factor helps to desensitize performance with respect to training, and provides a modest increase in capacity.

Analytical optimization of the training overhead for the model (4.2) with the ALS receiver has been considered in [52]. In that case, the ρ^{ALS} is given by (4.111), and

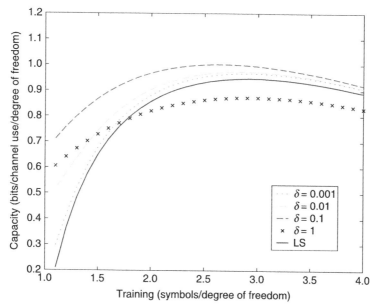

Figure 4.9. Large system capacity versus training length with an ALS linear receiver for the signal model (4.2). The normalized block size $\bar{B} = 10$, $\beta = 1$, and $SNR = 10\,dB$. Curves are shown for different values of the diagonal loading.

maximizing $\bar{C}$ with respect to η gives:

$$\eta^* = \sqrt{\frac{1}{4} + \frac{\bar{B} + \frac{1}{\rho^{\mathrm{MMSE}}} - 1}{\log\left(1 + \rho^{\mathrm{MMSE}}\right)}} + \frac{1}{2} - \frac{1}{\rho^{\mathrm{MMSE}}} + O\left(\frac{1}{\sqrt{\bar{B}}}\right) \qquad (4.122)$$

As $\bar{B} \to \infty$, the optimal training length increases as the square root of the block size, that is:

$$\lim_{B \to \infty} \frac{\eta^*}{\sqrt{\bar{B}}} = \sqrt{\frac{1}{\log\left(1 + \rho^{\mathrm{MMSE}}\right)}} \qquad (4.123)$$

This is in contrast to the optimal training length with a maximum-likelihood receiver, which is equal to the number of degrees of freedom [20]. The normalized training as a fraction of the block size therefore decreases to zero as $1/\sqrt{B}$.

From (4.122) it can be shown that:

- As $\rho^{\mathrm{MMSE}} \to \infty$, the optimal training length $\eta^* \to 1$. That is, as the transmit power becomes very large, the optimal training length decreases to the number of filter coefficients to be estimated, which is the minimum possible training length.

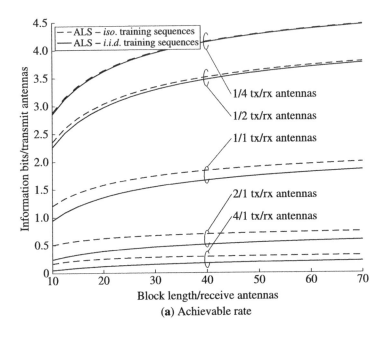

(a) Achievable rate

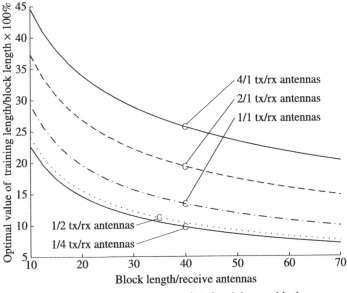

(b) Optimal percentage of *i.i.d.* training per block

Figure 4.10. Optimization of training overhead for the signal model (4.3). The ALS estimate is computed with rectangular windowing and no diagonal loading. The channel distribution is exponential, $\frac{E_b}{\sigma^2} = 10\,\text{dB}, \bar{B} = 15$, and the same transmit power is allocated to each transmitted stream (antenna).

- As $\rho^{\mathrm{MMSE}} \to 0$, $\eta^* \to (\bar{B} + 1)/2$. That is, at low SNRs about half the block should be devoted to training.[19]
- As $\bar{B} \to \infty$, $\bar{C}/C^{\mathrm{MMSE}} \to 1$ as $\left(1 - 1/\sqrt{C^{\mathrm{MMSE}}\bar{B}}\right)^2$, where $C^{\mathrm{MMSE}} = \log(1 + \rho^{\mathrm{MMSE}})$ is the large system capacity (per user and per degree of freedom) with the MMSE receiver.

For the system models (4.3) and (4.6) the optimal training in the large system limit can be found by maximizing $\bar{C}$ in (4.121) over η numerically. Figure 4.10a shows the growth in normalized capacity for the system model (4.3), optimized with respect to η, versus block length $\bar{B}$. For this example the ALS estimate is computed with rectangular windowing and no diagonal loading. The channel distribution is exponential, $\bar{B} = 15$, the same transmit power is allocated across transmit streams, and $\frac{E_b}{\sigma^2} = 10\,\mathrm{dB}$. (The labels in the figure correspond to the MIMO channel in which K is the number of transmit antennas and N is the number of receive antennas.) Results are shown for both *i.i.d.* and isometric training sequences. This example shows that the gain from orthogonal training sequences is more pronounced when there is a high ratio of transmit antennas to receive antennas (i.e., $\beta > 1$). Figure 4.10b shows the associated optimal training length η, expressed as a percentage of the block length $\bar{B}$, for *i.i.d.* training sequences. Although the model is different from that assumed to derive (4.122) and (4.123), Figure 4.10b shows that the optimized training still decreases roughly as $1/\sqrt{\bar{B}}$.

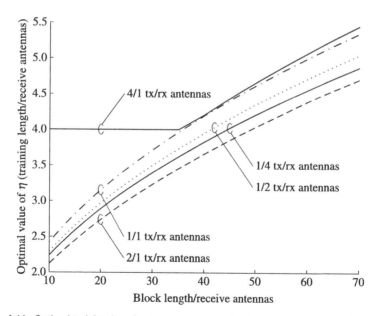

Figure 4.11. Optimal training length η^*, corresponding to the capacity curves in Figure 4.10a with orthogonal training sequences.

[19]A similar result has been observed in other contexts, e.g., [37].

Figure 4.11 shows the optimal value of η with orthogonal training sequences corresponding to the curves in Figure 4.10a. These curves are not normalized with respect to $\bar{B}$ in order to show that the optimal training value $\eta^* \geq \beta$. Namely, for $\eta < \beta$, **B** has orthogonal rows, and for $\eta > \beta$, **B** has orthogonal columns. Orthogonal columns reduce interference, and are therefore preferable. (If the y-axis were extended, we would see the same behavior in the other curves as for $\beta = 4$.)

4.9 OTHER MODELS AND EXTENSIONS

Large system analysis and applications of results from large random matrix theory have been used to evaluate the performance of many other types of receivers and communications systems. For example, the large system performance of *reduced-rank* filters, in which the incoming received signal vector is projected onto a lower-dimensional subspace before filtering, is analyzed in [26,31,34,55,71]. Rank reduction via the subspace projection has the advantage of reducing the complexity of the filter if the dimension of the lower-dimensional subspace is sufficiently small. Furthermore, it is found in [71] that reduced-rank least squares filters can converge faster than the analogous full-rank least squares filters, and are less sensitive to the choice of diagonal loading factor.

Large system analysis has also been used to evaluate the performance of decision feedback, or interference cancelling receivers in [6,24,25,38]. In a decision-feedback receiver, a linear filter is followed by a feedback loop in which symbol decisions are passed through a feedback filter to generate an estimate of the interference. The interference estimate is subtracted from the output of the first linear filter. It has been shown that this type of receiver is optimal in the sense that it can achieve the sum capacity [38]. Iterative, or turbo versions of this detector, in which soft tentative decisions used for cancellation are subsequently refined, have been proposed in [1,2,68]. The large system error rate performance of those detectors has been analyzed in [6,24].

There are still many extensions and modifications of the models discussed here yet to be considered. For example, the CDMA interpretation of the models considered in Section 4.2 assumes a single spatial dimension (i.e., one antenna at each transmitter and receiver). Extending those models to include multiple spatial dimensions could lead to interesting generalizations of the large system results presented here. Of course, other interesting modifications may arise by considering different types of transmission formats (e.g., multi-carrier with or without spreading) with multiple transmit dimensions across time, frequency and space, different multi-user networking configurations (e.g., cellular, peer-to-peer, and ad-hoc), and different statistical assumptions concerning the channel and noise models (e.g., accounting for correlation across channel gains and non-Gaussian noise).

Finally, we mention that evaluating the minimum *bit error rate* for the model (4.2), i.e., with the optimal *maximum a posteriori* detector, has been evaluated in the large system limit through an application of the *replica method* in statistical physics [19,53]. Although it is not rigorous, this method can also be used to derive some of the random matrix results presented here (e.g., see [3]), and may be useful in other related contexts. Other applications of large system analysis to signal processing and communications models are still emerging.

4.10 BIBLIOGRAPHICAL NOTES

Here we briefly review some of the background related to the discussion in this chapter. For a comprehensive overview of the relevant literature on random matrix theory, free probability, and its applications to CDMA and MIMO channels, see [60].

Large system analyses of CDMA were first presented in a series of three papers by different sets of authors [18,58,63]. Those references all assume the CDMA model (4.2), and the results rely on the Marčenko-Pastur a.e.d. [36] and the generalization derived by Silverstein and Bai [50]. That analysis was subsequently extended and applied to variations of the original CDMA model, accounting for asynchronous users [29], flat fading [48], variable-length signatures [4], frequency-selective channels [13,44], decision-feedback and iterative receivers [6,24,25,38], reduced-rank and multi-stage filtering [26,31,34,55], and multi-cell networks [70,73]. Much of that work relies on the same set of random matrix results needed to analyze the ideal model (4.2), although [29] uses random matrix results due to Girko [15], and [13] applies results from free probability.

Free probability was developed by Voiculescu [67] as a theory of non-commutative operators. Random matrices are a canonical example of such operators, and application of the theory leads to the R- and S-transforms in Appendix 4.10, which can be used to derive the a.e.d.'s of the sums and products of random unitary matrices. (See the monograph on free probability by Hiai and Petz [22] and related notes by Speicher [51].) Subsequent to the previous CDMA work mentioned, the R- and S-transforms have been used in [7,11,41,42] to analyze the model (4.3). Related results for Multi-Carrier CDMA are presented in [31], and are obtained using Girko's results [15]. (See also [16].)

In parallel with the large system analyses of CDMA, random matrix theory has been used to analyze the performance of MIMO channels, starting with the paper by Telatar [54]. (Both large system results and performance results for finite-size systems are presented in [54].) Results for MIMO channels, analogous to those presented for the CDMA model (4.2), are presented in [5]. This has been extended to MIMO channels with co-channel interference [35], and to the case where the MIMO channel gains are correlated (e.g., see [8,39,59]). We refer the reader to [60] for a more extensive overview of that work.

The matrix expansion approach presented here was originally motivated by the problem of evaluating performance metrics for the signal model (4.6). Namely, the matrices, which appear in the sum in (4.6) are not asymptotically free, so that the R-transform cannot be used to compute the a.e.d. of the received covariance matrix. The SINR results with *i.i.d.* signature matrices can also be obtained from [16, Ch. 16] (see [42]), although the derivation given here is more straightforward. Also, the results in [16] do not apply to unitary signature matrices.

The large system approach to analyzing the transient behavior of least squares filtering was introduced in [69,71]. Large system results with data windowing for the signal model (4.2) with *i.i.d.* signatures were expressed in terms of asymptotic moments of the sample covariance matrix. Computing those moments is a combinatorial problem which arises in the theory of non-crossing partitions, and has been used to derive and interpret results in free probability [51,60]. In other related work, large-system analysis has been applied to blind subspace-ALS receivers [28,72,75], where the subspace is formed from eigen-vectors of the covariance matrix. The ALS results presented here are derived in [43] using the matrix expansion approach, which can account for a wider range of models than considered in [71].

APPENDIX: PROOF SKETCH OF THEOREM 1

We are interested in computing $\gamma = G_{\bar{R}}(z)$. From the definition of Stieltjés transform (4.24), we can write:

$$\gamma^N = \frac{1}{M}\mathrm{tr}[\mathbf{R}^{-1}] = -z^{-1}\left(\frac{1}{M}\mathrm{tr}[(\mathbf{HSA})^{\ddagger}\mathbf{R}^{-1}] - 1\right) \tag{4.A.1}$$

where $\mathbf{R} = (\mathbf{HSA})^{\ddagger} - z\mathbf{I}_M$. We first apply the MIL to $\mathbf{R} = \mathbf{R}_{b_k} + P_k\mathbf{h}_k\mathbf{h}_k^{\dagger}$, where $\mathbf{h}_k = \mathbf{H}\mathbf{s}_k$, to obtain:

$$\mathbf{R}^{-1}\mathbf{h}_k = \frac{\mathbf{R}_{b_k}^{-1}\mathbf{h}_k}{1 + P_k\bar{\rho}_k^N} \tag{4.A.2}$$

where the asymptotic behavior of $\bar{\rho}_k^N$ is given by (4.50). From (4.A.2) and (4.50) we have:

$$\frac{1}{N}\mathrm{tr}[(\mathbf{HSA}^m)^{\ddagger}\mathbf{R}^{-1}] = \frac{1}{N}\sum_{k=1}^{K}\frac{\bar{\rho}_k^N}{1 + P_k\bar{\rho}_k^N} \asymp \beta\bar{\rho}^N\mathcal{E}_m^N \tag{4.A.3}$$

where:

$$\mathcal{E}_m^N = \frac{1}{K}\sum_{k=1}^{K}\frac{P_k^m}{1 + P_k\bar{\rho}^N} \tag{4.A.4}$$

converges to $\mathbf{E}[P^m/(1 + P\bar{\rho})]$ as $N \to \infty$. Note that $\mathcal{E}_1 = \mathcal{E}$ in (4.74).

To obtain the additional relations in the theorem we apply the results in Section 4.6.3. Namely, as stated in Proposition 1, we replace $\mathbf{H}$ by $\mathbf{VD}$. We emphasize that $\mathbf{VD}$ is an equivalent matrix, in the sense that this substition does not change the SINR, as opposed to a decomposition of $\mathbf{H}$. We denote the nth column of $\mathbf{V}$ and $\mathbf{S}^{\dagger}$ as $\mathbf{v}_n$ and $\tilde{\mathbf{s}}_n$, respectively, $1 \leq n \leq N$. The diagonal elements of $\mathbf{D}$ are $\{d_1, \ldots, d_{\alpha^*N}\}$, where $\alpha^*N = \min(M, N)$.

We wish to apply Lemma 1 to the expansion (4.60), so first check to make sure that the conditions (4.65–4.67) are satisfied. For any $n \leq \alpha^*N$, since $\mathbf{H}_{l_n}^{\dagger}\mathbf{v}_n = 0$, we have

$\mathbf{R}_{t_n}\mathbf{v}_n = -z\mathbf{v}_n$, $\mathbf{R}_{t_n}^{\dagger}\mathbf{v}_n = -z^{\dagger}\mathbf{v}_n$, and moreover:

$$\mathbf{v}_n^{\dagger}\mathbf{R}_{t_n}^{-1}\mathbf{v}_n = -z^{-1}, \tag{4.A.5}$$

$$\mathbf{u}_n^{\dagger}\mathbf{R}_{t_n}^{-1}\mathbf{v}_n = \mathbf{v}_n^{\dagger}\mathbf{R}_{t_n}^{-1}\mathbf{u}_n = 0, \tag{4.A.6}$$

where $\mathbf{v}_n^{\dagger}\mathbf{R}_{t_n}^{-1}\mathbf{v}_n$ corresponds to u_N in the Lemma, and (4.A.6) establishes condition (4.65). Since $|\mathbf{v}_n| = 1$, condition (4.66) is satisfied, and condition (4.67) is satisfied in the almost-sure sense.

Proceeding to apply Lemma 1 to (4.60) gives:

$$\mathbf{R}^{-1}\mathbf{v}_n \asymp \frac{\mathbf{R}_{t_n}^{-1}(\mathbf{v}_n + d_n z^{-1}\mathbf{u}_n)}{1 + d_n^2 z^{-1}(\tau_n^N - c_n)} \tag{4.A.7}$$

$$\mathbf{R}^{-1}\mathbf{u}_n \asymp \frac{\mathbf{R}_{t_n}^{-1}(-d_n \tau_n^N \mathbf{v}_n + (1 - d_n^2 c_n z^{-1})\mathbf{u}_n)}{1 + d_n^2 z^{-1}(\tau_n^N - c_n)} \tag{4.A.8}$$

for $1 \leq n \leq N$, where $\mathbf{u}_n$ is defined by (4.62), c_n is given by (4.63), and:

$$\tau_n^N = \mathbf{u}_n^{\dagger}\mathbf{R}_{t_n}^{-1}\mathbf{u}_n \tag{4.A.9}$$

$$c_n = \mathrm{tr}[\mathbf{P}\tilde{\mathbf{s}}_n\tilde{\mathbf{s}}_n^{\dagger}] \overset{n}{\asymp} \beta\mathbf{E}[P] \tag{4.A.10}$$

The last asymptotic equivalence is obtained by observing that $c_n \overset{n}{\asymp} \mathbf{E}[c_n]$. Following a similar argument as used to show (4.33) and (4.50), it can be shown that with *i.i.d.* sequences we have:

$$\tau_n^N \overset{n}{\asymp} \tau^N = \frac{1}{N}\mathrm{tr}[(\mathbf{HSA}^2)^{\ddagger}\mathbf{R}^{-1}] \tag{4.A.11}$$

The corresponding asymptotic relation with isometric signatures could be similarly obtained, in analogy with the one for $\bar{\rho}_k^N$ in (4.50), except that the matrix dimension K, which appears in (4.50), is replaced here by N. Hence the asymptotic value of τ_n^N is undefined. To avoid this problem, we redefine $\mathbf{D}$ and $\mathbf{S}$ with dimensions $M \times \bar{N}$ and $\bar{N} \times K$, respectively, where $0 < \bar{N} < N$. We then have, in analogy with (4.50) with isometric signatures:

$$\tau_n^N \overset{n}{\asymp} \frac{1}{N - \bar{N}}\mathrm{tr}[(\mathbf{HSA}^2\boldsymbol{\Xi})^{\ddagger}\mathbf{R}^{-1}] \tag{4.A.12}$$

where:

$$\boldsymbol{\Xi}^2 = \mathbf{I}_K - \mathbf{S}^{\dagger}\mathbf{S} \tag{4.A.13}$$

Also, refering to (4.50), we have $\Pi^2 = \mathbf{I}_{\bar{N}} - \mathbf{SS}^{\dagger}$. We can then take the limit $\bar{N} \to \infty$ with $\bar{N}/N \to \varsigma$ where $\varsigma \in (0, 1)$. With that modification diag $(\mathbf{D}) = \{d_1, \ldots, d_{\alpha^* N}\}$, where $\alpha^* = \min(\alpha, \varsigma)$, $\mathbf{S}^\dagger = [\tilde{\mathbf{s}}_1, \ldots, \tilde{\mathbf{s}}_{\bar{N}}]$, $\mathbf{s}_k$ is $\bar{N} \times 1$, and the maximum involved in $\overset{n}{\asymp}$ is over $n \leq \bar{N}$. The asymptotic results in the theorem are then obtained by letting $\varsigma \to 1^-$.

From (4.A.7)–(4.A.8) and (4.A.10)–(4.A.11) we get:

$$\frac{1}{N}\mathrm{tr}[\mathbf{H}^{\ddagger}\mathbf{R}^{-1}] = \frac{1}{N}\sum_{n=1}^{\alpha^* N} d_n^2 \mathbf{v}_n^{\dagger}\mathbf{R}^{-1}\mathbf{v}_n \asymp \frac{1}{N}\sum_{n=1}^{\alpha^* N} \frac{-z^{-1}d_n^2}{1 + d_n^2 z^{-1}(\tau_n^N - c_n)} \tag{4.A.14}$$

$$\asymp \alpha^* \mathcal{H}_1^N \tag{4.A.15}$$

and:

$$\frac{1}{N}\mathrm{tr}[(\mathbf{HSA}^2\mathbf{S}^\dagger)^{\ddagger}\mathbf{R}^{-1}] = \frac{1}{N}\sum_{n=1}^{N} \mathrm{tr}[(\mathbf{u}_n + d_n c_n \mathbf{v}_n)^{\ddagger}\mathbf{R}^{-1}] \tag{4.A.16}$$

$$\asymp \beta\bar{p} + (\tau^N - \beta\bar{p})[-\alpha^*(z\mathcal{H}_0^N + 1) + 1] \tag{4.A.17}$$

where $\bar{p} = \mathbf{E}[P]$ and:

$$\mathcal{H}_p^N = -\frac{1}{\alpha^* N}\sum_{n=1}^{\alpha^* N} \frac{d_n^{2p}}{z + d_n^2(\tau^N - \beta\bar{p})} \tag{4.A.18}$$

converges to $\mathbf{E}\left[\frac{H^p}{-z+\phi H}\right]$ as $N \to \infty$, where $\phi = \alpha\bar{p} - \tau^N$. Note that $\mathcal{H}_1 = \mathcal{H}$ in (4.75).

Combining the preceding relations gives:

$$\gamma^N \asymp -z^{-1}\left(\frac{\beta}{\alpha}\bar{\rho}^N \mathcal{E}^N - 1\right) \tag{4.A.19}$$

$$\bar{\rho}^N \asymp \begin{cases} \alpha^* \mathcal{H}_1^N, & \textit{i.i.d. } \mathbf{S}, \\ \dfrac{\alpha^* \mathcal{H}_1^N}{\beta(\mathcal{E}_0^N - 1) + 1}, & \text{iso. } \mathbf{S}, \end{cases} \tag{4.A.20}$$

$$\tau^N \asymp \begin{cases} \beta(\bar{p} - \mathcal{E}_1^N), & \textit{i.i.d. } \mathbf{S}, \\ \beta\bar{p} - \dfrac{\beta\mathcal{E}_1^N}{-\alpha^*(z\mathcal{H}_0^N + 1) + 1}, & \text{iso. } \mathbf{S}, \end{cases} \tag{4.A.21}$$

where in (4.A.21) we have used the equality $\mathbf{E}[P] = \mathcal{E}_1^N + \bar{\rho}^N \mathcal{E}_2^N$. Simplifying the preceding relations, using $\alpha(1 + z\gamma^N) = \beta(1 - \mathcal{E}_0^N) = \alpha^*(1 - \mathcal{H}_0^N)$, and letting $N \to \infty$ leads to the conclusion that $|\gamma^N - \gamma| \overset{a.s.}{\to} 0$, $|\bar{\rho}^N - \bar{\rho}| \overset{a.s.}{\to} 0$, and $|\tau^N - \tau| \overset{a.s.}{\to} 0$, where γ, $\bar{\rho}$, and $\phi = \alpha\bar{p} - \tau$ satisfy (4.72)–(4.75). It can be shown these limits exist and are unique (see [43]).

APPENDIX: FREE PROBABILITY TRANSFORMS

In this appendix, we apply the incremental matrix expansion approach to derive the a.e.d. of sums and products of independent, positive semi-definite, unitarily invariant large random matrices. These types of matrices are known to be *asymptotically free*. Free probability is concerned with non-commutative random variables, large random matrices being canonical examples. As mentioned in Section 4.10, free probability has been used to analyze several communications system models related to the ones considered in this chapter.

In free probability, the notion of independence (from commutative probability theory) is replaced by the notion of freeness. Namely, let $\{\mathbf{M}_N\}$ be a sequence of $N \times N$ random matrices, and define the operator:

$$\phi(\mathbf{M}) = \lim_{N \to \infty} \mathbf{E}\left[\frac{1}{N}\mathrm{tr}[\mathbf{M}_N]\right]. \tag{4.B.1}$$

If two sequences of random matrices $\mathbf{M}_N$ and $\mathbf{N}_N$ are asymptotically free, then $\phi(\mathbf{M}^k \mathbf{N}^m) = \phi(\mathbf{M}^k)\phi(\mathbf{N}^m)$. Furthermore, mixed moments, such as $\phi(\mathbf{M}^{k_1}\mathbf{N}^{k_2}\mathbf{M}^{k_3}\mathbf{M}^{k_4})$ can be expressed in terms of the asymptotic moments of $\mathbf{M}$ and $\mathbf{N}$, although computing the corresponding expansion is a nontrivial combinatorial problem (e.g., see [51]).

A common problem addressed by free probability is the calculation of the distribution of sums and products of random matrices. In particular, the a.e.d. of sums and products of asymptotically free random matrices can be computed, respectively, using the so-called R- and S-transforms, given the a.e.d. of each component term [65,66]. This is analogous to the way the Fourier transform is used to compute the distribution of a sum of independent scalar random variables. As such, the R- and S- transforms are often described as performing additive or multiplicative *free convolution* of the component distributions. For a more detailed introduction to free probability see [60, Sec. 2.4].

According to [22, Theorem 4.3.5], an independent family of $N \times N$ Hermitian positive semi-definite random matrices $\left(\mathbf{X}_j^{\ddagger}\right)_{j=1,\ldots,J}$ is almost surely asymptotically free as $N \to \infty$ if for each $j = 1, \ldots, J$, $\mathbf{X}_j$ is unitarily invariant and the e.e.d. of $\mathbf{X}_j^{\ddagger}$ converges almost surely in distribution to a compactly supported probability measure on $\mathbb{R}^*$ as $N \to \infty$. Hence the a.e.d.'s of sums and products of unitarily invariant, positive semi-definite Hermitian random matrices can be obtained using the R- and S-transforms, subject to the preceding condition on e.e.d.'s.

Here we use the incremental matrix expansion approach to derive the a.e.d.'s of sums and products of unitarily invariant matrices. This derivation does not explicitly require free probability results, and is therefore intended to be more accessible than the derivations of the R- and S-transforms found in [65,66]. We also note that similar, yet different, derivations of these results can be found in the mathematical physics literature [40,61].

4.B.1 Free Probability Transforms

DEFINITION 3 [65] *The* **R**-*Transform of the distribution of a random variable X (possibly non-cummutative) is*:

$$R_X(t) = \frac{1}{t} + G_X^{-1}(t) \qquad (4.B.2)$$

where $G_X(z)$ is the Stieltjés transform of the distribution of X, and $G_X^{-1}(\cdot)$ is the inverse with respect to composition. Also:

$$G_X(z) = \frac{1}{-z + \boldsymbol{R}_X \circ G_X(z)} \qquad (4.B.3)$$

where $\circ$ denotes composition.

DEFINITION 4 [66] *The* **S**-*Transform of the distribution of a random variable X (possibly non-commutative) is*:

$$S_X(x) = \frac{x+1}{x} Y_X^{-1}(x) \qquad (4.B.4)$$

$$Y_X(z) = \frac{G_X(z^{-1})}{-z} - 1 \qquad (4.B.5)$$

where $G_X(z)$ is the Stieltjés transform of the distribution of X, and $Y_X(z)$ is an auxiliary function. Note that $Y_X^{-1}(x)$ is the inverse with respect to composition.

Given a matrix $\mathbf{A}$, $R_\mathbf{A}(x)$ and $S_\mathbf{A}$ denote the R- and S-transforms of the a.e.d. of $\mathbf{A}$, respectively, assuming it exists and is well-defined.

IDENTITY 1 [65,66] *If the matrices $\mathbf{A}_1$, $\mathbf{A}_2$ are asymptotically free, then*:

$$R_{\mathbf{A}_1 + \mathbf{A}_2}(x) = R_{\mathbf{A}_1}(x) + R_{\mathbf{A}_2}(x) \qquad (4.B.6)$$
$$S_{\mathbf{A}_1 \mathbf{A}_2}(x) = S_{\mathbf{A}_1}(x) S_{\mathbf{A}_2}(x) \qquad (4.B.7)$$

We can express the preceding identities in terms of Stieltjés transforms as follows.

IDENTITY 2 *If $\mathbf{A}_j, j = 1, \ldots, J$ are asymptotically free and $\mathbf{A} = \sum_{j=1}^{J} \mathbf{A}_j$, then*:

$$G_\mathbf{A}(z) = \frac{J - 1}{z - \sum_{j=1}^{J} x_j} \qquad (4.B.8)$$

$$G_\mathbf{A}(z) = G_{\mathbf{A}_j}(x_j), \quad j = 1, \ldots, J \qquad (4.B.9)$$

where for $j = 1, \ldots, J$, $G_{A_j}(z)$ is the Stieltjés transform of A_j, and the x_j's are auxiliary variables.

IDENTITY 3 *If A_1, A_2 are asymptotically free, then*:

$$G_{A_1 A_2}(z) = \frac{1}{z(zz_1 z_2 - 1)} \tag{4.B.10}$$

$$G_{A_1 A_2}(z) = \frac{1}{zz_j} G_{A_j}(z_j^{-1}), \quad j = 1, 2. \tag{4.B.11}$$

where the z_j's are auxiliary variables.

4.B.2 Sums of Unitarily Invariant Matrices

We wish to determine the Stieltjés transform of the a.e.d. (i.e., as $N \to \infty$) of $\sum_{j=1}^{J} X_j^{\ddagger}$ where the $X_j^{\ddagger}$ are unitarily invariant, Hermitian, positive semi-definite $N \times N$ independent random matrices. That is, we seek $G_C(z) = \lim_{N \to \infty} G_C^N(z)$ where $G_C^N(z) = \frac{1}{N} \text{tr}[C^{-1}]$ and $C = -zI_N + \sum_{j=1}^{J} X_j^{\ddagger}$. To simplify the derivation, we assume $|z| < \infty$.

Denote the singular value decomposition of X_j as $V_j D_j^2 V_j^{\dagger}$. Since X_j is unitarily invariant, without loss of generality we can assume that V_j is $N \times N$ Haar unitary. In what follows, $v_{j,k}$ denotes the kth column of V_j and $D_{j,k}$ denotes the kth diagonal element of D_j^2 for $1 \leq j \leq J$ and $1 \leq k \leq N$.

Following the incremental matrix expansion approach, we start by expanding the identity $I_N = CC^{-1}$. The MIL can be used to remove column k of V_j from C, which gives the scalar term $v_{j,k}^{\dagger} C_{j,k}^{-1} v_{j,k}$, where $C_{j,k}$ is C with $v_{j,k}$ removed from $X_j^{\ddagger}$. To analyze this quadratic term asymptotically, we first write it as a matrix trace. Since $v_{j,k}$ is a column from a Haar distributed matrix, the large system limit can be evaluated in analogy with (4.52). Unfortunately, unlike the previous examples considered in this chapter, this approach does not yield any new information, i.e.:

$$v_{j,k}^{\dagger} C_{j,k}^{-1} v_{j,k} \asymp \frac{1}{N - (N-1)} \text{tr}[(I_N - V_j V_j^{\dagger} + v_{j,k} v_{j,k}^{\dagger}) C_{j,k}^{-1}] = v_{j,k}^{\dagger} C_{j,k}^{-1} v_{j,k} \tag{4.B.12}$$

This motivates an intermediate step in the derivation in which the rank of V_j is reduced to some value $K < N$. We then consider the asymptotic limit $(N, K) \to \infty$ with $K/N \to \beta$ where $\beta \in (0, 1)$. The result we seek is then obtained by letting $\beta \to 1$ from below, denoted $\beta \to 1^-$.

Formally, we have:

$$G_C(z) = \lim_{\beta \to 1^-} \acute{G}_C(z, \alpha) \tag{4.B.13}$$

where:

$$\acute{G}_C(z, \alpha) = \lim_{\substack{(N,K)\to\infty \\ K/N\to\beta}} \acute{G}_C^N(z, \alpha), \quad \beta \in (0, 1) \tag{4.B.14}$$

$$\acute{G}_C^N(z, \alpha) = \frac{1}{N}\text{tr}\left[\acute{\mathbf{C}}^{-1}\right] \tag{4.B.15}$$

$$\acute{\mathbf{C}} = -z\mathbf{I}_N + \sum_{j=1}^{J} \acute{\mathbf{X}}_j^{\ddagger}, \tag{4.B.16}$$

$$\acute{\mathbf{X}}_j = \acute{\mathbf{V}}_j \acute{\mathbf{D}}_j \tag{4.B.17}$$

where $\acute{\mathbf{V}}_j$ contains the first $K < N$ columns of $\mathbf{V}_j$, and $\acute{\mathbf{D}}_j$ is the corresponding $K \times K$ submatrix of $\mathbf{D}_j$.

Following the incremental matrix expansion approach, the next step is to remove column k of $\acute{\mathbf{V}}_j$ from $\acute{\mathbf{C}}$, i.e., we expand $\acute{\mathbf{C}} = \acute{\mathbf{C}}_{j,k} + D_{j,k}\mathbf{v}_{j,k}\mathbf{v}_{j,k}^{\dagger}$. We have:

$$\acute{\mathbf{C}}^{-1}\mathbf{v}_{j,k} = \frac{\acute{\mathbf{C}}_{j,k}^{-1}\mathbf{v}_{j,k}}{1 + D_{j,k}\rho_{j,k}^N} \tag{4.B.18}$$

from the MIL, where $\rho_{j,k}^N = \mathbf{v}_{j,k}^{\dagger}\acute{\mathbf{C}}_{j,k}^{-1}\mathbf{v}_{j,k}$. It can be shown that:

$$\rho_{j,k}^N \overset{k}{\asymp} \rho_j^N \tag{4.B.19}$$

in the limit considered, where:

$$\rho_j^N = \frac{1}{N-K}\text{tr}[\mathbf{Y}_j\acute{\mathbf{C}}^{-1}] \tag{4.B.20}$$

$$\mathbf{Y}_j = \mathbf{I}_N - \acute{\mathbf{V}}_j\acute{\mathbf{V}}_j^{\dagger} \tag{4.B.21}$$

Now expanding the identity $\mathbf{I}_N = \acute{\mathbf{C}}\acute{\mathbf{C}}^{-1}$ using (4.B.18) and (4.B.19) gives:

$$1 = \frac{1}{N}\text{tr}[\acute{\mathbf{C}}\acute{\mathbf{C}}^{-1}] \tag{4.B.22}$$

$$= -z\acute{G}_C^N(z, \alpha) + \frac{1}{N}\sum_{j=1}^{J}\sum_{k=1}^{K} D_{j,k}\mathbf{v}_{j,k}^{\dagger}\acute{\mathbf{C}}^{-1}\mathbf{v}_{j,k} \tag{4.B.23}$$

$$\asymp -z\acute{G}_C^N(z, \alpha) + \beta J - \beta\sum_{j=1}^{J}\frac{1}{K}\sum_{k=1}^{K}\frac{1}{1 + D_{j,k}\rho_j^N} \tag{4.B.24}$$

This gives one equation for the $J + 1$ unknowns $\acute{G}_C^N(z, \alpha)$ and ρ_j^N, $j = 1, \ldots, J$. The remaining J equations are found by expanding (4.B.20) as:

$$\rho_j^N = \frac{1}{N - K} \mathrm{tr}\left[\left(\mathbf{I}_N - \acute{\mathbf{V}}_j^{\ddagger}\right)\acute{\mathbf{C}}^{-1}\right] \tag{4.B.25}$$

$$\asymp \frac{1}{1 - \beta}\left(\acute{G}_C^N(z, \alpha) - \beta \frac{1}{K}\sum_{k=1}^{K} \frac{\rho_j^N}{1 + D_{j,k}\rho_j^N}\right) \tag{4.B.26}$$

It follows from (4.B.24) and (4.B.26) that $\acute{G}_C^N(z, \alpha) \asymp \acute{G}_C(z, \alpha)$ and $\rho_j^N \asymp \rho_j$, $j = 1, \ldots, J$, where $\acute{G}_C(z, \alpha)$ and $\rho_j, j = 1, \ldots, J$, satisfy:

$$\acute{G}_C(z, \alpha) = -z^{-1}\left(1 - \beta J + \beta \sum_{j=1}^{J} \mathcal{E}_j^{\mathrm{A}}\right) \tag{4.B.27}$$

$$\rho_j = \frac{\acute{G}_C(z, \alpha)}{\beta(\mathcal{E}_j^{\mathrm{A}} - 1) + 1}, \tag{4.B.28}$$

where $\mathcal{E}_j^{\mathrm{A}} = \mathbf{E}\left[\frac{1}{1 + X_j \rho_j}\right]$ and the distribution of the random variable X_j is the a.e.d. of $X_j^{\ddagger}$.

As previously discussed, we now let $\beta \to 1^-$ in (4.B.27) and (4.B.28), which gives the $J + 1$ simultaneous equations:

$$G_C(z) = \frac{J - 1}{z + \sum_{j=1}^{J} \rho_j^{-1}} \tag{4.B.29}$$

$$G_C(z) = \mathbf{E}\left[\frac{1}{X_j + \rho_j^{-1}}\right], \quad j = 1, \ldots, J \tag{4.B.30}$$

It can be shown that there exists a unique solution to these equations [9, Theorem 2.1], so that $G_C^N(z) \to G_C(z)$ with probability one. It is easily verified that this result matches that stated in Identity 2, obtained from the R-transform.

4.B.3   Products of Unitarily Invariant Matrices

The same approach discussed in the previous subsection can be applied to determine the a.e.d. of the product $(\prod_{j=1}^{J} \mathbf{X}_j)^{\ddagger}$, where the $\mathbf{X}_j$ are as defined previously. That is, we seek $G_B^N(z) = \lim_{N \to \infty} \frac{1}{N}\mathrm{tr}[\mathbf{B}^{-1}]$, where $\mathbf{B} = -z\mathbf{I}_N + (\prod_{j=1}^{J} \mathbf{X}_j)^{\ddagger}$. We start with $J = 2$ and note that $(\mathbf{X}_1\mathbf{X}_2)^{\ddagger}$ and $\mathbf{X}_1^{\ddagger}\mathbf{X}_2^{\ddagger}$ have the same eigen-values. As before, rather than applying an incremental matrix expansion to derive $G_B^N(z)$ directly, we consider an associated problem in which the rank of each $\mathbf{V}_j$, $j = 1, \ldots, J$, is reduced to K.

The desired solution is obtained by taking the asymptotic limit $(N, K) \to \infty$ with $K/N \to \infty$, and then taking $\beta \to 1^-$.

This results in three simultaneous equations in the variables $G_B(z)$, π_1, and π_2, given by:

$$G_B(z) = -z^{-1} \mathbf{E} \left[\frac{1}{1 + \pi_j X_j} \right], \quad j = 1, 2 \tag{4.B.31}$$

$$G_B(z) = \frac{1}{z(z\pi_1 \pi_2 - 1)} \tag{4.B.32}$$

The extension to $J > 2$ can be obtained by recursion. Since there exists a unique solution to these equations [9, Theorem 2.4], it follows that $G_B^N(z) \to G_B(z)$ with probability one. It can be easily verified that this result matches that shown in Identity 3, obtained using the S-transform.

REFERENCES

1. P. D. Alexander, A. J. Grant, and M. C. Reed. Iterative detection in code-division multiple-access with error control coding. *European Transactions on Telecommunications*, 9(5):419–425, September–October 1998.

2. P. D. Alexander, M. C. Reed, J. Asenstorfer, and C. B. Schlegel. Iterative multiuser interference reduction: Turbo CDMA. *IEEE Trans. on Communications*, 47(7):1008–1014, July 1999.

3. Z. D. Bai. Methodologies in spectral analysis of large dimensional random matrices, a review. *Statistica Sinica*, 9(3):611–677, 1999.

4. E. Biglieri, G. Caire, and G. Taricco. CDMA system design through asymptotic analysis. *IEEE Trans. on Communications*, 48(11):1882–1896, November 2000.

5. E. Biglieri, G. Taricco, and A. Tulino. Performance of space-time codes for a large number of antennas. *IEEE Trans. on Information Theory*, 48(7):1794–1803, July 2002.

6. J. Boutros and G. Caire. Iterative multiuser joint decoding: Unified framework and asymptotic analysis. *IEEE Trans. on Information Theory*, 48(7):1772–1793, July 2002.

7. J. M. Chaufray, W. Hachem, and P. Loubaton. Asymptotic analysis of optimum and suboptimum CDMA downlink MMSE receivers. *IEEE Trans. on Information Theory*, 50(11):2620–2638, November 2004.

8. C. Chuah, D. N. C. Tse, J. M. Kahn, and R. A. Valenzuela. Capacity scaling in MIMO wireless systems under correlated fading. *IEEE Trans. on Information Theory*, 48(3):637–650, March 2002.

9. G. P. Chystyakov and F. Götze. The arithmetic of distributions in free probability theory. *(preprint)* http://www.math.uni-bielefeld.de/~goetze/stochastics/WORK/DFG.html, 2005.

10. P. M. Crespo, M. L. Honig, and J. A. Salehi. Spread-time code-division multiple access. *IEEE Trans. on Communications*, 43(6):2139–2148, June 1995.

11. M. Debbah, W. Hachem, P. Loubaton, and M. de Courville. MMSE analysis of certain large isometric random precoded systems. *IEEE Trans. on Information Theory*, 49(5):1293–1311, May 2003.

12. E. Eleftheriou and D. Falconer. Tracking properties and steady-state performance of RLS adaptive filter algorithms. *IEEE Trans. on Acoustics, Speech, and Signal Processing*, 34(5):1097–1110, October 1986.

13. J. Evans and D. N. C. Tse. Large system performance of linear multiuser receivers in multipath fading channels. *IEEE Trans. on Information Theory*, 46(6):2059–2078, September 2000.

14. G. B. Giannakis and C. Tepedelenlioglu. Basis expansion models and diversity techniques for blind identification and equalization of time-varying channels. *Proceedings of the IEEE*, 86(10):1969–1986, October 1998.

15. V. L. Girko. *Theory of Random Determinants*. Kluwer Academic, 1990.

16. V. L. Girko. *Theory of Stochastic Canonical Equations*. Kluwer Academic, 2001.

17. A. Goldsmith. *Wireless Communication*. Cambridge University Press, 2005.

18. A. J. Grant and P. D. Alexander. Random sequence multisets for synchronous code-division multiple-access channels. *IEEE Trans. on Information Theory*, 44(7): 2832–2836, November 1998.

19. D. Guo and S. Verdú. Randomly spread CDMA: Asymptotics via statistical physics. *IEEE Trans. on Information Theory*, 51(6):1982–2010, June 2005.

20. B. Hassibi and B. M. Hochwald. How much training is needed in multiple-antenna wireless links? *IEEE Trans. on Information Theory*, 49(4):951–963, April 2003.

21. S. Haykin. *Adaptive Filter Theory*. Prentice Hall, 3rd edition, 1996.

22. F. Hiai and D. Petz. *The semicircle law, free random variables and entropy*. American Mathematics Society, Mathematical Surveys and Monographs, Vol. 77, 2000.

23. M. L. Honig and H. V. Poor. *Adaptive Interference Mitigation in Wireless Communications: A Signal Processing Perspective*, pages 64–128. H. V. Poor and G. Wornell, eds. Prentice Hall, New Jersey, 1998.

24. M. L. Honig and R. Ratasuk. Large-system performance of iterative multiuser decision-feedback detection. *IEEE Trans. on Communications*, 51(8):1368–1377, August 2003.

25. M. L. Honig and M. Tsatsanis. Adaptive techniques for multiuser cdma receivers. *Signal Processing Magazine*, 17(3):49–61, May 2000.

26. M. L. Honig and W. Xiao. Performance of reduced-rank linear interference suppression. *IEEE Trans. on Information Theory*, 47(5):1928–1946, July 2001.

27. R. A. Horn and C. R. Johnson. *Matrix Analysis*. Press Syndicate of the University of Cambridge, 1985.

28. A. Host-Madsen, X. Wang, and S. Bahng. Asymptotic analysis of blind multiuser detection with blind channel estimation. *IEEE Trans. on Signal Processing*, 52(6):1722–1738, June 2004.

29. Kiran and D. N. C. Tse. Effective interference and effective bandwidth of linear multiuser receivers in asynchronous CDMA systems. *IEEE Trans. on Information Theory*, 46(4):1426–1447, July 2000.

30. J. Lehnert. An asymptotic analysis of band-limited ds/ssma communication systems. *IEEE Trans. on Information Theory*, 52(2):759–766, February 2006.

31. L. Li, A. M. Tulino, and S. Verdú. Design of reduced-rank MMSE multiuser detectors using random matrix methods. *IEEE Trans. on Information Theory*, 50(6):986–1008, June 2004.

32. F. Ling and J. Proakis. Nonstationary learning characteristics of least squares adaptive estimation algorithms. In *IEEE ICASSP*, volume 9, pages 118–121, March 1984.

33. J. P. Linnartz. Performance analysis of synchronous MC-CDMA in mobile Rayleigh channel with both delay and doppler spreads. *IEEE Trans. on Vehicular Technology*, 50(6):1375–1387, November 2001.

34. P. Loubaton and W. Hachem. Asymptotic analysis of reduced-rank wiener filters. In *Proc. IEEE Information Theory Workshop*, pages pp. 329–331, Paris, France, April 2003.

35. A. Lozano and A. M. Tulino. Capacity of multiple-transmit multiple-receive antenna architectures. *IEEE Trans. on Information Theory*, 48(12):3117–3128, December 2002.

36. V. A. Marčenko and L. A. Pastur. Distributions of eigenvalues of some sets of random matrices. *Math. USSR-Sbornik*, 1(4):457–483, 1967.

37. S. Misra, A. Swami, and L. Tong. Optimal training over the Gauss-Markov fading channel: A cutoff rate analysis. In *Proc. IEEE Int. Conf. on Acoustics, Speech, and Signal Processing*, pages 809–12 vol. 3, Montreal, May 2004.

38. R. R. Müller. Multiuser receivers for randomly spread signals: Fundamental limits with and without decision-feedback. *IEEE Trans. on Information Theory*, 47(1):268–283, January 2001.

39. R. R. Müller. A random matrix model of communication via antenna arrays. *IEEE Trans. on Information Theory*, 48(9):2495–2506, September 2002.

40. L. Pastur and V. Vasilchuk. On the law of addition of random matrices. *Comm. Math. Phys.*, 214(21):249–286, November 2000.

41. M. J. M. Peacock, I. B. Collings, and M. L. Honig. Asymptotic analysis of MMSE multiuser receivers for multi-signature multicarrier CDMA in Rayleigh fading. *IEEE Trans. on Communications*, 52(6):964–972, June 2004.

42. M. J. M. Peacock, I. B. Collings, and M. L. Honig. Asymptotic spectral efficiency of multiuser multi-signature CDMA in frequency-selective channels. *IEEE Trans. on Info. Theory*, 52(3):1113–1129, March 2006.

43. M. J. M. Peacock, I. B. Collings, and M. L. Honig. Unified large system analysis of MMSE and adaptive least squares receivers for a class of random matrix channels. *IEEE Trans. on Information Theory*, 52(8):3567–3600, 2006.

44. W. G. Phoel and M. L. Honig. Performance of coded DS-CDMA with pilot-assisted channel estimation and linear interference suppression. *IEEE Trans. on Communications*, 50(5):822–832, May 2002.

45. H. V. Poor and X. Wang. Code-aided interference suppression for DS/CDMA communications—Part II: Parallel blind adaptive implementations. *IEEE Trans. on Communications*, 45(9):1112–1122, September 1997.

46. I. S. Reed, J. D. Mallet, and L. E. Brennan. Rapid convergence rate in adaptive arrays. *IEEE Trans. on Aerospace and Electronic Systems*, 10(6):853–863, September 1974.

47. M. Schwartz. *Mobile Wireless Communications*. Cambridge University Press, 2005.

48. S. Shamai and S. Verdú. The impact of frequency-flat fading on the spectral efficiency of CDMA. *IEEE Trans. on Information Theory*, 47(4):1302–1327, May 2001.

49. J. W. Silverstein. Strong convergence of the empirical distribution of eigenvalues of large dimensional random matrices. *Journal of Multivariate Analysis*, 55(2):331–339, 1995.

50. J. W. Silverstein and Z. D. Bai. On the empirical distribution of eigenvalues of a class of large dimensional random matrices. *Journal of Multivariate Analysis*, 54(2):175–192, 1995.

51. A. Nica and R. Speicher. *Lectures on the Combinatorics of Free Probability*. Cambridge University Press, 2006.

52. Y. Sun and M. L. Honig. Large system capacity of MIMO block fading channels with least squares linear adaptive receivers. In *Proc. of IEEE Globecom*, pages 1516–1519, St. Louis, MO, December 2005.

53. T. Tanaka. A statistical-mechanics approach to large-system analysis of CDMA multiuser detectors. *IEEE Trans. on Information Theory*, 48(11):2888–2910, November 2002.

54. E. Telatar. Capacity of multi-antenna gaussian channels. *Euro. Trans. Telecommun.*, 10:585–595, Nov.-Dec. 1999.

55. L. G. F. Trichard, J. S. Evans, and I. B. Collings. Large system analysis of linear multistage parallel interference cancellation. *IEEE Trans. on Communications*, 50(11):1778–1786, November 2002.

56. D. Tse and P. Viswanath. *Fundamentals of Wireless Communication*. Cambridge University Press, 2005.

57. D. Tse and O. Zeitouni. Performance of linear multiuser receivers in random environments. *IEEE Trans. on Information Theory*, 46(1):171–188, January 2000.

58. D. N. C. Tse and S. V. Hanly. Linear muliuser receivers: Effective interference, effective bandwidth and user capacity. *IEEE Trans. on Information Theory*, 45(2):641–657, March 1999.

59. A. M. Tulino, A. Lozano, and S. Verdú. Impact of antenna correlation on the capacity of multiantenna channels. *IEEE Trans. on Information Theory*, 51(7):2491–2509, July 2005.

60. A. M. Tulino and S. Verdú. Random matrix theory and wireless communications. *Foundations and Trends in Communications and Information Theory*, 1(1):1–182, 2004.

61. V. Vasilchuk. On the law of multiplication of random matrices. *Mathematical Physics, Analysis and Geometry*, 4(1):1–36, 2001.

62. S. Verdú. Minimum probability of error for asynchronous Gaussian multiple access channels. *IEEE Trans. on Information Theory*, 32(1):85–96, 1986.

63. S. Verdú and S. Shamai. Spectral efficiency of CDMA with random spreading. *IEEE Trans. on Information Theory*, 45(2):622–640, March 1999.

64. H. Vikalo, B. Hassibi, B. M. Hochwald, and T. Kailath. On the capacity of frequency-selective channels in training-based transmission schemes. *IEEE Trans. on Signal Processing*, 52(9):2572–2583, September 2004.

65. D. Voiculescu. Addition of certain non-commuting random variables. *J. Funct. Anal.*, 66:323–346, 1986.

66. D. Voiculescu. Multiplication of certain non-commuting random variables. *J. Operator Theory*, 18:223–235, 1987.

67. D. Voiculescu. *Free Probability Theory (Fields Institute Communications, Vol 12)*. American Mathematical Society, 1997.

68. X. Wang and H. V. Poor. Iterative (turbo) soft interference cancellation and decoding for coded CDMA. *IEEE Trans. on Communications*, 47(7):1046–1061, July 1999.

69. W. Xiao and M. L. Honig. Convergence analysis of adaptive reduced-rank linear filters for DS-CDMA. In *Conference on Information Sciences and Systems*, pages WP2-6–WP2-11, Princeton University, March 2000.

70. W. Xiao and M. L. Honig. Forward-link performance of satellite CDMA with linear interference suppression and one-step power control. *IEEE Trans. on Wireless Communications*, 1(4):600–610, October 2002.

71. W. Xiao and M. L. Honig. Large system transient analysis of adaptive least squares filtering. *IEEE Trans. on Information Theory*, 51(7):2447–2474, July 2005.

72. Z. Xu and X. Wang. Large-sample performance of blind and group-blind multiuser detectors: A perturbation perspective. *IEEE Trans. on Information Theory*, 50(10):2389–2401, October 2004.

73. B. M. Zaidel, S. Shamai, and S. Verdú. Multicell uplink spectral efficiency of coded DS-CDMA with random signatures. *IEEE Journal on Selected Areas in Communications*, 19(8):1556–1569, August 2001.

74. J. Zhang, E. K. P. Chong, and D. N. C. Tse. Output MAI distributions of linear MMSE multiuser receivers in DS-CDMA systems. *IEEE Trans. on Information Theory*, 47(3):1128–1144, March 2001.

75. J. Zhang and X. Wang. Large-system performance analysis of blind and group-blind multiuser receivers. *IEEE Trans. on Information Theory*, 48(9):2507–2523, September 2002.

5

GENERIC MULTIUSER DETECTION AND STATISTICAL PHYSICS

Dongning Guo and Toshiyuki Tanaka

5.1 INTRODUCTION

Fuelled by the advent and rapid development of cellular telephony, the problem of multiuser signal detection (or separation) has received great attention since the mid-1980s as one of the major avenues towards optimal error performance and spectrum usage in wireless communications. This chapter introduces the concept of *generic multiuser detection* and summarizes some recent advances in the analysis and design of generic detectors using statistical physics techniques.

5.1.1 Generic Multiuser Detection

Consider a multidimensional communication system in which each user randomly generates a "signature vector" and modulates its own (usually error-control coded) symbols onto the signature for transmission. The received signal is the superposition of all users' signals corrupted by Gaussian noise. With knowledge of all signature vectors, the goal of a multiuser receiver is to reliably recover the information intended for all or a subset of the users. The multiuser channel, best described by a vector model, is very versatile and is widely used in applications that include code-division multiple access (CDMA) as well as certain multiple-input multiple-output (MIMO) systems.

Advances in Multiuser Detection. Edited by Michael L. Honig
Copyright © 2009 John Wiley & Sons, Inc.

The maximum information rate through a multiuser channel is achieved by jointly optimal decoding, which is prohibitively complex for all but a small user population and codeword length. Hence, the tasks of untangling the mutually interfering streams and exploiting the redundancy in the error-control codes are often separated. Oftentimes, the multiuser detector plays the role of a front end, which provides individual stream of (hard or soft) decision statistics to independent single-user decoders.

The simplest meaningful detector ignores the presence of multiaccess interference (MAI) and carries out single-user matched filtering (SUMF), whose error performance is generally very poor. The best probability of error of an uncoded system is achieved by solving a hypothesis testing problem with an exponential number of hypotheses in the number of interfering users [1,2], which is in general NP-hard [3]. Optimal error performance in asynchronous channels is achieved by more involved sequence detection [4]. A large gap has been demonstrated between the probability of error of the naïve SUMF and that of individually optimal (IO) and jointly optimal (JO) detection. In order to explore the trade-offs between performance and computational complexity, numerous suboptimal detection schemes have been proposed, such as the decorrelator, linear minimum mean-square error (LMMSE) detector, and various interference cancelers.

All of the aforementioned detectors can be derived as some form of optimal detection with heuristic (but untrue) assumptions based on conventional wisdom and practical considerations. For example, the SUMF is optimal assuming the MAI to be Gaussian or absent; the decorrelator provides optimal detection assuming that there is no background noise; and the LMMSE detector maximizes the output signal-to-interference-and-noise ratio (SINR) assuming Gaussian inputs. With the exception of decorrelating receivers, the multiuser detector outputs are still contaminated by MAI.

The viewpoint taken in this chapter is that, in general, every suboptimal detector can be regarded as computing an "optimally" detected output given some "mismatched" system model. This perspective has its origin in the general theory of statistical inference or learning, where a "student" may adopt a probability model that is different from that of the "teacher" (see e.g., [5], for a discussion of such cases). In particular, the so-called *generic multiuser detector* computes the *posterior mean* of the transmitted symbols given the observation based on a postulated probability law of the system. More on this viewpoint will be discussed in Section 5.2.

5.1.2   Single-User Characterization of Multiuser Systems

Using techniques and methodologies originating in statistical physics, two fundamental questions about multiuser systems and generic detection are addressed in this chapter:

1) Given a multiuser detector, how to characterize the (single-user) subchannel between the input and output of each user?

2) Given a multiuser system, what are the achievable information rates by optimal joint decoding and suboptimal single-user decoding, respectively?

The preceding questions have been well studied for linear detection schemes. In fact, the analysis of multiuser communication systems is to a large extent the pursuit of a single-user characterization of the performance. A key performance measure, the *multiuser efficiency*, was introduced in [4,6] to refer to the signal-to-noise ratio (SNR) degradation of the multiuser detection output relative to single-user performance. The multiuser efficiencies of the SUMF, decorrelator, LMMSE detector and linear interference cancelers at any given SNR were found as functions of the correlation matrix of the spreading sequences (i.e., the signature vectors), which can also be written explicitly in terms of the eigen-values of the matrix.

The performance of finite-size multiuser systems is often not easy to evaluate, including when averaged over random sequences (e.g., [7–9]). An alternative paradigm for the analysis is to take the large-system limit instead, namely, to study the case where the number of users and the dimension of the channel both tend to infinity with a fixed ratio. A key consequence is that the dependence of performance measures on the spreading sequences vanishes as the system size increases. In the special case of linear detection, the output converges to a Gaussian statistic, which allows the performance to be solely quantified using the output SINR (e.g., [10–12]). It appears that the users are decoupled in such a way that each user experiences a single-user channel with SNR degradation in lieu of MAI.

A major spate of success of large-system analysis is achieved by using random matrix theory, the central dictate of which is that the empirical distributions of the eigen-values of a random matrix converge to a deterministic distribution as its dimension increases [13,14]. As a result, the multiuser efficiency of a sufficiently large system can be obtained as an integral with respect to the limiting eigen-value distribution. Indeed, this random matrix technique is applicable to any performance measure that can be expressed as a function of the eigen-values, e.g., the multiuser efficiency of the decorrelator [15–17] and the large-system capacity of CDMA channels [18,19] (see also [20,21]). Moreover, the large-system multiuser efficiency of the LMMSE detector is found to be the unique solution to the Tse-Hanly fixed-point equation [11] (see also [15] for the special equal-power case). The multiuser system with LMMSE detection admits a single-user characterization, as is also indicated by the notion of effective interference in [11]. It is important to note that such large-system results are often quite representative of the performance with a moderate to large user population.

Few explicit expressions of the efficiencies in terms of eigen-values are available beyond the above cases. Little success has been reported in the application of random matrix theory when the detector is nonlinear. It was not until statistical physics techniques were applied to the analysis of multiuser detection that a major breakthrough became possible. Using the so-called replica method, the large-system uncoded minimum bit-error-rate (BER) (hence the optimal multiuser efficiency) and spectral efficiency (the input–output mutual information per dimension) with equal-power binary inputs were first obtained in [22–26] and generalized to the case of arbitrary inputs and powers in [17,27]. Reference [28] studied the channel capacity under separate decoding and noted that the additive decomposition of the optimum spectral efficiency in [19] holds also for binary inputs. The same formula was conjectured to be valid regardless of the input distribution [29]. The most general framework to date is developed in [27,30–32], where both joint decoding and generic multiuser detection followed by separate decoding are studied assuming an arbitrary input distribution and flat fading.

The main results of the chapter are presented in Section 5.3. The centerpiece is a single-user characterization of the multiuser system, called the "decoupling principle," which states that the multiuser channel followed by generic detection is essentially equivalent to a bank of single-user channels, one for each user. The conjecture in [29] is also validated. The decoupling principle carries great practicality and finds convenient uses in finite-size systems where the analytical asymptotic results are a good approximation. It is also found to be applicable to multirate and multicarrier CDMA [33,34].

5.1.3 On the Replica Method

The replica method, which underlies most of the results in this chapter, was invented in 1975 by S. F. Edwards and P. W. Anderson to study the free energy of disordered magnetic systems, called spin glasses [35]. It has since become a standard technique in statistical physics [36]. Analogies between statistical physics and neural networks, image processing, and communications have gradually been noticed (e.g., [37,38]), on the basis of which the range of application of the replica method has been expanding. There have been many recent activities applying statistical physics wisdom and the replica method to sparse-graph error-control codes (e.g., [39–43]). The same techniques have also been used to study the capacity of MIMO channels [44,45]. Among other techniques, mean field theory is used to derive iterative detection algorithms [46,47].

For the purpose of analytical tractability, we will invoke several assumptions crucial to the replica method and common in the statistical physics literature (see Section 5.4.1.2). Unfortunately, these assumptions have not been fully justified. Thus although there has been some recent progress [48,49], the mathematical rigor of the general results in this chapter is pending on breakthroughs in those problems. Note, however, that the key results have been rigorously proved in the special case of relatively small load and where the spreading matrix is sparse in some sense by showing the optimality of belief propagation (BP) [50–52]. The technique paves a new avenue for the interpretation and justification of the general results. The replica analysis of generic detection is presented in Section 5.4. Some further discussions on the replica method is found in Section 5.5. Useful statistical physics concepts and methodologies are introduced in Section 5.6.

5.1.4 Statistical Inference Using Practical Algorithms

The input-output relationship of a multiuser systems can in general be fully described (probabilistically) using a bipartite factor graph. As the example shown in Figure 5.1, each edge connects a *symbol node* which represents an input symbol and a *chip node* which represents a component of the output signal. The problem of multiuser detection

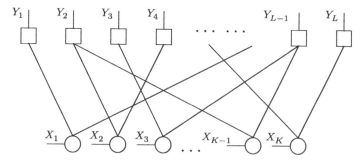

Figure 5.1. A bipartite factor graph describing the probability law of a multiuser system.

can be regarded as a statistical inference problem on the graph. As mentioned earlier, performing the inference exactly, e.g., in order to obtain the posterior mean, is computationally hard.

An important family of iterative algorithms for performing the computation approximately is known as *belief propagation*, the formulation of which is attributed to Pearl [53]. In fact, all multiuser detection algorithms discussed in this chapter can be regarded as BP on the factor graph with appropriate heuristic postulates. BP or its variations with linear complexity are especially appealing in practice. Section 5.7 discusses how to design low-complexity algorithms based on BP. In particular, it is shown that parallel interference cancellation (PIC) can be understood as a further simplification of BP.

5.1.5 Statistical Physics and Related Problems

This chapter can be regarded not only as an application of statistical physics ideas and techniques to the communication problem at hand, but also as progress in a much broader research trend, where large-scale problems in various fields are formulated using probability theory and analyzed using statistical physics. The trend can be traced back to the 1980s, when researchers of spin glasses, whose primary objective is to obtain macroscopic characterizations of large disordered systems on the basis of their microscopic specifications, became aware that their methodologies can also be applicable to problems outside statistical physics, such as constraint satisfaction problems [54,55] and neural networks [56]. Successes in these fields have triggered subsequent applications of statistical physics to various other fields, such as information and communication theory, computation theory, learning and artificial intelligence, etc. One example of the most exciting interplay between these multitude of disciplines is found in recent research activities of sparse-graph error-control codes, where "macroscopic" analysis and "microscopic" algorithm design are concurrently studied, revealing a deep relationship between statistical physics characterization of a problem and properties of inference algorithms for solving it, as well as demonstrating importance of statistical physics concepts such as phase transition and finite size scaling, in the context of error-control coding [57,58]. This fertile interdisciplinary field sets the stage for the unique treatment of the multidimensional communication problem described in this chapter.

5.2 GENERIC MULTIUSER DETECTION

In this section, we first describe the multiuser system model considered in this chapter. We then put forth a framework of generic multiuser detection and specialize it to several popular detectors.

5.2.1 CDMA/MIMO Channel Model

Consider the vector channel depicted in Figure 5.2, which in general models a MIMO system. In this chapter, the model describes a real-valued[1] fully-synchronous K-user CDMA system with spreading factor L. Each encoder maps its message into a sequence of channel symbols. All users employ the same type of signaling so that at each interval the K symbols are independent and identically distributed (*i.i.d.*) random variables with distribution (probability measure) P_X. Let $X = [X_1, \ldots, X_K]^\mathsf{T}$ denote the vector of input symbols from the K users in one symbol interval. For notational convenience in the analysis, it is assumed that either a probability density function (PDF) or a probability mass function (PMF) of the distribution P_X exists,[2] and is denoted by p_X. Let $p_X(x) = \prod_{k=1}^{K} p_X(x_k)$ denote the joint (product) distribution.[3]

Let the instantaneous SNR of user k be denoted by γ_k and $A = \mathrm{diag}$ $\{\sqrt{\gamma_1}, \ldots, \sqrt{\gamma_K}\}$. Denote the spreading sequence of user k by $S_k = \frac{1}{\sqrt{L}}$ $[S_{1k}, S_{2k}, \ldots, S_{Lk}]^\mathsf{T}$, where S_{nk} are *i.i.d.* random variables with zero mean, unit variance and finite moments. The realization of S_{lk} and S_k are denoted by s_{lk} and s_k to distinguish from their random counterparts. The $L \times K$ channel "state" matrix is denoted by $S = [\sqrt{\gamma_1} S_1, \ldots, \sqrt{\gamma_K} S_K]$. The synchronous flat-fading CDMA channel is described by:

$$Y = \sum_{k=1}^{K} \sqrt{\gamma_k} S_k X_k + N \tag{5.1}$$

$$= SX + N \tag{5.2}$$

where $N \sim \mathcal{N}(0, I)$ is a vector of independent standard Gaussian entries. Without loss of generality we assume P_X to have zero mean and unit variance.

5.2.2 Generic Posterior Mean Estimation

As depicted in Figure 5.2, the multiuser detector front end estimates the transmitted symbols given the received signal and the channel state without using any knowledge

---

[1] Extension to a complex-valued system is straightforward [27].
[2] Validity of the results in this chapter do not depend on the existence of a PDF or PMF.
[3] The main results of this chapter extend to cases where the entries of X are dependent: See [32].

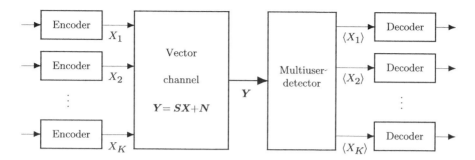

Figure 5.2. Single-user encoding, multiuser channel, and multiuser detection followed by independent single-user decoding.

of the error-control codes employed by the transmitters. Meanwhile, each single-user decoder only observes the sequence of decision statistics corresponding to one user, and ignores the existence of all other users. By adopting this separate decoding approach, the channel together with the multiuser detector front end is viewed as a bank of coupled single-user channels. Note that the detection output sequence for an individual user is in general not a sufficient statistic for decoding this user's own information.

To capture the intended suboptimal structure, we restrict the capability of the multiuser detector; otherwise the detector could in principle encode the channel state and the received signal (S, Y) into a single real number as its output to each user, which is a sufficient statistic for all users. A plausible choice is the (canonical) *posterior mean estimator* (PME), which computes the mean value of the posterior probability distribution $p_{X|Y,S}$, hereafter denoted by angle brackets $\langle \cdot \rangle$:

$$\langle X \rangle = \mathsf{E}\{X \mid Y, S\} \qquad (5.3)$$

The expectation is taken over the posterior probability distribution $p_{X|Y,S}$, which is induced from the input distribution p_X and the conditional Gaussian density function $p_{Y|X,S}$ of the channel (5.2) by Bayes' formula:[4]

$$p_{X|Y,S}(x \mid y, s) = \frac{p_X(x) p_{Y|X,S}(y \mid x, s)}{\int p_X(x) p_{Y|X,S}(y \mid x, s) \, dx} \qquad (5.4)$$

where the integral shall be replaced by a sum if X is discrete. Note that, although implicit in notation, $\langle X \rangle$ is a function of (Y, S), which is dependent on the input X through (5.2).

---

[4]Uppercase letters are usually used for matrices and random variates, while lowercase letters are used for deterministic scalars and vectors. As a compromise, the realization of the spreading matrix S is denoted as s. We keep the use of s to minimum.

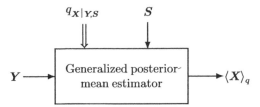

Figure 5.3. Generalized posterior mean estimator.

Also known as the *conditional mean estimator*, (5.3) achieves the minimum mean-square error for each user, and is therefore the (nonlinear) MMSE detector. We also regard it as a soft-output version of the individually optimal multiuser detector (assuming uncoded transmission). In case of binary antipodal transmission, the posterior mean estimate is consistent in its sign with the individually optimal hard decision. Although this consistency property does not hold for general *m*-ary constellation, (5.3) is optimal in mean-square sense, and is thus a sensible detection output to be used for further decoding.[5]

The PME can be understood as an "informed" optimal estimator which is supplied with the posterior distribution and then computes its mean. A generalization of the canonical PME is conceivable; instead of informing the estimator of the actual posterior $p_{X|Y,S}$, we can supply at will any well-defined conditional distribution $q_{X|Y,S}$ as depicted in Figure 5.3. Given (Y, S), the estimator can nonetheless perform "optimal" estimation based on this postulated measure q. We call this the *generalized posterior mean estimation*, which is conveniently denoted as:

$$\langle X \rangle_q = \mathsf{E}_q\{X \mid Y, S\} \tag{5.5}$$

where $\mathsf{E}_q\{\cdot\}$ stands for the expectation with respect to the postulated measure q. Suppose $q_{X|Y,S}$ is induced from a prior q_X and a conditional distribution $q_{Y|X,S}$, then the generalized PME can be expressed as the expectation of X taken over the postulated posterior probability distribution:

$$q_{X|Y,S}(x \mid y, s) = \frac{q_X(x) q_{Y|X,S}(y \mid x, s)}{\int q_X(x) q_{Y|X,S}(y \mid x, s)\, dx} \tag{5.6}$$

For brevity, we also refer to (5.5) as the PME. In view of (5.3), the subscript in (5.5) can be dropped if the postulated measure q coincides with the actual one p.

In general, postulating a mismatched measure $q \neq p$ causes degradation in detection performance. Such a strategy may be either due to lack of knowledge of the true statistics or a particular choice that anticipates benefits, such as reduction of

---

[5]A more sophisticated detector produces the posterior distribution about each input symbol, which generally contains much richer content than point estimates such as the posterior mean. However, as we shall see in this chapter, the posterior distribution is equivalent to a conditional Gaussian distribution in large systems so that the posterior mean suffices.

computational complexity. In principle, any deterministic estimation strategy can be regarded as a PME since we can always choose to put a unit mass at the desired estimation output given (Y, S), the fact which demonstrates that the concept of PME is generic and versatile. We will see in Section 5.2.3 that by postulating an appropriate measure q, the PME can be particularized to many popular multiuser detectors. The generic representation (5.5) is pivotal here because it allows a unified treatment of a large family of multiuser detectors which results in a simple single-user characterization for all of them.

In this chapter, the posterior $q_{X|Y,S}$ supplied to the PME is assumed to be the one that is induced from a postulated CDMA system, where the input follows a certain distribution q_X, and the input–output relationship of the postulated channel differs from the actual channel (5.2) by only the noise variance. Precisely, the postulated system is characterized by:

$$Y = SX' + \sigma N' \tag{5.7}$$

where S is the state matrix of the actual channel (5.2), the components of X' are *i.i.d.* with distribution q_X, and N' is statistically the same as the Gaussian noise N in (5.2). The postulated input distribution q_X is assumed to have zero mean and unit variance. The posterior $q_{X|Y,S}$ is determined by q_X and $q_{Y|X,S}$ according to Bayes' formula (5.6). The postulated noise level σ serves as a control parameter. Indeed, the PME so defined is the optimal detector for a postulated multiuser system with its input distribution and noise level different from the actual ones. In general, the postulated channel state could also be different from the actual instance S, but this is out of the scope of this chapter, as we limit ourselves to studying the (fairly rich) family of multiuser detectors that can be represented as PMEs parameterized by the postulated input and noise level (q_X, σ).

5.2.3 Specific Detectors as Posterior Mean Estimators

We identify specific choices of the postulated input distribution q_X and noise level σ under which the PME is particularized to well-known multiuser detectors.

The characteristic of the actual channel (5.2) is:

$$p_{Y|X,S}(y \mid x, s) = (2\pi)^{-\frac{L}{2}} \exp\left[-\frac{1}{2}\|y - sx\|^2\right] \tag{5.8}$$

and that of the postulated channel is:

$$q_{Y|X,S}(y|x, s) = (2\pi\sigma^2)^{-\frac{L}{2}} \exp\left[-\frac{1}{2\sigma^2}\|y - sx\|^2\right] \tag{5.9}$$

The posterior distribution can be obtained using Bayes' formula (cf. (5.6)) as:

$$q_{X|Y,S}(x \mid y, s) = \frac{(2\pi\sigma^2)^{-\frac{L}{2}}q_X(x)}{q_{Y|S}(y \mid s)} \exp\left[-\frac{\|y - sx\|^2}{2\sigma^2}\right] \tag{5.10}$$

where:

$$q_{Y|S}(y\,|\,s) = (2\pi\sigma^2)^{-\frac{L}{2}} \mathsf{E}_q \left\{ \exp\left[-\frac{\|y - SX\|^2}{2\sigma^2} \right] \middle| S = s \right\} \tag{5.11}$$

and the expectation in (5.11) is taken over $X \sim q_X$.

5.2.3.1 Linear Detectors Let the postulated input be Gaussian, i.e., q_X is $\mathcal{N}(0,1)$. The optimal detector (PME) for the postulated model (5.9) with this Gaussian input is a linear filtering of the received signal Y:

$$\langle X \rangle_q = [s^{\mathsf{T}}s + \sigma^2 I]^{-1} s^{\mathsf{T}} Y \tag{5.12}$$

which corresponds to the LMMSE detector and the decorrelator by choosing $\sigma = 1$ and $\sigma \to 0$ respectively. If $\sigma \to \infty$, (5.12) is consistent with the SUMF output:

$$\sigma^2 \langle X_k \rangle_q \longrightarrow s_k^{\mathsf{T}} Y, \qquad \text{in } L^2 \text{ as } \sigma \to \infty \tag{5.13}$$

5.2.3.2 Optimal Detectors Let the postulated prior distribution q_X be identical to p_X. Let $\sigma \to 0$, then the probability mass of the distribution $q_{X|Y,S}$ is concentrated on a vector that minimizes $\|y - sx\|$, which also maximizes the likelihood function $p_{Y|X,S}(y\,|\,x,s)$. The PME $\lim_{\sigma \to 0} \langle X \rangle_q$ is thus equivalent to that of jointly optimal (or maximum-likelihood) detection [15]. Alternatively, if $\sigma = 1$, then the postulated measure coincides with the actual measure, i.e., $q_{X|Y,S}(x\,|\,y,s) = p_{X|Y,S}(x\,|\,y,s)$. The PME output $\langle X \rangle$ is the mean of the posterior probability distribution, which is seen as the (soft) individually optimal detector. Also worth mentioning is that if $\sigma \to \infty$, the PME reduces to the SUMF.

5.2.3.3 Interference Cancelers Suppose all symbols but X_1 are revealed as $\hat{x}_2, \ldots, \hat{x}_K$. The detector can use:

$$q_{Y|X_1,S}(y\,|\,x_1,s) \propto p_{Y|X,S}\left(y\,|\,[x_1, \hat{x}_2, \ldots, \hat{x}_K]^{\mathsf{T}}, s\right) \tag{5.14}$$

as the postulated channel characteristics in order to estimate X_1. The resulting PME of X_1 is simply an estimate obtained by matched filtering the received signal after canceling the interference reconstructed from $\hat{x}_2, \ldots, \hat{x}_K$. This scheme can be used for all users in either a successive or a parallel manner as well as in multistage fashion (e.g., [59–62] and [Chapter Grant-Rasmussen this book]). As is shown in Section 5.7, interference cancellation is closely related to efficient approximate algorithms for statistical inference in Bayesian networks.

5.3 MAIN RESULTS: SINGLE-USER CHARACTERIZATION

Before burdening the reader with statistical physics concepts and methodologies, we introduce the main results of this chapter and describe the breakthrough in understanding large multiuser systems made possible by the replica analysis.

A *large system* in this chapter refers to the limit that both the number of users and the spreading factor tend to infinity but with their ratio, known as the *system load*, converging to a positive number, i.e., $K/L \to \beta > 0$. The load β may or may not be smaller than 1. It is also assumed that the SNRs of all users, $\{\gamma_k\}_{k=1}^{K}$, are *i.i.d.* with distribution P_γ, hereafter referred to as the *SNR distribution*. All moments of the SNR distribution are assumed to be finite. Clearly, the empirical distributions of the SNRs converge to the same distribution P_γ as $K \to \infty$. Note that this SNR distribution captures the (flat) fading characteristics of the channel.

Throughout this chapter we consider detection in one symbol interval assuming that the channel state is known by the receiver.

5.3.1 Is the Decision Statistic Gaussian?

Linear multiuser detectors are easy to analyze because of the simple structure of their decision statistics. In general, the detection output is the sum of three independent components: the desired signal, the MAI and the Gaussian noise, i.e., the (normalized) decision statistic for user k is expressed as:

$$\langle X_k \rangle = X_k + \sum_{i \neq k} I_i + N_k \qquad (5.15)$$

The error performance is determined by the statistics of the MAI and the noise. For a sufficiently large system, it is common to assume that the MAI is Gaussian conditioned on the SNRs, so that the performance is quantified as identical to that of a single-user Gaussian channel with the same input but enhanced noise (or, equivalently, degraded SNR). This simple single-user characterization is justified because the MAI converges weakly to a Gaussian random variable, independent of the noise, as $K \to \infty$ [12].

However, the above analysis does not apply beyond linear detection schemes. The problem here is inherent to nonlinear processing, where the detection output cannot be decomposed as a sum of independent components associated with the desired signal and the unwanted interference respectively. Moreover, the detection output is in general asymptotically non-Gaussian conditioned on the input (consider, e.g., the discrete output of the maximum-likelihood detector in case of binary transmission).

The above difficulty is largely overcome by applying statistical physics methodologies, and in particular the replica method, to the treatment of generic multiuser detection in the large-system regime. Although the output decision statistic of a nonlinear detector cannot be decomposed as (5.15), it converges in the large-system limit to a simple monotone function of a "hidden" Gaussian random variable conditioned on the input X_k, i.e.:

$$\langle X_k \rangle \longrightarrow f(Z_k) \qquad (5.16)$$

where $Z_k = X_k + W_k$ and W_k is Gaussian and independent of X_k. One may contend that it is always possible to monotonically map a non-Gaussian random variable to a Gaussian one. What is useful (and surprising) here is that:

1) the mapping f depends on neither the instantaneous spreading sequences, nor the transmitted symbols which we wish to estimate in the first place; and
2) the statistic Z_k is equal to the desired signal plus an independent Gaussian noise.

By applying an inverse of the function f (which can be readily determined) to $\langle X_k \rangle$, the equivalent conditionally Gaussian statistic Z_k is recovered, so that we are back to the familiar ground where the output SINR (defined for the equivalent Gaussian statistic Z_k) completely characterizes the performance for an individual user. We can thus define the multiuser efficiency as the ratio of the output SINR and the input SNR, which is consistent with its original notion in [15].

Example 1 *Figure 5.4a plots the approximate probability density functions obtained from the histogram of the output of the soft individually optimal detector conditioned on +1 being transmitted. Note that negative decision values correspond to decision error; hence the dark area on the negative half plane gives the BER. Since the distribution shown in Figure 5.4a is far from Gaussian, the usual notion of output SINR fails to capture the system performance. In fact, much work in the literature is devoted to evaluating the error performance by Monte Carlo simulation. Figure 5.4b plots the density of the conditionally Gaussian statistic obtained by applying f^{-1} to the non-Gaussian detection output in Figure 5.4a. The theoretically predicted Gaussian density function (the smooth curve) is also shown for comparison. The "fit" is remarkable considering that a relatively small system of eight users with spreading factor 12 is considered. Note that when the multiuser detector is linear, the mapping f is also linear, and (5.16) reduces to (5.15).*

The above example demonstrates the decoupling principle. The asymptotic normality of the decision statistic or its function allows the performance of multiuser systems to be simply characterized by the effective SNR, or SINR, of the detection output. The main claims are formally stated in the following, first for optimal detection (Section 5.3.2) and then for generic multiuser detection (Section 5.3.3). The analysis and discussion of statistical physics techniques are relegated to Sections 5.4 and 5.6.

5.3.2 The Decoupling Principle: Individually Optimal Detection

In order to describe the decoupling result, we first introduce a scalar channel:

$$Z = \sqrt{\gamma} X + \frac{1}{\sqrt{\eta}} N \tag{5.17}$$

where $X \sim p_X$, γ is the channel gain, $N \sim \mathcal{N}(0, 1)$ the additive Gaussian noise independent of X, and $\eta > 0$ the *inverse noise variance*, which is also understood as the

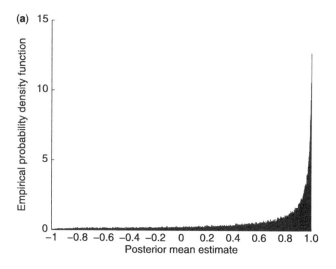

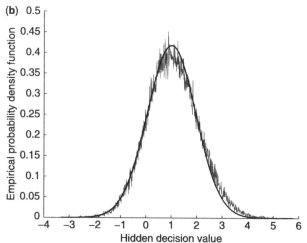

Figure 5.4. The empirical probability density functions of the decision statistics conditioned on +1 being transmitted. The system has eight users, the spreading factor is 12, and SNR = 2 dB. A total of 10,000 trials were recorded. (a) The soft individually optimal detection output. (b) The "hidden" equivalent Gaussian statistic. The asymptotic Gaussian distribution is also plotted for comparison.

degradation of the channel. The conditional distribution associated with the channel (5.17) is:

$$p_{Z|X,\gamma;\eta}(z \mid x, \ \gamma;\eta) = \sqrt{\frac{\eta}{2\pi}} \exp\left[-\frac{\eta}{2}\left(z - \sqrt{\gamma}x\right)^2\right] \tag{5.18}$$

where we generally treat γ as a random variable but η a deterministic parameter. Thus, (5.17) is a flat-fading channel. However, since we are interested in a single symbol

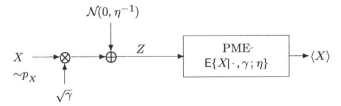

Figure 5.5. The single-user channel and PME.

interval with γ known to the receiver, it is more convenient to refer to (5.17) as a Gaussian channel (for given γ). With $X \sim p_X$ and given η and γ, the input–output mutual information of the channel (5.17) is denoted by $I(X; \sqrt{\eta\gamma}X + N)$. The posterior mean estimate of X given the output Z is:[6]

$$\langle X \rangle = \mathsf{E}\{X \,|\, Z, \gamma; \eta\} \tag{5.19}$$

which is an implicit function of Z. The Gaussian channel concatenated with the PME is depicted in Figure 5.5. Clearly, $\langle X \rangle$ is also the (nonlinear) MMSE estimate, since it achieves the minimum mean-square error:

$$\mathcal{E}_X(\eta\gamma) = \mathsf{E}\{(X - \langle X \rangle)^2 \,|\, \gamma; \eta\}. \tag{5.20}$$

Throughout this chapter, the (decreasing) function $\mathcal{E}_X(a)$ denotes the MMSE of estimating X in Gaussian noise with SNR equal to a.

Consider the individually optimal detection defined by (5.3) and also described in Section 5.2.3.2, where the detection output for user k is the posterior mean estimate $\mathsf{E}\{X_k \,|\, \boldsymbol{Y}, \boldsymbol{S}\}$. We claim[7] that, from a single user's perspective, the channel between the input and detection output is asymptotically equivalent to the scalar Gaussian channel (5.17) with an appropriate value of η that is interpreted as the multiuser efficiency.

CLAIM 1   *In the large-system limit, the distribution of the output $\langle X_k \rangle$ of the individually optimal detector for the multiuser channel (5.2) conditioned on $X_k = x$ being transmitted with SNR $\gamma_k = \gamma$ converges to the distribution of the posterior mean estimate $\langle X \rangle$ of the single-user Gaussian channel (5.17) conditioned on $X = x$ being transmitted, i.e., the posterior cumulative distribution function (CDF):*

$$P_{\langle X_k \rangle | X_k, \gamma_k}(\tilde{x} \,|\, x, \gamma) \;\longrightarrow\; P_{\langle X \rangle | X, \gamma}(\tilde{x} \,|\, x, \gamma) \tag{5.21}$$

---

[6]The posterior mean estimate is defined for both the single-user model [e.g., (5.17)] and the multiuser model [e.g., (5.2)] and denoted by the same notation $\langle \cdot \rangle$. The meaning of the notation should be clear from the context.

[7]Since as explained in Section 5.1, rigorous justification for some of the key statistical physics tools (essentially the replica method) is still pending, the key results in this chapter are referred to as claims. Proofs are provided in Section 5.4 based on several assumptions.

for all γ and all x, $\tilde{x}$ where the CDF P_X is continuous.[8] Here, the optimal multiuser efficiency η is determined from the following fixed-point equation:[9]

$$\eta^{-1} = 1 + \beta E\{\gamma \mathcal{E}_X(\eta \gamma)\} \tag{5.22}$$

where the expectation is taken over P_γ. In case (5.22) has more than one solution, η is chosen to minimize:[10]

$$C_{\text{joint}} = \beta I(X; \sqrt{\eta \gamma} X + N \mid \gamma) + \frac{1}{2}[(\eta - 1)\log e - \log \eta] \tag{5.23}$$

It is important to note that the efficiency η does not depend on any specific SNRs in the large system; rather, it depends only on the distributions P_γ and P_X. The conditional mutual information in (5.23) is obtained as an average over the SNR distribution:

$$I(X; \sqrt{\eta \gamma} X + N \mid \gamma) = \int_0^\infty I(X; \sqrt{\eta t} X + N) \, dP_\gamma(t) \tag{5.24}$$

The physical meaning of C_{joint} will be clear shortly. Note that the left-hand side (LHS) of (5.21) is a random CDF dependent on the matrix S. The convergence in (5.21) holds in probability.

The essence of Claim 1 is the following single-user characterization of multiuser systems: From an individual user's viewpoint, the input–output relationship of the multiuser channel and PME is increasingly similar to that under a simple single-user setting as the system becomes large. Indeed, given the (scalar) input and output statistics, it is impossible to distinguish whether the underlying system is in the (large) multiuser or the single-user setting. It is also interesting to note that the (asymptotically) equivalent single-user system takes an analogous structure as the multiuser one (compare Figures 5.2 and 5.5). Note that the conditionally Gaussian variable Z is not directly available in the multiuser system. Rather, one can process (Y, S) to obtain Z as a sufficient statistic for X (see e.g., [52]).

The single-user PME (5.19) is merely a decision function applied to the Gaussian channel output, which can be expressed explicitly as:

$$E\{X \mid Z = z, \gamma; \eta\} = \frac{p_1(z, \gamma, \eta)}{p_0(z, \gamma, \eta)} \tag{5.25}$$

---

[8]If X is a continuous random variable then the cdf is continuous on $(-\infty, \infty)$. If X is discrete, then the CDF is continuous at all but a finite or countable number of values.

[9]Because of the way the MMSE is defined, the fixed-point equation is true for arbitrary input distribution P_X, which need not have zero mean and unit variance.

[10]The base of logarithm is consistent with the unit of information measure throughout, unless stated otherwise.

where we define the following useful functions:

$$p_i(z, \gamma, \eta) = E\{X^i p_{Z|X,\gamma,\eta}(z \mid X, \gamma, \eta) \mid \gamma\} \qquad i = 0, 1, \dots \qquad (5.26)$$

where the expectation is taken over p_X. Note that $p_0(z, \gamma, \eta) = p_{Z|\gamma,\eta}(z \mid \gamma, \eta)$. The decision function (5.25) is in general nonlinear.

The MMSE can be computed as[11]

$$\mathcal{E}_X(\eta\gamma) = 1 - \int \frac{p_1^2(z, \gamma, \eta)}{p_0(z, \gamma, \eta)} \, dz \qquad (5.27)$$

Solutions to the fixed-point equation (5.22) can in general be found numerically. The conditional mutual information (over $\gamma \sim P_\gamma$) in (5.23) is also easy to compute.

Example 2   *Assume all users take binary antipodal input and the same SNR of 2 dB ($\gamma = 1.585$). Let $\beta = 2/3$. Solving the fixed-point equation (5.22) yields $\eta = 0.69$. Thus, from each user's point of view, if individually optimal detection is employed, the distribution of the decision statistic is identical to that of the posterior mean estimate of the input to a Gaussian channel with SNR equal to $\eta\gamma = 1.098$ (0.41 dB). The distribution of the detection output conditioned on $+1$ being transmitted is shown in Figure 5.4b, which is centered at 1 with a variance of $1/(\eta\gamma) = 0.911$.*

The fixed-point equations (5.22) may have multiple solutions. This is known as phase coexistence in statistical physics. Among those solutions, the (thermodynamically) dominant solution gives the smallest value of C_{joint}, which is in fact the optimal spectral efficiency as we shall discuss in Section 5.4.2.3. This is the solution that carries relevant operational meaning in the communication problem. In general, as the system parameters (such as the load) change, the dominant solution may switch from one of the coexisting solutions to another. This phenomenon is known as *phase transition*.

The decision function (5.25) is one-to-one because of the following, which is easily proved using the Cauchy-Schwartz inequality [27].

Proposition 1   *The decision function (5.25) is strictly monotone increasing in z for all γ, $\eta > 0$.*

This monotonicity result is not surprising because larger (smaller) values of channel output is likely to be caused by larger (smaller) values of the input.

In the large-system limit, given the detection output $\langle X_k \rangle$, one can apply the inverse of the decision function to recover an equivalent conditionally Gaussian statistic Z, which is centered at the actual input X_k scaled by $\sqrt{\gamma_k}$ with a variance of η^{-1}. Note that $\eta \in [0, 1]$ from (5.22). It is clear that the MAI is asymptotically equivalent to

---

[11]The integral with respect to z is from $-\infty$ to ∞. For notational simplicity, we omit integral limits in this chapter whenever they are clear from context.

an enhancement of the noise by η^{-1}, i.e., the effective SNR is reduced by a factor of η, hence the term *multiuser efficiency*. Indeed, in the large-system limit, the multiuser channel with the PME front end can be decoupled into a bank of independent single-user Gaussian channels with the same degradation in each user's SNR.

Corollary 1   *In the large-system limit, the mutual information between input symbol and the output of the individually optimal multiuser detector for each user is equal to the input–output mutual information of the equivalent single-user Gaussian channel with the same input and SNR degraded by η, which is the multiuser efficiency given by Claim 1. That is, conditioned on the input SNR being γ_k for user k:*

$$I(X_k; \langle X_k \rangle \mid S) \longrightarrow I(X; \sqrt{\eta\gamma_k}\, X + N) \tag{5.28}$$

where $X \sim p_X$ and $N \sim \mathcal{N}(0, 1)$ are independent.

The overall spectral efficiency under separate decoding is the average of all users' mutual information multiplied by the load:

$$C_{\text{sep}}(\beta) = \beta\, I(X; \sqrt{\eta\gamma}X + N \mid \gamma) \tag{5.29}$$

The optimal spectral efficiency under joint decoding is greater than that under separate decoding (5.29), where the increase is given by the following:

CLAIM 2   *The spectral efficiency gain of optimal joint decoding over individually optimal detection followed by separate decoding of the multiuser channel (5.2) is determined, in the large-system limit, by the optimal multiuser efficiency as:*

$$C_{\text{joint}}(\beta) - C_{\text{sep}}(\beta) = \frac{1}{2}[(\eta - 1)\log e - \log \eta] \tag{5.30}$$

$$= D(\mathcal{N}(0, \eta)\|\mathcal{N}(0, 1)) \tag{5.31}$$

Indeed, the spectral efficiency under joint decoding is given by (5.23).

As a by-product, Müller's conjecture on the mutual information loss [28,29] is true for arbitrary inputs and SNRs. Incidentally, the loss is identified as a Kullback-Leibler divergence [63] between two Gaussian distributions in (5.31) [27].

Interestingly, the spectral efficiencies under joint and separate decoding are also related by an integral equation, which was originally given in [19, (160)] for the special case of Gaussian inputs.

Theorem 1   *Regardless of the input and SNR distributions:*

$$C_{\text{joint}}(\beta) = \int_0^\beta \frac{1}{\beta'}\, C_{\text{sep}}(\beta')\, d\beta' \tag{5.32}$$

***Proof*:** Since $C_{\text{joint}}(0) = 0$ trivially, it suffices to show:

$$\beta \frac{d}{d\beta} C_{\text{joint}}(\beta) = C_{\text{sep}}(\beta) \tag{5.33}$$

By (5.31) and (5.23), it is enough to show:

$$\beta \frac{d}{d\beta} I(X; \sqrt{\eta\gamma} X + N \mid \gamma) + \frac{1}{2} \frac{d}{d\beta} [(\eta - 1) \log e - \log \eta] = 0 \tag{5.34}$$

As the efficiency η is a function of the system load β, (5.34) is equivalent to:

$$\frac{d}{d\eta} I(X; \sqrt{\eta\gamma} X + N \mid \gamma) + \frac{1}{2\beta} (1 - \eta^{-1}) \log e = 0 \tag{5.35}$$

The mutual information and the MMSE in Gaussian channels are related by the following formula [64, Theorem 1]:

$$\frac{1}{\log e} \frac{d}{dg} I(X; \sqrt{g} X + N) = \frac{1}{2} \mathcal{E}_X(g), \quad \forall_g. \tag{5.36}$$

Thus (5.35) holds as η satisfies the fixed-point equation (5.22).

Theorem 1 is an outcome of the chain rule of mutual information:

$$I(X; Y \mid S) = \sum_{k=1}^{K} I(X_k; Y \mid S, X_{k+1}, \ldots, X_K) \tag{5.37}$$

The LHS of (5.37) is the total mutual information of the multiuser channel. Each mutual information in the right-hand side (RHS) is a single-user mutual information over the multiuser channel conditioned on the symbols of previously decoded users. As argued below, the limit of (5.37) as $K \to \infty$ becomes the integral equation (5.32).

Consider a successive interference canceler with PME front ends against yet undecoded users in which reliably decoded symbols are used to reconstruct the interference for cancellation. Since the error probability of decoded symbols vanishes with code block-length, the interference from decoded users are asymptotically completely removed. Assume without loss of generality that the users are decoded in reverse order, then the PME for user k sees only $k-1$ interfering users. Hence the performance for user k under such successive decoding is identical to that under multiuser detection with separate decoding in a system with k instead of K users. Nonetheless, the equivalent single-user channel for each user is Gaussian by Claim 1. The multiuser efficiency experienced by user k, $\eta(k/L)$, is a function of the load k/L seen by the PME for user k. By Corollary 1, the single-user mutual information for user k is therefore:

$$I\left(X; \sqrt{\eta(k/L)\gamma_k} X + N\right) \tag{5.38}$$

The overall spectral efficiency under successive decoding converges almost surely by the law of large numbers:

$$\frac{1}{L}\sum_{k=1}^{K} I\left(X; \sqrt{\eta(k/L)\gamma_k}\,X + N\right) \longrightarrow \int_0^\beta I\left(X; \sqrt{\eta(\beta')\,\gamma}\,X + N \mid \gamma\right) d\beta' \quad (5.39)$$

which is the RHS of (5.32). This suggests that decoding and stripping users one-by-one in a large system is tantamount to increasing the SNR little-by-little in some intricate way.

Together with Theorem 1, the convergence in (5.39) implies the following:

Corollary 2 *In the large-system limit, successive decoding with an individually optimal detection front end against yet undecoded users achieves the optimal multiuser channel capacity under any constraint on the input.*

Corollary 2 is a generalization of the result that a successive canceler with a linear MMSE front end against undecoded users achieves the capacity of the CDMA channel under Gaussian inputs.[12]

5.3.3 Decoupling Principle: Generic Multiuser Detection

5.3.3.1 A Companion Channel
Consider a random transformation $p_{Y|X}$ that characterizes a memoryless channel $X \to Y$. The problem of Bayesian inference is in general to infer about X given Y based on the posterior probability law $p_{X|Y}$. Under many circumstances, e.g., when $p_{X|Y}$ is not exactly known, inference may be carried out using an alternative law $q_{X|Y}$. For all estimation purposes, it suffices to know the joint probability distribution of (X, Y, X') where $Y \to X'$ is characterized by $q_{X|Y}$ and X' is independent of X conditioned on Y. Precisely, $p_{XX'|Y}(x, x'|y) = p_{X|Y}(x|y)q_{X|Y}(x'|y)$ for all (x, x', y). We call the random transformation $q_{X|Y}$ a *companion channel* of the channel $p_{Y|X}$.

The above can be specialized to the current problem. Let $q_{Z|X,\gamma;\xi}$ represent the input–output relationship of a Gaussian channel akin to (5.17), the only difference being that the inverse noise variance is ξ instead of η:

$$q_{Z|X,\gamma;\xi}(z \mid x, \gamma; \xi) = \sqrt{\frac{\xi}{2\pi}}\exp\left[-\frac{\xi}{2}\left(z - \sqrt{\gamma}x\right)^2\right] \quad (5.40)$$

Throughout, we choose to explicitly associate η with distribution p and ξ with distribution q for clarity. Similar to that in the multiuser setting, by postulating the input distribution to be q_X, a posterior probability distribution $q_{X|Z,\gamma;\xi}$ is induced by q_X and $q_{Z|X,\gamma;\xi}$ using Bayes' formula [cf. (5.6)]. Thus we have a single-user companion

---

[12]This principle, originally discovered in [65], has been shown with other proofs and in other settings [18,66–70].

channel defined by $q_{X|Z,\gamma,\xi}$, which outputs a random variable X' given the channel output Z (Figure 5.6b). A (generalized) single-user PME is defined naturally as:

$$\langle X \rangle_q = \mathsf{E}_q\{X \,|\, Z, \gamma, \xi\} = \frac{q_1(Z, \gamma, \xi)}{q_0(Z, \gamma, \xi)} \tag{5.41}$$

where the following functions are defined akin to (5.26):

$$q_i(z, \gamma, \xi) = \mathsf{E}_q\{X^i q_{Z|X,\gamma,\xi}(z \,|\, X, \gamma, \xi) \,|\, \gamma\}, \quad i = 0, 1, \ldots \tag{5.42}$$

where the expectation is taken over q_X. The probability law of the composite system depicted by Figure 5.6b is determined by γ and two parameters η and ξ.

Let us define the mean squared error of the PME as

$$\mathcal{E}(\gamma, \eta, \xi) = \mathsf{E}\left\{ \left(X - \langle X \rangle_q \right)^2 \,|\, \gamma, \eta, \xi \right\}, \tag{5.43}$$

and also define the variance of the companion channel as

$$\mathcal{V}(\gamma, \eta, \xi) = \mathsf{E}\left\{ \left(X' - \langle X \rangle_q \right)^2 \,\middle|\, \gamma, \eta, \xi \right\}. \tag{5.44}$$

Note that $\xi = \eta$ if X and X' are *i.i.d.* given Z.

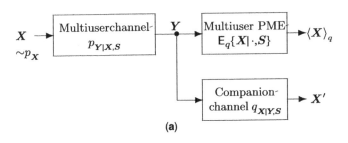

(a)

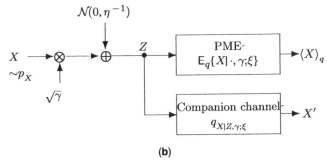

(b)

Figure 5.6. (a) The multiuser channel, the (multiuser) PME, and the (multiuser) companion channel. (b) The equivalent single-user Gaussian channel, PME and companion channel.

5.3.3.2 Main Results Consider the multiuser channel (5.2) with input distribution p_X and SNR distribution P_γ. Let its output be fed into the posterior mean estimator (5.5) and a companion channel $q_{X|Y,S}$, both parameterized by the postulated input q_X and noise level σ (refer to Figure 5.6a). Let X_k, X'_k, and $\langle X_k \rangle_q$ be the input, the companion channel output and the posterior mean estimate for user k with input SNR γ_k.

Fix $(\beta, P_\gamma, p_X, q_X, \sigma)$. Consider also the single-user Gaussian channel (5.18) with inverse noise variance η and its companion channel depicted in Figure 5.6b. Let $X \sim p_X$ be the input to the single-user Gaussian channel, X' be the output of the single-user companion channel parameterized by (q_X, ξ), and $\langle X \rangle_q$ is the corresponding posterior mean estimate (5.41), with $\gamma = \gamma_k$.

CLAIM 3   *Consider the multiuser and single-user systems described above (also Figure 5.6).*

(a) The joint distribution of $(X_k, X'_k, \langle X_k \rangle_q)$ conditioned on the channel state S converges in probability as $K \to \infty$ and $K/L \to \beta$ to the joint distribution of $(X, X', \langle X \rangle_q)$ with $\gamma = \gamma_k$, i.e., the posterior CDF:

$$P_{X_k, X'_k, \langle X'_k \rangle_q | \gamma_k}(x, x', \tilde{x} | \gamma) \longrightarrow P_{X, X', \langle X \rangle | \gamma}(x, x', \tilde{x} | \gamma) \qquad (5.45)$$

in probability for every $x, x', \tilde{x}$ where the CDF P_X is continuous.

(b) The parameter η, known as the multiuser efficiency, satisfies together with ξ the coupled equations:

$$\eta^{-1} = 1 + \beta E\{\gamma \cdot \mathcal{E}(\gamma; \eta, \xi)\} \qquad (5.46a)$$

$$\xi^{-1} = \sigma^2 + \beta E\{\gamma \cdot \mathcal{V}(\gamma; \eta, \xi)\} \qquad (5.46b)$$

where the expectations are taken over P_γ. In case of multiple solutions to (5.46), (η, ξ) is chosen to minimize the free energy expressed as:

$$\mathcal{F} = -E\left\{ \int p_{Z|\gamma,\eta}(z|\gamma;\eta) \log q_{Z|\gamma,\xi}(z|\gamma;\xi) dz \right\} + \frac{1}{2\beta}[(\xi - 1)\log e - \log \xi]$$

$$- \frac{1}{2}\log\frac{2\pi}{\xi} - \frac{\xi}{2\eta}\log e + \frac{\sigma^2\xi(\eta - \xi)}{2\beta\eta}\log e + \frac{1}{2\beta}\log(2\pi) + \frac{\xi}{2\beta\eta}\log e \qquad (5.47)$$

Claim 3 reveals that, from an individual user's viewpoint, the input–output relationship of the multiuser channel, PME and companion channel is increasingly similar to that under a simple single-user setting as the system becomes large.

Finally, it is straightforward to verify that the decoupling result for individually optimal detection (Claim 1) is a special case of the results for generic detection (Claim 3) with the postulated distribution q identical to the actual distribution p as well as symmetry assumption $\xi = \eta$.

5.3.4 Justification of Results: Sparse Spreading

Claims 1–3 have not been rigorously proved because the underlying replica method is until now an unjustifiable technique. In the following, we provide an interpretation of the central fixed-point equation which was first discussed in [31]. In particular, we derive (5.22) under the assumption that interference cancellation based on posterior mean estimates of interfering users is optimal, along with some additional independence assumptions.

Suppose we construct without loss of generality an estimator for User 1 using interference cancellation as depicted in Figure 5.7 (see also a discussion of interference cancellation in Section 5.7). Let $\langle X_2 \rangle, \ldots, \langle X_K \rangle$ be the generalized PME estimates for User 2 through User K. A decision statistic for User 1 can be generated by first subtracting the reconstructed interferences using those estimates and then matched filtering with respect to user 1's spreading sequence:

$$Z_1 = \sqrt{\gamma_1} X_1 + \sum_{k=2}^{K} \mathbf{S}_1^{\mathsf{T}} \mathbf{S}_k \sqrt{\gamma_k} \left(X_k - \langle X_k \rangle \right) + N_1 \tag{5.48}$$

where N_1 is standard Gaussian. We make two specious assumptions:

1) The desired symbol X_1, the Gaussian noise N_1, and the residual errors $(X_k - \langle X_k \rangle)$ are independent;
2) The statistic Z_1 is sufficient for achieving the MMSE for X_1.

By the first assumption, the sum of the residual MAI and Gaussian noise converges to a Gaussian random variable as $K \to \infty$ by virtue of the central limit theorem. Let the variance of $X_k - \langle X_k \rangle$ be denoted by $V(\gamma_k)$, which depends on γ_k. Then variance of the total interference in (5.48) is:

$$1 + \beta \, \mathsf{E}\{\gamma \, V(\gamma)\} \tag{5.49}$$

which implies that the efficiency for User 1 is:

$$\eta_1 = \frac{1}{1 + \beta \, \mathsf{E}\{\gamma \, V(\gamma)\}} . \tag{5.50}$$

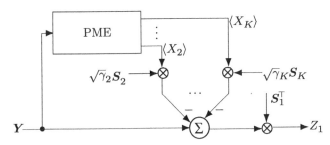

Figure 5.7. A canonical interference canceler equivalent to the single-user channel.

Evidently, the efficiency is not dependent on the user number and hence identical for all users; the subscript of the efficiency can be dropped. By the second assumption, the mean squared error based on the statistic Z_1 should be equal to the MMSE, which is $\mathcal{E}_X(\eta\gamma_1)$. Note that the same applies to all users, hence $V(\gamma_k) = \mathcal{E}_X(\eta\gamma_k)$. Therefore, formula (5.50) becomes exactly the fixed-point equation (5.22).

Therfore, if the above two assumptions were valid, we would have recovered the fixed-point equation (5.22) in Claim 1. Moreover, we would have constructed a degraded Gaussian channel for User 1 equivalent to the single-user channel as shown in Figure 5.6b. We can also argue that every user enjoys the same efficiency since otherwise users with worse efficiency may benefit from users with better efficiency until an equilibrium is reached. Roughly speaking, the PME output is a "fixed-point" of a parallel interference canceler. The multiuser efficiency, in a sense, is the outcome of such an equilibrium.

The above interpretation does not hold in general due to the unjustifiable independence assumption.[13] In particular, $S_1^{\mathsf{T}}S_k$ are not independent, albeit uncorrelated, for all k. Also, $\langle X_k \rangle$ are dependent on the desired signal X_1 and the noise N_1, which is evident in the special case of linear MMSE detection.

Interestingly, the above argument can be made rigorous in the special case where the spreading matrix S is *sparse* (or extremely diluted) in some sense (see also [72,73]). In [50–52], the general formula (5.22) has been justified for binary inputs and arbitrary inputs and SNR respectively with sparse spreading and a relatively small load, which is the first partial proof of (5.22) without resorting to the replica method. The key observation is that, the residual errors are independent over almost all choices of the spreading matrix if the posterior mean estimates are replaced by the (asymptotically equivalent) estimates supplied by parallel interference cancellation, or belief propagation. Interference cancellation and belief propagation are the subject of Section 5.7 where some practical multiuser detection schemes are discussed.

5.3.5 Well-Known Detectors as Special Cases

As shown in Section 5.2.3, several well-known multiuser detectors can be regarded as appropriately parameterized PMEs. Thus many previously known results can be recovered as special cases of the findings in Sections 5.3.2 and 5.3.3.

5.3.5.1 *Linear Detectors*   Let the postulated prior q_X be standard Gaussian so that the multiuser PME represents a linear detector. Since the input Z and output X of the companion channel are jointly Gaussian (refer to Figure 5.6b), the single-user PME is simply a linear attenuator:

$$\langle X \rangle_q = \frac{\xi\sqrt{\gamma}}{1 + \xi\gamma} Z. \tag{5.51}$$

---

[13]It should be noted, however, that a similar independence argument can also be found in statistical physics literature (see, e.g., [36]). Such independence property is called the "cluster property" in statistical physics [71].

From (5.43), the mean squared error is:

$$\mathcal{E}(\gamma, \eta, \xi) = \mathsf{E}\left\{\left[X_0 - \frac{\xi\sqrt{\gamma}}{1 + \xi\gamma}\left(\sqrt{\gamma}X_0 + \frac{N}{\sqrt{\eta}}\right)\right]^2\right\} \tag{5.52}$$

$$= \frac{\eta + \xi^2\gamma}{\eta(1 + \xi\gamma)^2} \tag{5.53}$$

Meanwhile, the variance of X conditioned on Z is independent of Z. Hence the variance (5.44) of the companion channel output is independent of η:

$$\mathcal{V}(\gamma, \eta, \xi) = \frac{1}{1 + \xi\gamma} \tag{5.54}$$

From Claim 3, one finds that ξ is the solution to:

$$\xi^{-1} = \sigma^2 + \beta\mathsf{E}\left\{\frac{\gamma}{1 + \xi\gamma}\right\} \tag{5.55}$$

and the multiuser efficiency is determined as:

$$\eta = \xi + \xi(\sigma^2 - 1)\left[1 + \beta\mathsf{E}\left\{\frac{\gamma}{(1 + \xi\gamma)^2}\right\}\right]^{-1} \tag{5.56}$$

which is independent of the input distribution p_X.

Let $\sigma \to \infty$ so that the PME becomes the matched filter. One finds $\xi\sigma^2 \to 1$ by (5.55) and consequently, the multiuser efficiency of the matched filter is [15]:

$$\eta^{(\mathrm{mf})} = \frac{1}{1 + \beta\mathsf{E}\{\gamma\}} \tag{5.57}$$

In case $\sigma = 1$, one has the linear MMSE detector. By (5.56), $\eta = \xi$ and by (5.55), the efficiency $\eta^{(\mathrm{lm})}$ is the unique solution to the Tse-Hanly equation [11,18]:

$$\eta^{-1} = 1 + \beta\mathsf{E}\left\{\frac{\gamma}{1 + \eta\gamma}\right\} \tag{5.58}$$

By letting $\sigma \to 0$ one obtains the decorrelator. If $\beta < 1$, then (5.55) gives $\xi \to \infty$ and $\xi\sigma^2 \to 1 - \beta$, and the multiuser efficiency is found as $\eta = 1 - \beta$ by (5.56) regardless of the SNR distribution (as shown in [15]). If $\beta > 1$, and assuming the generalized form of the decorrelator as the Moore-Penrose inverse of the correlation matrix [15], then ξ is the unique solution to:

$$\xi^{-1} = \beta\mathsf{E}\left\{\frac{\gamma}{1 + \xi\gamma}\right\} \tag{5.59}$$

and the multiuser efficiency is found by (5.56) with $\sigma = 0$. In the special case of identical SNRs, an explicit expression is found [16,17]:

$$\eta^{(\text{dec})} = \frac{\beta - 1}{\beta + \gamma(\beta - 1)^2}, \quad \beta > 1 \tag{5.60}$$

By Claim 3, the mutual information with input distribution p_X for a user with SNR given as γ under linear multiuser detection is $I(X; \langle X \rangle_q) = I(X; \sqrt{\eta\gamma}X + N)$ where $N \sim \mathcal{N}(0, 1)$ and η depends on which type of linear detector is in use. By Claim 2, the total spectral efficiency, which is achieved by Gaussian inputs, is expressed in terms of the LMMSE efficiency [19]:

$$C_{\text{joint}}^{(\text{G})} = \frac{\beta}{2} \, \mathsf{E}\big\{\log\big(1 + \eta^{(\text{lm})}\gamma\big)\big\} + \frac{1}{2}\big[\big(\eta^{(\text{lm})} - 1\big)\log e - \log \eta^{(\text{lm})}\big] \tag{5.61}$$

5.3.5.2 *Optimal Detectors*

Using the actual input distribution p_X as the postulated prior of the PME results in optimum multiuser detectors. As discussed in Section 5.2.3.2, in case of the jointly optimal detector, the postulated noise level σ is 0, and (5.46) becomes:

$$\eta^{-1} = 1 + \beta\mathsf{E}\{\gamma \cdot \mathcal{E}(\gamma; \eta, \xi)\} \tag{5.62a}$$

$$\xi^{-1} = \beta\mathsf{E}\{\gamma \cdot \mathcal{V}(\gamma; \eta, \xi)\} \tag{5.62b}$$

The parameters can then be solved numerically.

In case of the individually optimal detector, $\sigma = 1$ and $q = p$. The optimal efficiency η is the solution to the fixed-point equation (5.22) given in Claim 1.

It is of practical interest to find the spectral efficiency under the constraint that the input symbols are antipodally modulated as in the popular BPSK. In this case, equally likely prior maximizes the mutual information. The MMSE is:

$$\mathcal{E}^{(b)}(\gamma) = 1 - \int \frac{e^{-\frac{z^2}{2}}}{\sqrt{2\pi}} \tanh\big(\gamma - z\sqrt{\gamma}\big)\mathrm{d}z \tag{5.63}$$

where the superscript (b) stands for binary inputs. By Claim 1, the multiuser efficiency $\eta^{(b)}$ is a solution to the fixed-point equation [17]:

$$\frac{1}{\eta} = 1 + \beta\mathsf{E}\left\{\gamma\left[1 - \int \frac{e^{-\frac{z^2}{2}}}{\sqrt{2\pi}} \tanh\big(\eta\gamma - z\sqrt{\eta\gamma}\big)\mathrm{d}z\right]\right\} \tag{5.64}$$

The channel capacity for a user with binary input, SNR equal to γ and separate decoding is given by [28]:

$$C^{(b)}(\gamma) = -\int \frac{e^{-\frac{z^2}{2}}}{\sqrt{2\pi}} \log \cosh\big(\eta^{(b)}\gamma - z\sqrt{\eta^{(b)}\gamma}\big)\mathrm{d}z + \eta^{(b)}\gamma \log e \tag{5.65}$$

The joint-decoding spectral efficiency with binary inputs is thus:

$$C_{joint}^{(b)} = \beta E\{C^{(b)}(\gamma)\} + \frac{1}{2}\left[(\eta^{(b)} - 1)\log e - \log \eta^{(b)}\right] \qquad (5.66)$$

which is also a generalization of an implicit result in [26].

5.4 THE REPLICA ANALYSIS OF GENERIC MULTIUSER DETECTION

This section introduces the replica method and presents the replica analysis of generic multiuser detection which leads to Claims 1–3. We first describe the procedure of the replica method and demonstrate its use with a simple example. We then apply the method to the analysis of the multiuser system and present the calculation of the mutual information in some detail.

5.4.1 The Replica Method

Before describing the replica method, we first revisit the key measures used to characterize the multiuser system, including in particular the input–output mutual information. For convenience, natural logarithms are assumed from this point on.

5.4.1.1 Spectral Efficiency and Detection Performance
Consider the multiuser channel, the PME and the companion channel as depicted in Figure 5.6a. Fix the input distribution p_X. The key quantity is the spectral efficiency:

$$C = \frac{1}{L}I(X; Y|S) \qquad (5.67)$$

which we wish to evaluate. In some cases, one may want to evaluate the mutual information $I(X; Y | S = s)$, which is a function of the realization s of S. In such cases one assumes the self-averaging property, in which the random quantity $(1/L)I(X; Y | S = s)$ is assumed to converge to C as $L \to \infty$ for almost all realizations of S. This property has been justified in the special case where q_X is Gaussian [11,18] as well as in the case of Gaussian spreading sequence and binary input [74].

The spectral efficiency is expressed as:

$$C = \frac{1}{L}E\left\{\log \frac{p_{Y|X,S}(Y|X, S)}{p_{Y|S}(Y|S)}\right\} \qquad (5.68)$$

$$= -\beta E\left\{\frac{1}{K}\log p_{Y|S}(Y|S)\right\} - \frac{1}{2}\log(2\pi e) \qquad (5.69)$$

where the simplification to (5.69) is because $p_{Y|X,S}$ given by (5.8) is an L-dimensional Gaussian density. In Section 5.4.1.2 we show that the replica method can be used to calculate the normalized conditional differential entropy in (5.69):

$$E\left\{\frac{1}{K}\log p_{Y|S}(Y | S)\right\} \qquad (5.70)$$

which is also referred to as the *free energy* using the physics terminology.

In case of a multiuser detector front end, one is interested in the quality of the detection output for each user, which is completely described by the distribution of the detection output conditioned on the input. Let us focus on an arbitrary user k, and let X_k, $\langle X_k \rangle_q$ and X'_k be the input, the PME output, and the companion channel output, respectively (cf. Figure 5.6a). Instead of the conditional distribution $P_{\langle X_k \rangle_q | X_k}$, we solve a somewhat more ambitious problem: the joint distribution of $(X_k, \langle X_k \rangle_q, X'_k)$ conditioned on the channel state S in the large-system limit. The replica approach calculates the joint moments:

$$\mathsf{E}\left\{ X_k^i (X'_k)^j \langle X_k \rangle_q^l \right\}, \quad i, j, l = 0, 1, \ldots \tag{5.71}$$

by studying a free-energy-like quantity, as will be discussed in Section 5.4.3. The joint distribution becomes clear once all the moments (5.71) are determined, so does the relationship between the detection output $\langle X_k \rangle_q$ and the input X_k. It turns out that, as stated in Claim 3, the large-system joint distribution of $(X_k, \langle X_k \rangle_q, X'_k)$ is identical to that of the input, PME output and companion channel output associated with a single-user Gaussian channel with the same input distribution but with a degradation in the SNR.

We have distilled the problems under both joint and separate decoding to finding some ensemble averages, namely, the free energy (5.70) and the joint moments (5.71). In order to calculate these quantities, we resort to a powerful technique, the heart of which is sketched in the following.

5.4.1.2 The Replica Method

Direct calculation of the differential entropy (free energy) (5.70) is hard. The replica method can be described as the following procedure to that effect:

1. Reformulate the free energy (5.70) as:

$$\mathcal{F} = -\lim_{K \to \infty} \frac{1}{K} \lim_{u \to 0} \frac{\partial}{\partial u} \log \mathsf{E}\left\{ p_{Y|S}^u(Y|S) \right\} \tag{5.72}$$

The equivalence of (5.70) and (5.72) can be verified by noticing that for all positive random variable Θ:

$$\lim_{u \to 0} \frac{\partial}{\partial u} \log \mathsf{E}\{\Theta^u\} = \lim_{u \to 0} \frac{\mathsf{E}\{\Theta^u \log \Theta\}}{\mathsf{E}\{\Theta^u\}} = \mathsf{E}\{\log \Theta\} \tag{5.73}$$

2. For an arbitrary positive integer u, calculate:

$$-\lim_{K \to \infty} \frac{1}{K} \log \mathsf{E}\left\{ p_{Y|S}^u(Y|S) \right\} \tag{5.74}$$

by introducing u replicas of the system (hence the name "replica" method).

3. Assuming the resulting expression from Step 2 to be valid for all real-valued u in the vicinity of $u = 0$, take its derivative at $u = 0$ to obtain the free energy (5.72). It is also assumed that the limits in (5.72) can be interchanged.

We note that when analyzing general suboptimal estimators, the free energy is defined as (5.72) with $p_{Y|S}(Y|S)$ replaced by some alternative distribution $q_{Y|S}$ $(Y|S)$ while the expectation remains over the joint probability measure $p_{Y,S}$.

The rigorous mathematical minds will immediately question the validity of taking Step 3. In particular, the expression obtained for integer values may not be valid for real values in general [75]. In fact, the continuation of the expression to real values is not unique, e.g., $f(u) + \sin(\pi u)$ and $f(u)$ coincide at all integer u for every function f. Nevertheless, as we shall see, the the replica method simply takes the same expression derived for integer values of u, which is natural and straightforward in the problem at hand. The rigorous justification for Step 3 is still an open problem. Surprisingly, this continuation assumption, along with other assumptions—sometimes very intricate—on symmetries of solutions, if necessary (see Section 5.5.1), leads to correct results in all non-trivial cases where the results are known through other rigorous methods. In other cases, the replica method produces results that match well with numerical studies.

5.4.1.3 A Simple Example

Before applying the replica method to the much more involved multiuser detection problem, we give a simple example of its application to the analysis of a single-user system. Let:

$$Y = \sqrt{\frac{\gamma}{L}}SX + N \qquad (5.75)$$

where $X \sim p_X$, $N \sim \mathcal{N}(0, I)$, and $S = [S_1, \ldots, S_L]^\mathsf{T}$ is a column vector with *i.i.d.* entries of mean 0 and unit variance. It is easy to see that the channel is equivalent (via matched filtering) to a single-user Gaussian channel with the same SNR. In the following, we obtain the mutual information for $L \to \infty$ using the replica method as a warm-up exercise of the technique.

Similar to (5.69), conditioned on the channel state matrix, the input–output mutual information of (5.75) is:

$$I(X; Y|S) = -\mathsf{E}\{\log p_{Y|S}(Y|S)\} - \frac{L}{2}\log(2\pi e) \qquad (5.76)$$

where $p_{Y|S}(y|S) = \mathsf{E}\{p_{Y|S,X}(y|S, X)\}$. In the following, we evaluate (5.76) using the replica method. The calculation is rather lengthy, while the outcome is quite simple.

The differential entropy can be obtained from:

$$\mathsf{E}\{\log p_{Y|S}(Y|S)\} = \lim_{u \to 0} \frac{\partial}{\partial u} \log \mathsf{E}\left\{ p_{Y|S}^u(Y|S)\right\} \qquad (5.77)$$

For any positive integer u, one can introduce u replicas of the original system, and evaluate the moment as follows:

$$\mathsf{E}\left\{ p_{Y|S}^u(Y|S)\right\} = \mathsf{E}\left\{ \int p_{Y|S}^{u+1}(y\,|\,S)\mathrm{d}y\right\} \qquad (5.78)$$

$$= \mathsf{E}\left\{ \int \prod_{a=0}^u p_{Y|S,X}(y\,|\,S, X_a)\mathrm{d}y\right\} \qquad (5.79)$$

where the integral is over all entries of the vector $\mathbf{y}$ from $-\infty$ to ∞. Plugging in the Gaussian densities $p_{Y|S,X}$, the RHS of (5.79) becomes $(2\pi)^{-(u+1)L/2}$ times:

$$
E\left\{ \int \prod_{a=0}^{u} \exp\left[-\frac{1}{2} \sum_{l=1}^{L} \left(y_l - \sqrt{\frac{\gamma}{L}} S_l X_a \right)^2 \right] d\mathbf{y} \right\}
$$

$$
= E\left\{ E\left\{ \int \prod_{l=1}^{L} \exp\left[-\frac{1}{2} \sum_{a=0}^{u} \left(y_l - \sqrt{\frac{\gamma}{L}} S_l X_a \right)^2 \right] d\mathbf{y} \middle| X \right\} \right\} \tag{5.80}
$$

$$
= E\left\{ \left[E\left\{ \int \exp\left[-\frac{1}{2} \sum_{a=0}^{u} \left(y - \sqrt{\frac{\gamma}{L}} S X_a \right)^2 \right] dy \middle| X \right\} \right]^L \right\} \tag{5.81}
$$

where in (5.80) and (5.81) the inner expectations is with respect to the spreading chip(s) conditional on the symbols X, and (5.81) is due to symmetry and independence of the L chips. The integral in (5.81) is simply over a Gaussian density, which can be evaluated as:

$$
\int \exp\left[-\frac{1}{2} \sum_{a=0}^{u} \left(y - \sqrt{\frac{\gamma}{L}} S X_a \right)^2 \right] dy
$$

$$
= \sqrt{\frac{2\pi}{u+1}} \exp\left[\frac{\gamma}{2(u+1)L} \left(S \sum_{a=0}^{u} X_a \right)^2 - \frac{\gamma}{2L} \sum_{a=0}^{u} (S X_a)^2 \right] \tag{5.82}
$$

By (5.81) and (5.82), the RHS of (5.79) becomes:

$$
\frac{(2\pi)^{-uL/2}}{(u+1)^{L/2}} E\left\{ \left(E\left\{ \exp\left[\frac{\gamma S^2 \left(\sum\limits_{a=0}^{u} X_a \right)^2}{2(u+1)L} - \frac{\gamma S^2}{2L} \sum_{a=0}^{u} X_a^2 \right] \middle| X \right\} \right)^L \right\} \tag{5.83}
$$

Note that the exponent in (5.83) vanishes as $L \to \infty$. Using $E\{S\}^2 = 1$, we have

$$
[(2\pi)^u (u+1)]^{\frac{L}{2}} E\left\{ p_{Y|S}^u (Y|S) \right\} \longrightarrow E\left\{ \exp\left[\frac{\gamma \left(\sum\limits_{a=0}^{u} X_a \right)^2}{2(u+1)} - \frac{\gamma}{2} \sum_{a=0}^{u} X_a^2 \right] \right\} \tag{5.84}
$$

as $L \to \infty$. The RHS of (5.84) can be rearranged using the unit area property of Gaussian density:[14]

$$e^{x^2} = \sqrt{\frac{\eta}{2\pi}} \int \exp\left[-\frac{\eta}{2}z^2 + \sqrt{2\eta}\,xz\right] dz, \quad \forall x, \eta \tag{5.85}$$

with $\eta = u+1$ and $x = \sqrt{\frac{\gamma}{(2u+2)}} \sum_{a=0}^{u} X_a$. The RHS of (5.84) becomes:

$$\sqrt{\frac{u+1}{2\pi}} \mathsf{E}\left\{ \int \exp\left[-\frac{1}{2}(u+1)z^2 + \sqrt{\gamma}z \sum_{a=0}^{u} X_a - \frac{\gamma}{2}\sum_{a=0}^{u} X_a^2\right] dz \right\}$$

$$= \sqrt{\frac{u+1}{2\pi}} \int \left[\mathsf{E}\left\{\exp\left[-\frac{1}{2}(z - \sqrt{\gamma}X)^2\right]\right\}\right]^{u+1} dz \tag{5.86}$$

It is convenient to define a random variable $Z = \sqrt{\gamma}X + N$ where N is standard Gaussian. Let us define:

$$p_{Z|X}(z \mid x) = \frac{1}{\sqrt{2\pi}} \exp\left[-\frac{1}{2}(z - \sqrt{\gamma}x)^2\right] \tag{5.87}$$

and:

$$p_Z(z) = \frac{1}{\sqrt{2\pi}} \mathsf{E}\left\{\exp\left[-\frac{1}{2}(z - \sqrt{\gamma}X)^2\right]\right\} \tag{5.88}$$

From (5.84) and (5.86):

$$[(2\pi)^u (u+1)]^{\frac{L-1}{2}} \mathsf{E}\left\{p_{Y|S}^u(Y|S)\right\} \to \int p_Z^{u+1}(z)\,dz = \mathsf{E}\left\{p_Z^u(Z)\right\} \tag{5.89}$$

Therefore, from (5.76) and (5.89):

$$I(X; Y|S) = -\lim_{u \to 0} \frac{\partial}{\partial u} \log \mathsf{E}\left\{p_{Y|S}^u(Y|S)\right\} - \frac{L}{2}\log(2\pi e) \tag{5.90}$$

$$\to -\int p_Z(z) \log p_Z(z)\,dz - \frac{1}{2}\log(2\pi e) \tag{5.91}$$

$$= h(Z) - h(N) \tag{5.92}$$

$$= I(X; \sqrt{\gamma}X + N) \tag{5.93}$$

It has thus been shown that in the large-dimension limit, the multi-dimensional channel (5.75) has the same mutual information as the scalar Gaussian channel with the same input and SNR as we initially expected.

---

[14] Equation (5.85) is a variant of the Hubbard-Stratonovich transform [76].

5.4.2 Free Energy

In the remainder of this section, we present major steps of the replica analysis that lead to Claims 1–3. The outline of this development is as follows. We calculate the free energy (5.70) using (5.72) so that the spectral efficiency under joint decoding is immediate from (5.69). In Section 5.4.3, we show a sketch for calculating the joint moments (5.71) that lead to the decoupling of the multiuser channel. Some of the calculations are tedious, so we omit some details but provide enough clues and intuition so that the reader can connect the dots. For more details, we refer the reader to [26,27,31].

For an arbitrary positive integer u, we introduce u independent replicas of the companion channel with the same received signal Y and channel state S as depicted in Figure 5.8. The *partition function* of the replicated system, from which we evaluate the free energy (see Section 5.4.1.2), is:

$$q_{Y|S}^u(y|s) = \mathsf{E}_q\left\{\prod_{a=1}^u q_{Y|X,S}(y|X_a, s)\right\} \tag{5.94}$$

where the expectation is taken over the replicated variables $\{X_{ak}|a = 1, \ldots, u, k = 1, \ldots, K\}$. In (5.94), $X_{ak} \sim q_X$ are *i.i.d.* since $(Y, S)=(y, s)$ are given. From (5.94):

$$\mathsf{E}\left\{q_{Y|S}^u(Y, S)\right\} = \mathsf{E}\left\{\int p_{Y|X,S}(y|X_0, S)\prod_{a=1}^u q_{Y|X,S}(y|X_a, S)dy\right\} \tag{5.95}$$

where the expectations are taken over the channel state matrix S, the original symbol vector X_0 (*i.i.d.* entries with distribution p_X), and the replicated symbols X_a, $a = 1, \ldots, u$. For convenience, let $\sigma_0 = 1$ and $\sigma_a = \sigma$ for $a = 1, 2, \ldots$. Plugging in (5.8) and (5.9), we have:

$$\mathsf{E}\left\{q_{Y|S}^u(Y, S)\right\} = \mathsf{E}\left\{\int \frac{(2\pi\sigma^2)^{-\frac{uL}{2}}}{(2\pi)^{\frac{L}{2}}}\prod_{a=0}^u \exp\left[-\frac{\|y - SX_a\|^2}{2\sigma_a^2}\right]dy\right\} \tag{5.96}$$

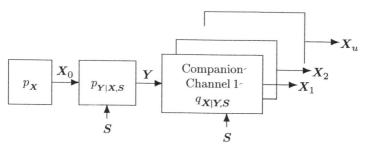

Figure 5.8. The replices of the companion channel.

Note that S and X_a are independent in (5.96). Let $\underline{X} = [X_0, \ldots, X_u]$. The fact that the L dimensions of the multiuser channel are independent and statistically identical allows the RHS of (5.96) to be written as:

$$\mathsf{E}\left\{ \left[(2\pi\sigma^2)^{-\frac{u}{2}} \int \mathsf{E}\left\{ \prod_{a=0}^{u} \exp\left[-\frac{(y - \overline{S}AX_a)^2}{2\sigma_a^2} \right] \middle| A\underline{X} \right\} \frac{dy}{\sqrt{2\pi}} \right]^L \right\} \tag{5.97}$$

where the inner expectation is taken over $\overline{S} = [S_1, \ldots, S_K]$, a row vector of *i.i.d.* random variables each taking the same distribution as the random chips S_{nk}. It is clear that the original expectation over the growing chip dimension L is replaced by the fixed dimension u of the replicas.

Define the following variables:

$$V_a = \frac{1}{\sqrt{K}} \sum_{k=1}^{K} \sqrt{\gamma_k}\, S_k X_{ak}, \quad a = 0, 1, \ldots, u \tag{5.98}$$

Clearly, (5.97) can be rewritten as:

$$\mathsf{E}\left\{ q_{Y|S}^u(Y, S) \right\} = \mathsf{E}\left\{ \exp\left[K G_K^{(u)}(A\underline{X}) \right] \right\} \tag{5.99}$$

where:

$$G_K^{(u)}(A\underline{X}) = -\frac{u}{2\beta} \log (2\pi\sigma^2)$$
$$+ \frac{1}{\beta} \log \int \mathsf{E}\left\{ \prod_{a=0}^{u} \exp\left[-\frac{(y - \sqrt{\beta}V_a)^2}{2\sigma_a^2} \right] \middle| A\underline{X} \right\} \frac{dy}{\sqrt{2\pi}} \tag{5.100}$$

Note that given A and $\underline{X}$, each V_a is a sum of K weighted *i.i.d.* random chips. Due to a vector version of the central limit theorem, $V = [V_0, V_1, \ldots, V_u]^{\mathsf{T}}$ converges to a zero-mean Gaussian random vector as $K \to \infty$. For $a, b = 0, 1, \ldots, u$, define:

$$Q_{ab} = \mathsf{E}\{ V_a V_b \mid A\underline{X} \} = \frac{1}{K} \sum_{k=1}^{K} \gamma_k X_{ak} X_{bk} \tag{5.101}$$

Although implicit in notation, Q_{ab} is a function of $\{\gamma_k, X_{ak}, X_{bk}\}_{k=1}^{K}$. The random vector V can essentially be replaced by a zero-mean Gaussian vector with covariance matrix $Q = (1/K)X^{\mathsf{T}}A^2 X$. As a result:

$$\exp\left[G_K^{(u)}(A\underline{X}) \right] = \exp\left[G^{(u)}(Q) + \mathcal{O}(K^{-1}) \right] \tag{5.102}$$

where the integral of the Gaussian density in (5.100) can be simplified to obtain:

$$G^{(u)}(Q) = -\frac{1}{2\beta} \log \det(I + \Sigma Q) - \frac{1}{2\beta} \log\left(1 + \frac{u}{\sigma^2}\right) - \frac{u}{2\beta} \log(2\pi\sigma^2) \quad (5.103)$$

where Σ is a $(u+1)\times(u+1)$ matrix:[15]

$$\Sigma = \frac{\beta}{\sigma^2 + u} \left[\begin{array}{c|c} u & -e^{\mathsf{T}} \\ \hline -e & \left(1 + \frac{u}{\sigma^2}\right)I - \frac{1}{\sigma^2}ee^{\mathsf{T}} \end{array} \right] \quad (5.104)$$

where e is a $u\times 1$ column vector whose entries are all 1. It is clear that Σ is invariant if two non-zero indexes are interchanged, i.e., Σ is symmetric in the replicas.

By (5.99) and (5.102):

$$\frac{1}{K} \log \mathsf{E}\left\{ q_{Y|S}^u(Y, S) \right\} = \frac{1}{K} \log \mathsf{E}\left\{ \exp\left[K(G^{(u)}(Q) + \mathcal{O}(K^{-1})) \right] \right\} \quad (5.105)$$

$$= \frac{1}{K} \log \int \exp\left[KG^{(u)}(Q) \right] d\mu_K^{(u)}(Q) + \mathcal{O}\left(\frac{1}{K}\right) \quad (5.106)$$

where the expectation over the replicated symbols is rewritten as an integral over the probability measure of the covariance matrix Q.

5.4.2.1 *Large Deviations and Saddle Point*   Since Q_{ab} given by (5.101) is a sum of independent random variables for each pair (a, b), the probability measure $\mu_K^{(u)}$ satisfies the large deviations property. By Cramér's Theorem [77, Theorem II.4.1], there exists a rate function $I^{(u)}$ such that the measure $\mu_K^{(u)}$ satisfies:

$$-\lim_{K\to\infty} \frac{1}{K} \log \mu_K^{(u)}(\mathcal{A}) = \inf_{Q\in\mathcal{A}} I^{(u)}(Q) \quad (5.107)$$

for all measurable sets $\mathcal{A}$ of $(u+1)\times(u+1)$ matrices.

Let the moment generating function be defined as:

$$M^{(u)}(\tilde{Q}) = \mathsf{E}\left\{ \exp\left[\gamma X^{\mathsf{T}} \tilde{Q} X \right] \right\} \quad (5.108)$$

where $\tilde{Q}$ is a $(u+1)\times(u+1)$ symmetric matrix, $X = [X_0, X_1, \ldots, X_u]^{\mathsf{T}}$, and the expectation in (5.108) is taken over independent random variables $\gamma \sim P_\gamma$, $X_0 \sim p_X$ and $X_1, \ldots, X_u \sim q_X$. The rate of the measure $\mu_K^{(u)}$ is given by the Legendre-Fenchel

---

[15]The indexes of all $(u+1)\times(u+1)$ matrices in this chapter start from 0.

transform of the cumulant generating function [77]:

$$I^{(u)}(\mathbf{Q}) = \sup_{\tilde{\mathbf{Q}}} \left[\mathrm{tr}\{\tilde{\mathbf{Q}}\mathbf{Q}\} - \log M^{(u)}(\tilde{\mathbf{Q}}) \right] \tag{5.109}$$

where the supremum is taken with respect to the symmetric matrix $\tilde{\mathbf{Q}}$.

As the exponential factor in (5.106) is proportional to K, and since we are taking the limit $K \to \infty$, the integral is dominated by the maximum of the overall effect of the exponent and the rate of the measure on which the integral takes place (the saddle-point method). Precisely, by Varadhan's theorem [77, Theorem II.7.1], the free energy for a given replica number u is:

$$\mathcal{F}_u = \lim_{K \to \infty} \frac{1}{K} \log \mathsf{E}\left\{ q^u_{Y|S}(Y, S) \right\} = -\sup_{\mathbf{Q}} \left[G^{(u)}(\mathbf{Q}) - I^{(u)}(\mathbf{Q}) \right] \tag{5.110}$$

where the supremum is over all valid covariance matrices. Plugging in (5.103), (5.108) and (5.109):

$$\mathcal{F}_u = \inf_{\mathbf{Q}} \sup_{\tilde{\mathbf{Q}}} T^{(u)}(\mathbf{Q}, \tilde{\mathbf{Q}}) \tag{5.111}$$

with:

$$T^{(u)}(\mathbf{Q}, \tilde{\mathbf{Q}}) = \frac{1}{2\beta} \log \det(\mathbf{I} + \mathbf{\Sigma}\mathbf{Q}) + \mathrm{tr}\{\tilde{\mathbf{Q}}\mathbf{Q}\} - \log \mathsf{E}\left\{ \exp\left[\gamma X^{\mathsf{T}} \tilde{\mathbf{Q}} X \right] \right\}$$
$$+ \frac{1}{2\beta} \log\left(1 + \frac{u}{\sigma^2} \right) + \frac{u}{2\beta} \log\left(2\pi\sigma^2 \right) \tag{5.112}$$

For an arbitrary $\mathbf{Q}$, we first seek the point of zero gradient with respect to $\tilde{\mathbf{Q}}$ and find that for any given $\mathbf{Q}$, the exremum in $\tilde{\mathbf{Q}}$, satisfies:

$$\mathbf{Q} = \frac{\mathsf{E}\left\{ \gamma X X^{\mathsf{T}} \exp\left[\gamma X^{\mathsf{T}} \tilde{\mathbf{Q}} X \right] \right\}}{\mathsf{E}\left\{ \exp\left[\gamma X^{\mathsf{T}} \tilde{\mathbf{Q}} X \right] \right\}} \tag{5.113}$$

Let $\tilde{\mathbf{Q}}^*(\mathbf{Q})$ denote the solution to (5.113). We then seek the point of zero gradient of $T^{(u)}(\mathbf{Q}, \tilde{\mathbf{Q}}^*(\mathbf{Q}))$ with respect to $\mathbf{Q}$.[16] By virtue of the zero-gradient condition with respect to $\tilde{\mathbf{Q}}$, one finds that the derivative of $\tilde{\mathbf{Q}}^*(\mathbf{Q})$ with respect to $\mathbf{Q}$ is multiplied by 0 and hence inconsequential. Therefore, the extremum in $\mathbf{Q}$ satisfies:

$$\tilde{\mathbf{Q}} = -\beta^{-1}(\mathbf{I} + \mathbf{\Sigma}\mathbf{Q})^{-1}\mathbf{\Sigma} \tag{5.114}$$

---

[16]The following identifies are useful:

$$\frac{\partial \log \det \mathbf{Q}}{\partial x} = \mathrm{tr}\left\{ \mathbf{Q}^{-1} \frac{\partial \mathbf{Q}}{\partial x} \right\}, \quad \frac{\partial \mathbf{Q}^{-1}}{\partial x} = -\mathbf{Q}^{-1} \frac{\partial \mathbf{Q}}{\partial x} \mathbf{Q}^{-1}$$

It is interesting to note from the resulting joint equations (5.113)–(5.114) that the order in which the supremum and infimum are taken in (5.111) can be exchanged. The solution $(Q^*, \tilde{Q}^*)$ is in fact a saddle point of $T^{(u)}$. Notice that (5.113) can also be expressed as:

$$Q = E\{\gamma XX^\mathsf{T} | \tilde{Q}\} \qquad (5.115)$$

where the expectation is over an appropriately defined conditional measure $P_\gamma \times p_{X|\tilde{Q},\gamma}$ where:

$$p_{X|\tilde{Q},\gamma}(x|\tilde{Q}, \gamma) = p_X(x) \frac{\exp\left[\gamma x^\mathsf{T} \tilde{Q} x\right]}{E\{\exp\left[\gamma X^\mathsf{T} \tilde{Q} X\right]\}} \qquad (5.116)$$

which is evidently a PDF or PMF. Let $Q^*(u)$ and $\tilde{Q}(u)$ be the solution to (5.113)–(5.114) as functions of u. The free energy is then found by (5.72) and (5.111).

5.4.2.2 Replica Symmetry Solution

Solving joint equations (5.113)–(5.114) directly is prohibitive except in the simplest cases such as q_X being Gaussian. In the general case, suggested by the symmetry in the matrix Σ (5.104), we postulate that the solution to the joint equations satisfies *replica symmetry*, namely, both $Q^*(u)$ and $\tilde{Q}^*(u)$ are invariant if two (nonzero) replica indexes are interchanged. In other words, the extremum can be written as:

$$Q^*(u) = \begin{bmatrix} r & m & m & \cdots & m \\ m & p & q & \cdots & q \\ m & q & p & \ddots & \vdots \\ \vdots & \vdots & \ddots & \ddots & q \\ m & q & \cdots & q & p \end{bmatrix}, \quad \tilde{Q}^*(u) = \begin{bmatrix} c & d & d & \cdots & d \\ d & g & f & \cdots & f \\ d & f & g & \ddots & \vdots \\ \vdots & \vdots & \ddots & \ddots & f \\ d & f & \cdots & f & g \end{bmatrix} \qquad (5.117)$$

where r, m, p, q, c, d, f, g are some real functions of u. The validity of the replica symmetry assumption is discussed in Section 5.5.1. Under this symmetry assumption, the problem of seeking the extremum (5.111) over a $(u+1)^2$-dimensional space (with u also a variable) is reduced to seeking the extremum over several parameters.

The eight parameters (r, m, p, q, c, d, f, g) can be solved from the joint equations (5.113)–(5.114) under replica symmetry assumption. The detailed calculation is omitted. It is interesting to note that the u-dependence of the parameters obtained from the joint equations (5.133)–(5.114) does not contribute to the free energy (5.111) due to the zero gradient conditions. Thus for the purpose of the free energy (5.111), it suffices to find the derivative in (5.111) with $Q^*(u)$ and $\tilde{Q}^*(u)$ replaced by their values at $u = 0$, which we simply denote by Q^* and $\tilde{Q}^*$. From this point on, with slight abuse of notation, let r, m, p, q, c, d, f, g represent their values at $u = 0$.

Using (5.114) and (5.117), it can be shown that at $u = 0$:

$$c = 0 \tag{5.118a}$$

$$d = \frac{1}{2[\sigma^2 + \beta(p - q)]} \tag{5.118b}$$

$$f = \frac{1 + \beta(r - 2m + q)}{2[\sigma^2 + \beta(p - q)]^2} \tag{5.118c}$$

$$g = f - d \tag{5.118d}$$

The parameters r, m, p, q can be determined from (5.115) by studying the measure $P_{X,\gamma|\tilde{Q}}$ under replica symmetry and $u \to 0$. For that purpose, define two useful parameters with a modest amount of foresight:

$$\eta = \frac{2d^2}{f} \quad \text{and} \quad \xi = 2d \tag{5.119}$$

The moment generating function (5.108) is evaluated using the property (5.85) with $\eta = 2d^2/f$ and noticing that $c = 0$, $g - f = -d$ to obtain:

$$M^{(u)}(\tilde{Q}^*) = \mathsf{E}\left\{ \sqrt{\frac{\eta}{2\pi}} \int \exp\left[-\frac{\eta}{2}(z - \sqrt{\gamma}X_0)^2\right] \right.$$
$$\left. \times \left[\mathsf{E}_q\left\{ \exp\left[-\frac{\xi}{2}z^2 - \frac{\xi}{2}(z - \sqrt{\gamma}X)^2\right] \middle| \gamma \right\} \right]^u dz \right\} \tag{5.120}$$

It is clear that the limit of (5.120) as $u \to 0$ is 1, i.e.:

$$\lim_{u \to 0} \mathsf{E}\left\{ \exp\left[\gamma X^{\mathsf{T}} \tilde{Q}^* X \right] \right\} = 1 \tag{5.121}$$

Hence (5.113) implies that, as $u \to 0$, the limit of $Q_{ab}^* = \mathsf{E}\left\{ \gamma X_a X_b | \tilde{Q}^* \right\}$ is identical to:

$$\lim_{u \to 0} \mathsf{E}\left\{ \gamma X_a X_b \exp\left[\gamma X^{\mathsf{T}} \tilde{Q}^* X \right] \right\} \tag{5.122}$$

We apply the transform (5.85) to decouple the cross terms of the form $X_c X_d$ in the exponent in (5.122). In fact all terms unrelated to X_a and X_b integrate to 1 that do not contribute to the limit. More details are found in [27].

5.4.2.3 Single-User Channel Interpretation   We now give a useful representation for the parameters r, m, p, q defined in (5.117). Consider $a = 0$ and

$b = 1$, for instance. Expanding (5.122), as $u \to 0$:

$$Q_{01}^* = \mathsf{E}\left\{\gamma X_0 X_1 \exp\left[\gamma X^\mathsf{T} \tilde{Q}^* X\right]\right\} \tag{5.123}$$

$$\to \mathsf{E}\left\{\gamma X_0 \int \sqrt{\frac{\eta}{2\pi}} \exp\left[-\frac{\eta}{2}(z - \sqrt{\gamma} X_0)^2\right]\right. \tag{5.124}$$

$$\left. \times \frac{X_1 \sqrt{\frac{\xi}{2\pi}} \exp\left[-\frac{\xi}{2}(z - \sqrt{\gamma} X_1)^2\right]}{\mathsf{E}_q\left\{\sqrt{\frac{\xi}{2\pi}} \exp\left[-\frac{\xi}{2}(z - \sqrt{\gamma} X_1)^2\right]\bigg| \gamma\right\}} \, dz\right\} \tag{5.125}$$

Let two single-user Gaussian channels be defined as in Section 5.3.2, i.e., the input–output relationship of the two channels are described by $p_{Z|X,\gamma,\eta}$ given by (5.18) and $q_{Z|X,\gamma,\xi}$ by (5.40). Assuming that the input distribution to the channel $q_{Z|X,\gamma,\xi}$ is q_X, a posterior probability distribution $q_{X|Z,\gamma,\xi}$ is induced, which defines a companion channel. Let X_0 be the scalar input to the channel $p_{Z|X,\gamma,\eta}$ and $X = X_1$ be the output of the companion channel $q_{X|Z,\gamma,\xi}$. The posterior mean with respect to the measure q, denoted by $\langle X \rangle_q$, is given by (5.41). The Gaussian channel $p_{Z|X,\gamma,\eta}$, the companion channel $q_{X|Z,\gamma,\xi}$ and the PME, all in the single-user setting, are depicted in Figure 5.6b. Then, (5.125) can be understood as an expectation over X_0, X, and Z to obtain:

$$\mathsf{E}\left\{\gamma X_0 \int \mathsf{E}_q\{X \mid Z = z, \gamma, \xi\} p_{Z|X,\gamma,\eta}(z \mid X_0, \gamma, \eta) \, dz\right\} = \mathsf{E}\{\gamma X_0 \langle X \rangle_q\}. \tag{5.126}$$

Similarly, (5.122) can be evaluated for all (a, b) yielding together with (5.117):

$$r = \lim_{u \to 0} Q_{00}^* = \mathsf{E}\{\gamma X_0^2\} = \mathsf{E}\{\gamma\} \tag{5.127a}$$

$$m = \lim_{u \to 0} Q_{01}^* = \mathsf{E}\{\gamma X_0 \langle X \rangle_q\} \tag{5.127b}$$

$$p = \lim_{u \to 0} Q_{11}^* = \mathsf{E}\{\gamma X^2\} \tag{5.127c}$$

$$q = \lim_{u \to 0} Q_{12}^* = \mathsf{E}\{\gamma (\langle X \rangle_q)^2\} \tag{5.127d}$$

In summary, the parameters c, d, f, g are given by (5.118) as functions of r, m, p, q that are in turn determined by the statistics of the two channels (5.18) and (5.40) parameterized by $\eta = 2d^2/f$ and $\xi = 2d$ respectively. It is not difficult to see that:

$$r - 2m + q = \mathsf{E}\left\{\gamma (X_0 - \langle X \rangle_q)^2\right\} \tag{5.128a}$$

$$p - q = \mathsf{E}\left\{\gamma (X - \langle X \rangle_q)^2\right\} \tag{5.128b}$$

Using (5.118) and (5.119), it can be checked that:

$$r - 2m + q = \frac{1}{\beta}\left(\frac{1}{\eta} - 1\right), \quad \text{and} \quad p - q = \frac{1}{\beta}\left(\frac{1}{\xi} - \sigma^2\right) \tag{5.129}$$

Under replica symmetry, $G^{(u)}(Q^*)$ is evaluated using (5.103) and expressed in η and ξ. Together with (5.111) and (5.120), the free energy is found as (5.47), whereby (5.128) and (5.129), (η, ξ) satisfies:

$$\eta^{-1} = 1 + \beta E\left\{\gamma(X_0 - \langle X \rangle_q)^2\right\} \tag{5.130a}$$

$$\xi^{-1} = \sigma^2 + \beta E\left\{\gamma(X - \langle X \rangle_q)^2\right\} \tag{5.130b}$$

Because of the saddle-point evaluation in (5.110), in case of multiple solutions to (5.130), (η, ξ) is chosen as the solution that gives the minimum free energy $\mathcal{F}$. By defining $\mathcal{E}(\gamma; \eta, \xi)$ and $\mathcal{V}(\gamma; \eta, \xi)$ as in (5.43) and (5.44), the coupled equations (5.118) and (5.127) can be summarized to establish the key fixed-point equations (5.46). It will be shown in Section 5.4.3 that, from an individual user's viewpoint, the multiuser PME and the multiuser companion channel, parameterized by arbitrary (q_X, σ), have an equivalence as a single-user PME and a single-user companion channel.

5.4.2.4 Spectral Efficiency and Multiuser Efficiency

Finally, for the purpose of the total spectral efficiency, we set the postulated measure q to be identical to the actual measure p [i.e., $(q_X, \sigma) = (p_X, 1)$]. The inverse noise variances (η, ξ) satisfy joint equations but we choose the replica-symmetric solution $\eta = \xi$. Using the identity:

$$C_{\text{joint}} = \beta \mathcal{F}\big|_{q=p} - \frac{1}{2}\log(2\pi e), \tag{5.131}$$

the total spectral efficiency is:

$$C_{\text{joint}} = -\beta E\left\{\int p_{Z\,|\,\gamma,\eta}(z|\gamma;\eta)\log p_{Z\,|\,\gamma,\eta}(z|\gamma;\eta)dz\right\}$$
$$- \frac{\beta}{2}\log\frac{2\pi e}{\eta} + \frac{1}{2}(\eta - 1 - \log \eta) \tag{5.132}$$

where η satisfies:

$$\eta + \eta\beta E\left\{\gamma\left[1 - \int \frac{[p_1(z, \gamma; \eta)]^2}{p_{Z|\gamma,\eta}(z|\gamma;\eta)}dz\right]\right\} = 1 \tag{5.133}$$

The optimal spectral efficiency of the multiuser channel is thus found.

We remark that the essence of the replica method here is its capability of converting a difficult expectation (e.g., of a logarithm) with respect to a given large system to an expectation of a simpler form with respect to the replicated system. Quite different from conventional techniques is the emphasis of large systems and symmetry from the beginning, where the central limit theorem and large deviations help to calculate the otherwise intractable quantities.

5.4.3 Joint Moments

Consider the multiuser Gaussian channel, the PME and the companion channel depicted in Figure 5.6a. The joint moments (5.71) are of interest here. For simplicity, we first study joint moments of the input and the companion channel output, which can be obtained as expectations under the replicated system [31, Lemma 3.1]:

$$\mathsf{E}\{X_{0k}^i X_k^j\} = \mathsf{E}\{X_{0k}^i X_{mk}^j\}, \quad m = 1, \ldots, u \quad (5.134)$$

It is then straightforward to calculate (5.71) by following the same procedure.

In [27], it is shown that the moments (5.134) can be obtained as:

$$\lim_{u \to 0} \frac{\partial}{\partial h} \frac{1}{\alpha_1 K} \log \mathsf{E}\{Z^{(u)}(Y, S, X_0; h)\}\big|_{h=0} \quad (5.135)$$

where $\alpha_1 \in (0, 1)$ and:

$$Z^{(u)}(y, s, x_0; h) = \mathsf{E}_q\left\{\exp\left[h \sum_{k=1}^{k_1} x_{0k}^i X_{mk}^j\right] \prod_{a=1}^u \exp\left[-\frac{\|y - sX_a\|^2}{2\sigma^2}\right]\right\} \quad (5.136)$$

where $K_1 = \alpha_1 K$ and we assume that the SNRs of the first K_1 users are equal to γ. Regarding (5.136) as a partition function for some random system allows the same techniques in Section 5.4.2 to be used to write:

$$\lim_{K \to \infty} \frac{1}{K} \log \mathsf{E}\{Z^{(u)}(Y, S, X_0; h)\} = \sup_Q \left[\beta^{-1} G^{(u)}(Q) - I^{(u)}(Q; h)\right] \quad (5.137)$$

where $G^{(u)}(Q)$ is given by (5.103) and the rate $I^{(u)}(Q; h)$ is found as:

$$I^{(u)}(Q; h) = \sup_{\tilde{Q}} \left[\operatorname{tr}\{\tilde{Q}Q\} - (1-\alpha_1) \log M^{(u)}(\tilde{Q}) - \alpha_1 \log M^{(u)}(\tilde{Q}, \gamma; h)\right] \quad (5.138)$$

where $M^{(u)}(\tilde{Q})$ is defined in (5.108), and:

$$M^{(u)}(\tilde{Q}; \gamma; h) = \mathsf{E}\{\exp[h X_0^i X_m^j] \exp[\gamma X^{\mathsf{T}} \tilde{Q} X] \mid \gamma\} \quad (5.139)$$

From (5.137) and (5.138), taking the derivative in (5.135) with respect to h at $h = 0$ leaves only one term:

$$\frac{\partial}{\partial h} \log M^{(u)}(\tilde{Q}, \gamma, h)\Big|_{h=0} = \frac{\mathsf{E}\{X_0^i X_m^j \exp[\gamma X^\mathsf{T} \tilde{Q} X]\,|\,\gamma\}}{\mathsf{E}\{\exp[\gamma X^\mathsf{T} \tilde{Q} X]\,|\,\gamma\}} \tag{5.140}$$

Since:

$$Z^{(u)}(Y, S, X_0; h)\big|_{h=0} = q_{Y|S}^u(Y|S) \tag{5.141}$$

the $\tilde{Q}^*$ which satisfies (5.140) and gives the supremum in (5.138) at $h \to 0$ is exactly the $\tilde{Q}^*$ which gives the supremum of (5.109), which is replica-symmetric by assumption. By introducing the parameters (η, ξ) as in Section 5.4.2, and by definition of q_i and p_i in (5.42) and (5.26) respectively, (5.140) can be further evaluated as:

$$\frac{\displaystyle\int \left(\sqrt{\frac{2\pi}{\xi}} e^{\frac{\xi z^2}{2}}\right)^u p_i(z, \gamma, \eta) q_0^{u-1}(z, \gamma, \xi) q_j(z, \gamma, \xi)\,\mathrm{d}z}{\displaystyle\int \left(\sqrt{\frac{2\pi}{\xi}} e^{\frac{\xi z^2}{2}}\right)^u p_0(z, \gamma, \eta) q_0^u(z, \gamma, \xi)\,\mathrm{d}z} \tag{5.142}$$

Taking the limit $u \to 0$, one has from (5.134)–(5.142) that as $K \to \infty$:

$$\frac{1}{K_1} \sum_{K=1}^{K_1} \mathsf{E}\{X_{0k}^i X_{mk}^j\} \to \int p_i(z, \gamma, \eta) \frac{q_j(z, \gamma, \xi)}{q_0(z, \gamma, \xi)}\,\mathrm{d}z \tag{5.143}$$

Let $X_0 \sim p_X$ be the input to the scalar Gaussian channel $p_{Z|X,\gamma,\eta}$ and Z be its output (see Figure 5.6b). Let X be the output of the companion channel with Z as its input. Then $X_0 - Z - X$ is a Markov chain. The RHS of (5.143) is:

$$\int p_0(z, \gamma, \eta) \frac{p_i(z, \gamma, \xi)}{p_0(z, \gamma, \xi)} \frac{q_j(z, \gamma, \xi)}{q_0(z, \gamma, \xi)}\,\mathrm{d}z = \mathsf{E}\{\mathsf{E}\{X_0^i\,|\,Z\}\mathsf{E}\{X^j\,|\,Z\}\} \tag{5.144}$$

Letting $K_1 \to 1$ (thus $\alpha_1 \to 0$)[17] so that the requirement that the first K_1 users take the same SNR becomes unnecessary, we have shown by (5.134), (5.143) and (5.144) that for every SNR distribution and every user $k \in \{1, \dots, K\}$:

$$\mathsf{E}\{X_{0k}^i X_k^j\} \longrightarrow \mathsf{E}\{X_0^i X^j\} \quad \text{as } K \to \infty \tag{5.145}$$

---

[17]To be precise, this step requires a more delicate treatment, since the saddle-point evaluation involved in our calculation only captures terms of order $\mathcal{O}(K)$ in the exponent. It has been shown that the result remains the same even if we take $\mathcal{O}(1)$ terms into consideration [78].

We assume that the joint distribution $P_{X_{0k}X_k}$ is determinate, i.e., uniquely determined by the joint moments.[18] Therefore, for every user k, the joint distribution of the input X_{0k} to the multiuser channel and the output X_k of the multiuser companion channel converges to the joint distribution of the input X_0 to the single-user Gaussian channel $p_{Z|X,\gamma,\eta}$ and the output X of the single-user companion channel $q_{X|Z,\gamma,\xi}$.

Applying the same methodology as developed thus far, one can also calculate the joint moments (5.71) to obtain[19]:

$$\mathsf{E}\left\{X_{0k}^i X_k^j \langle X_k \rangle_q^l\right\} \quad \longrightarrow \quad \mathsf{E}\left\{X_0^i X^j \langle X \rangle_q^l\right\} \tag{5.146}$$

where $\langle X \rangle_q$ is the single-user PME output as seen in Figure 5.6b, which is a function of the Gaussian channel output Z. Again, assuming the determinacy, the joint distributions of $(X_{0k}, X_k, \langle X_k \rangle_q)$ converge to that of $(X_0, X, \langle X \rangle_q)$. Indeed, from the viewpoint of user k, the multiuser setting is equivalent to the single-user setting in which the SNR suffers a degradation η (compare Figures 5.6b and 5.6a). Hence we have justified the decoupling principle and Claim 3.

In the large-system limit, the transformation from the input X_{0k} to the multiuser detection output $\langle X_k \rangle_q$ is nothing but a single-user Gaussian channel $p_{Z|X,\gamma,\eta}$ concatenated with a decision function (5.41). The decision function is one-to-one due to Proposition 1 and hence inconsequential from both detection- and information-theoretic viewpoints.

We now conclude that the equivalent single-user channel is an additive Gaussian noise channel with input SNR γ and noise variance η^{-1} as depicted in Figure 5.6b. Claim 3 follows, which implies Claims 1 and 2 in the special case that the postulated measure q is identical to the actual measure p. Curiously, this decoupling result is identical to what is obtained using parallel interference cancellation in Section 5.3.4 with invalid independence assumption, or using belief propagation in the special case of sparse spreading matrix [52].

5.5 FURTHER DISCUSSION

5.5.1 On Replica Symmetry

The validity of the replica symmetry assumption can be checked by calculating the Hessian of $[G^{(u)}(\mathbf{Q}) - I^{(u)}(\mathbf{Q})]$ at the replica symmetric supremum [37]. If all the

---

[18]Note that the determinacy does not necessarily hold in general (the moment problem [79, p. 227] [80]), even though all distributions of finite support and most discrete and continuous distributions of practical interest (e.g., Gaussian distribution) are determinate. Sufficient conditions for a multidimensional distribution to be determinate are given in [81,82]. In particular, if the marginals are determinate, the joint distribution is also determinate [81].

[19]Note the change of notation: X is replaced by X_0 which corresponds to the 0-th replica and X' is replaced by X.

eigen-values of the Hessian associated with modes that break replica symmetry are negative at the replica symmetric supremum, then the solution is stable against perturbations which break replica symmetry. If not, then the solution is said to suffer from the de Almeida-Thouless (AT) instability, which has been named after two physicists who first performed the stability analysis on a spin glass model [83].

For the basic prescription of the AT stability analysis, we ask the reader to see [26], where one will find detailed description of the analysis for the equal-power binary input case. Essentially the same analysis can be performed in the generic case discussed in this chapter, and the result is summarized as follows.

CLAIM 4 *(AT stability) A replica-symmetric solution is stable against replica symmetry breaking (RSB) if the following inequality holds*:

$$-\beta\xi^2 + \left[\mathsf{E}\{\gamma(\langle X^2\rangle_q - (\langle X\rangle_q)^2)^2\}\right]^{-1} < 0 \qquad (5.147)$$

The LHS of (5.147) is the eigen-value of the perturbation modes (the so-called "replicon" modes) that determines the AT stability. One can numerically check the AT stability condition in order to see if a particular numerical solution of the replica symmetric fixed-point equations (5.46) is stable against replica symmetry breaking. In the equal-power binary case, AT instability may actually be observed, although not always, when the postulated noise level σ is less than 1.

When the AT stability is violated for a solution with replica symmetry, it means that Q at the true supremum (5.110) should lack the replica symmetry, and Q with broken symmetry will give us even larger values of $[G^{(u)} - I^{(u)}]$. A systematic way of improving replica-symmetric solutions has been proposed for spin glass models, which consists of considering a series of symmetry breaking schemes, the so-called 1-step RSB, 2-step RSB, etc. A preliminary study on the equal-power binary CDMA problem suggests that consideration of the 1-step RSB alters the solutions only slightly [84]. We thus expect that the analysis with replica symmetry assumption provides us with quantitatively accurate enough picture even if the assumption is not valid.

5.5.2 On Metastable Solutions

As we have briefly mentioned in Section 5.4.2.3, the fixed-point equations (5.46) determining (η, ξ) may have multiple solutions. Since we are interested in obtaining the true supremum of $[G^{(u)} - I^{(u)}]$, what we have to do is to compare the values of the free energy $\mathcal{F}$ of those solutions, and to pick up the one that minimizes the free energy. Then we can safely discard the other solutions, since they seem not to have any operational meaing. Or, do they?

They do, if practical (suboptimal) schemes for obtaining the PME are considered. In order to understand benefits of considering solutions other than the one giving the true supremum, it should be a good idea to exploit an analogy with a "magnet." Typical magnetic materials respond to externally applied magnetic field by expressing magnetization. Magnetization curves (Figure 5.9a) represent how the magnetization depends

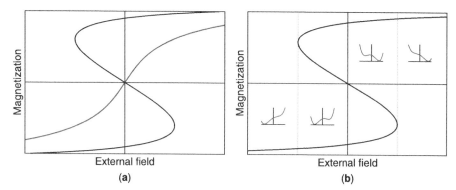

Figure 5.9. Typical magnetization curves of ferromagnets. (a) Magnetization curve at high (monotone curve) and low (S-shaped curve) temperatures. (b) Magnetization curve at low temperature. Insets show the free energy profiles in the corresponding regions.

on the external magnetic field. At high enough temperature the magnetization depends monotonically on, and uniquely determined by, the external field (Figure 5.9a, monotone curve). On the other hand, at low temperature (lower than the so-called Curie temperature), the magnetization may take multiple values within a certain range of the external field (Figure 5.9a, the S-shaped curve). In particular, non-zero magnetization will be observed even when the external field is absent. It is called spontaneous magnetization, and is the theory underlying the phenomenon such as seen about a magnet. The structural change of the magnetization curve with temperature is an example of the phase transition.

A simple mathematical model of magnetism can explain the phase transition analytically. The analysis consists of evaluating the free energy in essentially the same manner as the analysis in this chapter, that is, with the fixed-point method. At the low-temperature ("ferromagnetic") phase, the fixed-point equations may have three solutions. When it is the case, the free energy, as a function of magnetization, has a structure as shown in the insets of Figure 5.9b. Thus the true solution in the mathematical sense is the one that minimizes the free energy, and is given by the topmost and the lowermost branches in Figure 5.9b when the external field is positive and negative, respectively. The solution is called a (globally) stable solution. The solution in the topmost branch with negative external fields, and the one in the lowermost branch with positive external fields, only locally minimizes the free energy. They are called "metastable" solutions. The last solution in the middle branch, called an unstable solution, locally maximizes the free energy.

The significance of the metastable solutions is manifested in a phenomenon called *hysteresis*. In the low-temperature condition, the magnetization we observe may not be the one corresponding to the true solution (the globally stable solution). Depending on history, the system may take a state corresponding to a metastable solution, the fact that explains the hysteresis.

Essentially the same description applies to the multiuser detection problem as well, as can be seen in Figure 5.10. Whereas the globally optimum solution switches from

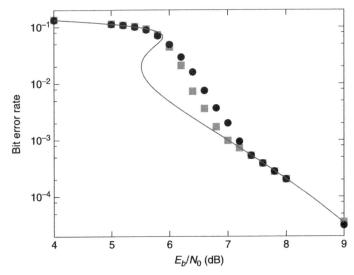

Figure 5.10. Comparison of BER between replica analysis of the individually optimal detector (solid curve) and numerical experiments on the BP-based multiuser detection algorithm described in Section 5.7.2.3 (dots: $L = 2000$, squares: $L = 4000$). System load β is 1.6. BPSK data modulation, as well as perfect power control is assumed.

the uppermost branch to the lowermost one, the numerical results obtained using suboptimal belief propagation algorithms (see Section 5.7.2) seem to follow the uppermost branch even though it is not the true solution on the shoulder at the RHS of the S-shaped curve. The phenomenon can be ascribed to the fact that the multiuser detection algorithm does not know the true detection results initially: Indeed, the system may easily get trapped in the metastable solution with large error probability, due to the "history" effect caused by the initial configurations being far away from the true detection results.

5.6 STATISTICAL PHYSICS AND THE REPLICA METHOD

So far, we have worked only with the mathematical aspect of the replica method and avoided physical concepts. In retrospect, it is enlightening to draw an equivalence between multiuser communications and many-body problems in statistical physics, which also provides the underlying rationale for applying the replica theory in the first place.

5.6.1 A Note on Statistical Physics

Consider the physics of a macroscopic system, which typically consists of 10^{20} or more particles. Let the microscopic configuration of the system be described by a vector x. The configuration of the system evolves over time according to the laws of physics. However, in view of the enormous degrees of freedom of such a macroscopic system, it is practically impossible to track the time evolution of the configuration. On the other hand, we are most interested in the macroscopic properties of the system, not the detailed configuration of the humongous number of particles. Elegantly, statistical physics introduces a probabilistic description of the system. Let $p(x)$ denote the probability that the system is found in configuration x. Assuming the system is at thermal equilibrium in contact with a heat bath, statistical physics states that $p(x)$ is given by the so-called *Gibbs–Boltzmann distribution*:

$$p(x) = Z^{-1} \exp\left[-\frac{1}{T}H(x)\right] \tag{5.148}$$

where $H(x)$ denotes the *Hamiltonian*, i.e., the function that associates each configuration x to its energy, where:

$$Z = \sum_x \exp\left[-\frac{1}{T}H(x)\right] \tag{5.149}$$

is the *partition function* normalizing $p(x)$, and where the parameter $T > 0$ denotes the *temperature* of the system.

The Gibbs–Boltzmann distribution can also be characterized as the solution to a constrained optimization problem, in which the *entropy* (disorder) of the system:

$$S = -\sum_x p(x)\log p(x) \tag{5.150}$$

is maximized under the constraint that the *energy* of the system is fixed to be:

$$\mathcal{E} = \sum_x p(x)H(x) \tag{5.151}$$

This optimization problem can be solved using the Lagrange multiplier method. Using (5.150) and (5.151), the probability distribution $p(x)$ is found to be equal to (5.148), where the Lagrange multiplier $1/T$, serving as the inverse temperature, is determined by the energy constraint (5.151).

Generally speaking, statistical physics is a theory that studies macroscopic properties (e.g., pressure, magnetization) of such a system starting from the Hamiltonian by taking the above probabilistic viewpoint. For the system shown above, it is not the most probable configuration (the ground state which has the minimum energy), but those configurations with energy close to $\mathcal{E}$ that contribute to the physics of the system. Indeed, such configurations form the "typical set," which determines macroscopic properties of the system. From the mathematical point of view, one can regard that statistical physics provides a framework of statistical theory regarding probability models with huge degrees of freedom. This view is fundamental in linking statistical physics with various problems in information and communication theory. One particularly useful macroscopic quantity of the thermodynamic system is the *free energy*:

$$\mathcal{F} = \mathcal{E} - T\mathcal{S}. \tag{5.152}$$

Using (5.148)–(5.151), the free energy at equilibrium can also be expressed as:

$$\mathcal{F} = -T \log Z \tag{5.153}$$

Indeed, at thermal equilibrium, the temperature and energy of the system remain constant, the entropy is the maximum possible, and the free energy is at its minimum. The free energy is often the starting point for calculating macroscopic properties of a thermodynamic system. For example, the energy $\mathcal{E}$ and the entropy $\mathcal{S}$ are obtained by differentiating $\mathcal{F}$; $\mathcal{E} = \partial(\mathcal{F}/T)/\partial(1/T)$ and $\mathcal{S} = -\partial\mathcal{F}/\partial T$ hold, respectively.

5.6.2 Multiuser Communications and Statistical Physics

5.6.2.1 Equivalence of Multiuser Systems and Spin Glasses
In order to take advantage of the statistical physics methodologies, we equate the multiuser communication problem to an artificial thermodynamic system, called spin glass. A *spin glass* is a system consisting of many directional spins, in which the interaction of the spins is determined by the so-called *quenched random variables* whose values are determined by the realization of the spin glass. An example is a system consisting of molecules with magnetic spins that evolve over time, while the positions of the molecules that determine the amount of interactions are random (disordered) but remain fixed for each concrete instance. In the probabilistic context, the quenched variables are simply the random variables we condition on to calculate the expectation values of the performance measures. We then average over the quenched variables in order to obtain the average performance. Let the quenched random variables be denoted by (Y, S). The system can be understood as K random spins sitting in quenched randomness $(Y, S) = (y, s)$, and its statistical physics described as in Section with a parameterized Hamiltonian $H_{y,s}(x)$.

Suppose the temperature $T = 1$ and that the Hamiltonian is defined as:

$$H_{y,s}(x) = \frac{\|y - sx\|^2}{2\sigma^2} - \log q_X(x) + \frac{L}{2}\log\left(2\pi\sigma^2\right) \tag{5.154}$$

then the Gibbs–Boltzmann distribution, the configuration distribution of the spin glass at equilibrium, is given by (5.10) and its corresponding partition function by (5.11) [cf. (5.148) and (5.149)]. Precisely, the probability that the transmitted symbol is $X = x$ under the postulated model, given the observation $Y = y$ and the channel state $S = s$, is equal to the probability that the spin glass is found at configuration x, given quenched random variables $(Y, S) = (y, s)$.

The characteristics of the system is encoded in the quenched randomness (Y, S). In the communication channel described by (5.2), (Y, S) takes a specific distribution, i.e., the distributions of the received signal and channel state matrix according to the prior and conditional distributions that underlie the "original" spins.

The free energy of the thermodynamic (or communication) system normalized by the number of users is:

$$-\frac{T}{K}\log Z(\mathbf{Y}, \mathbf{S}) = -\frac{1}{K}\log q_{\mathbf{Y}|\mathbf{S}}(\mathbf{Y} \mid \mathbf{S}) \qquad (5.155)$$

where we assume $T = 1$. If we assume self-averaging for the per-user free energy (5.155), it converges in probability to its expected value over the distribution of the quenched random variables $(\mathbf{Y}, \mathbf{S})$ in the large-system limit $K \to \infty$, which is denoted by $\mathcal{F}$:

$$\mathcal{F} = -\lim_{K \to \infty} \mathsf{E}\left\{\frac{1}{K}\log q_{\mathbf{Y}|\mathbf{S}}(\mathbf{Y} \mid \mathbf{S})\right\} \qquad (5.156)$$

Hereafter, by the free energy we refer to its large-system limit (5.156).

The reader should be cautioned that for disordered systems, thermodynamic quantities may or may not be self-averaging [85]. Buttressed by numerical examples and associated results using random matrix theory, as well as recent progress [74], the self-averaging property is assumed to hold in this work.

The self-averaging property resembles the asymptotic equipartition property (AEP) in information theory [63]. An important consequence is that a macroscopic quantity of a thermodynamic system, which is a function of a large number of random variables, may become increasingly predictable from merely a few parameters independent of the realization of the quenched randomness as the system size grows without bound.

In view of (5.69) and (5.156), the large-system spectral efficiency of the multiuser system is affine in the free energy with a postulated measure q identical to the actual measure p:

$$\mathsf{C} \to \beta \mathcal{F}\big|_{q=p} - \frac{1}{2}\log(2\pi e). \qquad (5.157)$$

Indeed, the replica analysis presented in Section 5.4 was developed based on this observation.

5.7 INTERFERENCE CANCELLATION

5.7.1 Conventional Parallel Interference Cancellation

So far we have discussed, via the single-user characterization, theoretical structure and information transmission capability of the CDMA channels, equipped with various multiuser posterior mean estimators. Straightforward computation of the PME requires K-dimensional integration (or summation if X_k's are discrete), the computational complexity of which generally grows exponentially in K. This is practically hard (see Section 5.1.1).[20]

The interference cancellation is a heuristic idea for reducing complexity by suboptimal processing. Suppose we are given a realization s of the channel state matrix S. The basic observation is that the matched filter output for user k is decomposed into three terms:

$$s_k^\mathsf{T} Y = \|s_k\|^2 \sqrt{\gamma_k} X_k + \sum_{k' \neq k} s_k^\mathsf{T} s_{k'} \sqrt{\gamma_{k'}} X_{k'} + s_k^\mathsf{T} N \qquad (5.158)$$

The first is the "signal" term that is proportional to the desired symbol X_k. The second is the MAI consisting of the symbols of the remaining users. The third is the noise term. In order for a good estimation, we want the interference and noise terms to be small. Whereas essentially nothing can be done in order to eliminate the noise term, one could think of reducing the interference because, in the context of multiuser detection, we want to estimate not only X_k but $\{X_{k'}, k' \neq k\}$ as well, which means that we will certainly have some estimates for the latter, and that these estimates could be used to reconstruct and then cancel the interference. For example, if the estimates $\{\hat{x}_{k'}, k' \neq k\}$ are good enough, the "interference cancellation":

$$z_k = s_k^\mathsf{T} \left(y - \sum_{k' \neq k} s_{k'} \sqrt{\gamma_{k'}} \, \hat{x}_{k'} \right) \qquad (5.159)$$

would give us a quantity that is almost free of the interference.

The parallel interference cancellation (PIC), also referred to as the multistage detector [61], is the idea of performing the interference cancellation in stages and in parallel. It is formulated as:

$$z_k^t = s_k^\mathsf{T} y - \sum_{k' \neq k} s_k^\mathsf{T} s_{k'}' \sqrt{\gamma_{k'}} \, \hat{x}_{k'}^t \qquad (5.160)$$

$$\hat{x}_k^{t+1} = f_k(z_k^t) \qquad (5.161)$$

where $f_k(\cdot)$ is a decision function for user k, which may be defined on the basis of a postulated channel characteristics such as (5.14). Initialization of PIC is typically done by setting $\{\hat{x}_k^0\}$ with a computationally simple estimator, such as a linear detector.

5.7.2 Belief Propagation

5.7.2.1 Application of Belief Propagation to Multiuser Detection   We next turn our focus to a systematic method for approximate computation of the posterior means. Posterior mean estimation is also important in researches of artificial

---

[20]The Gaussian-prior case is an exception, in which the PMEs are calculated algebraically, yielding a linear detector (5.12).

intelligence; how one can represent uncertainties surrounding an intelligent agent is an important issue in artificial intelligence, and one might think it natural to use probability models to handle the uncertainties. Pearl's proposal of BP [53] has provided us with a unified approach to calculating posterior means, provided that a probability model is represented as a graphical model. As mentioned in Section 5.1.4, the multiuser system can be described by a bipartite graph shown in Figure 5.1 where a symbol node X_k and a chip node Y_l are connected by an edge if $s_{lk} \neq 0$. The task of a multiuser detector is to infer the symbols based on the observation at the chip nodes, to which the framework of BP is applicable. However, BP gives exact posterior means only for a limited class of probability models (i.e., those that do not contain cycles), and it generally provides approximate posterior means. BP has nevertheless been regarded as very important because the decoding algorithms of many capacity-achieving error-control coding, e.g., the turbo decoding algorithm for turbo codes and the sum-product algorithm for sparse-graph codes, turn out to be instances of BP [86,87]. In view of such outstanding success of BP in error-control coding, one might think it worthwhile to consider application of BP to the multiuser detection.

It is possible, at least in principle, to apply BP to the multiuser detection problem, which yields the following procedure iterating the "Horizontal" and "Vertical" steps until convergence is achieved:

Input: Channel output y, channel state s, prior $q_X(x)$.

Initialization: Set $t := 0$, and

$$\pi_{lk}^0(x_k) = q_X(x_k), \quad l = 1, \dots, L; \quad k = 1, \dots, K. \tag{5.162}$$

Main Iterations:

for $t = 0$ to maximum number of iterations **do**
  "Horizontal" step:

$$\rho_{lk}^{t+1}(x_k) = \int \frac{1}{\sqrt{2\pi\sigma^2}} \exp\left[-\frac{(y_l - (sx)_l)^2}{2\sigma^2}\right] \prod_{k' \neq k} [\pi_{lk'}^t(x_{k'}) \, dx_{k'}] \tag{5.163}$$

  "Vertical" step:

$$\pi_{lk}^{t+1}(x_k) = \alpha_{lk} \, q_X(x_k) \prod_{l' \neq l} \rho_{l'k}^{t+1}(x_k) \tag{5.164}$$

  where α_{lk} is the normalization coefficient so that $\pi_{lk}^{t+1}(x_k)$ is a PMF.
end for

Output: After convergence is achieved, calculate

$$\pi_k(x_k) = \alpha_k q_X(x_k) \prod_{l=1}^{L} \rho_{lk}(x_k), \tag{5.165}$$

which gives an approximate marginal distribution of X_k.
return $\pi_k(x)$, for all $k = 1, \dots, K$ and all x.

The difficulty in applying BP to the user detection resides in the Horizontal step, where one has to perform $(K-1)$-dimensional integration, whose computational complexity grows exponentially in K.

5.7.2.2 Conventional Parallel Interference Cancellation as Approximate BP
Here we discuss a simple heuristic approximation [89,90] to alleviate the computational difficulty, which, interestingly, leads to the conventional parallel interference cancellation scheme of Section 5.7.1. Their heuristics consists of two basic ideas: The first idea is to replace the quantities $\{\pi^t_{lk}(x_k); l = 1, \ldots, N\}$, computed in the vertical steps, with:

$$\pi^t_k(x_k) = \alpha_k q_X(x_k) \prod_{l=1}^{N} \rho^t_{lk}(x_k) \propto \pi^t_{lk}(x_k)\rho^t_{lk}(x_k) \qquad (5.166)$$

The approximation of $\pi^t_{lk}(x_k)$ with $\pi^t_k(x_k)$ is expected to be quite good asymptotically, because the factor $\rho^t_{lk}(x_k)$ is one among the N factors. The second idea is, instead of evaluating the expectation with respect to $\prod_{k'\neq k}\pi^t_{lk'}(x_{k'})$ in the horizontal steps (5.163), to plug the expectations $m^t_{k'}$ of $X_{k'}$ with respect to $\pi^t_{k'}(x_{k'})$ into the exponent of the integrand. The approximation gives:

$$\rho^{t+1}_{lk}(x_k) \propto \exp\left[\frac{1}{\sigma^2}\sqrt{\frac{\gamma_k}{L}}u^t_l s_{lk}x_k - \frac{\gamma_k}{2\sigma^2}\frac{s^2_{lk}}{L}x^2_k\right] \qquad (5.167)$$

and:

$$\pi^{t+1}_{lk}(x_k) \propto q_X(x_k)\exp\left[\sqrt{\frac{\gamma_k x_k}{\sigma^2}}s^T_k u^t - \frac{\gamma_k x^2_k}{2\sigma^2}\|s_k\|^2\right] \qquad (5.168)$$

where $u_t = [u_1, \ldots, u_L]^T$, $u^t_l = y_l - L^{-1/2}\sum_{k'\neq k}\sqrt{\gamma_{k'}}s_{lk'}m^t_{k'}$. Under the random spreading assumption and in the large-system regime, $\|s_k\|^2 = L^{-1}\sum_{l=1}^{L}s^2_{lk}$ can safely be regarded as being equal to 1 due to the law of large numbers, so that (5.168) is represented as an update rule in terms of $\{m^t_k\}$, as:

$$m^{t+1}_k = f\left(s^T_k y - \sum_{k'\neq k}s^T_k s_{k'}\sqrt{\gamma_{k'}}m^t_{k'}, \gamma_k; \frac{1}{\sigma^2}\right) \qquad (5.169)$$

The function $f(z, \gamma; \xi)$ is the decision function $\mathsf{E}_q\{X|Z = z, \gamma; \xi\}$ induced by the single-user channel with q_X and $q_{Z|X,\gamma;\xi}$ as defined in (5.40). Equation (5.169) is nothing but the conventional PIC algorithm discussed in Section 5.7.1, demonstrating that the conventional PIC algorithm can be regarded as an approximation to the BP algorithm.

5.7.2.3 BP-Based Parallel Interference Cancellation Algorithm
We now discuss a more sophisticated approximation scheme [46]. Let us first rewrite the quantity $(sx)_l$ as:

$$(sx)_l = \frac{1}{\sqrt{L}} \sum_{k' \neq k} \sqrt{\gamma_{k'}} s_{lk'} x_{k'} + \sqrt{\frac{\gamma_k}{L}} s_{lk} x_k \equiv (sx)_{l \setminus k} + \sqrt{\frac{\gamma_k}{L}} s_{lk} x_k \qquad (5.170)$$

Noting that the second term of the rightmost side of (5.170) can be regarded as small compared with the first term as L becomes large, one can expand the integrand of (5.163) as:

$$\exp\left[-\frac{(y_l - (sx)_l)^2}{2\sigma^2} \right] = \exp\left[-\frac{(y_l - (sx)_{l \setminus k})^2}{2\sigma^2} \right]$$

$$\times \left(1 + \frac{1}{\sigma^2} \sqrt{\frac{\gamma_k}{L}} s_{lk} x_k [y_l - (sx)_{l \setminus k}] \right.$$

$$\left. + \frac{\gamma_k}{2\sigma^4 L} \left(s_{lk} x_k [y_l - (sx)_{l \setminus k}] \right)^2 + O(L^{-3/2}) \right) \qquad (5.171)$$

The key observation is that the distributions $\{ \pi_{lk'}^t(x_{k'}) \}$ affects the LHS of (5.163) only through the distribution of $(sX)_{l \setminus k}$, which can be regarded as a Gaussian random variable in the large-system regime due to the central-limit theorem, since it is a weighted sum of the independent random variables $X_{k'} \sim \pi_{lk'}^t$; $k' \neq k$. The mean and variance are:

$$\mu_{lk}^t = \frac{1}{\sqrt{L}} \sum_{k' \neq k} \sqrt{\gamma_{k'}} s_{lk'} m_{lk'}^t \qquad (5.172)$$

and:

$$C_{lk}^t = \frac{1}{L} \sum_{k' \neq k} \gamma_{k'} s_{lk'}^2 V_{lk'}^t \qquad (5.173)$$

respectively, where m_{lk}^t and V_{lk}^t are the mean and the variance of $X_k \sim \pi_{lk}^t$. Note that μ_{lk}^t is an estimate, based on $\{ \pi_{lk'}^t \}$, of the MAI component in y_l for user k, and that C_{lk}^t represents uncertainty of the estimate, quantifying magnitude of residual MAI component after the cancellation of MAI with μ_{lk}^t. Retaining terms up to order of L^{-1}, and calculating the Gaussian integral, the horizontal step can be represented as follows:

$$\rho_{lk}^{t+1}(x_k) \propto \exp\left[\sqrt{\frac{\gamma_k}{L}} \frac{s_{lk}(y_l - \mu_{lk}^t)}{\sigma^2 + C_{lk}^t} x_k - \frac{\gamma_k s_{lk}^2}{2L(\sigma^2 + C_{lk}^t)} x_k^2 \right] \qquad (5.174)$$

Introducing the parametrization:

$$\rho_{lk}^t(x_k) \propto \exp\left[\sqrt{\frac{\gamma_k}{L}}\theta_{lk}^t x_k - \frac{\gamma_k}{2L}\Xi_{lk}^t x_k^2\right] \tag{5.175}$$

the horizontal and vertical steps are represented as:

$$\theta_{lk}^{t+1} = \frac{s_{lk}(y_l - \mu_{lk}^t)}{\sigma^2 + C_{lk}^t} \quad \text{and} \quad \Xi_{lk}^{t+1} = \frac{s_{lk}^2}{\sigma^2 + C_{lk}^t} \tag{5.176}$$

and:

$$\pi_{lk}^t(x_k) = \alpha_{lk}q_X(x_k)\exp\left[\sqrt{\gamma_k}x_k\left(\frac{1}{\sqrt{L}}\sum_{l'\neq l}\theta_{l'k}^t\right) - \frac{\gamma_k x_k^2}{2}\left(\frac{1}{L}\sum_{l'\neq l}\Xi_{l'k}^t\right)\right] \tag{5.177}$$

respectively. The mean m_{lk}^t and the variance V_{lk}^t are to be calculated from $\pi_{lk}^t(x_k)$. Equations (5.172), (5.173), (5.176), (5.177) define an approximate BP algorithm for user detection. This algorithm has polynomial-order computational complexity per iteration, as opposed to the exponential-order complexity of the original BP. The final results are to be read out, after convergence is achieved, as statistics of the distributions:

$$\pi_k(x_k) = \alpha_k q_X(x_k)\exp\left[\sqrt{\gamma_k}x_k\left(\frac{1}{\sqrt{L}}\sum_{l=1}^{L}\theta_{lk}^*\right) - \frac{\gamma_k x_k^2}{2}\left(\frac{1}{L}\sum_{l=1}^{L}\Xi_{lk}^*\right)\right] \tag{5.178}$$

where θ_{lk}^* and Ξ_{lk}^* denote the respective quantities at the equilibrium.

One might ask how well the algorithm performs. The key observations to the question are that one can regard $\pi_{lk}^t(x_k)$, as given by (5.177), as a posterior distribution with the prior q_X and the Gaussian channel $q_{Z|X,\gamma,\xi}(z_{lk}^t|x_k, \gamma_k; \xi_{lk}^t)$ [see (5.40)] with appropriately chosen parameters:

$$\xi_{lk}^t = \frac{1}{L}\sum_{l'\neq l}\Xi_{l'k}^t \quad \text{and} \quad z_{lk}^t = \frac{1}{\xi_{lk}^t\sqrt{L}}\sum_{l'\neq l}\theta_{l'k}^t \tag{5.179}$$

and that one can apply the density evolution [90] idea to analyze macroscopic dynamical behaviors of the BP-based algorithm [89,91], which is motivated by its great success in the analysis of BP-based decoding algorithms of sparse-graph codes. Basically, the density evolution describes time evolution of the distributions of the "messages" $(\theta_{lk}^t, \Xi_{lk}^t)$ and (m_{lk}^t, V_{lk}^t). When applied to the approximate BP algorithm introduced above, it turns out that the distributions of z_{lk}^t and ξ_{lk}^t are relevant. Relying on a heuristic argument (which can be justified only in case of sparse spreading [50,51]), one finds, under random spreading and in the large-system limit, that ξ_{lk}^t becomes deterministic and independent of l or k, and that:

$$(\xi^{t+1})^{-1} = \sigma^2 + \beta E\{\gamma V^t\} \tag{5.180}$$

holds, where $\mathcal{V}^t \approx V^t_{lk}$ denotes the variance of π^t_{lk}, and where we dropped the indexes lk from ξ^t_{lk} due to the asymptotic independence. As for z^t_{lk}, one can regard it as following a zero-mean Gaussian distribution, and the variance, denoted here by $(\eta^t)^{-1}$, turns out to satisfy:

$$(\eta^{t+1})^{-1} = 1 + \beta \mathsf{E}\{\gamma \mathcal{E}^t\} \tag{5.181}$$

where $\mathcal{E}^t$ denotes the mean squared error of the estimate m^t_{lk}. Comparing the density evolution formulas (5.180) and (5.181) with the fixed-point equations of the replica analysis (5.46), one observes that stationarity condition of the density evolution formulas coincide with the fixed-point equations for arbitrary inputs, which has been proven only in case of sparse spreading [51], as is pointed out in Sections 5.1.3 and 5.3.4. On the theoretical side, the coincidence suggests an interesting and not yet fully understood link between the replica analysis and the BP-based algorithm. As for the application side, on the other hand, it suggests that, under the random spreading, the BP-based algorithm performs as predicted by the replica analysis in the large-system limit, and thus can be "asymptotically optimal."

5.8 CONCLUDING REMARKS

This chapter presents a simple characterization of the large-system performance of multiuser detection under arbitrary input and SNR distribution (and/or flat fading). A broad family of multiuser detectors is studied under the umbrella of posterior mean estimators, which includes well-known detectors such as the matched filter, decorrelator, linear MMSE detector, maximum likelihood (jointly optimal) detector, and the individually optimal detector.

A key conclusion is the decoupling of a multiuser channel concatenated with a generic multiuser detector front end. It is found that the detection output for each user is a deterministic function of a "hidden" Gaussian statistic centered at the transmitted symbol. Hence the single-user channel seen at the multiuser detection output is equivalent to a Gaussian channel conditioned on the input SNR in which the overall effect of MAI is a degradation in the effective SNR. The degradation factor, known as the multiuser efficiency, is the solution to a pair of coupled fixed-point equations, and can be easily computed numerically if not analytically.

Another set of results, tightly related to the decoupling principle, lead to general formulas for the large-system spectral efficiency of multiuser channels expressed in terms of the multiuser efficiency, both under joint and separate decoding.

Turning to algorithmic issues, the chapter also discusses application of belief propagation. It is shown that the conventional parallel interference cancellation is an approximate BP, and that more systematic approximation leads to a variant of PIC, whose performance is expected to be asymptotically optimal in the large-system limit. The suggested relation between density evolution formulas of the algorithm and the fixed-point equations obtained by the replica analysis, which has not been fully explored yet, might be of help in further interpreting the replica results.

From a practical viewpoint, this chapter presents results on the efficiency of CDMA communication under arbitrary user powers and input signaling such as PSK and QAM. The results in this chapter allow the performance of multiuser detection to be characterized by a single parameter, the multiuser efficiency. Thus, the results offer convenient performance measures and valuable insights in the design and analysis of multiuser systems, e.g., in power control [92].

The linear system in our study also models MIMO channels under various circumstances. The results can thus be used to evaluate the output SINR or spectral efficiency of high-dimensional MIMO channels (such as multiple-antenna systems) with arbitrary signaling and various detection techniques. Some of the results in this chapter have been generalized to MIMO channels with spatial correlation at both transmitter and receiver sides [93], as well as to MIMO-CDMA channels [94].

ACKNOWLEDGMENTS

We are grateful to the editor and anonymous reviewer for their helpful comments. Portions of the results included in this chapter originated during DG's Ph.D. study under the advice of Professor Sergio Verdú, to whom DG is very grateful. DG would also like to thank Dr. Chih-Chun Wang for useful discussions. TT would like to thank Mr. Keigo Takeuchi of Kyoto University, for his helpful discussion and constructive comments on the drafts, and Professor Mihai Putinar, University of California at Santa Barbara, for providing information about recent development of the moment problem.

This work is supported in part by the National Science Foundation CAREER Award CCF-0644344, the DARPA IT-MANET Grant W911NF-07-1-0028, and Grant-in-Aid for Scientific Research on Priority Areas (Nos. 14084209 and 18079010), MEXT, Japan.

REFERENCES

1. D. Horwood and R. Gagliardi. "Signal design for digital multiple access communications," *IEEE Trans. Commun.*, vol. 23, pp. 985–995, May 1975.

2. K. S. Schneider. "Optimum detection of code division multiplexed signals," *IEEE Transactions on Aerospace and Electronic Systems*, vol. 15, no. 1, pp. 181–185, Jan. 1979.

3. S. Verdú. "Computational complexity of optimum multiuser detection," *Algorithmica*, vol. 4, no. 3, pp. 303–312, 1989.

4. S. Verdú. "Minimum probability of error for asynchronous Gaussian multiple-access channels," *IEEE Trans. Inform. Theory*, vol. 32, no. 1, pp. 85–96, Jan. 1986.

5. M. Opper and W. Kinzel. "Statistical mechanics of generalization," in *Models of Neural Networks III: Association, Generalization, and Representation*, E. Domany, J. L. van Hemmen, and K. Schulten, Eds. Springer, 1996.

6. S. Verdú. "Minimum probability of error for asynchronous multiple access communication systems," in *Proc. IEEE Military Communications Conference*, vol. 1, Nov. 1983, pp. 213–219.

7. J. S. Lehnert and M. B. Pursley. "Error probabilities for binary direct-sequence spread-spectrum communications with random signature sequences," *IEEE Trans. Commun.*, vol. 35, pp. 87–98, Jan. 1987.

8. D. Guo, L. K. Rasmussen, and T. J. Lim. "Linear parallel interference cancellation in long-code CDMA multiuser detection," *IEEE J. Selected Areas Commun.*, vol. 17, pp. 2074–2081, Dec. 1999.

9. U. Madhow and M. L. Honig. "On the average near-far resistance for MMSE detection of direct sequence CDMA signals with random spreading," *IEEE Trans. Inform. Theory*, vol. 45, no. 6, pp. 2039–2045, Sep. 1999.

10. M. B. Pursley. "Performance evaluation for phase-coded spread-spectrum multiple-access communication—Part I: System analysis," *IEEE Trans. Commun.*, vol. 25, no. 8, pp. 795–799, Aug. 1977.

11. D. N. C. Tse and S. V. Hanly. "Linear multiuser receivers: Effective interference, effective bandwidth and user capacity," *IEEE Trans. Inform. Theory*, vol. 45, no. 2, pp. 641–657, Mar. 1999.

12. D. Guo, S. Verdú, and L. K. Rasmussen. "Asymptotic normality of linear multiuser receiver outputs," *IEEE Trans. Inform. Theory*, vol. 48, no. 12, pp. 3080–3095, Dec. 2002.

13. Z. D. Bai. "Methodologies in spectral analysis of large dimensional random matrices, a review," *Statistica Sinica*, vol. 9, no. 3, pp. 611–677, Jul. 1999.

14. A. M. Tulino and S. Verdú. "Random matrix theory and wireless communications," *Foundations and Trends in Communications and Information Theory*, vol. 1, no. 1, pp. 1–182, 2004.

15. S. Verdú. *Multiuser Detection*. Cambridge University Press, 1998.

16. Y. C. Eldar and A. M. Chan. "On the asymptotic performance of the decorrelator," *IEEE Trans. Inform. Theory*, vol. 49, no. 9, pp. 2309–2313, Sep. 2003.

17. D. Guo and S. Verdú. "Multiuser detection and statistical mechanics," in *Communications, Information and Network Security*, V. Bhargava, H. V. Poor, V. Tarokh, and S. Yoon, Eds. Kluwer Academic Publishers, 2002, ch. 13, pp. 229–277.

18. S. Verdú and S. Shamai. "Spectral efficiency of CDMA with random spreading," *IEEE Trans. Inform. Theory*, vol. 45, no. 2, pp. 622–640, Mar. 1999.

19. S. Shamai and S. Verdú. "The impact of frequency-flat fading on the spectral efficiency of CDMA," *IEEE Trans. Inform. Theory*, vol. 47, no. 4, pp. 1302–1327, May 2001.

20. A. J. Grant and P. D. Alexander. "Random sequence multisets for synchronous code-division multiple-access channels," *IEEE Trans. Inform. Theory*, vol. 44, no. 7, pp. 2832–2836, Nov. 1998.

21. P. B. Rapajic and D. Popescu. "Information capacity of a random signature multiple-input multiple-output channel," *IEEE Trans. Commun.*, vol. 48, no. 8, pp. 1245–1248, Aug. 2000.

22. T. Tanaka. "Analysis of bit error probability of direct-sequence CDMA multiuser demodulators," in *Advances in Neural Information Processing Systems*, T. K. Leen et al., Ed. The MIT Press, 2001, vol. 13, pp. 315–321.

23. T. Tanaka. "Average-case analysis of multiuser detectors," in *Proc. IEEE Int. Symp. Inform. Theory*. Washington, D.C. USA, Jun. 2001, p. 287.

24. T. Tanaka. "Performance analysis of neural CDMA multiuser detector," in *Proc. INNS-IEEE International Joint Conference on Neural Networks*. Washington, DC, USA, Jul. 2001, pp. 2832–2837.

25. T. Tanaka. "Statistical mechanics of CDMA multiuser demodulation," *Europhysics Letters*, vol. 54, no. 4, pp. 540–546, 2001.

26. T. Tanaka. "A statistical mechanics approach to large-system analysis of CDMA multiuser detectors," *IEEE Trans. Inform. Theory*, vol. 48, no. 11, pp. 2888–2910, Nov. 2002.

27. D. Guo and S. Verdú. "Randomly spread CDMA: Asymptotics via statistical physics," *IEEE Trans. Inform. Theory*, vol. 51, no. 6, pp. 1982–2010, Jun. 2005.

28. R. R. Müller and W. H. Gerstacker. "On the capacity loss due to separation of detection and decoding," *IEEE Trans. Inform. Theory*, vol. 50, no. 8, pp. 1769–1778, Aug. 2004.

29. R. R. Müller. "On channel capacity, uncoded error probability, ML-detection and spin glasses," in *Proc. Workshop on Concepts in Information Theory*. Breisach, Germany, 2002, pp. 79–81.

30. D. Guo and S. Verdú. "Replica analysis of CDMA spectral efficiency," in *Proc. IEEE Inform. Theory Workshop*. Paris, France, 2003.

31. D. Guo. "Gaussian channels: Information, estimation and multiuser detection," Ph.D. dissertation, Department of Electrical Engineering, Princeton University, 2004.

32. T. Tanaka. "Replica analysis of performance loss due to separation of detection and decoding in CDMA channels," in *Proc. IEEE Int. Symp. Inform. Theory*. Seattle, WA, USA, 2006, pp. 2368–2372.

33. D. Guo. "Performance of synchronous multirate CDMA via statistical physics," in *Proc. IEEE Int. Symp. Information Theory*. Adelaide, Australia, Sep. 2005.

34. D. Guo. "Performance of multicarrier CDMA in frequency-selective fading via statistical physics," *IEEE Trans. Inform. Theory*, vol. 52, no. 4, pp. 1765–1774, Apr. 2006.

35. S. F. Edwards and P. W. Anderson. "Theory of spin glasses," *Journal of Physics F: Metal Physics*, vol. 5, pp. 965–974, 1975.

36. M. Mézard, G. Parisi, and M. A. Virasoro. *Spin Glass Theory and Beyond*. World Scientific, 1987.

37. H. Nishimori. *Statistical Physics of Spin Glasses and Information Processing: An Introduction*, ser. Number 111 in International Series of Monographs on Physics. Oxford University Press, 2001.

38. N. Sourlas. "Spin-glass models as error-correcting codes," *Nature*, vol. 339, no. 6227, pp. 693–695, Jun. 1989.

39. Y. Kabashima and D. Saad. "Statistical mechanics of error-correcting codes," *Europhysics Letters*, vol. 45, no. 1, pp. 97–103, 1999.

40. A. Montanari. "Turbo codes: The phase transition," *European Physical Journal B*, vol. 18, pp. 121–136, 2000.

41. T. Murayama, Y. Kabashima, D. Saad, and R. Vicente. "Statistical physics of regular low-density parity-check error-correcting codes," *Physical Review E*, vol. 62, no. 2, pp. 1577–1591, 2000.

42. A. Montanari and N. Sourlas. "The statistical mechanics of turbo codes," *European Physical Journal B*, vol. 18, no. 1, pp. 107–119, 2000.

43. Y. Kabashima and D. Saad. "Statistical mechanics of low-density parity-check codes," *J. Phys. A: Math. Gen.*, vol. 37, pp. R1–R43, 2004.

44. A. L. Moustakas, S. H. Simon, and A. M. Sengupta. "MIMO capacity through correlated channels in the presence of correlated interferers and noise: A (not so) large N analysis," *IEEE Trans. Inform. Theory*, vol. 49, no. 10, pp. 2545–2561, Oct. 2003.

45. R. R. Müller. "Channel capacity and minimum probability of error in large dual antenna array systems with binary modulation," *IEEE Trans. Signal Processing*, vol. 51, pp. 2821–2828, Nov. 2003.

46. Y. Kabashima. "A CDMA multiuser detection algorithm on the basis of belief propagation," *J. Phys. A: Math. Gen.*, vol. 36, 11111–11121, 2003.

47. T. Fabricius and O. Winther. "Correcting the bias of subtractive interference cancellation in CDMA: Advanced mean field theory," Informatics and Mathematical Modelling, Technical University of Denmark, Tech. Rep., 2003.

48. M. Talagrand. "Rigorous results for mean field models for spin glasses," *Theoretical Computer Science*, vol. 265, pp. 69–77, Aug. 2001.

49. M. Talagrand. *Spin Glasses: A Challenge for Mathematicians.* Springer, 2003.

50. A. Montanari and D. Tse. "Analysis of belief propagation for non-linear problems: The example of CDMA (or: How to prove Tanaka's formula)," in *Proc. IEEE Inform. Theory Workshop.* Punta del Este, Uruguay, Mar. 2006, pp. 122–126.

51. D. Guo and C.-C. Wang. "Multiuser detection of sparsely spread CDMA," *IEEE J. Select. Areas Commun., Special Issue on Multiuser Detection for Advanced Communication Systems and Networks*, vol. 26, pp. 412–431, April 2008.

52. D. Guo and C.-C. Wang. "Random sparse linear systems observed via arbitrary channels: A decoupling principle," in *Proc. IEEE Int. Symp. Inform. Theory.* Nice, France, Jun. 2007.

53. J. Pearl. *Probabilistic Reasoning in Intelligent Systems: Networks of Plausible Inference.* Morgan Kaufmann, 1988, revised 2nd printing.

54. M. Mézard and G. Parisi. "Replicas and optimization," *J. Physique Lett.*, vol. 46, no. 17, pp. L–771–L–778, Sep. 1985.

55. Y. Fu and P. W. Anderson. "Application of statistical mechanics to NP-complete problems in combinatorial optimisation," *J. Phys. A: Math. Gen.*, vol. 19, no. 9, pp. 1605–1620, Jun. 1986.

56. D. J. Amit, H. Gutfreund, and H. Sompolinsky. "Storing infinite numbers of patterns in a spin-glass model of neural networks," *Phys. Rev. Lett.*, vol. 55, no. 14, pp. 1530–1533, Sep. 1985.

57. M. Mézard and A. Montanari. *Information, Physics and Computation.* Oxford Univ. Press, 2008.

58. R. Richardson and R. Urbanke. *Modern Coding Theory.* Cambridge Univ. Press, 2008.

59. P. Patel and J. Holtzman. "Analysis of simple successive interference cancellation scheme in a DS/CDMA," *IEEE J. Select. Areas Commun.*, vol. 12, pp. 796–807, Jun. 1994.

60. L. K. Rasmussen, T. J. Lim, and A.-L. Johansson. "A matrix-algebraic approach to successive interference cancellation in CDMA," *IEEE Trans. Commun.*, vol. 48, no. 1, pp. 145–151, Jan. 2000.

61. M. K. Varanasi and B. Aazhang. "Multistage detection in asynchronous code-division multiple-access communications," *IEEE Trans. Commun.*, vol. 38, pp. 509–519, Apr. 1990.

62. D. Guo, L. K. Rasmussen, S. Sun, and T. J. Lim. "A matrix-algebraic approach to linear parallel interference cancellation in CDMA," *IEEE Trans. Commun.*, vol. 48, pp. 152–161, Jan. 2000.

63. T. M. Cover and J. A. Thomas. *Elements of Information Theory*, 2nd ed. Wiley, 2006.

64. D. Guo, S. Shamai, and S. Verdú. "Mutual information and minimum mean-square error in Gaussian channels," *IEEE Trans. Inform. Theory*, vol. 51, no. 4, pp. 1261–1282, Apr. 2005.

65. M. K. Varanasi and T. Guess. "Optimum decision feedback multiuser equalization with successive decoding achieves the total capacity of the Gaussian multiple-access channel,"

in *Proc. Asilomar Conf. on Signals, Systems and Computers.* Monterey, CA, USA, Nov. 1997, pp. 1405–1409.

66. P. Rapajic, M. Honig, and G. Woodward. "Multiuser decision-feedback detection: Performance bounds and adaptive algorithms," in *Proc. IEEE Int. Symp. Inform. Theory.* Cambridge, MA, USA, Aug. 1998, p. 34.

67. S. L. Ariyavisitakul. "Turbo space-time processing to improve wireless channel capacity," *IEEE Trans. Commun.*, vol. 48, no. 8, pp. 1347–1359, Aug. 2000.

68. R. R. Müller. "Multiuser receivers for randomly spread signals: Fundamental limits with and without decision-feedback," *IEEE Trans. Inform. Theory*, vol. 47, no. 1, pp. 268–283, Jan. 2001.

69. T. Guess and M. K. Varanasi. "An information-theoretic framework for deriving canonical decision-feedback receivers in Gaussian channels," *IEEE Trans. Inform. Theory*, vol. 51, no. 1, pp. 173–187, Jan. 2005.

70. G. D. Forney, Jr. "Shannon meets Wiener II: On MMSE estimation in successive decoding schemes," in *Proc. 42nd Allerton Conf. Commun., Control, and Computing.* Monticello, IL, USA, 2004. [Online]. Available: arXiv:cs/0409011 [cs.IT]

71. D. Ruelle. *Statistical Mechanics: Rigorous Results.* Imperial College Press and World Scientific Publishing, 1999.

72. M. Yoshida and T. Tanaka. "Analysis of sparsely-spread CDMA via statistical mechanics," in *Proc. IEEE Int. Symp. Inform. Theory.* Seattle, WA, USA, 2006, pp. 2378–2382.

73. J. Raymond and D. Saad. "Sparsely spread CDMA – a statistical mechanics-based analysis," *J. Phys. A: Math. Theor.*, vol. 40, pp. 12,315–12,333, 2007.

74. S. B. Korada and N. Macris. "On the concentration of the capacity for a code division multiple access system," in *Proc. IEEE Int. Symp. Inform. Theory.* Nice, France, Jun. 2007.

75. T. Tanaka. "Moment problem in replica method," *Interdisciplinary Information Sciences*, vol. 13, no. 1, pp. 17–23, 2007.

76. J. Hubbard. "Calculation of partition functions," *Physics Review Letters*, vol. 3, no. 2, pp. 77–78, 1959.

77. R. S. Ellis. *Entropy, Large Deviations, and Statistical Mechanics*, ser. A series of comprehensive studies in mathematics. Springer-Verlag, 1985, vol. 271.

78. K. Nakamura and T. Tanaka. "Microscopic analysis for decoupling principle of linear vector channel," in *Proc. IEEE Int. Symp. Inform. Theory.* Toronto, Canada, pp. 519–523, July 2008.

79. W. Feller. *An Introduction to Probability Theory and Its Applications*, 2nd ed., John Wiley & Sons, Inc., 1971, vol. II.

80. N. I. Akhiezer. *The Classical Moment Problem.* Oliver and Boyd Ltd., 1965.

81. L. C. Petersen. "On the relation between the multidimensional moment problem and the one-dimensional moment problem," *Math. Scand.*, vol. 51, pp. 361–366, 1982.

82. B. Fuglede. "The multidimensional moment problem," *Expo. Math.*, vol. 1, pp. 47–65, 1983.

83. J. R. L. de Almeida and D. J. Thouless. "Stability of the Sherrington-Kirkpatrick solution of a spin glass model," *Journal of Physics A: Mathematical and Physical*, vol. 11, pp. 983–990, 1978.

84. T. Uezu, M. Yoshida, T. Tanaka, and M. Okada. "Statistical mechanical analysis of CDMA multiuser detectors—AT stability and entropy of the RS solution, and 1RSB solution," *Progress of Theoretical Physics Supplement*, vol. 157, pp. 254–257, 2005.

85. F. Comets. "The martingale method for mean-field disordered systems at high temperature," in Mathematical Aspects of Spin Glasses and Neural Networks, A. Bovier and P. Picco, Eds. Birkhäuser, 1998, pp. 91–113.

86. R. J. McEliece, D. J. C. MacKay, and J.-F. Cheng. "Turbo decoding as an instance of Pearl's belief propagation' algorithm," *IEEE J. Select. Areas Commun.*, vol. 16, no. 2, pp. 140–152, Feb. 1998.

87. D. J. C. MacKay. "Good error-correcting codes based on very sparse matrices," *IEEE Trans. Inform. Theory*, vol. 45, no. 2, pp. 399–431, Mar. 1999, errata, 47(5): 2101, July 2001.

88. T. Tanaka. "Density evolution for multistage CDMA multiuser detector," in *Proc. IEEE Int. Symp. Inform. Theory*. Lausanne, Switzerland, 2002, p. 23.

89. T. Tanaka and M. Okada. "Approximate belief propagation, density evolution, and statistical neurodynamics for CDMA multiuser detection," *IEEE Trans. Inform. Theory*, vol. 51, no. 2, pp. 700–706, Feb. 2005.

90. T. J. Richardson and R. L. Urbanke. "The capacity of low-density parity-check codes under message-passing decoding," *IEEE Trans. Inform. Theory*, vol. 47, pp. 599–618, Feb. 2001.

91. T. Ikehara and T. Tanaka. "Decoupling principle in belief-propagation-based CDMA multiuser detection algorithm," in *Proc. IEEE Int. Symp. Inform. Theory*. Nice, France, 2007, pp. 2081–2085.

92. F. Meshkati, D. Guo, H. V. Poor, and S. C. Schwartz. "A unified approach to energy-efficient power control in large CDMA systems," *IEEE Trans. Wireless Commun.*, vol. 7, pp. 1208–1216, April 2008.

93. C. K. Wen, P. Ting, and J.-T. Chen. "Asymptotic analysis of mimo wireless systems with spatial correlation at the receiver," *IEEE Trans. Commun.*, vol. 54, no. 2, pp. 349–363, Feb. 2006.

94. K. Takeuchi, T. Tanaka, and T. Yano. "Asymptotic analysis of general multiuser detectors in MIMO DS-CDMA channels," *IEEE J. Select. Areas Commun., Special Issue on Multiuser Detection for Advanced Communication Systems and Networks*, vol. 26, pp. 486–496, April 2008.

6

JOINT DETECTION FOR MULTI-ANTENNA CHANNELS

Antonia Tulino, Matthew R. McKay, Jeffrey G. Andrews,
Iain B. Collings, and Robert W. Heath, Jr.

6.1 INTRODUCTION

Multiuser and multiantenna communication systems have many similarities. This is particularly true when the transmit antenna array sends multiple independent data streams that interfere in time and frequency, a paradigm that is commonly referred to as multiple-input multiple-output (MIMO) or more specifically, spatial multiplexing. In the case of spatial multiplexing, each data stream must be separated and decoded at the receiver, which creates a scenario that is formally nearly identical to the case of jointly decoding multiple interfering users. Indeed, many of the theoretical advances in multiantenna receivers have closely followed previous advances in multiuser detectors. At the present time, innovation in both realms is synergistic since advances in either the spatial or user domain can be used to spur progress in the other.

The goal of this chapter is to provide a contemporary overview of joint detection techniques in MIMO systems and to generate insights into parallels with multiuser systems. We begin by developing a general matrix channel model for MIMO systems, and note how this model is applicable to many multiuser scenarios, as well as highlighting the areas of departure for single user MIMO versus multiuser single antenna systems.

We then review the key capacity results for multi-antenna systems under varying assumptions of the amount of channel knowledge and other practical constraints like antenna correlation. Next, we describe receiver architectures for MIMO systems with particular emphasis on linear detectors, i.e., zero-forcing (ZF) and minimum mean-square error (MMSE). We present fairly recent results that quantify the exact performance of such receivers. We also discuss the challenges posed by multiuser MIMO systems, where the interference can be either non-coherent (e.g., background interference from neighboring cells) or coherent (strong interference from within the same cell or nearby interfering base stations). Finally, we discuss how MIMO systems can balance raw throughput and reliability by using the spatial dimensions for both diversity and multiplexing; here we contrast recent theory results with practical switching schemes that are useful in practice.

6.2   WIRELESS CHANNELS: THE MULTI-ANTENNA REALM

A wireless system spans multiple complementary dimensions: time, frequency, space, and users. The system model used throughout this chapter is a general linear model that accommodates most single- and multiuser communication scenarios. This MIMO model is given by:

$$\mathbf{y} = \sqrt{g}\,\mathbf{H}\mathbf{x} + \mathbf{n} \tag{6.1}$$

where $\mathbf{x}$ is a K-dimensional vector of the signal input, $\mathbf{y}$ is the N-dimensional vector of the signal output, and the N-dimensional vector $\mathbf{n}$ is complex additive Gaussian noise, whose real and imaginary components are typically modelled as independent Gaussian random variables with zero mean and variance $\sigma^2/2$ (i.e., circularly distributed). The channel is described by $\mathbf{H}$, which is an $N \times K$ complex (possibly random) matrix while g is a deterministic scalar that represents the average channel gain.

Although simple, the model in (6.1) encompasses a wide variety of channels of interest in wireless communications, including multiple access channels, dispersive and selective linear channels in both the time and frequency domains, spatial channels, and other multidimensional signalling scenarios. In each of these cases, the values N, K and $\mathbf{H}$ assume different physical meanings. For example, in a synchronous direct-sequence code division multiple access (DS-CDMA) additive noise broadcast channel, K and N are respectively the number of users and the spreading gain while the columns of $\mathbf{H}$ are the spreading codes for the K users. Fading can easily be integrated by multiplying $\mathbf{H}$ by a diagonal matrix $\tilde{\mathbf{H}}_k = \operatorname{diag}(h_1 \ h_2 \ \cdots \ h_N)$ where h_n is the fading coefficient to the kth user for the nth coded symbol. Note that each user would have a unique noise vector $\mathbf{n}_k$ as well.

In a single-antenna and single user block transmission system over a broadband (time-dispersive/frequency-selective) channel, $\mathbf{H}$ is typically used to model the inter-symbol interference. In this case, the K input symbols refer to the block length, and the $K + \nu$ output symbols are a result of convolution of an input vector of length K with a channel of $\nu + 1$ non-negligible taps. In orthogonal frequency division multiplexing (OFDM), a cyclic prefix is appended to the transmit vector to make $\mathbf{x}$ have length $K' > K + \nu$, which allows $\mathbf{H}$ to appear circulant and hence decomposable into a diagonal (interference-free) matrix with the Inverse Discrete Fourier Transform (IDFT) and Discrete Fourier Transform (DFT) matrices applied at the transmitter and receiver, respectively.

In the single-user narrowband channel with multiple antennas at the transmitter and receiver, respectively, we identify K and N with the number of transmit and receive antennas while $\mathbf{H}$ models the propagation coefficients between each pair of transmit and receive antennas.

Naturally, the same linear model can be adapted to problems that incorporate many of the above features in combination; in such cases, it may be more convenient to use matrices or higher-dimensional objects to represent the input and output quantities. For example, for a channel with J users each transmitting with n_T antennas using spread-spectrum with spreading gain G and a receiver with n_R antennas, $K = n_T J$ and $N = n_R G$. In short, the simple model $\mathbf{y} = \sqrt{g}\,\mathbf{Hx} + \mathbf{n}$ is surprisingly general and we will use variants of it throughout this chapter to demonstrate means for detecting $\mathbf{x}$ in a spatial context. Because of the generality of the model, particulary its relations to a CDMA signalling, it is clear that many of the techniques developed in the last decade by the MIMO community are directly applicable to multiuser systems in general and CDMA receivers in particular.

Although, fundamentally, the single-user multiantenna problem is analogous to that of an unfaded synchronous CDMA channel with random spreading [135] there are, nonetheless, some important differences.

1. Joint coding of the transmit signals is typically more feasible in single user multiantenna systems than in CDMA. Particulary in the uplink, it is not possible to coordinate the transmitted waveforms of CDMA users. In a single user multiantenna system, it is possible to advantageously correlate the transmitted waveforms.

2. The transmit power constraint applies to the sum of powers in the multiantenna case whereas in CDMA it applies to each individual user.

3. The spatial signatures for the various transmit antennas are imposed by the physics of the channel, whereas in a CDMA system the user sequences can be designed.

4. Accurate channel estimation is easier to achieve in a single $n_T \times n_R$ link than in a $K = n_T n_R$ user system with multiuser detection, despite these channels having the same dimensionality. This is primarily due to advantages from the antennas being co-located, which allows for efficient use of pilot signals and channel state feedback, and less variance in the channel dynamics.

Despite these differences, many of the results on random CDMA are very relevant to the multiantenna problem considered in this chapter, especially when the channel matrix $\mathbf{H}$ is unknown to the multiantenna transmitter. In the rest of the chapter, we will focus on a single-user channel where multiple transmit and receive antennas are employed, before revisiting the combination of multiple antennas and multiple users in Section 6.7.

6.3 DEFINITIONS AND PRELIMINARIES

Let us now elaborate on the model for a single-user MIMO channel where the transmitter has n_T antennas and the receiver has n_R antennas. The early contributions on this topic are [36,123], and we refer readers to [12,31,42,44,131] for recent articles of a more tutorial nature.

With reference to the general model in (6.1), $\mathbf{x}$ contains the symbols transmitted from the n_T transmit antennas and $\mathbf{y}$ the symbols received by the n_R receive antennas. The general model given in (6.1) becomes:

$$\mathbf{y} = \sqrt{g}\mathbf{H}\mathbf{x} + \mathbf{n} \tag{6.2}$$

with $\mathbf{H}$ denoting the channel matrix, whose entries represent the fading coefficients between each transmit and each receive antenna, typically modelled as zero-mean complex Gaussian random variables. The scalar factor g, is defined such that:

$$g\mathbb{E}[\text{tr}(\mathbf{H}\mathbf{H}^\dagger)] = n_T n_R \tag{6.3}$$

The average signal-to-noise ratio (SNR) observed over the n_R receive antennas, with the condition in (6.3) and in (6.5):

$$\text{SNR} = g\frac{\mathbb{E}\left[\|\mathbf{x}\|^2\right]}{\frac{1}{n_R}\mathbb{E}\left[\|\mathbf{n}\|^2\right]} \tag{6.4}$$

Thus any difference in average power gain is factored out of $\mathbf{H}$ and absorbed into the SNR through g. In contrast with the multiaccess scenarios, in this case the signals transmitted by different antennas can be advantageously correlated and thus the covariance of $\mathbf{x}$ becomes relevant. Normalized by its energy per dimension, the input covariance is denoted by:

$$\mathbf{\Phi_x} = \frac{\mathbb{E}[\mathbf{x}\mathbf{x}^\dagger]}{\frac{1}{n_T}\mathbb{E}\left[\|\mathbf{x}\|^2\right]} \tag{6.5}$$

where the normalization ensures that $\mathbb{E}[\text{tr}\{\mathbf{\Phi_x}\}] = n_T$. We assume that the receiver is aware of the channel realization, i.e., coherent reception.

It is worth mentioning that particularly in the context of a multiuser system, this is a very significant assumption. In fact, obtaining accurate channel knowledge for the K interfering users is in many cases the most important impediment to the viability of most multiuser detectors [6,8]. However, it should be noted that channel estimation is considerably easier in a single user $n_T \times n_R$ antenna system than for a K user single-antenna system. This is primarily due to difficulties in measuring K distributed channels that are by definition at very low signal to interference plus noise ratio (SINR) and generally heavily dependent on coding; if the multiuser signals are not at low

SINR then the system is either underloaded or the spreading factor is very small. On the other hand, it is considerably easier to coordinate high-fidelity multiantenna channel estimation by using time-frequency-space orthogonal pilot patterns at the transmitter [50,147]. Therefore, the assumption of perfect channel state information at the receiver is at least somewhat realistic in a multiantenna system.

In the next section, we focus on multiantenna channel *capacity* as the basic performance measure. The capacity provides the maximum fundamental limit of error-free communication rate achievable by any scheme. For the fading channel, there are several capacity measures, relevant for different scenarios. Two distinct scenarios provide particular insight:

- the *slow fading* channel (nonergodic channel), where the channel stays constant (at a random value) over the entire time scale of communication, e.g., for a packet-time.
- the *fast fading* channel (ergodic channel), where the channel varies significantly over the time scale of communication.

In the slow fading channel, the key event of interest is outage: this is the situation where the channel is so poor that no scheme can communicate reliably at a certain target data rate. The largest rate for reliable communication at a certain outage probability is called the *outage capacity*; a coding scheme that achieves the outage capacity is said to be *universal*, since it communicates reliably over all slow fading channels that are not in outage.

In the fast fading channel, outages can be avoided due to the ability to average over the time variations of the channel, e.g., with coding and interleaving or adaptive modulation. In this case, the *ergodic capacity* is of interest. This is the maximum average rate at which arbitrarily reliable communication is possible.

6.4   MULTI-ANTENNA CAPACITY: ERGODIC REGIME

In this section, as well as in Section 6.5, we catalog and elaborate on a number of results in the multiantenna capacity literature. Specifically, we address several results on the optimum input distribution and on the impact of model features such as antenna correlation on multiantenna capacity. Particular attention is given to the large-dimensional regime (as the number of transmit and receive antennas go to infinity). In this section, we focus on the ergodic channels, while in Section 6.5, we analyze the non-ergodic regime. Both sections, however, deal with coherent communication.

In the coherent regime, where the channel realization is available at the receiver, if the input $\mathbf{x}$ in (6.1) is circularly symmetric complex Gaussian, the mutual information conditioned on the channel matrix is:

$$\mathcal{I}(\text{SNR}, \boldsymbol{\Phi_x}) = \log \det\left(\mathbf{I} + \frac{\text{SNR}}{n_{\text{T}}} \mathbf{H}\boldsymbol{\Phi_x}\mathbf{H}^{\dagger}\right) \tag{6.6}$$

where $\mathbf{I}$ is the identity matrix. The formula (6.6) has been derived specifically for multiantenna channels in [37,123]. However, previous appearances of this celebrated log-determinant formula and its generalizations can be found in [15,28,99,125,134].

Most information-theoretic studies of multiantenna channels have focused on the ergodic regime, in which during transmission a long enough code word experiences essentially all states of the channel, and hence it averages out the channel randomness. We defer until Section 6.5 the examination of non-ergodic channels.

In the ergodic regime, the fundamental operational limit is the mutual information averaged over the fading coefficients, i.e.:

$$\mathbb{E}[\mathcal{I}(\text{SNR}, \mathbf{\Phi_x})] = \mathbb{E}\left[\log\det\left(\mathbf{I} + \frac{\text{SNR}}{n_{\text{T}}}\mathbf{H}\mathbf{\Phi_x}\mathbf{H}^\dagger\right)\right] \qquad (6.7)$$

In order to achieve capacity, the input covariance $\mathbf{\Phi_x}$ must be properly determined depending on the channel-state information (CSI) available to the transmitter. In this respect, there are three main regimes of interest:

- The transmitter has full CSI, i.e., access to $\mathbf{H}$ instantaneously.
- The transmitter has only statistical CSI, i.e., access to the distribution of $\mathbf{H}$ but not to its realization.
- The transmitter has no CSI whatsoever.

6.4.1 Input Optimization and Capacity-Achieving Transceiver Architectures

For the optimization of the input in each of the aforementioned regimes, it is convenient to decompose this input covariance into its eigen-vectors and eigen-values, $\mathbf{\Phi} = \mathbf{V}\mathbf{P}\mathbf{V}^\dagger$. Each eigen-value represents the (normalized) power allocated to the corresponding signalling eigen-vector.

If the channel realization is known at the transmitter, the capacity is achieved with an input covariance whose eigen-vector matrix, $\mathbf{V}$, coincides with that of $\mathbf{H}^\dagger\mathbf{H}$. For the eigen-value matrix, $\mathbf{P}$, which of course will be a function of the channel realizations, several solutions have been obtained depending on the time horizon over which the power is allowed to be averaged and the temporal dynamics of the problem. Specifically, if the power constrain is on a per channel-use basis (i.e., $\sum_{j=1}^{n_{\text{T}}}\mathsf{P}_j(\mathbf{H}) = n_{\text{T}}$) then $\mathbf{P}$ is obtained via a waterfill process on the eigen-values of $\mathbf{H}^\dagger\mathbf{H}$, thus the resulting jth diagonal entry of $\mathbf{P}$ is:

$$\mathsf{P}_j(\mathbf{H}) = \left(\nu - \frac{1}{\text{SNR}\,\lambda_j^2(\mathbf{H})}\right)^+ \qquad (6.8)$$

where $\lambda_j(\mathbf{H})$ denotes the j-th singular value of $\mathbf{H}$, and the cutoff value ν is chosen such that $\sum_{j=1}^{n_{\text{T}}}\mathsf{P}_j(\mathbf{H}) = n_{\text{T}}$. The waterfill policy devised by Shannon, in its original frequency-domain waterfill approach [112], was formalized in the 1960s [125] for deterministic matrix channels and later rederived by several authors in the 1990s, specifically for multiantenna communication [102,123].

If the power constraint applies over a time interval long enough for the fading distribution to be revealed (*long-term power constraint*, i.e., $\mathbb{E}[\sum_{j=1}^{n_T} \mathsf{P}_j(\mathbf{H})] = n_T$, with the expectation over the channel distribution), then [45] shows how ergodic capacity is achieved by adapting the transmit power and data rate using a "time-water-filling" procedure that depends on the channel statistics only through the cutoff value ν chosen such that long-term power constraint is satisfied.

Finally if the power constraint applies over a limited time interval (*short-term power constraint*), such that the channel variations thereon do not fully reveal the fading distribution, then the optimal transmission strategy is realized via dynamic programming [94]. For the short-term constraint, it was observed in [94] that whereas optimizing the transmitted power does not significantly increase capacity at high values of SNR, it provides a substantial gain at low SNR.

Instantaneous CSI at the transmitter is an operational regime that applies, for example, to fixed wireless access systems (backhaul, local loop, broadband residential) and to low-mobility systems (local-area networks, pedestrians) and it is particularly appealing whenever uplink and downlink are reciprocal (time-duplexed systems) [29]. The capacity-achieving transceiver architecture for this regime follows naturally from the input optimization. Information bits are split into $t = \min\{n_T, n_R\}$ parallel streams, each coded separately, and then enlarged by $n_T - t$ streams of zeros. The symbols across the streams form the input vector $\mathbf{s} = [s_1, \ldots, s_{n_T}]^T$ which is rotated into the directions corresponding to the right singular vectors of the channel matrix $\mathbf{H}$. The power allocated to each stream is time-dependent and obtained, depending on the time horizon, via either temporal waterfilling or dynamic programming as discussed above, and the rates are dynamically allocated accordingly. At the receiver, the left singular vectors of the channel matrix $\mathbf{H}$ are used for reception. Specifically the output is post-multiplied by the matrix of left singular vectors of $\mathbf{H}$ to extract the independent streams, which are then separately decoded.

While with full CSI at the transmitter the coordinate system in which the independent data streams are multiplexed is obviously channel-dependent, with only statistical CSI available at the transmitter, in order to achieve capacity the choice of this coordinate system has to be fixed *a priori* and should be set, for several important classes of channels (Rayleigh fading with certain correlation structures and uncorrelated Ricean fading), to coincide with the eigen-vectors of $\mathbb{E}[\mathbf{H}^\dagger\mathbf{H}]$. (Note that the aforementioned principle fails to hold in general [2]). The capacity-achieving power allocation, $\mathbf{P}$, can be, instead, found iteratively [128]. This architecture, known in the literature as V-BLAST [37,38], consists of a space-time encoder that at each use of the channel outputs a set of independent unit-variance Gaussian symbols, $s_1, \ldots, s_{n_T}$, each of which, before being simultaneously radiated out of the n_T transmit antennas, is assigned a certain transmit power, $\{\mathsf{P}_1, \ldots, \mathsf{P}_{n_T}\}$ (which may be zero) and rotated into the directions of $\mathbf{V}$.

The earliest statement on the optimization of the input covariance statistical CSI available at the transmitter can be found in [123] where it was shown that, for channel matrices with zero-mean independent identically distributed (IID) Gaussian entries, the capacity-achieving input is isotropic. For the region of low SNR, a complete characterization of the limiting covariance, valid for arbitrary channels, was provided in [136]. Subsequent findings on the eigen-vectors of the capacity-achieving input covariance can be found in [57,62,75,106,139] for correlated multiantenna channels with zero-mean Gaussian entries and in [54,133,139] for channels with IID Gaussian entries with arbitrary mean. In every case, the eigen-values were left to numerical optimization.[1] A complete characterization of capacity-achieving input covariance has been addressed in [130]: introducing a general channel model, [130] unifies all previous results on the eigen-vectors of the input covariance and also addresses the characterization of the corresponding eigen-values.

Theorem 6.4.1 [130] *Consider the following channel:*

$$\mathbf{H} = \mathbf{U_R} \tilde{\mathbf{H}} \mathbf{U_T} \tag{6.9}$$

where $\mathbf{U_R}$ and $\mathbf{U_T}$ are $n_R \times n_R$ and $n_T \times n_T$ deterministic unitary matrices respectively and $\tilde{\mathbf{H}}$ is a $n_R \times n_T$ random matrix with independent columns the distribution of whose entries is jointly symmetric with respect to zero. Denoting by $\mathbf{\Phi} = \mathbf{VPV}^\dagger$ the capacity-achieving input covariance matrix, normalized such that $\mathrm{Tr}\{\mathbf{P}\} = n_T$, we have that:

- $\mathbf{V} = \mathbf{U_T}$
- *while the jth diagonal entry of $\mathbf{P}$ is:*

$$\mathsf{P}_j = 0 \qquad\qquad \frac{\mathrm{SNR}}{n_T} E\left[\tilde{\mathbf{h}}_j^\dagger \mathbf{B}_j \tilde{\mathbf{h}}_j\right] \le \frac{1}{n_T} \sum_{\ell=1}^{n_T} (1 - \overline{\mathrm{MMSE}}_\ell)$$

$$\mathsf{P}_j = \frac{1 - \overline{\mathrm{MMSE}}_j}{\frac{1}{n_T}\sum_{\ell=1}^{n_T}\left(1 - \overline{\mathrm{MMSE}}_\ell\right)} \qquad \text{otherwise} \tag{6.10}$$

where $\tilde{\mathbf{h}}_j$ denotes the j-th column of $\tilde{\mathbf{H}}$:

$$\mathbf{B}_j \triangleq \left(\mathbf{I} + \frac{\mathrm{SNR}}{n_T}\sum_{i \neq j}\tilde{\mathbf{h}}_i \mathsf{P}_i \tilde{\mathbf{h}}_i^\dagger\right)^{-1} \tag{6.11}$$

and, finally, $\overline{\mathrm{MMSE}}_j$ indicates the expectation with respect to $\tilde{\mathbf{H}}$, of:

$$\mathrm{MMSE}_j = \frac{1}{1 + p_j \dfrac{\mathrm{SNR}}{n_T}\tilde{\mathbf{h}}_j^\dagger \mathbf{B}_j^{-1} \tilde{\mathbf{h}}_j} \tag{6.12}$$

which is restricted to the interval [0,1].

---

[1] Some relationships between the eigen-values have been uncovered in [52] where a search procedure based on the Blahut-Arimoto algorithm has been given.

Because of the unitary-independent-unitary structure of (6.9), this model is refered to as the UIU model [129]. The UIU framework encompasses most of the zero-mean channel models that have been treated in the multiantenna literature and it is advocated and experimentally supported in [96,142]. Relevant special cases of (6.9) when $\tilde{\mathbf{H}}$ having independent zero-mean Gaussian entries include:

- canonical model [36,123] if $\tilde{\mathbf{H}}$ has IID Rayleigh-faded entries.
- separable correlation model [22,98,114] if the entries of $\tilde{\mathbf{H}}$ are Rayleigh-faded with variances that conform to a certain pattern (cf. Section 6.4.3.1). This model can be, equivalently, expressed as:

$$\mathbf{H} = \Theta_R^{1/2} \mathbf{W} \Theta_T^{1/2} \qquad (6.13)$$

where $\mathbf{W}$ is $n_R \times n_T$ IID Rayleigh-faded complex matrix while Θ_R and Θ_T are deterministic receive and transmit correlation matrices.

- polarization diversity and/or pattern diversity [10,76,120], if $\mathbf{U}_R$ and $\mathbf{U}_T$ are constrained to be identity matrices and $\tilde{\mathbf{H}}$ has independent nonidentically distributed (IND) entries.
- *virtual representation* [108] if $\mathbf{U}_R$ and $\mathbf{U}_T$ are Fourier matrices while $\tilde{\mathbf{H}}$ is IND Rayleigh-faded.

A notable exception that falls outside the representation in (6.9) is that of the keyhole (or pinhole) channel, where $\mathbf{H} = \mathbf{c}_R \mathbf{c}_T^\dagger$ with $\mathbf{c}_R$ and $\mathbf{c}_T$ being random column vectors. In the hypothesis that $\mathbf{c}_R$ and $\mathbf{c}_T$ have independent random entries whose distribution is symmetric with respect to zero, the capacity-achieving input covariance for this channel is diagonal (i.e., $\mathbf{V} = \mathbf{I}$) [130].

With no CSI, the most reasonable strategy is to transmit an isotropic signal ($\Phi = \mathbf{I}$) [97,141]. In fact, because of its simplicity and because many space-time coding schemes conform to it, this strategy may be appealing even if some degree of CSI is available.

Having analyzed the transceiver architectures that achieve the capacity, we now shift our focus towards the study of the ergodic capacity for various channel models of interest. In particular, we start with the very simple canonical model and proceed beyond it onto the so-called separable correlation model able to account for correlations between antennas at both transmitter and receiver.[2]

Note that the characterization of the capacity can be mathematically involved when correlations between channel entries are allowed and for some channel models, the capacity becomes analytically tractable only as the SNR approaches zero or infinity or else in the large-dimensional regime (as n_T and n_R and n_R go to infinity). Fortunately, the number of antennas required for the asymptotics to be closely

---

[2]Further results on the ergodic capacity of very general models, able to account for very broad range of correlations including indoor propagation environments [96] and several diversity mechanisms [10,76,84,120] like polarization diversity, pattern diversity etc, can be found in [129].

approached are usually very small and therefore the expressions derived therein are valid almost universally.

Accordingly, in the next sections we seek to characterize the ergodic capacity of several channels of interest in the coherent regime with and without CSI at the transmitter asymptotically in the number of antennas for arbitrary SNR, and for arbitrary numbers of antennas when it is possible.

6.4.2 Random Matrix Theory

Asymptotic analyses (as n_T and n_R go to infinity with a constant ratio) of the fundamental limits of wireless communication channels provide valuable engineering insights and are often feasible using results in random matrix theory [131]. Thus, before embarking on the study of capacity of several channel models, it is worth reviewing some definitions and results on random matrix theory from [131] that will be useful in the sequel.

6.4.2.1 Eigen-Value Distributions Given an $n \times n$ Hermitian matrix $\mathbf{A}$, the empirical cumulative distribution function (CDF) of the eigen-values (also referred to as the empirical spectral distribution (ESD)) of $\mathbf{A}$ is defined as:

$$\mathsf{F}_{\mathbf{A}}^n(x) = \frac{1}{n} \sum_{i=1}^{n} 1\{\lambda_i(\mathbf{A}) \le x\} \tag{6.14}$$

where $\lambda_1(\mathbf{A}), \ldots, \lambda_n(\mathbf{A})$ are the eigen-values of $\mathbf{A}$ and $1\{\cdot\}$ is the indicator function. If $\mathsf{F}_{\mathbf{A}}^n(\cdot)$ converges almost surely as $n \to \infty$, then the corresponding limit (asymptotic ESD) is denoted by $\mathsf{F}_{\mathbf{A}}(\cdot)$.[3]

Example 6.1 [83] Let $\mathbf{H}$ be an $n_R \times n_T$ random matrix whose entries are independent random variables with identical mean and variance. The asymptotic ESD of $\frac{1}{n_R}\mathbf{H}\mathbf{H}^\dagger$ has a density:

$$\tilde{\mathsf{f}}_\beta(x) = (1 - \beta)^+ \delta(x) + \frac{\sqrt{(x - a)^+ (b - x)^+}}{2\pi x} \tag{6.15}$$

where $\beta = \lim_{n_R \to \infty} \frac{n_T}{n_R}$, while:

$$a = (1 - \sqrt{\beta})^2, \qquad b = (1 + \sqrt{\beta})^2$$

6.4.2.2 Transforms For our purposes, it is advantageous to make use of the η-transform and the Shannon transform, which were motivated by the application of random matrix theory to various problems in the information theory of noisy communication channels [131]. These transforms, intimately related to each other

---

[3] A slight abuse of notation, customary in asymptotic random matrix theory, is to avoid a dimension subscript in $\mathbf{A}$, which, depending on the context, stands for a matrix of a given dimension or for a sequence of matrices.

and to the Stieltjes transform traditionally used in random matrix theory [122], characterize the spectrum of a random matrix while carrying certain engineering intuition, as explained in [131].

DEFINITION 6.4.2 [131] *Given a nonnegative definite random matrix* $\mathbf{A}$, *its η-transform is:*

$$\eta_{\mathbf{A}}(\gamma) = \mathbb{E}\left[\frac{1}{1 + \gamma X}\right] \tag{6.16}$$

where X is a nonnegative random variable whose distribution is the asymptotic ESD of $\mathbf{A}$ while γ is a nonnegative real number. Thus, $0 < \eta_X(\gamma) \leq 1$.

DEFINITION 6.4.3 [131] *Given a nonnegative definite random matrix* $\mathbf{A}$, *its Shannon transform is defined as:*

$$\mathcal{V}_{\mathbf{A}}(\gamma) = \mathbb{E}[\log(1 + \gamma X)] \tag{6.17}$$

where X is a nonnegative random variable whose distribution is the asymptotic ESD of $\mathbf{A}$ while γ is a nonnegative real number.

Note that the expected mutual information per receive antenna[4] can be rewritten as function of the ESD of $\mathbf{A} = \frac{1}{n_{\mathrm{T}}}\mathbf{H}\boldsymbol{\Phi}_x\mathbf{H}^{\dagger}$ via:

$$\frac{\mathbb{E}[I(\mathrm{SNR}, \boldsymbol{\Phi})]}{n_{\mathrm{R}}} = \int \log(1 + \mathrm{SNR}\,\xi)\,dF_{\mathbf{A}}^n(\xi) \tag{6.18}$$

which, as $n_{\mathrm{R}}, n_{\mathrm{T}} \to \infty$ while $\frac{n_{\mathrm{T}}}{n_{\mathrm{R}}} \to \beta$ converges to:

$$\frac{\mathbb{E}[I(\mathrm{SNR}, \boldsymbol{\Phi})]}{n_{\mathrm{R}}} \to \int \log(1 + \mathrm{SNR}\,\xi)\,dF_{\mathbf{A}}(\xi) \tag{6.19}$$

Thus according to (6.17) and (6.19):

$$\frac{\mathbb{E}[I(\mathrm{SNR}, \boldsymbol{\Phi})]}{n_{\mathrm{R}}} \to \mathcal{V}_{\mathbf{A}}(\mathrm{SNR}) \tag{6.20}$$

6.4.3 Canonical Model (IID Channel)

The earliest works by [36,123] started with the canonical model where the entries of the channel matrix $\mathbf{H}$ are modelled as IID zero-mean complex Gaussian random variables (all antennas are assumed be co-polarized).

---

[4]Since our analysis focus on both asymptotic (in the number of antennas) and non-asymptotic regime, it is more convenient to operate with the mutual information per receive antenna, sure to remain finite.

Following these pioneering works, the capacity with full transmit CSI for this canonical model is studied in [5,58]. In [58], in particular, an explicit expression is given although as function of a parameter that must be solved for numerically.

A closed-form expression of the capacity with full CSI can be however obtained asymptotically in the number of antennas [26,47,90,107]. In fact as n_R, $n_T \to \infty$ while $\frac{n_T}{n_R} \to \beta$:

$$\frac{C(\text{SNR})}{n_R} \longrightarrow \int_{\max\{a,\nu^{-1}\}}^{b} \log\left(\frac{\nu\,\text{SNR}}{\beta}\lambda\right)\tilde{f}_\beta(\lambda)\,d\lambda \tag{6.21}$$

where ν satisfies:

$$\int_{\max\{a,\nu^{-1}\}}^{b} \left(\nu - \frac{\beta}{\text{SNR}\,\lambda}\right)^+ \tilde{f}_\beta(\lambda)\,d\lambda = 1 \tag{6.22}$$

with β, a, b and $\tilde{f}_\beta(\cdot)$ given as in Example 6.1. Exploiting (6.21) and (6.22) the following result is obtained:

Theorem 6.4.4 [127] *For*:

$$\text{SNR} \geq \frac{2\min\{1, \beta^{3/2}\}}{|1 - \sqrt{\beta}||1 - \beta|} \tag{6.23}$$

the capacity of the canonical channel with full CSI at the transmitter converges almost surely to:

$$\frac{C(\text{SNR})}{n_R} \to \begin{cases} \beta\log\left(\frac{\text{SNR}}{\beta} + \frac{1}{1-\beta}\right) + (1 - \beta)\log\frac{1}{1-\beta} - \beta\log e & \beta < 1 \\ \log\left(\beta\,\text{SNR} + \frac{\beta}{\beta-1}\right) + (\beta - 1)\log\frac{\beta}{\beta-1} - \log e & \beta > 1 \end{cases}$$

If only statistical CSI is available at the transmitter, from Theorem 6.4.1 it follows that the capacity-achieving input covariance matrix is $\mathbf{\Phi_x} = \mathbf{I}$ (This result was first shown in [123]). Consequently the erogodic capacity per receive antenna is:

$$\frac{C(\text{SNR})}{n_R} = \frac{1}{n_R}\mathbb{E}\left[\log\det\left(\mathbf{I} + \frac{\text{SNR}}{n_T}\mathbf{HH}^\dagger\right)\right] \tag{6.24}$$

For finite number of antennas, [123] gave an integral expression for (6.24) as a function of n_R, n_T (number of transmit and receive antennas respectively) and the signal-to-noise ratio. Denoting t and r as the minimum and maximum between n_R and n_T, an explicit expression for (6.24) can be found in [32]:

$$\frac{C(\text{SNR})}{n_R} = \frac{\log e}{n_R}\sum_{k=0}^{t-1}\sum_{\ell_1=0}^{k}\sum_{\ell_2=0}^{k}\binom{k}{\ell_1}\frac{(k+r-t)!(-1)^{\ell_1+\ell_2}I_{\ell_1+\ell_2+r-t}(\frac{\text{SNR}}{n_T})}{(k-\ell_2)!(r-t+\ell_1)!(r-t+\ell_2)!\ell_2!} \tag{6.25}$$

where $I_0(\gamma) = -e^{\frac{1}{\gamma}}E_i(-\frac{1}{\gamma})$ with $E_i(\cdot)$ denoting the exponential integral

$$E_i(z) = -\int\limits_{-z}^{\infty} \frac{e^{-t}}{t}\, dt$$

while

$$I_n(\gamma) = nI_{n-1}(\gamma) + (-\gamma)^{-n}\left(I_0(\gamma) + \sum_{k=1}^{n}(k-1)!(-\gamma)^k\right) \qquad (6.26)$$

A more simple and compact expression of the capacity per antenna was obtained for this canonical model in [103,137] resorting, again, to the asymptotic regime. As the number of transmit and receiver antennas grows to infinity while $\frac{n_T}{n_R} \to \beta$, the capacity per antenna converges to:

$$\frac{C(\text{SNR})}{n_R} \to \beta \log\left[1 + \frac{\text{SNR}}{\beta} - \frac{1}{4}\mathcal{F}\left(\frac{\text{SNR}}{\beta}, \beta\right)\right] + \log\left[1 + \text{SNR} - \frac{1}{4}\mathcal{F}\left(\frac{\text{SNR}}{\beta}, \beta\right)\right]$$

$$-\beta\frac{\log e}{4\text{SNR}}\mathcal{F}\left(\frac{\text{SNR}}{\beta}, \beta\right) \qquad (6.27)$$

with

$$\mathcal{F}(x, z) = \left(\sqrt{x(1+\sqrt{z})^2 + 1} - \sqrt{x(1-\sqrt{z})^2 + 1}\right)^2 \qquad (6.28)$$

6.4.3.1 *Separable Correlation Model*

While very simple, the canonical model enabled a number of fundamental observations on the benefits that accrue as a function of the numbers of antennas. However, more recently [26,81,93,126], the analysis has been extended in order to characterize the impact of antenna correlation but using a very simple model, the so-called separable correlation model proposed by several authors [22,98,114], where the $n_R \times n_T$ matrix with IID entries is pre-multiplied and post-multiplied by two deterministic positive-semidefinite matrices Θ_T and Θ_R which account for the correlation between the transmit antennas and the receiver antennas:

$$\mathbf{y} = \Theta_R^{1/2}\mathbf{W}\Theta_T^{1/2}\mathbf{x} + \mathbf{n} \qquad (6.29)$$

with $\mathbf{W}$ an $n_R \times n_T$ complex matrix whose entries are independent complex Gaussian random variables with zero-mean and unit variance. Recall here that, denoting by Λ_T and Λ_R the diagonal eigen-value matrices of Θ_T and Θ_R, respectively, the separable model in (6.29) admits a UIU expression as in (6.9) where $\mathbf{U}_R$ and $\mathbf{U}_T$ coincide with the eigen-vector matrices of Θ_R and Θ_R respectively, while $\tilde{\mathbf{H}}$ is a zero-mean IND Rayleigh-faded channel whose (i, j)th entry has variance $\mathbb{E}[|(\tilde{\mathbf{H}})_{i,j}|^2] = (\Lambda_R)_{ii}(\Lambda_T)_{jj}$ with $(\Lambda_R)_{ii}$ and $(\Lambda_T)_{jj}$ denoting the i-th diagonal entry and the j-th diagonal entry of Λ_R and Λ_T, respectively.

With full CSI at the transmitter, no explicit expression of the capacity for arbitrary number of antennas is available. Fortunately the problem becomes analytically tractable in the large-dimensional regime where the asymptotic capacity is obtained as [26]:

$$\frac{\mathcal{C}(\text{SNR})}{n_R} \longrightarrow \beta \int_0^\infty (\log(\nu\text{SNR}\lambda))^+ \, dG(\lambda) \tag{6.30}$$

where ν satisfies:

$$\int_0^\infty \left(\nu - \frac{1}{\text{SNR}\,\lambda}\right)^+ dG(\lambda) = 1 \tag{6.31}$$

and $G(\cdot)$ represents the asymptotic spectrum of $\frac{1}{n_T}\mathbf{H}\mathbf{H}^\dagger$. The capacity in (6.30) can be evaluated as follows.

Theorem 6.4.5 [127] *Let Λ_R and Λ_T be independent random variables whose distributions are the asymptotic spectra of the full-rank matrices Θ_R and Θ_T respectively. Further define:*

$$\Lambda_1 = \begin{cases} \Lambda_T & \beta < 1, \\ \Lambda_R & \beta > 1, \end{cases} \qquad \Lambda_2 = \begin{cases} \Lambda_R & \beta < 1, \\ \Lambda_T & \beta > 1, \end{cases} \tag{6.32}$$

and let κ be the infimum (excluding any mass point at zero) of the support of the asymptotic spectrum of $\frac{1}{n_T}\mathbf{H}^\dagger\mathbf{H}$. For:

$$\text{SNR} \geq \frac{1}{\kappa} - \delta\mathbb{E}\left[\frac{1}{\Lambda_1}\right] \tag{6.33}$$

with δ satisfying:

$$\eta_{\Lambda_2}(\delta) = 1 - \min\left\{\beta, \tfrac{1}{\beta}\right\},$$

the asymptotic capacity of a channel with separable correlations and full CSI at the transmitter is:

$$\frac{\mathcal{C}(\text{SNR})}{n_R} \longrightarrow \begin{cases} \beta\,\mathbb{E}[\log\frac{\Lambda_T}{e\vartheta}] + \mathcal{V}_{\Lambda_R}(\vartheta) + \beta\log\left(\text{SNR} + \vartheta\mathbb{E}[\frac{1}{\Lambda_T}]\right) & \beta < 1 \\ \mathbb{E}[\log\frac{\Lambda_R}{\alpha e}] + \beta\,\mathcal{V}_{\Lambda_T}(\alpha) + \log\left(\text{SNR} + \alpha\mathbb{E}[\frac{1}{\Lambda_R}]\right) & \beta > 1 \end{cases}$$

with α and ϑ the solutions to:

$$\eta_{\Lambda_T}(\alpha) = 1 - \frac{1}{\beta} \qquad \eta_{\Lambda_R}(\vartheta) = 1 - \beta$$

As for the canonical channel, no asymptotic characterization of the asymptotic capacity with full CSI at the transmitter is known for $\beta = 1$ and arbitrary SNR.

When the correlation is present only at either the transmit or receive ends of the link, the solutions in Theorem 6.4.5 sometimes become explicit:

Corollary 6.4.6 *With correlation at the end of the link with the fewest antennas, the capacity per antenna with full CSI at the transmitter converges to:*

$$
\frac{\mathcal{C}(\mathrm{SNR})}{n_{\mathrm{R}}} \longrightarrow
\begin{cases}
\beta\, \mathbb{E}[\log \frac{\Lambda_{\mathrm{T}}}{e}] + \log \frac{1}{1-\beta} + \beta \log\left(\mathrm{SNR}\frac{1-\beta}{\beta} + \mathbb{E}[\frac{1}{\Lambda_{\mathrm{T}}}]\right) & \begin{array}{l}\beta < 1 \\ \Lambda_{\mathrm{R}} = 1\end{array} \\[2ex]
\mathbb{E}[\log \frac{\Lambda_{\mathrm{R}}}{e}] - \beta \log \frac{\beta-1}{\beta} + \log\left(\mathrm{SNR}(\beta - 1) + \mathbb{E}[\frac{1}{\Lambda_{\mathrm{R}}}]\right) & \begin{array}{l}\beta > 1 \\ \Lambda_{\mathrm{T}} = 1\end{array}
\end{cases}
$$

The usually rapid convergence to these limits renders the expressions illustrated above relevant to a wide range of nonasymptotic scenarios.

Having analyzed the full CSI regime, we now turn our attention to the realm of statistical CSI at the transmitter, for which we can give explicit expressions of the capacity even for finite number of antennas.

With statistical CSI at the transmitter, using the UIU representation of the separable model from Theorem 6.4.1 it is immediate to see that achieving capacity requires that the eigen-vectors of the input covariance, $\boldsymbol{\Phi}_{\mathbf{x}} = \mathbf{V}\mathbf{P}\mathbf{V}^{\dagger}$, coincide with those of $\boldsymbol{\Theta}_{\mathrm{T}}$ (This result was first derived in [57,139]). Consequently the capacity per antenna is:

$$
\frac{\mathcal{C}(\mathrm{SNR})}{n_{\mathrm{R}}} = \frac{1}{n_{\mathrm{R}}} \log \det\left(\mathbf{I} + \frac{\mathrm{SNR}}{n_{\mathrm{T}}} \Lambda_{\mathrm{R}}^{1/2} \mathbf{W} \Lambda_{\mathrm{T}}^{1/2} \mathbf{P} \Lambda_{\mathrm{T}}^{1/2} \mathbf{W}^{\dagger} \Lambda_{\mathrm{R}}^{1/2} \right)
$$

where $\mathbf{P}$ denotes the diagonal eigen-value matrix of the capacity-achieving input covariance $\boldsymbol{\Phi}_{\mathbf{x}}$, whose diagonal elements satisfy (6.10) (cf. [128]).

Analytical non-asymptotic expressions for the capacity of the separable correlation model with statistical CSI at the transmitter, have been reported by several authors [4,21,64,70,71,104,105,116,118]. References [70,71,116] compute the two-sided correlated MIMO channel capacity constraining the eigen-values of the correlation matrix to be distinct. Specifically, denoting by ϕ_i the i-th diagonal element of $\Lambda_{\mathrm{T}}^{1/2}\mathbf{P}\Lambda_{\mathrm{T}}^{1/2}$, for $n_{\mathrm{R}} \leq n_{\mathrm{T}}$ we have that:

$$
\frac{\mathcal{C}(\mathrm{SNR})}{n_{\mathrm{R}}} = \frac{(-1)^{\frac{d(d-1)}{2}} \prod_{i=1}^{n_{\mathrm{T}}-1} \frac{i!}{i^{n_{\mathrm{T}}}}}{\frac{1}{\log e}\left(\frac{\mathrm{SNR}}{n_{\mathrm{T}}}\right)^{\frac{n_{\mathrm{T}}(n_{\mathrm{T}}-1)}{2}}} \prod_{i<j}^{n_{\mathrm{T}}} \frac{1}{\phi_i - \phi_j} \prod_{i<j}^{n_{\mathrm{R}}} \frac{1}{(\Lambda_{\mathrm{R}})_{i,i} - (\Lambda_{\mathrm{R}})_{j,j}} \sum_{\ell=1}^{n_{\mathrm{R}}} \det\left(\begin{bmatrix} \mathbf{X}_\ell \\ \mathbf{Y} \end{bmatrix}\right) \quad (6.34)
$$

where $d = n_{\mathrm{T}} - n_{\mathrm{R}}$, $\mathbf{X}_\ell$ is a $n_{\mathrm{R}} \times n_{\mathrm{T}}$ matrix whose (i,j)th entry, for $i \in \{1,\ldots,n_{\mathrm{R}}\}$ and $j \in \{1,\ldots,n_{\mathrm{T}}\}$, is:

$$
(\mathbf{X}_\ell)_{i,j} =
\begin{cases}
-(n_{\mathrm{R}} - 1)! \dfrac{\left(\frac{\mathrm{SNR}}{n_{\mathrm{T}}}\phi_j\right)^{n_{\mathrm{R}}-1}}{(\Lambda_{\mathrm{R}})_{i,i}^{1-n_{\mathrm{R}}}} e^{\frac{n_{\mathrm{T}}}{\mathrm{SNR}\phi_j(\Lambda_{\mathrm{R}})_{i,i}}} E_i\left(-\frac{n_{\mathrm{T}}}{\mathrm{SNR}\phi_j(\Lambda_{\mathrm{R}})_{i,i}}\right) & i = \ell \\[3ex]
\displaystyle\sum_{k=n_{\mathrm{T}}-n_{\mathrm{R}}}^{n_{\mathrm{T}}-1} \frac{\left(-\frac{\mathrm{SNR}}{n_{\mathrm{T}}}\phi_j(\Lambda_{\mathrm{R}})_{i,i}\right)^k}{(\Lambda_{\mathrm{R}})_{i,i}^{n_{\mathrm{T}}-n_{\mathrm{R}}}}[1 - n_{\mathrm{R}}]_k & i \neq \ell
\end{cases}
$$

and $\mathbf{Y}$ is an $(n_T - n_R) \times n_T$ matrix whose (i,j)th entry, for $j \in \{1, \ldots, n_T\}$ and $i \in \{1, \ldots, n_T - n_R\}$, is:

$$(\mathbf{Y})_{i,j} = [1 - n_T]_{i-1} \left(-\frac{\text{SNR}}{n_T} \phi_j \right)^{i-1}.$$

While explicit, the above expression of the capacity is very involved and very limited insight on how multiantenna links are impacted by correlation can be derived from (6.34). However, the establishment of sound design principles often requires a more mature and deep understanding. To that end, we shall develop our analysis in the asymptotic regime which yields valuable insight—that could not be drawn from (6.34)—on the impact of correlation and on the capacity-achieving input at high and low SNR. Results on the asymptotic capacity and mutual information of channels that obey the separable model can be found in [26,91,93,127,129,131].

In particular we have that:

Theorem 6.4.7 *Given a correlated multiantenna channel as in (6.29), its asymptotic capacity per antenna converges to:*

$$\frac{C(\text{SNR})}{n_R} \rightarrow \beta \, \mathbb{E}[\log(1 + \text{SNR} \, \Lambda_T \Gamma(\text{SNR}))] + \mathbb{E}[\log(1 + \text{SNR} \, \Lambda_R Y(\text{SNR})]$$

$$- \beta \, \text{SNR} \, \Gamma(\text{SNR}) Y(\text{SNR}) \log e \qquad (6.35)$$

where:

$$\Gamma(\text{SNR}) \triangleq \frac{1}{\beta} \mathbb{E}\left[\frac{\Lambda_R}{1 + \text{SNR} \, \Lambda_R Y(\text{SNR})} \right] \qquad Y(\text{SNR}) \triangleq \mathbb{E}\left[\frac{\Lambda_T}{1 + \text{SNR} \, \Lambda_T \Gamma(\text{SNR})} \right]$$

with expectation over Λ and Λ_R whose distributions are given by the asymptotic empirical eigen-value distributions of $\Lambda_T \mathbf{P}$ and Θ_R, respectively.

For correlation only at the receiver (i.e., $\Theta_T = \mathbf{I}$), from Theorem 6.4.4 it follows that isotropic inputs achieve the capacity and the capacity per antenna converges to:

$$\frac{C(\text{SNR})}{n_R} \rightarrow \beta \log(1 + \text{SNR} \, \Gamma(\text{SNR})) + \mathbb{E}[\log(1 + \text{SNR} \, \Lambda_R Y(\text{SNR})]$$

$$- \beta \, \text{SNR} \, \Gamma(\text{SNR}) Y(\text{SNR}) \log e \qquad (6.36)$$

where:

$$\Gamma(\text{SNR}) \triangleq \frac{1}{\beta} \mathbb{E}\left[\frac{\Lambda_R}{1 + \text{SNR} \, \Lambda_R Y(\text{SNR})} \right] \qquad Y(\text{SNR}) \triangleq \frac{1}{1 + \text{SNR} \, \Gamma(\text{SNR})}$$

From (6.35) and (6.36), using Jensen's inequalities, we can easily identify two distinct behaviors: If isotropic inputs achieve the capacity of the correlated channel (as for $\Theta_T = \mathbf{I}$), then it holds that correlation is detrimental at any SNR. Otherwise, the capacity curves may intersect with correlation being pernicious above the intersection SNR but beneficial below it. In [129] using (6.35) and its asymptotic expansions for high and low SNR it is shown how receive correlation reduces the effective dimensionality of the receiver without increasing the captured power and thus it is always detrimental. On the other hand, transmit correlation reduces the effective dimensionality of the transmitter, but it also enables focusing power. The net effect is an advantage at low SNR where the minimum energy per bit required for reliable communication sees a reduction given precisely by the largest eigen-value of Θ_T. At high SNR, an advantage can also be realized if $n_T > n_R$.

This model describes with extreme accuracy a linear array, where the diversity is based only on antenna spacing. However, it has some limitations. In fact, the separable correlation model usually suffices to represent the correlation that arises with spatial diversity, due to antenna proximity; however it does not accommodate, for instance, other diversity mechanisms that are becoming increasingly popular such as polarization [10,76,84,120] and pattern diversity or some pathological phenomena like keyhole and pinhole [23,41]. A richer model, able to encompass a broader range of channel structures, is the UIU-model (cf. [129]) illustrated in (6.9), where the channel matrix **H** is modeled as a complex Gaussian matrix and the correlation between the (i, j) and (i', j') entries of **H** is, in general, a joint function of (i, j, i', j'). Results on the ergodic capacity of this very general model can be found in [129].

Note also that the UIU model only accommodates zero-mean channels. As it turns out, nonetheless, for large numbers of antennas the addition to the model of an unfaded (i.e, deterministic) matrix becomes immaterial in terms of the capacity as long as such matrix does not have full rank. For arbitrary numbers of antennas, capacity characterizations for nonzero-mean channels are available in the specific case of the IID Ricean distribution in [131] and references therein. Finally, for even more general transmit and/or receive correlated Rician scenarios, the sum capacity does not admit an exact closed-form solution, however accurate closed-form approximations can be found in [89].

6.5 MULTI-ANTENNA CAPACITY: NON-ERGODIC REGIME

We now turn our attention to the slow fading MIMO channel. In the non-ergodic regime, where the fading is not such that the statistics of the channel are revealed to the receiver during the span of a codeword, a more suitable performance measure is the outage capacity, which coincides with the classical Shannon-theoretic notion of ε-capacity [30], namely the maximal rate for which block error probability ε is attainable. Under certain conditions, the outage capacity can be obtained through the probability that $\mathcal{I}(\text{SNR}, \mathbf{\Phi_x})$ in (6.6) (whose distribution is induced by **H**) falls below the transmission rate R [11,37,123]:

$$\mathcal{P}(I(\text{SNR}, \mathbf{\Phi_x}) \leq R)$$

In order to maximize the rate supported at some chosen outage level, the input covariance, $\mathbf{\Phi_x}$, must be properly determined depending on the SNR level and on distribution of $\mathcal{I}(\text{SNR}, \mathbf{\Phi_x})$. At very low SNR relative to the target rate, it is optimal to use just one transmit antenna [136]. For arbitrary SNR, establishing $\mathbf{\Phi_x}$ is a problem not easily tackled analytically. Some results on the eigen-vectors of $\mathbf{\Phi_x}$ can be found in [115]. A long-standing conjecture is that the optimal strategy is to restrict to a subset of the antennas and then transmit isotropically among the antennas used: the lower the SNR level relative to the target rate, the smaller the number of antennas used. At typical outage probability levels, the SNR is high relative to the target rate and it is expected that using all the antennas is a good strategy.

Concerning the distribution of $\mathcal{I}$, for $n_R = 1$ the distribution of $\mathcal{I}$ has been found for correlated Rayleigh-faded channels in [61,92] and for uncorrelated Ricean channels in [92,117].[5] For an arbitrary number of transmit and receive antennas, the distribution of $\mathcal{I}$ can be obtained via its moment-generating function $M(\zeta) = \mathbb{E}[e^{\zeta \mathcal{I}}]$ which for the canonical channel with $\mathbf{\Phi_x} = \mathbf{I}$ has been derived in [20,140]. In the case of separable correlation models findings on $M(\cdot)$ can be found in [4,64,70,71]. Finally, for uncorrelated Ricean channels with $\mathbf{\Phi_x} = \mathbf{I}$, $M(\cdot)$ is provided in [63] in terms of the integral of hypergeometric functions.

As the number of antennas grows, the distribution of $\mathcal{I}$ for many multiantenna channels of interest can be approximated as Gaussian. In the engineering literature, two distinct approaches have been followed in order to explore this property:

1. The mean and variance of $\mathcal{I}(\text{SNR}, \mathbf{\Phi_x})$ are obtained through the moment generating function (for fixed number of antennas). A Gaussian distribution with such mean and variance is then compared, through Monte Carlo simulation, to the empirical distribution of $\mathcal{I}(\text{SNR}, \mathbf{\Phi_x})$.

2. Starting from:

$$\lim_{n_R \to \infty} \frac{\mathbb{E}[\mathcal{I}(\text{SNR}, \mathbf{\Phi_x})]}{n_R} = \mathcal{V}_{\frac{1}{n_T} \mathbf{H}\mathbf{\Phi_x}\mathbf{H}^\dagger}(\text{SNR}),$$

the random fluctuations of the mutual information in the asymptotic regime are analyzed. Specifically the random variable:

$$\Delta_{n_R} = \mathcal{I}(\text{SNR}, \mathbf{\Phi_x}) - n_R \mathcal{V}_{\frac{1}{n_T} \mathbf{H}\mathbf{\Phi_x}\mathbf{H}^\dagger}(\text{SNR}) \qquad (6.37)$$

is either shown or conjectured to converge to a zero-mean Gaussian random variable as $n_R \to \infty$ and its variance for arbitrary SNR characterized using some sharper mathematical tools like random matrix theory or replica method (cf. [131] and reference therein).

Following the first approach, the computation of mean and variance of the mutual information of the canonical channel (IID zero-mean Gaussian entries) was carried out

---

[5]The input covariance is constrained to be $\mathbf{\Phi_x} = \mathbf{I}$ in [117], which also gives the corresponding distribution of $\mathcal{I}$ for $\min(n_T, n_R) = 2$ and arbitrary $\max(n_T, n_R)$ although in the form of an involved integral expression.

in [13,119,140], which laid out numerical evidence of an excellent Gaussian fit. These results were extended to Ricean fading in [66] and to correlated antennas (at the array with the most antennas) in [118]. Although, in every case, the match is excellent, no proof of asymptotic Gaussianity is provided. Only for SNR $\to \infty$ with $\mathbf{\Phi}_x = \mathbf{I}$ and with $\mathbf{H}$ being a real Gaussian matrix with IID entries has it been shown that $\mathcal{I} - \mathbb{E}[\mathcal{I}]$ converges to a Gaussian random variable [43].

Concerning the second approach, the following result on the mutual information of a correlated multiantenna channel with isotropic inputs can be provided:

Theorem 6.5.1   [132] *Consider a $n_{\mathrm{R}} \times n_{\mathrm{T}}$ multiantenna channel as in (6.29) where the correlation takes place only at transmitter[6] ($\mathbf{\Theta}_{\mathrm{R}} = \mathbf{I}$). Assume that the $n_{\mathrm{R}} \times n_{\mathrm{T}}$ complex random matrix, $\mathbf{W}$, in (6.29) is such that the entries are IID with unit variance. As $n_{\mathrm{T}}, n_{\mathrm{R}} \to \infty$ with $\frac{n_{\mathrm{T}}}{n_{\mathrm{R}}} \to \beta$,*

$$\lim_{n_{\mathrm{R}} \to \infty} \frac{1}{n_{\mathrm{R}}} \log \det \left(\mathbf{I} + \frac{\mathrm{SNR}}{n_{\mathrm{T}}} \mathbf{W} \mathbf{\Theta}_{\mathrm{T}} \mathbf{W}^{\dagger} \right) = \mathcal{V}_{\frac{1}{n_{\mathrm{T}}} \mathbf{W} \mathbf{\Theta}_{\mathrm{T}} \mathbf{W}^{\dagger}} (\mathrm{SNR})$$

$$= \beta \mathcal{V}_{\mathbf{\Theta}_{\mathrm{T}}} \left(\eta \frac{\mathrm{SNR}}{\beta} \right) + \log \frac{1}{\eta} + (\eta - 1) \log e$$

where η, denoting η-transform of $\frac{1}{n_{\mathrm{T}}} \mathbf{W} \mathbf{\Theta}_{\mathrm{T}} \mathbf{W}^{\dagger}$, satisfies:

$$\beta = \frac{1 - \eta}{1 - \eta_{\mathbf{\Theta}_{\mathrm{T}}}(\frac{\mathrm{SNR}}{\beta} \eta)} \tag{6.38}$$

Furthermore, the random variable:

$$\Delta_{n_{\mathrm{R}}} = \log \det \left(\mathbf{I} + \frac{\mathrm{SNR}}{n_{\mathrm{T}}} \mathbf{W} \mathbf{\Theta}_{\mathrm{T}} \mathbf{W}^{\dagger} \right) - n_{\mathrm{R}} \mathcal{V}_{\frac{1}{n_{\mathrm{T}}} \mathbf{W} \mathbf{\Theta}_{\mathrm{T}} \mathbf{W}^{\dagger}} (\mathrm{SNR}) \tag{6.39}$$

is asymptotically zero-mean Gaussian with variance:

$$\mathbb{E}[\Delta^2] = -\log \left(1 - \beta \mathbb{E} \left[\left(\frac{\Lambda_{\mathrm{T}} \mathrm{SNR}\, \eta}{\beta + \Lambda_{\mathrm{T}} \mathrm{SNR}\, \eta} \right)^2 \right] \right)$$

where η satisfies (6.38) while the expectation is over the nonnegative random variable Λ_{T} whose distribution is given by the asymptotic ESD of $\mathbf{\Theta}_{\mathrm{T}}$.

With both transmit and receive correlations, the asymptotic Gaussianity of $\Delta_{n_{\mathrm{R}}}$ is conjectured in [93,109] by observing the behavior of the second- and third-order moments obtained via the replica method from statistical physics (which has yet to find a rigorous justification).

---

[6]The result is valid also in the case that the correlation takes place only at the receiver ($\mathbf{\Theta}_{\mathrm{T}} = \mathbf{I}$).

6.6 RECEIVER ARCHITECTURES AND PERFORMANCE

In this section, we turn our attention to practical MIMO receiver structures. As previously mentioned, the MIMO channel model is intimately related to the multiuser CDMA model, and as such, the same fundamental approaches to receiver design can be taken. The main practical receivers that have been proposed and studied include ZF, MMSE, and successive interference cancelation (SIC) receivers, as well as suboptimal sphere-detection approaches for approximating maximum likelihood (ML) detectors. Of course, each of these practical receivers incurs a capacity loss in comparison to the calculations made in the previous two sections, which inherently assumed the use of optimal receivers (which by their nature have prohibitive implementation complexity). In this section, we will examine the performance of the practical receivers for the MIMO channel model, as well as the capacity loss in comparison to the optimal receiver.

Of course, since the statistics of the MIMO channel differ from the statistics of the equivalent CDMA matrix of spreading signature vectors, the respective receiver performances will also differ (as was the case for capacity, discussed in the previous section). While many of the analysis techniques can be carried over from CDMA receiver analysis, it is important to note that for practical MIMO applications, the comparatively small system dimensionality means that it is not always appropriate to do so. For example, typical MIMO systems will have two to four antennas, compared with the length of CDMA spreading codes, which are usually orders of magnitude greater. The main implication is that for MIMO we seek explicit finite-dimensional analysis results. This contrasts with much of the CDMA performance analysis which is based on asymptotic techniques (as discussed in detail in Chapters 4 and 5). The large-system asymptotic approach provides very good approximations for CDMA, since the spreading codes are long, and it often results in simple expressions yielding practical insights. As such, exact finite-dimensional analysis has been largely avoided for CDMA systems; partly also due to its difficulty, especially since the practical spreading codes are discrete. In contrast, for MIMO the channel matrix is Gaussian, and for these matrices a number of important finite-dimensional statistical techniques are available.

In this section, we will present results for three important performance metrics for the practical linear ZF and MMSE receivers; namely SNR/SINR distribution, capacity, and symbol error rates (SER). We will consider equal-power spatial multiplexing transmission, and assume that the channel is estimated perfectly at the receiver.

6.6.1 Linear Receivers

Within the context of MIMO, the task of the linear receiver is to provide an estimate for each of the multiple transmitted symbols by performing only linear operations to the measured samples at the receive antennas. Any linear receiver can therefore be represented by a matrix $\mathbf{A}$, for which the $n_T \times 1$ vector of receiver outputs is given as follows:

$$\hat{\mathbf{x}} = \mathbf{A}\mathbf{y} = \mathbf{A}(\sqrt{g}\,\mathbf{H}\mathbf{x} + \mathbf{n}) \tag{6.40}$$

It is important to note that the estimate for a given symbol (i.e., a given element of $\hat{\mathbf{x}}$) will generally be corrupted by residual interference from the other transmitted symbols as well as noise, and different receivers are characterized by their ability to suppress these two forms of disturbance. In this section, we focus on two important practical linear receivers; namely ZF and MMSE.

6.6.1.1 Zero-Forcing Receiver
The ZF receiver (also known as the decorrelator) has the following structure:

$$\mathbf{A} = \frac{1}{\sqrt{g}} (\mathbf{H}^\dagger \mathbf{H})^{-1} \mathbf{H}^\dagger \tag{6.41}$$

Clearly for the matrix $\mathbf{A}$ to be well-defined, it is required that $\mathbf{H}^\dagger \mathbf{H}$ be invertible. This, in turn, implies the condition $n_R \geq n_T$, which will be assumed throughout this section. Note that for systems with $n_R < n_T$, the same receiver structure can be employed in conjunction with antenna selection at the transmitter, where a subset of no more than n_R transmitters is used.

From (6.40), the ZF output can be written as:

$$\hat{\mathbf{x}} = \mathbf{x} + \tilde{\mathbf{n}} \tag{6.42}$$

where $\tilde{\mathbf{n}} = \mathbf{A}\mathbf{n}$.

The primary advantage of the ZF receiver is that it provides estimates for each transmitted symbol which contain no interference from the other transmitted symbols, thereby decoupling the linear model (6.40) into the n_T parallel non-interacting scalar subchannels described by (6.42). The main tradeoff is that, in doing so, it enhances the noise.

Conditioned on the channel realization $\mathbf{H}$, the SNR for the k-th subchannel is given by:

$$\mathrm{SNR}_k = \frac{\mathrm{SNR}}{n_T} \frac{1}{[(\mathbf{H}^\dagger \mathbf{H})^{-1}]_{k,k}} \tag{6.43}$$

where $[\cdot]_{k,k}$ represents the kth diagonal element. Note that in this ZF case, the SNR is also the SINR, since the interference is totally removed.

One of the main challenges in the performance analysis of any MIMO receiver is to characterize the statistics of the SNR for finite numbers of antennas and different fading channel scenarios. For ZF, this problem has been well studied in the large-antenna regime (see, e.g., [124,137]) where SNR_k is shown to converge to a deterministic value.

For the finite-antenna case, when $\mathbf{H}$ exhibits IID Rayleigh fading it is well-known that SNR_k follows a Gamma distribution with probability density function (PDF) [143]

$$f(\mathrm{SNR}_k) = \frac{n_T}{\mathrm{SNR}} \frac{\exp\left(-\frac{\mathrm{SNR}_k n_T}{\mathrm{SNR}}\right)}{(n_R - n_T)!} \left(\frac{\mathrm{SNR}_k n_T}{\mathrm{SNR}}\right)^{n_R - n_T} \tag{6.44}$$

A similar result has also been found for Rayleigh channels with transmit correlation [46]. For other practical MIMO channel scenarios, such as Rayleigh fading with both transmit and receiver correlation, as well as Rician fading, obtaining exact closed-form expressions for the SNR PDF is much more difficult. In the former case, however, the statistics have been characterized in terms of the moment generating function (m.g.f.) [73], whereas for the latter case accurate Gamma approximations have been proposed [146].

Sum Capacity: Assuming that the ZF subchannels are treated as separate parallel SISO channels, then single-user encoding/decoding with Gaussian inputs on each subchannel achieves the capacity. Summing the capacities of the individual ZF subchannels normalized by the number of recieve antennas, we obtain:

$$\frac{\mathcal{C}_{\text{ZF}}(\text{SNR})}{n_{\text{R}}} = \frac{1}{n_{\text{R}}} \sum_{k=1}^{n_{\text{T}}} \mathbb{E}_{\text{SNR}_k}[\log(1 + \text{SNR}_k)]. \tag{6.45}$$

In some cases, closed-form solutions can be obtained by directly averaging over the SNR distribution $f(\text{SNR}_k)$. For example, when $\mathbf{H}$ exhibits IID Rayleigh fading, the capacity (6.45) is evaluated using (6.44) to give:

$$\frac{\mathcal{C}_{\text{ZF}}(\text{SNR})}{n_{\text{R}}} = \frac{n_{\text{T}}}{n_{\text{R}}} \log e \, e^{n_{\text{T}}/\text{SNR}} \sum_{\ell=1}^{n_{\text{R}}-n_{\text{T}}+1} \text{E}_\ell(n_{\text{T}}/\text{SNR}) \tag{6.46}$$

where $\text{E}_\ell(\cdot)$ denotes the ℓth order Exponential Integral function [1]. A closed-form expression for more general Rayleigh channels with transmit correlation can be found in [34].

Figure 6.1 compares the ZF sum capacity (6.46) with the MIMO channel capacity (6.25) as a function of SNR (in dB), in IID Rayleigh fading with $n_{\text{R}} = n_{\text{T}} = 4$. The corresponding sum capacity curve for the linear MMSE receiver (considered in the next section) is also shown. It is clearly seen that in the high SNR region the slope of the ZF curve is the same as for the MIMO channel capacity. There is, however, a fixed offset between the curves, which we now examine.

To quantify the high SNR offset between the MIMO channel capacity and the ZF sum capacity, it is useful to derive analytic expressions specifically for the high SNR regime. For IID Rayleigh fading, this can be achieved by taking SNR large in (6.46). More generally, though, we can apply results from [110] to express the ZF sum capacity (6.45) at high SNR according to the following affine expansion:

$$\frac{\mathcal{C}_{\text{ZF}}(\text{SNR})}{n_{\text{R}}} = \frac{n_{\text{T}}}{n_{\text{R}}} \log \text{SNR} + \frac{1}{n_{\text{R}}} \sum_{k=1}^{n_{\text{T}}} \mathbb{E}_{\mathbf{H}}\left[\log\left(\frac{1/n_{\text{T}}}{[(\mathbf{H}^\dagger\mathbf{H})^{-1}]_{k,k}}\right)\right] + o(1) \tag{6.47}$$

where the $o(1)$ term vanishes as SNR $\rightarrow \infty$.

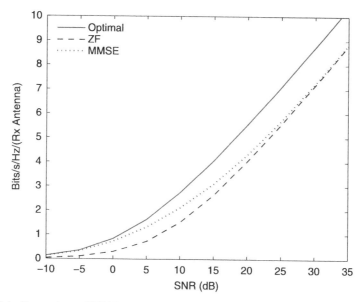

Figure 6.1. Comparison of MIMO channel capacity per receive antenna and normalized (by n_R) sum capacity for ZF and MMSE receivers in IID Rayleigh fading. Results are shown for $n_R = n_T = 4$.

Now, it is interesting to compare (6.47) with the high SNR affine expansion for the MIMO mutual information (6.7) with isotropic inputs[7]

$$\frac{\mathcal{C}(\text{SNR})}{n_R} = \frac{n_T}{n_R} \log \text{SNR} + \frac{1}{n_R} \mathbb{E}_{\mathbf{H}}[\log \det((1/n_T)\mathbf{H}^\dagger \mathbf{H})] + o(1) \tag{6.48}$$

where $\frac{1}{n_R}\mathbb{E}_{\mathbf{H}}[\log \det((1/n_T)\mathbf{H}^\dagger \mathbf{H})] = \mathcal{L}_\infty$, represents the *power offset*, in 3-dB units, with respect to a reference channel whose dimensions are unfaded and orthogonal. This measure was first introduced in [110]. An extensive analytical characterization of the *power offset* (in dB) for several MIMO channel of interest has been found in where insightful closed-form expressions for $\mathcal{L}_\infty$ have been derived and the impact of various features (correlation, unfaded terms, interference, etc) has be consequently assessed.

We can immediately make two important observations. First, the leading (first-order) terms in (6.47) and (6.48) are identical, confirming that the ZF receiver achieves the optimal capacity scaling in the high SNR regime (i.e., it scales linearly with the minimum number of antennas). Second, the non-vanishing (zeroth order) terms (6.47) and (6.48) are different. In fact, it is precisely these terms that account for the

---

[7]Note that here we have given the mutual information under the assumption of isotropic inputs, for any channel distribution. For IID Rayleigh channels, recall that isotropic inputs are capacity achieving, and as such the mutual information in (6.48) is the capacity. At high SNR, isotropic inputs also become capacity-achieving for certain other channels/assumptions [130].

fixed rate offset (or equivalently, fixed power offset [82]) observed between the MIMO capacity and ZF curves in Figure 6.1. To examine this further, let us define the asymptotic rate offset as follows:

$$
\begin{aligned}
\Delta C^\infty &\triangleq \lim_{\mathrm{SNR}\to\infty} \left(\frac{C(\mathrm{SNR})}{n_\mathrm{R}} - \frac{C_\mathrm{ZF}(\mathrm{SNR})}{n_\mathrm{R}} \right) \\
&= \frac{1}{n_\mathrm{R}} \mathbb{E}_\mathbf{H}[\log \det(\mathbf{H}^\dagger\mathbf{H})] - \frac{1}{n_\mathrm{R}} \sum_{k=1}^{n_\mathrm{T}} \mathbb{E}_\mathbf{H}\left[\log\left(\frac{1}{[(\mathbf{H}^\dagger\mathbf{H})^{-1}]_{k,k}} \right) \right]
\end{aligned}
\tag{6.49}
$$

It is difficult to directly evaluate the expectations involving matrix inverses in (6.49), although some results do exist. However, it is possible to employ a matrix identity from [53] to circumvent this problem. Using the identity:

$$
\frac{1}{[(\mathbf{H}^\dagger\mathbf{H})^{-1}]_{k,k}} = \frac{\det(\mathbf{H}^\dagger\mathbf{H})}{\det(\mathbf{H}_k^\dagger\mathbf{H}_k)}
\tag{6.50}
$$

where $\mathbf{H}_k$ corresponds to $\mathbf{H}$ but with the kth column removed, the rate offset in (6.49) can be expressed as:

$$
\Delta C^\infty = \frac{1}{n_\mathrm{R}} \sum_{k=1}^{n_\mathrm{T}} \mathbb{E}_{\mathbf{H}_k}[\log \det(\mathbf{H}_k^\dagger\mathbf{H}_k)] - \frac{n_\mathrm{T}-1}{n_\mathrm{R}} \mathbb{E}_\mathbf{H}[\log \det(\mathbf{H}^\dagger\mathbf{H})]
\tag{6.51}
$$

Importantly, ΔC^∞ can now be evaluated in closed-form for many different MIMO channel scenarios of interest, including IID and correlated Rayleigh and Rician fading, by exploiting a number of recent random matrix theory results for expectations of the form $\mathbb{E}[\log \det(\cdot)]$ (see, e.g., [82,87,95,151]). For example, for the case of IID Rayleigh fading, the rate offset is evaluated as:

$$
\Delta C^\infty = \frac{\log e}{n_\mathrm{R}} \sum_{k=1}^{n_\mathrm{T}-1} \frac{k}{n_\mathrm{R}-k} \quad (\mathrm{b/s/Hz})
\tag{6.52}
$$

Note that this result was first established in [59] via direct evaluation of (6.49). For square systems (i.e., with $n_\mathrm{R} = n_\mathrm{T} = n$), (6.52) reduces further to:

$$
\Delta C^\infty = \log e \sum_{k=2}^{n} \frac{1}{k} \quad (\mathrm{b/s/Hz})
\tag{6.53}
$$

which, interestingly, shows the rate offset per receive antenna increasing with n. This implies that the overall rate loss incurred by using suboptimal ZF receivers, as opposed to optimal joint ML decoders, can be quite large even for moderate numbers of antennas. We can also see from (6.52) that the rate penalty due to ZF is reduced by increasing n_R (for fixed n_T). These points are illustrated further in Figure 6.2, where the rate offset ΔC^∞ is plotted for different antenna configurations.

Error Performance: In addition to capacity, it is also important to examine the error performance, which is in many ways a more practical measure. The focus here will be on characterizing the average uncoded SER; however, the analysis with coding is also possible (see, e.g., [85,86,88]).

For the ZF receiver, the average SER is given by:

$$\text{SER}_{\text{ZF}}(\text{SNR}) = \frac{1}{n_T} \sum_{k=1}^{n_T} \text{SER}_{\text{ZF},k}(\text{SNR}) \tag{6.54}$$

where $\text{SER}_{\text{ZF},k}(\cdot)$ is the average SER for the kth ZF subchannel. This can be expressed for a number of modulation formats of interest as follows:

$$\text{SER}_{\text{ZF},k}(\text{SNR}) = \mathbb{E}_{\text{SNR}_k}\left[a\,Q(\sqrt{2b\,\text{SNR}_k})\right] \tag{6.55}$$

where $Q(\cdot)$ is the standard Gaussian Q-function, and a and b are constellation-specific constants. Specific modulation formats which conform to (6.55) include binary phase-shift keying (BPSK) ($a = 1$, $b = 1$); binary frequency shift keying (BFSK) with orthogonal signaling ($a = 1$, $b = 0.5$), or minimum correlation ($a = 1$, $b = 0.715$); and M-ary pulse amplitude modulation (PAM) ($a = 2(M - 1)/M, b = 3/(M^2 - 1)$).

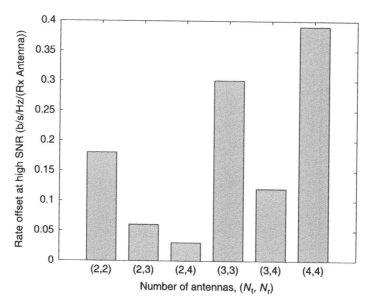

Figure 6.2. High SNR rate offset between normalized (by n_R) MIMO channel capacity and normalized ZF sum capacity for different antenna configurations.

The expression (6.55) also gives an accurate approximation for M-ary PSK ($a = 2$, $b = \sin^2(\pi/M)$) [101, Eq. 5.2-61].

In some cases, the average SER can be evaluated in closed-form by directly averaging (6.55) over the distribution of SNR_k. For example, for IID Rayleigh fading, averaging (6.55) over (6.44) and substituting the result into (6.54) yields the closed-form solution:

$$\text{SER}_{\text{ZF}}(\text{SNR}) = \frac{a}{2}\left(1 - \sqrt{\theta(\text{SNR})}\sum_{k=0}^{n_{\text{R}}-n_{\text{T}}}\binom{2k}{k}\left(\frac{1-\theta(\text{SNR})}{4}\right)^k\right) \qquad (6.56)$$

where:

$$\theta(\text{SNR}) = \frac{1}{1 + n_{\text{T}}/(b\,\text{SNR})} \qquad (6.57)$$

Note that closed-form SER expressions have also been derived for more general channels in [46], which considered Rayleigh fading with transmit correlation, and in [73], which considered Rayleigh fading with both transmit and receive correlation.

Figure 6.3 plots the average SER expression (6.56) against SNR (in dB) for different antenna configurations, considering BPSK modulation. We clearly see the benefit from having more receive antennas than transmit antennas, especially in the high SNR regime where the curves are linear and the slope varies with the difference between n_{R} and n_{T}.

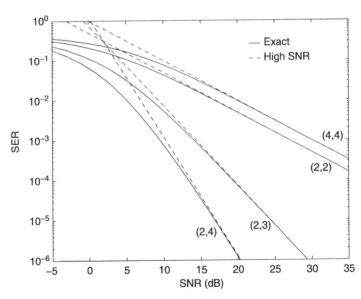

Figure 6.3. Exact and asymptotic SER for MIMO systems employing a ZF receiver in IID Rayleigh fading. Results are shown for different antenna configurations (n_{T}, n_{R}), and for BPSK modulation.

It is also interesting to observe that the curve for (4, 4) is shifted by 3 dB from the (2, 2) curve. While it is not obvious that this would be the case, it is not totally unexpected since the total transmit power is shared across twice as many antennas.

These observations motivate us to investigate the high SNR regime more closely. To this end, letting SNR $\to \infty$, it can be shown that the SER (6.56) is well-approximated by:

$$\text{SER}_{\text{ZF}}^{\infty}(\text{SNR}) = (G_a \, \text{SNR})^{-G_d} \tag{6.58}$$

where:

$$G_d = n_\text{R} - n_\text{T} + 1 \tag{6.59}$$

$$G_a = \frac{2b}{n_\text{T}} \left(\frac{a \, (2(n_\text{R} - n_\text{T}) + 1)!!}{2 \, (n_\text{R} - n_\text{T} + 1)!} \right)^{-1/(n_\text{R} - n_\text{T} + 1)} \tag{6.60}$$

where $N!! = 1 \cdot 3 \cdot \ldots \cdot N$, for odd $N > 0$.

Note that G_a gives an offset term on a SER-SNR plot (in the log domain), and as such is commonly referred to as an array gain. The negative slope is given by G_d, which is commonly referred to as the diversity order. Intuitively, the diversity order gives the number of equivalent independent received versions of each transmitted symbol. For example, consider the simple case when $n_\text{T} = 1$, and there are therefore n_R diversity paths. Clearly, from (6.59), $G_d = n_\text{R}$ in this case.

The SER result in (6.58)–(6.60) is shown in Figure 6.3 and is clearly seen to converge at high SNR , whilst yielding accurate results for most SERs of practical interest (e.g., $<10^{-3}$).

For the special case $n_\text{R} = n_\text{T} = n$, (6.58)–(6.60) admits the particularly simple form:

$$\text{SER}_{\text{ZF}}^{\infty}(\text{SNR}) = \left(\frac{4b}{an} \text{SNR} \right)^{-1} \tag{6.61}$$

From this, we clearly see that as n increases, the diversity order does not change, but the array gain includes a $1/n$ term. We can now see that the shift we observed in the plot between the (2, 2) and (4, 4) curves is precisely 3 dB. Naturally, it would be preferable to have the SER decrease much more quickly than as SNR^{-1}, and in Section 6.8 we'll see how some of the spatial channels can be sacrificed to increase the rate of SER falloff.

6.6.1.2 Minimum Mean-Square Error Receiver
The MMSE receiver has the following structure:

$$\mathbf{A} = \frac{1}{\sqrt{g}} \left((n_\text{T}/\text{SNR}) \mathbf{I}_{n_\text{T}} + \mathbf{H}^\dagger \mathbf{H} \right)^{-1} \mathbf{H}^\dagger \tag{6.62}$$

Note that the MMSE receiver is well-defined for both $n_\text{R} \geq n_\text{T}$ and $n_\text{R} < n_\text{T}$. We will see, however, that the performance in the latter case is undesirable for most practical scenarios.

In contrast to the ZF receiver that completely removed all interference at the expense of noise enhancement, the MMSE receiver trades off interference suppression and noise enhancement. Therefore, in this case we are interested in quantifying the SINR, instead of the SNR. The MMSE receiver is the optimal linear receiver in the sense of maximizing the SINR.

The SINR for the estimate of the symbol transmitted on the kth antenna is given by:

$$\text{SINR}_k = \frac{1}{[(\mathbf{I}_{n_T} + (\text{SNR}/n_T)\mathbf{H}^\dagger\mathbf{H})^{-1}]_{k,k}} - 1 \tag{6.63}$$

The distribution of SINR_k has been very well studied in the large-antenna regime. For finite-dimensional systems, when $\mathbf{H}$ exhibits IID Rayleigh fading, an exact solution has been derived for the distribution of SINR_k in [40] (in the context of multiple-access channels with optimal linear diversity combining), and is given by:

$$f(\text{SINR}_k) = -\frac{\mathrm{d}}{\mathrm{d}\gamma}\bar{F}(\gamma)\Big|_{\gamma=\text{SINR}_k} \tag{6.64}$$

where $\bar{F}(\gamma) = \Pr(\text{SINR}_k > \gamma)$ is the complementary cumulative distribution function (CCDF) of SINR_k given by:

$$\bar{F}(\gamma) = \frac{e^{-\gamma n_T/\text{SNR}} \sum_{i=0}^{n_R-1} \beta_i \gamma^i}{(1+\gamma)^{n_T-1}} \tag{6.65}$$

Here, β_i is the coefficient of z^i in the series expansion of $e^{zn_T/\text{SNR}}(1+z)^{n_T-1}$. A simple recursive method for numerically computing these coefficients is presented in [40]. After some algebra, however, it can be also shown that β_i admits the following closed-form solution:

$$\beta_i = \sum_{k=p}^{i} \binom{n_T-1}{i-k} \frac{(n_T/\text{SNR})^k}{k!} \tag{6.66}$$

where $p = 0$ for $i < n_T$, and $p = i - (n_T - 1)$ for $i \geq n_T$.

In contrast to the case of the ZF receiver the MMSE SINR PDF given by (6.64)–(6.66) is not well-known. This result comes about by observing a duality between MIMO MMSE receivers, and traditional multi-user optimum combining systems with smart antennas. In what follows, we provide a sketch of the derivation, highlighting the use of finite-dimensional random matrix theory. Such a derivation is instructive in that it involves hypergeometric functions of matrix arguments, which often arise in MIMO analysis.

Sketch of derivation: The first step is to exploit standard properties of inverses of partitioned matrices and the matrix inversion lemma to write the conditional kth subchannel SINR (6.63) in the form:

$$\text{SINR}_k = \mathbf{h}_k^\dagger \left((n_T/\text{SNR})\mathbf{I}_{n_R} + \sum_{i\neq k} \mathbf{h}_i\mathbf{h}_i^\dagger \right)^{-1} \mathbf{h}_k \tag{6.67}$$

where $\mathbf{h}_i$ denotes the ith column of $\mathbf{H}$. Now, noting that from the law of large numbers:

$$\mathbf{I}_{n_R} = \lim_{K \to \infty} \sum_{i=1}^{K} \frac{1}{K} \mathbf{u}_i \mathbf{u}_i^\dagger, \quad a.s. \tag{6.68}$$

for independent random vectors $\mathbf{u}_i \sim \mathcal{CN}_{n_R}(\mathbf{0}, \mathbf{I}_{n_R})$, (6.67) can be further expressed as:

$$\mathrm{SINR}_k = \lim_{K \to \infty} \mathrm{SINR}_{k,K}, \quad a.s. \tag{6.69}$$

where:

$$\mathrm{SINR}_{k,K} = \mathbf{h}_k^\dagger \left(\mathbf{H}_\omega \mathbf{L} \mathbf{H}_\omega^\dagger \right)^{-1} \mathbf{h}_k \tag{6.70}$$

Here:

$$\mathbf{L} = \mathrm{diag}\left(\mathbf{I}_{n_T - 1}, \frac{n_T}{\mathrm{SNR} K} \mathbf{I}_K \right) \tag{6.71}$$

and $\mathbf{H}_\omega$ is given by:

$$\mathbf{H}_\omega = [\mathbf{H}_k, \ \mathbf{U}] \tag{6.72}$$

with $\mathbf{H}_k$ defined as in (6.50), and $\mathbf{U} = [\mathbf{u}_1, \ldots, \mathbf{u}_K]$.

The structure (6.70) is a quadratic form in Gaussian variables; the properties of which have been studied for many years in the multivariate statistics literature. Importantly, for a given K, the distribution of $\mathrm{SINR}_{k,K}$ in (6.70) is found as a special case of a general quadratic form result in [67], and has PDF:

$$f(\mathrm{SINR}_{k,K}) = n_R \binom{n_T + K - 1}{n_R} \left(\frac{q^{n_R}}{1 + q\mathrm{SINR}_{k,K}} \right)^{n_T + K} \mathrm{SINR}_{k,K}^{n_R - 1} \tag{6.73}$$
$$\times \det(\mathbf{L})^{-n_R} {}_1F_0(n_T + K; \mathbf{T}, \mathbf{S})$$

where $q > 0$ is an arbitrary constant, and ${}_1F_0(\cdot)$ is a hypergeometric function of the two matrix arguments:

$$\mathbf{T} = \mathbf{I}_{n_T + K - 1} - q\mathbf{L}^{-1}$$
$$\mathbf{S} = \mathrm{diag}((1 + q\mathrm{SINR}_{k,K})^{-1}, \mathbf{I}_{n_R - 1}) \tag{6.74}$$

In principle, the desired PDF $f(\mathrm{SINR}_k)$ could now be obtained by directly taking the limit of (6.73) as $K \to \infty$. This is a difficult task, however, due to the complexity of the ${}_1F_0(\cdot)$ hypergeometric function, which was traditionally expressed as an infinite series of zonal polynomials with an inner sum over partitions, and was therefore completely impractical for numerical computation purposes. In fact, only very recently has an efficient numerical algorithm been developed for approximating these series [74]. It turns out, however, that alternative determinant representations have been derived for this function (as well as other hypergeometric functions of matrix arguments), independently in [48,68], which involve only scalar functions, and therefore permit efficient computation. It appears that the first application of these determinant representations to problems in communications and information theory was made in the context of the current proof in [40], and this has since become a standard technique (see e.g., [65,72,87]).

The determinant representation for the hypergeometric function $_1F_0(\cdot)$ is given by:

$$_1F_0(L+1; \mathbf{X}, \mathbf{Y}) = \frac{1}{L!} \frac{\det\left(\{_1\mathcal{F}_0(2; y_i x_j)\}_{i,j=1,\ldots,L}\right)}{\prod_{1 \leq i < j \leq L}(x_i - x_j) \prod_{1 \leq i < j \leq L}(y_i - y_j)} \tag{6.75}$$

for certain matrices $\mathbf{X}$ and $\mathbf{Y}$, with eigen-values $x_1 < \cdots < x_L$ and $y_1 < \cdots < y_L$ respectively, and $_1\mathcal{F}_0(2; x) = (1-x)^{-2}$ is the classical scalar hypergeometric function.

Note that, as discussed in [40], there are some difficulties in directly applying the representation (6.75) to (6.73). This includes, for example, the requirement that all of the eigen-values of both of the matrix arguments in (6.75) be distinct, which is clearly not met in (6.73). This can be alleviated, however, by first perturbing the equal-valued eigen-values in $\mathbf{T}$ and $\mathbf{S}$ prior to application of the identity (6.75), and subsequently taking limits as these eigen-values reapproach one another. For full details of this approach, as well as some algebraic simplifications which follow, see [40].

The proof is then completed upon applying the limit $K \to \infty$ to the resulting expression.

Beyond the IID Rayleigh case, there appear to be few results on the distribution of the SINR for finite numbers of antennas. One notable exception is presented in [77], where simple approximations are derived, applying for both IID and transmit-correlated Rayleigh fading channels. For more general Rayleigh fading channels with both transmit and receive correlation, the SINR distribution has been characterized in [69] in terms of an integral expression for the m.g.f.

Sum Capacity: As for the ZF receiver, the MMSE receiver can be viewed as separating the MIMO channel into an equivalent set of parallel SISO channels. Assuming we treat these SISO channels separately and employ capacity-achieving codes on each, then the sum capacity is given by:

$$\frac{\mathcal{C}_{\mathrm{MMSE}}(\mathrm{SNR})}{n_{\mathrm{R}}} = \frac{1}{n_{\mathrm{R}}} \sum_{k=1}^{n_{\mathrm{T}}} \mathbb{E}_{\mathrm{SINR}_k}[\log(1 + \mathrm{SINR}_k)] \tag{6.76}$$

As mentioned above, the distribution of SINR_k is not available in closed-form for many channels of interest, which therefore precludes direct evaluation of (6.76). By using a simple manipulation however, the MMSE sum capacity (6.76) can be expressed in a form which can be easily evaluated, exactly in closed-form, for most general fading channels of interest (e.g., correlated Rayleigh/Rician channels); without requiring explicit statistical characterization of SINR_k. This is obtained by substituting (6.63) into (6.76), and using an identity from [53] [as also used in (6.50)] to yield:

$$\frac{\mathcal{C}_{\text{MMSE}}(\text{SNR})}{n_{\text{R}}} = \frac{n_{\text{T}}}{n_{\text{R}}} \mathbb{E}_{\mathbf{H}}\left[\log\det\left(\mathbf{I}_{n_{\text{T}}} + \frac{\text{SNR}}{n_{\text{T}}}\mathbf{H}^{\dagger}\mathbf{H}\right)\right]$$

$$- \frac{1}{n_{\text{R}}}\sum_{k=1}^{n_{\text{T}}}\mathbb{E}_{\mathbf{H}_k}\left[\log\det\left(\mathbf{I}_{n_{\text{T}-1}} + \frac{\text{SNR}}{n_{\text{T}}}\mathbf{H}_k^{\dagger}\mathbf{H}_k\right)\right] \quad (6.77)$$

Recognizing that the expectations in (6.77) are in the same form as (6.7) with $\mathbf{\Phi_x} = \mathbf{I}$, we can directly evaluate the MMSE sum capacity simply by substituting into (6.77) those expressions discussed and illustrated in Section 6.4. For example, let us assume an IID Rayleigh fading channel. Then, defining:

$$n = \min(n_{\text{R}}, n_{\text{T}}), \quad n' = \min(n_{\text{R}}, n_{\text{T}} - 1)$$
$$m = \max(n_{\text{R}}, n_{\text{T}}), \quad m' = \max(n_{\text{R}}, n_{\text{T}} - 1)$$

$$\frac{\mathcal{C}_{\text{MMSE}}(\text{SNR})}{n_{\text{R}}} = \frac{n_{\text{T}}}{n_{\text{R}}}\log_2 e$$

$$\times \left(\sum_{k=0}^{n-1}\sum_{\ell_1=0}^{k}\sum_{\ell_2=0}^{k}\binom{k}{\ell_1}\frac{(k+m-n)!(-1)^{\ell_1+\ell_2}I_{\ell_1+\ell_2+m-n}(\text{SNR}/n_{\text{T}})}{(k-\ell_2)!(m-n+\ell_1)!(m-n+\ell_2)!\ell_2!}\right.$$

$$\left. - \sum_{k=0}^{n'-1}\sum_{\ell_1=0}^{k}\sum_{\ell_2=0}^{k}\binom{k}{\ell_1}\frac{(k+m'-n')!(-1)^{\ell_1+\ell_2}I_{\ell_1+\ell_2+m'-n'}(\text{SNR}/n_{\text{T}})}{(k-\ell_2)!(m'-n'+\ell_1)!(m'-n'+\ell_2)!\ell_2!}\right) \quad (6.78)$$

where $I(\cdot)$ is defined as in (4.17).

Figure 6.1 shows $\frac{\mathcal{C}_{\text{MMSE}}(\text{SNR})}{n_{\text{R}}}$ in bits/s/Hz for IID Rayleigh channels, where it is seen to converge to the MIMO channel capacity for low SNR and the ZF sum capacity for high SNR. Clearly, at low SNR the noise dominates the interference, and from (6.62) it can be seen that the MMSE receiver reduces to the matched filter, which is optimal in the absence of interference. At high SNR, the interference dominates and as such the MMSE receiver primarily concentrates on suppressing the interference. This phenomenon motivates us to investigate the high SNR regime more closely.

For the case $n_{\text{R}} \geq n_{\text{T}}$, the sum capacity (6.77) can be expressed at high SNR as follows:

$$\mathcal{C}_{\text{MMSE}}(\text{SNR}) = n_{\text{T}}\log\left(\frac{\text{SNR}}{n_{\text{T}}}\right) + n_{\text{T}}\mathbb{E}_{\mathbf{H}}[\log\det(\mathbf{H}^{\dagger}\mathbf{H})]$$

$$- \sum_{k=1}^{n_{\text{T}}}\mathbb{E}_{\mathbf{H}_k}\left[\log\det(\mathbf{H}_k^{\dagger}\mathbf{H}_k)\right] + o(1) \quad (6.79)$$

where the $o(\cdot)$ term vanishes as $\text{SNR} \to \infty$. As for the ZF sum capacity and the MIMO channel capacity considered previously, we see that the MMSE sum capacity scales linearly with the minimum number of antennas (i.e., n_T). It is also interesting to compare (6.79) with the high-SNR ZF sum capacity expression given in (6.47). To this end, using the identity (6.50), it is easily verified that:

$$\lim_{\text{SNR}\to\infty} (\mathcal{C}_{\text{MMSE}}(\text{SNR}) - \mathcal{C}_{\text{ZF}}(\text{SNR})) = 0 \tag{6.80}$$

confirming that the sum capacity of the MIMO ZF and MMSE receivers are asymptotically equal at high SNR.

For the case $n_R < n_T$, the MMSE sum capacity (6.77) can be expressed as:

$$\mathcal{C}_{\text{MMSE}}(\text{SNR}) = n_T \mathbb{E}_{\mathbf{H}}[\log\det(\mathbf{H}\mathbf{H}^\dagger)]$$
$$- \sum_{k=1}^{n_T} \mathbb{E}_{\mathbf{H}_k}\left[\log\det(\mathbf{H}_k\mathbf{H}_k^\dagger)\right] + o(1) \tag{6.81}$$

which shows that whenever $n_R < n_T$, the sum capacity no longer scales linearly with the minimum number of antennas in the high SNR regime, but rather, converges to a deterministic limit given by the right-hand side of (6.81) (excluding the $o(\cdot)$ term).

Note that the expectations in (6.79) and (6.81) can be evaluated in closed-form for many channels of interest by using results from [82,87,95,151].

Error Performance: For the MMSE receiver, the average SER is given by:

$$\text{SER}_{\text{MMSE}}(\text{SNR}) = \frac{1}{n_T} \sum_{k=1}^{n_T} \text{SER}_{\text{MMSE},k}(\text{SNR}) \tag{6.82}$$

where $\text{SER}_{\text{MMSE},k}(\cdot)$ is the average SER for the kth MMSE subchannel, given by:

$$\text{SER}_{\text{MMSE},k}(\text{SNR}) = \mathbb{E}_{\text{SINR}_k}\left[a\,Q(\sqrt{2b\,\text{SINR}_k})\right] \tag{6.83}$$

where a and b are constellation-specific constants, defined as in (6.55).

For the case of IID Rayleigh fading, closed-form expressions can be derived for the SER. To this end, it is convenient to first apply integration by parts to (6.83) to express the subchannel SER in terms of the CCDF $\bar{F}(\cdot)$ as follows:

$$\text{SER}_{\text{MMSE},k}(\text{SNR}) = \frac{a}{2}\left(1 - \sqrt{\frac{b}{\pi}}\int_0^\infty \frac{e^{-bu}}{\sqrt{u}}\bar{F}(u)\,du\right) \tag{6.84}$$

Now substituting (6.65) into (6.84), integrating, and then substituting the result into (6.82) yields the closed-form solution:

$\text{SER}_{\text{MMSE}}(\text{SNR})$

$$= \frac{a}{2}\left(1 - \sqrt{b}\sum_{i=0}^{n_R-1}\beta_i \frac{(2i-1)!!}{2^i}U\left(i+\frac{1}{2}, i+\frac{5}{2}-n_T, b+\frac{n_T}{\text{SNR}}\right)\right) \quad (6.85)$$

where $(\cdot)!!$ is defined as in (6.60), with $(-1)!! = 1$, and $U(\cdot)$ is the Kummer-U function [1].

Figure 6.4 plots the average SER (6.85) against SNR for different antenna configurations. Results are shown for BPSK modulation. We clearly see that, as for the ZF receiver, the performance is significantly improved as n_R is increased for fixed n_T. It is also interesting to note that for the (5, 3) antenna case (i.e., when there are more transmitters than receivers), the SER exhibits an irreducible error floor at high SNR. In this scenario, regardless of the SNR, the receiver antenna array does not have enough degrees of freedom to completely cancel out the interferers. The plot indicates that for $n_T < n_R$, a system designer would be better off using only a subset of the transmit antennas.

Once again, we are motivated to examine the high SNR regime. In this case, it can be shown that the SER (6.85) is well-approximated by:

$$\text{SER}_{\text{MMSE}}^{\infty}(\text{SNR}) = \begin{cases} (G_a\,\text{SNR})^{-G_d}, & \text{for } n_R \geq n_T \\ \kappa, & \text{for } n_R < n_T \end{cases} \quad (6.86)$$

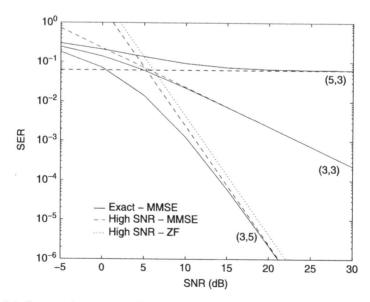

Figure 6.4. Exact and asymptotic SER for MMSE and ZF receivers with different antenna configurations (n_T, n_R), in IID Rayleigh fading. Results are shown for BPSK modulation.

where, for the case $n_R \geq n_T$, G_d and G_a are the diversity order and array gain respectively for the MMSE receiver, given by [73]:

$$G_d = n_R - n_T + 1$$

$$G_a = \frac{2b}{n_T} \left[\frac{a}{2^{n_T}} \frac{(2n_R - 1)!!}{(n_R - n_T + 1)!} U\left(n_T - 1, n_T - n_R - \frac{1}{2}, b \right) \right]^{-1/(n_R - n_T + 1)} \tag{6.87}$$

and, for the case $n_R < n_T$, κ is a constant independent of SNR, given by:

$$\kappa = \frac{a\, n_R \binom{n_T - 1}{n_R}}{2} \sum_{\ell=0}^{n_T - n_R} \frac{\binom{n_T - n_R}{\ell}}{(-1)^{\ell+1}} \sum_{i=1}^{n_R + \ell - 1} \frac{(2i - 1)!!}{i\, 2^i} U(i, 1/2, b) \tag{6.88}$$

Considering the case $n_R \geq n_T$, it is interesting to compare the high SNR result in (6.86)–(6.87) with the corresponding ZF SER result given in (6.58)–(6.60). Clearly, both ZF and MMSE have the same diversity order, as shown in Figure 6.4. We see, however, that there is a fixed power offset between the ZF and MMSE curves. Interestingly, this behavior is in contrast to the sum capacity considered previously, where the ZF and MMSE curves were shown to converge in the high SNR regime. This fixed power offset can be expressed as:

$$\Delta\text{SNR}_{\text{MMSE–ZF}} = 10 \log_{10} (\text{SNR}_{\text{ZF}} - \text{SNR}_{\text{MMSE}}) \quad (\text{dB}) \tag{6.89}$$

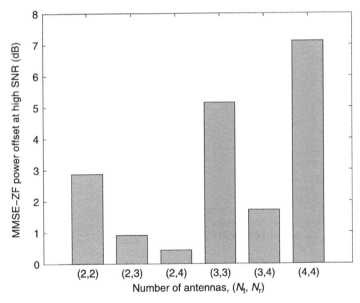

Figure 6.5. MMSE–ZF High SNR SER power offset. Results are shown for different antenna configuration (n_T, n_R), and for BPSK modulation.

where SNR_{MMSE} and SNR_{ZF} are chosen to satisfy:

$$\text{SER}_{\text{MMSE}}^{\infty}(\text{SNR}_{\text{MMSE}}) = \text{SER}_{\text{ZF}}^{\infty}(\text{SNR}_{\text{ZF}}) \tag{6.90}$$

This can be easily calculated in closed-form as:

$$\Delta\text{SNR}_{\text{MMSE}-\text{ZF}} = \frac{10\log_{10}\left(\dfrac{[2(n_R - n_T) + 1]!!\,2^{n_T - 1}}{(2n_R - 1)!!\,U(n_T - 1, n_T - n_R - 1/2, b)}\right)}{n_R - n_T + 1} \quad (\text{dB}) \tag{6.91}$$

In Figure 6.5, the power offset (6.91) is shown for different antenna configurations. Clearly, the relative performance depends not only on the number of antennas, but also on the difference between n_R and n_T. From the figure we can conclude that for systems with $n_R = n_T$, there is a considerable advantage in using a MMSE receiver compared with ZF, but for other scenarios the advantage is much less.

6.7 MULTIUSER MULTI-ANTENNA SYSTEMS

Although this chapter has focused on single user multiantenna systems, it can be readily pointed out that all wireless systems are at least in some sense inherently *multiuser*: at the outset, they must share bandwidth resources with all other wireless devices, and more specifically any deployed system generally must serve a number of users in overlapping geographic areas. Therefore, the question is not whether the system is multiuser or single user, but rather how the multiple users are accommodated. In this sense, the single user model explored thus far in this chapter can be viewed as the scenario in which a higher layer networking protocol has allocated the channel and the entire transmit and receive antenna array to a single user for a certain period of time.

The single user model, which is implicitly TDMA/FDMA, immediately raises other questions. Why shouldn't users share resources in the spatial domain in addition to the time-frequency domain? What about the interference from users who are allocated the same time-frequency slot in neighboring cells or systems? In this section of the chapter, we explore these questions, and in particular consider interference precancellation techniques that are conceptually like performing multiuser detection at the transmitter.

We begin by observing that the two scenarios we have just mentioned—same-cell cooperation and other-cell interference—have quite different characteristics, as shown for a hypothetical downlink MIMO system in Figure 6.6. In the case of same-cell cooperation and interference, there are several options. First and most simply, a scheduler can restrict each base station to serve a single mobile station (MS) at any point in time. This is the dominant approach in current and emerging wireless broadband standards. Second, the base station can apply sophisticated pre-cancellation algorithms in order to allow each MS to receive its desired signal with straightforward signal processing. These multiuser precoding techniques typically require the base station to know

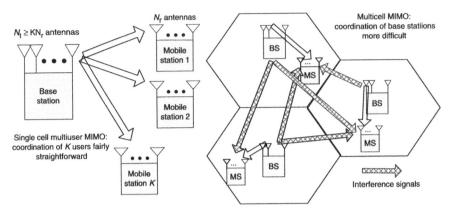

Figure 6.6. In-cell multiuser interference (left) and other-cell multiuser interference (right) in a MIMO cellular system. The former is easy to deal with than the latter, since other-cell interference suggests the need for base-station cooperation.

the data and channels to each user. The third option is for the base station to simply share the total of n_T data streams among the multiple users at the same time, leaving the spatial interference cancellation to the MSs.

Other-cell interference poses additional problems. All well-designed cellular systems are by nature interference-limited: if they were not, it would be possible to increase the spectral efficiency by lowering the frequency reuse or increasing the average loading per cell. Therefore, MIMO systems will operate at a delicate balancing point: most of the gains promised by spatial multiplexing are at high SINR (15 dB and up), yet most cellular users in today's systems experience low SINR. In fact, a significant percentage get an SINR below 3 dB due to aggressive frequency reuse. Therefore, this type of multiuser interference is very important, regardless of the adopted approach for same-cell interference suppression. We overview emerging methods for managing other-cell interference in Section 6.7.2: the better of these techniques rely in essence on cooperation between spatially-dispersed transmitters.

6.7.1 Same-Cell Interference and Cooperation

As we noted above, there are three principle approaches to managing same-cell interference in the spatial domain. First, the network can circumvent the interference problem with an orthogonal allocation of time-frequency resources so that within each cell, communication takes the form of single user MIMO. This is the simplest and most common approach, but suboptimal. The third approach mentioned above is interference cancellation in downlink, where streams are simply transmitted to each user with the burden on the receiver to decode them. This is not expected to be a viable approach, particularly in the downlink, since it requires each MS to have on the order of n_T sufficiently uncorrelated receive antennas and a tolerance for considerable computational complexity. On the other hand, receiver multiuser interference cancellation for K mobile stations is theoretically possible in the uplink if the base station has at least KN antennas, where each MS transmits N data streams. The most attractive and interesting option is to use precoding at the base station transmitter in order to allow multiple users to be serviced simultaneously.

6.7.1.1 Downlink: Precoding

In the downlink, we assume that the number of base station antennas is $n_T \geq n_R$, where for simplicity of notation each mobile has n_R receive antennas. From straightforward linear algebra arguments, MIMO receivers are able to decode parallel data streams by suppressing the spatial interference between the signals sent from the n_T transmit antennas (using linear signal processing techniques), as long as the number of receive antennas $n_R \geq n_T$. Therefore, a K user MIMO system appears to require $n_R \geq Kn_T$ receive antennas in order to fully suppress other-cell interference (OCI) with a linear receiver. However, instead it is possible to view a K user system as having Kn_R receive antennas, and so as long as $Kn_R \geq n_T$, an encoding scheme is possible that will allow each unit to decode their desired n_R streams of data. The key distinction from the normal single user model is that each set of n_R antennas is not allowed to cooperate in the decoding process with the other sets, since they are geographically dispersed.

To be concrete, the multiuser MIMO model is similar to before:

$$\mathbf{y} = \sqrt{g}\mathbf{H}\mathbf{x} + \mathbf{n}$$

but now $\mathbf{y}$ is a Kn_R vector while $\mathbf{x}$ is still $n_T \times 1$. Furthermore, each receiver only has access to a subset of size n_R of $\mathbf{y}$. What we will consider now is how to best design $\mathbf{x}$ at the base station in order to make this task possible for the distributed receivers.

Dirty Paper Coding: From an information theory standpoint, the optimal approach to this multiuser problem is to adapt the so-called "dirty paper coding" of Costa [27] to the MIMO interference channel. The philosophy of this technique is to construct codewords for each user that use the spatial interference due to the other users to each user's advantage (e.g., see the discussion in Chapter 8). In practice, such an approach would require non-causal knowledge of the channels to each user, and a prohibitive successive encoding process [16,138,149]. Like many breakthroughs in information theory, although the optimal scheme is impractical, it provides insight and an upper bound to what is possible.

Block Diagonalization: A natural alternative approach is to design a linear precoder that suppresses the interference from concurrent transmissions, in a manner similar to linear multiuser detection performed at the transmitter. In this case, the base station transmits to K users simultaneously. For simplicity each user has $n_R^{(k)}$ receive antennas, and for simplicity we will assume that the number of streams transmitted to that user is also $n_R^{(k)}$, which will allow for a linear receiver.

In this case, the received signal at each user k can be represented as:

$$\mathbf{y}_k = \sqrt{g_k}\mathbf{H}_k\mathbf{T}_k\mathbf{x}_k + \sqrt{g_k}\mathbf{H}_k \sum_{j=1, j \neq k}^{K} \mathbf{T}_j\mathbf{x}_j + \mathbf{n}_k$$

where $\mathbf{T}_k$ is a precoding matrix for user k and now both $\mathbf{y}_k$ and $\mathbf{x}_k$ are $n_R^{(k)}$ vectors. The objective of block diagonalization is to find precoding matrices $\{\mathbf{T}_k\}_{k=1}^K$ such that $\mathbf{H}_k\mathbf{T}_j = 0$, $\forall_j \neq k$. In other words, the interference for all other active users is nulled by the precoder for user k, which effectively diagonalizes the channel, giving the technique its name. The zero-interference constraint is expressed as:

$$\bar{\mathbf{H}}_k\mathbf{T}_k = \mathbf{0}, \quad k = 1,\dots,K \tag{6.92}$$

where $\bar{\mathbf{H}}_k = \left(\mathbf{H}_1^\dagger \quad \cdots \quad \mathbf{H}_{k-1}^\dagger \quad \mathbf{H}_{k+1}^\dagger \quad \cdots \quad \mathbf{H}_K^\dagger\right)^\dagger$.

Denoting the SVD of $\bar{\mathbf{H}}_k$ as $\bar{\mathbf{H}}_k = \bar{\mathbf{U}}_k\left(\bar{\boldsymbol{\Sigma}}_k \quad \mathbf{0}\right)\left(\bar{\mathbf{V}}_k^1 \quad \bar{\mathbf{V}}_k^0\right)^\dagger$, where $\bar{\boldsymbol{\Sigma}}_k$ is the $\bar{r}_k \times \bar{r}_k$ diagonal matrix containing the $\bar{r}_k$ non-zero singular values of $\bar{\mathbf{H}}_k$, and $\bar{\mathbf{V}}_k^1$ and $\bar{\mathbf{V}}_k^0$ contain the singular vectors corresponding to the non-zero and zero singular values, respectively. Since the columns of $\bar{\mathbf{V}}_k^0$ span the null space of $\bar{\mathbf{H}}_k$, constructing $\mathbf{T}_k$ with n_R columns of $\bar{\mathbf{V}}_k^0$ will automatically satisfy the zero-interference constraint. Assuming the channel is full-rank, the required singular vectors exists as long as $n_T \geq \sum_{j=1}^K n_R^{(j)}$. Further details on block diagonalization can be found in [19,24,121,144].

Performance Comparison. In Figure 6.7 from [113], we compare the capacity of DPC, block diagonalization, and the conventional TDMA approach. Clearly, block diagonalization is an attractive approach, since it approaches the DPC bound, while requiring only linear processing. The capacity gain versus a time-sharing approach

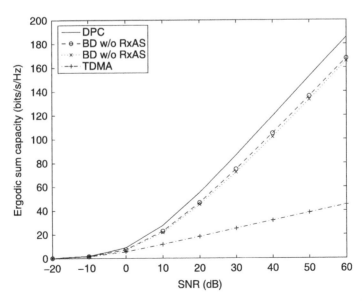

Figure 6.7. Ergodic sum capacity of Dirty Paper Coding, Block Diagonalization with and without receive antenna selection, and TDMA for $n_T = 10$, $n_R = 2$, $K = 5$. If antenna selection is used, $n_R = 3$ but only the best two are used at a time.

is very significant. Since it is expected that the number of antennas on a base station can be quite a bit larger than what is possible on an MS, an approach like BD appears to be quite attractive in practice, assuming that suitable channel state information can be provided to the transmitter for each of the K users. Limited feedback techniques such as those developed for single-user MIMO systems [79,80] have been extended to multiuser MIMO systems [55,148], with the conclusion that multiuser precoding does not necessitate a prohibitive amount of feedback.

6.7.1.2 Uplink: Interference Cancellation In the uplink, spatial interference cancellation even for multiple users can be performed by the base station. Naturally, the performance decreases fairly rapidly when $n_R < Kn_T$, where n_R is now the size of the base station's receive antenna array and each MS sends n_T streams of data. Conceptually, once the K users are selected for transmission the problem is no different than a $Kn_T \times n_R$ single user MIMO system, with the following differences. First, the channel gain matrix $\mathbf{H}$ and its correlation matrix $\mathbb{E}[\mathbf{HH}^\dagger]$ will change since uplink users often have very different channel gains due to large differences in path loss and shadowing, and naturally antennas on the same MS would have a fair amount of correlation while antennas on different mobiles would have none. Second, for the simpler spatial multiplexing scenarios, each mobile would encode its n_T streams independently and so stream rotation techniques such as that used in D-BLAST would not be possible, or more generally, any cooperative precoding among the Kn_T transmit elements (e.g., for precoding) would require extensive real-time cooperation between all the distributed units, which is probably not viable in the foreseeable future. Overall, however, the signal processing involved is substantially similar to the single-user MIMO channel, so further discussion of multiuser uplink interference cancellation designs is deferred here.

6.7.2 Other-Cell Interference and Cooperation

In contrast to the single user MIMO scenario, cellular MIMO systems are subject to a large amount of interference. Particularly near the cell edges in a downlink system with aggressive frequency reuse patterns, the total amount of interference dictates operating points below 5 dB or even 0 dB SINR. As the number of interfering sources increases, it is not typically possible to suppress all the OCI with spatial signal processing, and instead the interference is generally treated as noise. As the number of interfering sources becomes large, the interference becomes increasingly Gaussian due to the Central Limit Theorem, which is worst case according to information theory. In view of this, it is not surprising that recent research on spatial multiplexing in cellular systems has shown that adding data streams at the base station can actually decrease the throughput owing to the increased dimensionality of spatial interference [14,17]. Instead, at low SINR, higher capacity is achieved by focusing power on the best eigen-value(s) in the channel matrix. Therefore, if spatial multiplexing is to be viable in a cellular system, new means for increasing the SINR to active users need to be developed.

Base station cooperation holds considerable promise as a means of managing (or suppressing) other-cell interference. This implies that the base stations are networked together and can share information with each other. Communication between base stations already occurs in practice to coordinate handoffs and other network-level operations; additional cooperative operations can result in increased system capacity. Naturally, the amount of cooperation has to be balanced against the gain it provides.

6.7.2.1 Joint Encoding

One possibility is to extend to the multiuser precoding techniques of Section 6.7.1 to the cellular scenario [56,100,111]. If the received interference signals are known to each base station, then each base station can form part of a macroscopic MIMO array, sometimes called a "super base station". For example, if each base station has n_T antennas, and L of them cooperate, then at least in theory, this can be viewed as a single base station with Ln_T antennas. From such a viewpoint, the same mathematical theory of block diagonalization from the previous section can be applied, with however the significant following changes.

First, each set of n_T antennas has its own transmit power constraint. Second, the base stations now have to be able to share both channel state and user data information and collaboratively decide which subset of users to transmit at each time instance. In addition to the obvious strain on the back-end network, this imposes the requirement that the base stations be precisely synchronized and capable of real-time information exchange. Third, the signals to each user will usually become extremely unbalanced, since each set of n_T antennas is geographically isolated. Typically, only two or at most three of the L base stations can be heard by each user, while the other base stations will be close to the noise floor. This further complicates the encoding process for each user.

6.7.2.2 Base Station Cooperative Scheduling

Since the aforementioned challenges are fairly daunting, a simpler approach is to allow the base stations to simply take turns transmitting, or to schedule transmissions in a way that causes minimal interference to other users. Recent work, e.g., [25,39,150], has investigated the possibility of neighboring base stations scheduling their transmissions in a cooperative fashion. This is a generalization of the concept of spatial frequency reuse in which simple intercell coordination is used to select appropriate users on a system wide level for each time slot. For example, two base stations should not simultaneously transmit at full power to mobile stations on their mutual cell boundaries. On the contrary, it would be desirable to transmit to a near-in user in one cell, and a cell-edge user in an adjacent cell [60,145].

To make this concept more concrete, consider two classes of users: cell edge users (exterior) and users close to the base station (interior). Base stations could then alternate between transmitting to their interior and exterior users, with the two most significant neighboring two base stations alternating on opposite time slots. Because of the path gain differences between the interior and exterior classes, the achievable rates at each time slot for the exterior and interior users can be written as [60]:

$$R_e = \log\left(1 + \frac{Q_e}{1 + \alpha^2 Q_i}\right)$$
$$R_i = \log\left(1 + \beta^2 Q_i\right), \tag{6.93}$$

where α characterizes the attenuation from the interfering base station to the exterior users, and β is the attenuation to the interior users from its desired base station, and we assume that the interior users are close enough to the desired base station that interference from other cells is negligible. Additionally, Q_e and Q_i are related to the transmitted power and satisfy $Q_i + Q_e = 2\text{SNR}$, and can be optimized as a function of α and β. The key point is that this scheme, known as "cell breathing," provides a framework for achieving acceptable SINR to both interior and exterior users since $\beta \gg \alpha$. This could also allow spatial multiplexing techniques to be deployed in a cellular system, since these require higher SINR, as previously noted. In reality, it is likely that while interior users will be able to make use of spatial multiplexing, exterior users will likely have to be more conservative in their use of MIMO techniques. The next section describes how a MIMO system can adaptively select the mode of operation for the different users depending on their SINR conditions.

6.8 DIVERSITY-MULTIPLEXING TRADEOFFS AND SPATIAL ADAPTATION

In this section we turn our attention to the roles of diversity and multiplexing in a MIMO communication channel. It turns out that this discussion is one of the departure points between MIMO communication and joint detection. It is important, however, as it provides insight into MIMO communication link engineering.

Thus far the role of MIMO in this chapter has been a source of increased spectral efficiency without bandwidth or power enhancement. This can be achieved in many cases by sending independent data streams over each transmit antenna and thus is known as the spatial multiplexing mode of operation. Antennas, though, can play other roles in wireless communication; they can be a source of diversity to mitigate fading. Space-time codes can take advantage of the diversity from multiple transmit antennas, while techniques such as maximum ratio combining can exploit the diversity from multiple receive antennas. MIMO communication links using spatial multiplexing also benefit from diversity. For example, in Section 6.6 the diversity order of spatial multiplexing with a zero forcing receiver was shown to be $n_R - n_T + 1$ for the case $n_R \geq n_T$. Maximizing diversity requires $n_T = 1$ but this would not be a MIMO system since it only has one transmit antenna. Maximizing the number of streams would result in $n_R = n_T$ but the resulting diversity gain would be 1 (the lowest possible). Clearly, there are some tradeoffs between the multiplexing rate and the diversity gain.

In this section, we will present results on diversity and multiplexing tradeoffs from two perspectives. First, we will summarize some theoretical results on the diversity-multiplexing tradeoff curve. These results provide a certain perspective on robust system operation with minimal channel state information required at the transmitter. Second, we will discuss spatial adaptation through diversity-multiplexing switching. The idea is to switch between different possible transmit configurations based on transmit channel state information. This approach varies the number of transmit streams but maximizes the diversity order even with ZF receivers.

6.8.1 Diversity-Multiplexing Tradeoff

The diversity-multiplexing tradeoff from a rate-distortion perspective was developed by Zheng and Tse [152]. For a space-time coding scheme, a diversity gain $d(r)$ is said to be achieved at multiplexing rate r if the data rate scales as $R = r \log \text{SNR}$ while the error probability scales as $p_e \approx \text{SNR}^{-d}$. Essentially, this means that $\lim_{\text{SNR} \to \infty} \log p_e / \log \text{SNR} = -d$. The optimum tradeoff for any transmission scheme is denoted $d^*(r)$ and is derived from the outage probability $p_{out} = \mathbb{P}\{\log \det(\mathbf{I} + \text{SNR}\mathbf{H}\mathbf{H}^\dagger) < r \log \text{SNR}\}$. Specifically, a diversity gain of $d^*(r)$ is achieved with multiplexing gain r if $R = r\text{SNR}$ and $p_{out}(R) \approx \text{SNR}^{-d^*(r)}$.

The diversity-multiplexing tradeoff curve has been computed for many different space-time coding schemes and the optimum tradeoff curve has been calculated in some cases, notably for IID Rayleigh fading channels. To understand the diversity-multiplexing tradeoff curve, it is prudent to study an example. We consider Rayleigh fading channels with $n_T = n_R = 2$. For this case, spatial multiplexing with an ML receiver achieves $d_{sm}(r) = 2 - r$ for $r \in [0, 2]$. The Alamouti space-time code [3] achieves $d_a(r) = 4 - 4r$ for $r \in [0, 4]$. The optimum tradeoff curve is the piecewise linear curve connecting points defined by $(0, 4)$, $(1, 1)$, and $(2, 0)$. The curve is illustrated in Figure 6.8. First consider the multiplexing and Alamouti curves. As intuition would suggest, for higher data rates, multiplexing is optimum while for larger diversity requirements the Alamouti code is appropriate. Something that is somewhat less intuitive, though, is this fact: the two curves cross at $r = 2/3$. This means that spatial multiplexing is optimum at least part of the time for $r < 1$, which is where the Alamouti code should be better. This can be explained by noting that while spatial multiplexing overlaps with the optimum tradeoff curve for a piece, the Alamouti code does not. This means that there are other coding approaches that can improve on Alamouti, in terms of the diversity-multiplexing tradeoff.

The optimum tradeoff for Rayleigh fading channels was found to be the piecewise linear curve connecting points defined by $(k, (n_T - k)(n_R - k))$ for

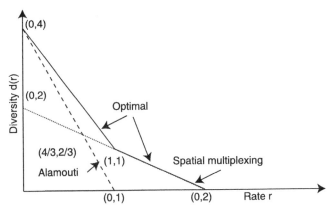

Figure 6.8. Diversity-multiplexing tradeoff curve for $n_R = n_T = 2$. Shown are spatial multiplexing with an ML receiver, the Alamouti space-time code, and the optimal tradeoff.

$k = 0, 1, \ldots, \min(n_T, n_R)$ [152]. Spatial multiplexing is always optimum for the last piece, from rates $r \in [\min(n_T, n_R) - 1, \min(n_T, n_R)]$. Recently, space-time codes have been developed that can achieve other portions of the curve. Well-known examples include the Golden code [9] and cyclic-division-algebra based codes [33]. These codes generally increase performance at the expense of additional receiver complexity.

To summarize, the diversity-multiplexing tradeoff curve provides some insight into the relationship between rate and diversity for large SNR. It justifies the use of spatial multiplexing for high rates and motivated the discovery of tradeoff-optimal space-time codes to fill in the gaps at other rates.

6.8.2 Mode Adaptation: Switching Between Diversity and Multiplexing

While the tradeoff curve is useful, it does not capture all the leverage from diversity and multiplexing in practical systems. One important omission is that it applies only for large SNR. For example, it is known that beamforming with a single vector is optimal at low SNR (from waterfilling), yet this insight is not captured in the tradeoff curve. Another dimension not included is space-time adaptation. The tradeoff curve assumes that the same space-time code family is used for all channel realizations. Practically, though, the rate or space-time code may change as the channel changes thanks to channel state information at the transmitter [18]. These limitations make the tradeoff curve less useful in the design of practical MIMO links with some spatial adaptation (like WiMax, for example [7]).

Including the possibility of adapting between diversity and multiplexing modes of operation exposes the importance of this capability. This concept was first proposed by Heath and Paulraj who addressed switching between the Alamouti space-time code and spatial multiplexing [51]. This concept can be generalized by allowing the transmitter to send a different number of data streams, or modes, based on the channel state [78]. Intuitively, when the channel is full rank spatial multiplexing is used, while when it is low rank some form of beamforming on the dominant eigen-modes is preferred.

To illustrate the concept of spatial adaptation and its advantages, consider $n_T = n_R$ and a Rayleigh fading channel. Suppose that the transmitter can use either equal power allocation and spatial multiplexing with a zero forcing receiver or transmit beamforming. Spatial multiplexing with a zero forcing receiver was studied in Section 6.6. With transmit beamforming, the receiver observes:

$$\mathbf{y} = \sqrt{g}\mathbf{H}\mathbf{w}x + \mathbf{n} \tag{6.94}$$

or equivalently after matched filtering with $\mathbf{Hw}$ and rescaling:

$$y = g\lambda_{max}^2(\mathbf{H})x + v \tag{6.95}$$

Beamforming only uses the dominant right singular vector of $\mathbf{H}$ and thus intuitively is only a good idea if $\mathbf{H}$ is low rank. Clearly $r = 1$ for transmit beamforming and $r = n_T$ for spatial multiplexing. In terms of diversity, transmit beamforming and receive combining obtains $d = n_T n_R$ while spatial multiplexing achieves only $d = 1$ from (6.59).

To make the comparison fair, assume that the constellations with each approach are chosen such that the total spectral efficiency is the same. Thus if spatial multiplexing sends two bits on each of n_T streams, then transmit beamforming would need to send $2n_T$ bits per symbol to keep the same spectral efficiency. Consider an adaptive algorithm that chooses the approach that has the lowest probability of error for a given channel realization. Since transmit beamforming obtains a diversity of $n_T n_R$ (the maximum achievable), then this algorithm also obtains a diversity order of $n_T n_R$, no matter the size of n_T! The implication of this result is illustrated in Figure 6.9 for $n_T = n_R = 4$. The adaptive curve enjoys a huge performance improvement over spatial multiplexing and some improvement over beamforming. For reference, note that with an optimum ML receiver, spatial multiplexing would obtain $d = n_R$. Thus spatial adaptation provides more diversity than spatial multiplexing alone (with the best receiver) and has low complexity.

There are many other studies of diversity-multiplexing switching and space-time adaptation. Combining with antenna subset selection [50] or limited feedback [78] shows that these benefits can be achieved with as little as $\log_2 n_T$ bits of feedback per channel realization. Spatial adaptation has also been combined with adaptive modulation to form space-time adaptive modulation [35]. In such a configuration, both the modulation/coding scheme and the space-time transmission scheme are varied with the objective of maximizing the rate at a given SNR.

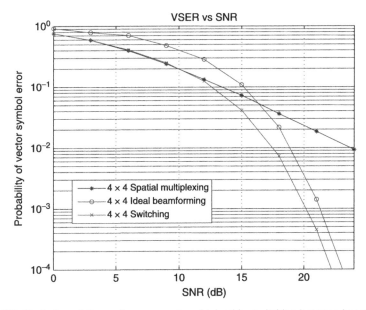

Figure 6.9. Illustration of the performance gains obtained by switching between beamforming and spatial multiplexing for $n_R = n_T = 4$, 4-QAM for spatial multiplexing for each stream and 256-QAM for beamforming. VSER is the symbol error rate (vector symbol error rate for spatial multiplexing, the probability that one or more symbols is in error).

To summarize, by adapting between diversity and multiplexing schemes based on channel state information it is possible to achieve full diversity of $d = n_T n_R$ and a maximum rate of $\min(n_T, n_R)$ in practical systems with low complexity feedback.

6.9 CONCLUSIONS

This chapter has intended to provide an overview of state-of-the-art MIMO communications from three complementary perspectives, all of which parallel multiuser detectors in some distinctive ways, while diverging in others. First, we overviewed recent results on the fundamental information theoretic limits of MIMO channels, characterized by the ergodic and outage capacity, for fast-varying and slow-varying channels, respectively. The key results diverge from multiuser detection primarily due to (i) the different structure of the signature matrix, which is (approximately) complex Gaussian for MIMO, but generally discrete in CDMA, and (ii) the ability of the transmitter to jointly encode the transmit vectors.

Next, we summarized key results on the error-rate performance of practical (linear) receiver architectures. In this section, the connection to multiuser detection is the most concrete, with clear parallels between the ZF and MMSE receivers for MIMO channels and CDMA systems.

Finally, we discussed the important system-level considerations for MIMO systems, namely how to simultaneously accommodate multiple users and how to tradeoff between MIMO's diversity and multiplexing benefits. Although CDMA systems are relatively robust to SINR low by design, the performance of MIMO systems is very sensitive to SINR. We overviewed recently proposed techniques for controlling performance in challenging environments, namely through base station cooperation and MIMO mode adaptation.

REFERENCES

1. M. Abramowitz and I. A. Stegun. *Handbook of Mathematical Functions with Formulas, Graphs, and Mathematical Tables.* Dover Publications, New York, 9th edition, 1970.

2. M. Abramowitz and I. A. Stegun. *Multiantenna capacity: Myths and Realities, Chapter 8, Space-Time Wireless Systems: From Array Processing to MIMO Communications.* H. Boelcskei, D. Gesbert, C. Papadias, and A. J. van der Veen, Eds., Cambridge University Press, 2006.

3. S. M. Alamouti. A simple transmit diversity technique for wireless communications. *IEEE J. on Selected Areas in Communications,* 16(8):1451–1458, Oct. 1998.

4. G. Alfano, A. M. Tulino, A. Lozano, and S. Verdú. Capacity of MIMO channels with one-sided correlation. In *IEEE Int. Symp. on Spread Spec. Tech. and Appl. (ISSSTA),* pages 515–519, Sydney, Australia, Aug. 30–Sept. 02, 2004.

5. J. B. Andersen. Array gain and capacity for known random channels with multiple element arrays at both ends. *IEEE J. on Selected Areas in Communications,* 18(11):2172–2178, Nov. 2000.

6. J. G. Andrews. Interference cancellation for cellular systems: A contemporary overview. *IEEE Wireless Communications Magazine*, 12(2):19–29, Apr. 2005.

7. J. G. Andrews, A. Ghosh, and R. Muhamed. *Fundamentals of WiMAX*. Prentice-Hall, 2007.

8. J. G. Andrews and T. Meng. Optimum power control for successive interference cancellation with imperfect channel estimation. *IEEE Trans. on Wireless Communications*, 2(2):375–383, Mar. 2003.

9. J. C. Belfiore, G. Rekaya, and E. Viterbo. The golden code: A 2×2 full-rate space-time code with nonvanishing determinants. *IEEE Trans. on Information Theory*, 51(4): 1432–1436, Apr. 2005.

10. S. A. Bergmann and H. W. Arnold. Polarization diversity in portable communications environment. *IEE Electronics Letters*, 22(11):609–610, May 1986.

11. E. Biglieri, J. Proakis, and S. Shamai. Fading channels: Information-theoretic and communications aspects. *IEEE Trans. on Information Theory*, 44(6):2619–2692, Oct. 1998.

12. E. Biglieri and G. Taricco. Transmission and reception with multiple antennas: Theoretical foundations. *Foundations and Trends in Communications and Information Theory*, 1(2), 2004.

13. E. Biglieri, A. M. Tulino, and G. Taricco. How far away is infinity? Using asymptotic analyses in multiple antenna systems. *Proc. IEEE Int. Symp. on Spread Spectrum Techn. and Applications (ISSSTA'02)*, 1:1–6, Sep. 2002.

14. R. Blum. MIMO capacity with interference. *IEEE Journal on Sel. Areas in Communications*, 21(5):793–801, June 2003.

15. L. H. Brandenburg and A. D. Wyner. Capacity of the Gaussian channel with memory: The multivariate case. *Bell System Technical Journal*, 53(5):745–778, May–June 1974.

16. G. Caire and S. Shamai. On the achievable throughput of a multi-antenna gaussian broadcast channel. *IEEE Trans. on Info. Theory*, 49(7):1691–1706, July 2003.

17. S. Catreux, P. Driessen, and L. Greenstein. Attainable throughput of an interference-limited multiple-input multiple-output (MIMO) cellular system. *IEEE Trans. on Communications*, 49(8):1307–1311, Aug. 2001.

18. S. Catreux, V. Erceg, D. Gesbert, and R. W. Heath Jr. Adaptive modulation and MIMO coding for broadband wireless data networks. *IEEE Communications Magazine*, 40(6):108–115, 2002.

19. R. Chen, R. W. Heath, and J. G. Andrews. Transmit selection diversity for multiuser spatial division multiplexing wireless systems. *IEEE Trans. on Signal Processing*, Mar. 2007.

20. M. Chiani. Evaluating the capacity distribution of MIMO Rayleigh fading channels. *Proc. IEEE Int. Symp. on Advances in Wireless Communications*, pages 3–4, Victoria, Canada, Sep. 23–24, 2002.

21. M. Chiani, M. Z. Win, and A. Zanella. On the capacity of spatially correlated MIMO Rayleigh-fading channels. *IEEE Trans. on Information Theory*, 49(10):2363–2371, Oct. 2003.

22. D. Chizhik, F. R. Farrokhi, J. Ling, and A. Lozano. Effect of antenna separation on the capacity of BLAST in correlated channels. *IEEE Communications Letters*, 4(11):337–339, Nov. 2000.

23. D. Chizhik, G. J. Foschini, and R. A. Valenzuela. Capacities of multi-element transmit and receive antennas: Correlations and keyholes. *IEEE Electronic Letters*, pages 1099–1100, June 2000.

24. R. L. Choi and R. D. Murch. A transmit processing technique for multiuser MIMO systems using a decomposition approach. *IEEE Trans. on Wireless Communications*, 3(1):20–24, Jan. 2004.

25. W. Choi and J. G. Andrews. The capacity gain from base station cooperative scheduling in a MIMO DPC cellular system. In *Proc., IEEE Intl. Symposium on Information Theory*, Seattle, WA, June 2006.

26. C. Chuah, D. Tse, J. Kahn, and R. Valenzuela. Capacity scaling in dual-antenna-array wireless systems. *IEEE Trans. on Information Theory*, 48(3):637–650, Mar. 2002.

27. M. Costa. Writing on dirty paper. *IEEE Trans. on Info. Theory*, 29(3):439–441, May 1983.

28. T. M. Cover and J. A. Thomas. *Elements of Information Theory*. John Wiley and Sons, Inc., 1991.

29. D. Cox. Universal digital portable radio communications. *Proc. IEEE*, 75(4):436–477, Apr. 1987.

30. I. Csiszár and J. Körner. *Information Theory: Coding Theorems for Discrete Memoryless Systems*. Academic, New York, 1981.

31. S. N. Diggavi, N. Al-Dhahir, A. Stamoulis, and A. R. Calderbank. Great expectations: The value of spatial diversity in wireless networks. *Proc. IEEE*, 92(2):219–270, Feb. 2004.

32. M. Dohler. *Virtual Antenna Array*. PhD thesis, King's College London, University of London, London, 2003.

33. P. Elia, K. R. Kumar, S. A. Pawar, P. V. Kumar, and H. F. Lu. Explicit space–time codes achieving the diversity–multiplexing gain tradeoff, *IEEE Trans. on Information Theory*, 52(9):3869–3884, Sept. 2006.

34. A. Forenza, M. R. McKay, A. Pandharipande, R. W. Heath Jr., and I. B. Collings. Adaptive MIMO transmission for exploiting the capacity of spatially correlated channels. *IEEE Trans. Veh. Technol.*, 56(2):619–630, Mar. 2007.

35. A. Forenza, A. Pandharipande, H. Kim, and R. W. Heath Jr. Adaptive MIMO transmission scheme: exploiting the spatial selectivity of wireless channels. In *Proc. IEEE Vehicular Technology Conference*, volume 5, pages 3188–3192, 2005.

36. G. J. Foschini and M. J. Gans. On limits of wireless communications in fading environment when using multiple antennas. *Wireless Personal Communications*, 6(3): 311–335, Mar. 1998.

37. G. J. Foschini. Layered space-time architecture for wireless communication in a fading environment when using multi-element antennas. *Bell Labs Technical Journal*, 1:41–59, 1996.

38. G. J. Foschini, G. D. Golden, R. A. Valenzuela, and P. W. Wolniansky. Simplified processing for high spectral efficiency wireless communication employing multi-element arrays. *IEEE J. on Selected Areas in Communications*, 17(11):1841–1852, Nov. 1999.

39. G. J. Foschini, H. Huang, K. Karakayali, R. A. Valenzuela, and S. Venkatesan. The value of coherent base station coordination. In *Proc., Conference on Information Sciences and Systems (CISS)*, Johns Hopkins University, Mar. 2005.

40. H. Gao, P. J. Smith, and M. V. Clark. Theoretical reliability of MMSE linear diversity combining in Rayleigh-fading additive interference channels. *IEEE Trans. Commun.*, 46(5):666–672, May 1998.

41. D. Gesbert, H. Bolcskei, D. Gore, and A. J. Paulraj. MIMO wireless channels: Capacity and performance prediction. *Proc. IEEE GLOBECOM'2000*, San Francisco, CA, USA, Dec. 2000.

42. D. Gesbert, M. Shafi, D. Shiu, P. J. Smith, and A. Naguib. From theory to practice: An overview of MIMO Space-Time coded wireless systems. *IEEE J. on Selected Areas in Communications*, 21(3):281–302, Apr. 2003.

43. V. L. Girko. A refinement of the central limit theorem for random determinants. *Theory Prob. Appl.*, 42(1):121–129, 1997.

44. A. Goldsmith, S. A. Jafar, N. Jindal, and S. Vishwanath. Capacity limits of MIMO channels. *IEEE J. on Selected Areas on Communications*, 21(5):684–702, June 2003.

45. A. Goldsmith and P. Varaiya. Capacity of fading channels with channel side information. *IEEE Trans. on Information Theory*, 43:1986–1992, Nov. 1997.

46. D. J. Gore, R. W. Heath, Jr., and A. J. Paulraj. Transmit selection in spatial multiplexing systems. *IEEE Commun. Lett.*, 6(11):491–493, Nov. 2002.

47. A. Grant. Rayleigh fading multi-antenna channels. *EURASIP J. on Applied Signal Processing*, 3:316–329, Mar. 2002.

48. K. I. Gross and D. S. P. Richards. Total positivity, spherical series, and hypergeometric functions of a matrix argument. *J. Approximation Theory*, 59(2):224–246, 1989.

49. B. Hassibi and B. Hochwald. How much training is needed in multiple-antenna wireless links? *IEEE Trans. on Info. Theory*, 49(4):951–963, Apr. 2003.

50. R. W. Heath Jr. and D. Love. Multimode antenna selection for spatial multiplexing systems with linear receivers. *IEEE Trans. on Signal Processing*, 53(8):3042–3056, Aug. 2005.

51. R. W. Heath Jr. and A. Paulraj. Switching between diversity and multiplexing in MIMO systems. *IEEE Trans. on Communications*, 53(6):962–968, June 2005.

52. D. Hoesli and A. Lapidoth. How good is an isotropic input on a MIMO Rician channel? *Proc. of IEEE Int. Symp. on Inform. Theory (ISIT'04)*, July 2004.

53. R. A. Horn and C. R. Johnson. *Matrix Analysis*. University of Cambridge Press, New York, 4th edition, 1990.

54. D. Hösli and A. Lapidoth. The capacity of a MIMO Rician channel is monotonic in the singular values of the mean. *5th Int. ITG Conf. on Source and Channel Coding*, Jan. 2004.

55. K. Huang, R. W. Heath, and J. G. Andrews. Joint beamforming and scheduling for SDMA systems with limited feedback. *IEEE Trans. on Communications*, submitted, available at: http://arxiv.org/abs/cs.IT/0606121.

56. S. Jafar, G. Foschini, and A. Goldsmith. Phantomnet: Exploring optimal multicellular multiple antenna systems. In *Proc., IEEE Veh. Technology Conf.*, pages 24–28, Sept. 2002.

57. S. A. Jafar, S. Vishwanath, and A. J. Goldsmith. Channel capacity and beamforming for multiple transmit and receive antennas with covariance feedback. *Proc. IEEE Int. Conf. on Communications (ICC'01)*, 7:2266–2270, 2001.

58. S. K. Jayaweera and H. V. Poor. Capacity of multiple-antenna systems with both receiver and transmitter channel state information. *IEEE Trans. on Information Theory*, 49(10):2697–2709, Oct. 2003.

59. N. Jindal. High SNR analysis of MIMO broadcast channels. In *IEEE Int. Symp. on Info. Theory (ISIT)*, pages 2310–2314, 4–9 Sept. 2005.

60. S. Jing, D. N. C. Tse, J. Hou, J. B. Soriaga, J. E. Smee, and R. Padovani. Multi-cell downlink capacity with coordinated processing. In *Proc., Information Theory and its Applications (ITA)*, San Diego, CA, Jan. 2007.

61. E. Jorswieck and H. Boche. On transmit diversity with imperfect channel state information. *Proc. IEEE Conf. Acoustics, Speech, and Signal Processing (ICASSP'02)*, 3:2181–2184, May 2002.

62. E. Jorswieck and H. Boche. Channel capacity and capacity-range of beamforming in MIMO wireless systems under correlated fading with covariance feedback. *IEEE Trans. Wireless Communications*, 3(5):1543–1553, Sept. 2004.

63. M. Kang and M. S. Alouini. On the capacity of MIMO Rician channels. In *Proc. 40th Annual Allerton Conference on Communication, Control, and Computing*, Monticello, IL, Oct. 2002.

64. M. Kang and M. S. Alouini. Impact of correlation on the capacity of MIMO channels. In *Proc. IEEE Int. Conf. in Communications (ICC'03)*, May 2003.

65. M. Kang and M. S. Alouini. Capacity of MIMO Rician channels. *IEEE Trans. Wireless Commun.*, 5(1):112–122, Jan. 2006.

66. M. Kang, L. Yang, and M. S. Alouini. Capacity of MIMO Rician channels with multiple correlated Rayleigh co-channel interferers. In *Proc. 2003 IEEE Globecom*, San Francisco, CA, Dec. 2003.

67. C. G. Khatri. On certain distribution problems based on positive definite quadratic functions in normal vectors. *Ann. Math. Statist.*, 37:468–479, 1966.

68. C. G. Khatri. On the moments of traces of two matrices in three situations for complex multivariate normal populations. *Sankhya, The Indian J. Statist., Ser. A*, 32:65–80, 1970.

69. M. Kiessling and J. Speidel. Analytical performance of MIMO MMSE receivers in Rayleigh fading environments. In *IEEE Veh. Technol. Conf. (VTC)*, pages 1738–1742, Orlando, USA, Oct. 2003.

70. M. Kiessling and J. Speidel. Exact ergodic capacity of MIMO channels in correlated Rayleigh fading environments. In *Proc. Int. Zurich Seminar on Communications (IZS)*, Zurich, Switzerland, Feb. 2004.

71. M. Kiessling and J. Speidel. Mutual information of MIMO channels in correlated Rayleigh fading environments—a general solution. In *Proc. of IEEE Int. Conf. in Communications. (ICC'04)*, Paris, France, June 2004.

72. M. Kiessling and J. Speidel. Mutual information of MIMO channels in correlated Rayleigh fading environments—a general solution. In *IEEE Int. Conf. on Commun. (ICC)*, pages 814–818, Paris, France, Jun 2004.

73. M. Kiessling and J. Speidel. Unifying performance analysis of linear MIMO receivers in correlated Rayleigh fading environments. In *IEEE Int. Symp. on Spread Spec. Tech. and Appl. (ISSSTA)*, pages 634–638, Sydney, Australia, Aug. 30–Sept. 02 2004.

74. P. Koev and A. Edelman. The efficient evaluation of the hypergeometric function of a matrix argument. *Math. Comp.*, 75:833–846, 2006.

75. J. Kotecha and A. Sayeed. On the capacity of correlated MIMO channels. *Proc. of IEEE Intern. Symp. on Inform. Theory (ISIT'03)*, July 2003.

76. W. C. Y. Lee and Y. S. Yeh. Polarization diversity system for mobile radio. *IEEE Trans. on Communications*, 20(5):912–923, Oct 1972.

77. P. Li, D. Paul, R. Narisimhan, and J. Cioffi. On the distribution of SINR for the MMSE MIMO receiver and performance analysis. *IEEE Trans. Inform. Theory*, 52(1): 271–286, Jan. 2006.

78. D. Love and R. W. Heath Jr. Multimode precoding for MIMO wireless systems, 53:3674–3687, 2005.

79. D. J. Love and R. W. Heath. Limited feedback unitary precoding for spatial multiplexing systems. *IEEE Trans. on Info. Theory*, 51(8):1967–1976, Aug. 2005.

80. D. J. Love, R. W. Heath, W. Santipach, and M. L. Honig. What is the value of limited feedback for MIMO channels. *IEEE Communications Magazine*, 42(10):54–59, Oct. 2004.

81. A. Lozano, A. M. Tulino, and S. Verdú. Correlation number: A new design criterion in multi-antenna communication. *Proc. IEEE Vehicular Technology Conf. (VTC'2003) Spring*, Apr. 2003.

82. A. Lozano, A. M. Tulino, and S. Verdú. High-SNR power offset in multiantenna communication. *IEEE Trans. Inform. Theory*, 51(12):4134–4151, Dec. 2005.

83. V. A. Marčenko and L. A. Pastur. Distributions of eigenvalues for some sets of random matrices. *Math. USSR-Sbornik*, 1:457–483, 1967.

84. T. L. Marzetta. Fundamental limitations on the capacity of wireless links that use polarimetric antenna arrays. *Proc. IEEE Int. Symp. on Information Theory (ISIT'02)*, page 51, July 2002.

85. M. R. McKay, I. Collings, A. Forenza, and R. W. Heath Jr. Adaptive coded-MIMO in spatially-correlated channels based on closed-form BER expressions. *IEEE Trans. Veh. Technol.*, 56(5):2555–2567, Sept. 2007.

86. M. R. McKay and I. B. Collings. Capacity and performance of MIMO-BICM with zero forcing receivers. *IEEE Trans. Commun.*, 53(1):74–83, Jan. 2005.

87. M. R. McKay and I. B. Collings. General capacity bounds for spatially correlated Rician MIMO channels. *IEEE Trans. Inform. Theory*, 51(9):3121–3145, Sept. 2005.

88. M. R. McKay and I. B. Collings. Error performance of MIMO-BICM with zero-forcing receivers in spatially-correlated Rayleigh channels. *IEEE Trans. Wireless Commun.*, 6(3):787–792, Mar. 2007.

89. M. R. McKay, P. J. Smith, and I. B. Collings. New properties of complex noncentral quadratic forms and bounds on MIMO mutual information. In *Proc. of IEEE Int. Symp. on Info. Theory (ISIT)*, pages 1209–1213, Seattle, USA, July 2006.

90. X. Mestre. *Space processing and channel estimation: performance analysis and asymptotic results*. PhD thesis, Dept. de Teoria del Senyal i Comunicacions, Universitat Politècnica de Catalunya, Barcelona, Catalonia, Spain, 2002.

91. X. Mestre, J. R. Fonollosa, and A. Pages-Zamora. Capacity of MIMO channels: asymptotic evaluation under correlated fading. *IEEE J. on Selected Areas in Communications*, 21(5):829-838, June 2003.

92. A. L. Moustakas and S. H. Simon. Optimizing multiple-input single-output (MISO) communication systems with general Gaussian channels: nontrivial covariance and non-zero mean. *IEEE Trans. on Information Theory*, 49(10):2770–2780, Oct. 2003.

93. A. L. Moustakas, S. H. Simon, and A. M. Sengupta. MIMO capacity through correlated channels in the presence of correlated interferers and noise: A (not so) large N analysis. *IEEE Trans. on Information Theory*, 49(10):2545–2561, Oct. 2003.

94. R. Negi and J. M. Cioffi. Delay-constrained capacity with causal feedback. *IEEE Trans. on Inform. Theory*, 48:2478–2494, September 2002.

95. O. Oyman, R. U. Nabar, H. Bölcskei, and A. J. Paulraj. Characterizing the statistical properties of mutual information in MIMO channels. *IEEE Trans. Signal Processing*, 51(11):2784–2795, Nov 2003.

96. H. Ozcelik, M. Herdin, W. Weichselberger, G. Wallace, and E. Bonek. Deficiencies of the Kronecker MIMO channel model. *IEE Electronics Letters*, 39:209–210, Aug. 2003.

97. D. P. Palomar, J. M. Cioffi, and M. A. Lagunas. Uniform power allocation in MIMO channels: A game theoretic approach. *IEEE Trans. on Information Theory*, 49(7): 1707–1727, July 2003.

98. K. I. Pedersen, J. B. Andersen, J. P. Kermoal, and P. E. Mogensen. A stochastic multiple-input multiple-output radio channel model for evaluations of space-time coding algorithms. *Proc. IEEE Vehicular Technology Conf. (VTC'2000 Fall)*, pages 893–897, Sep. 2000.

99. M. S. Pinsker. *Information and Information Stability of Random Variables and Processes*. Holden-Day, San Francisco, CA, 1964.

100. O. Popescu and C. Rose. Sum capacity and tsc bounds in collaborative multi-base wireless systems. *IEEE Transactions on Information Theory*, 50(10):2433–2438, October 2004.

101. J. G. Proakis. *Digital Communications*. McGraw-Hill, New York, 4th edition, 2001.

102. G. Raleigh and J. M. Cioffi. Spatio-temporal coding for wireless communications. *IEEE Trans. on Communications*, 46(3):357–366, Mar. 1998.

103. P. Rapajic and D. Popescu. Information capacity of a random signature multiple-input multiple-output channel. *IEEE Trans. on Communications*, 48(8):1245–1248, Aug. 2000.

104. T. Ratnarajah, R. Vaillancourt, and M. Alvo. Complex random matrices and applications. *Math. Rep. of the Acad. of Sci. of the Royal Soc. of Canada*, 25(4):114–120, Dec. 2003.

105. T. Ratnarajah, R. Vaillancourt, and M. Alvo. Complex random matrices and Rayleigh channel capacity. *Communications in Information and Systems*, (2):119–138, Oct. 2003.

106. W. Rhee and J. M. Cioffi. On the capacity of multiuser wireless channels with multiple antennas. *IEEE Trans. Inform. Theory*, 49(10):2580–2595, Oct. 2003.

107. U. Sacoglu and A. Scaglione. Asymptotic capacity of space-time coding for arbitrary fading: A closed form expression using Girko's law. In *Proc. Intl. Conf. on Acoust. Speech and Signal Proc., ICASSP(2001)*, Salt Lake City, UT, May 7–12, 2001.

108. A. Sayeed. Deconstructing multi-antenna channels. *IEEE Trans. on Signal Processing*, 50(10):2563–2579, Oct. 2002.

109. A. M. Sengupta and P. P. Mitra. Capacity of multivariate channels with multiplicative noise: I. Random matrix techniques and large-N expansions for full transfer matrices. *LANL arXiv:physics*, Oct. 2000.

110. S. Shamai and S. Verdú. The impact of frequency-flat fading on the spectral efficiency of CDMA. *IEEE Trans. Inform. Theory*, 47(4):1302–1327, May 2001.

111. S. Shamai and B. Zaidel. Enhancing the cellular downlink capacity via co-processing at the transmitting end. In *Proc., IEEE Veh. Technology Conf.*, pages 1745–1749, May 2001.

112. C. Shannon. Communication in the presence of noise. *Proc. IRE*, 37, January 1949.

113. Z. Shen, R. Chen, J. G. Andrews, R. W. Heath, and B. L. Evans. Sum capacity of multiuser MIMO broadcast channels with block diagonalization. *IEEE Trans. on Wireless Communications*, to appear 2007.

114. D.-S. Shiu, G. J. Foschini, M. J. Gans, and J. M. Kahn. Fading correlation and its effects on the capacity of multi-element antenna systems. *IEEE Trans. on Communications*, 48(3):502–511, Mar. 2000.

115. S. H. Simon and A. L. Moustakas. Optimizing MIMO systems with channel covariance feedback. *IEEE J. on Selected Areas in Communications*, 21(3):406–417, Apr. 2003.

116. S. H. Simon, A. L. Moustakas, and L. Marinelli. Capacity and character expansions: Moment generating function and other exact results for MIMO correlated channels. *Bell Labs Technical Memorandum ITD-04-45211T*, Mar. 2004.

117. P. J. Smith and L. M. Garth. Exact capacity distribution for dual MIMO systems in Ricean fading. *IEEE Communications Letters*, 8(1):18–20, Jan. 2004.

118. P. J. Smith, S. Roy, and M. Shafi. Capacity of MIMO systems with semi-correlated flat-fading. *IEEE Trans. on Information Theory*, 49(10):2781–2788, Oct. 2003.

119. P. J. Smith and M. Shafi. On a Gaussian approximation to the capacity of wireless MIMO systems. *Proc. IEEE Int. Conf. in Communications. (ICC'02)*, pages 406–410, Apr. 2002.

120. P. Soma, D. S. Baum, V. Erceg, R. Krishnamoorthy, and A. Paulraj. Analysis and modelling of multiple-input multiple-output (MIMO) radio channel based on outdoor measurements conducted at 2.5 GHz for fixed BWA application. *Proc. IEEE Int. Conf. on Communications (ICC'02), New York City, NY*, pages 272–276, 28 Apr.–2 May 2002.

121. Q. Spencer, A. Swindlehurst, and M. Haardt. Zero-forcing methods for downlink spatial multiplexing in multi-user MIMO channels. *IEEE Trans. on Signal Processing*, 52:461–471, Feb. 2004.

122. T. J. Stieltjes. Recherches sur les fractions continues. *Annales de la Faculte des Sciences de Toulouse*, 8 (9)(A (J)):1–47 (1–122), 1894 (1895).

123. E. Telatar. Capacity of multi-antenna Gaussian channels. *Euro. Trans. Telecommunications*, 10(6):585–595, Nov.–Dec. 1999.

124. D. N. C. Tse and S. V. Hanly. Linear multiuser receivers: Effective interference, effective bandwidth and user capacity. *IEEE Trans. Inform. Theory*, 45(2):641–657, 1999.

125. B. S. Tsybakov. The capacity of a memoryless Gaussian vector channel. *Problems of Information Transmission*, 1(1):18–29, 1965.

126. A. M. Tulino, A. Lozano, and S. Verdú. Power-bandwidth tradeoff of multi-antenna systems in the low-power regime. In G. Foschini and S. Verdú, editors, *Multiantenna channels: Capacity, Coding and Signal Processing*, pages 15–42. American Mathematical Society Press, 2003.

127. A. M. Tulino, A. Lozano, and S. Verdú. MIMO capacity with channel state information at the transmitter. In *Proc. IEEE Int. Symp. on Spread Spectrum Tech. and Applications (ISSSTA'04)*, Aug. 2004.

128. A. M. Tulino, A. Lozano, and S. Verdú. Power allocation in multi-antenna communication with statistical channel information at the transmitter. In *Proc. IEEE Int. Conf.*

on Personal, Indoor and Mobile Radio Communications. (PIMRC'04), Barcelona, Catalonia, Spain, Sep. 2004.

129. A. M. Tulino, A. Lozano, and S. Verdú. Impact of correlation on the capacity of multi-antenna channels. *IEEE Trans. on Information Theory*, 51(7):2491–2509, July 2005.

130. A. M. Tulino, A. Lozano, and S. Verdú. Capacity-achieving input covariance for single-user multi-antenna channels. *IEEE Trans. on Wireless Communications*, 5(3):662–671, Mar. 2006.

131. A. M. Tulino and S. Verdú. Random matrix theory and wireless communications. *Foundations and Trends in Communications and Information Theory*, 1(1):1–182, 2004.

132. A. M. Tulino and S. Verdú. Asymptotic outage capacity of multiantenna channels. In *Proc. IEEE Int. Conf. Acoustics, Speech and Signal Processing (ICASSP'05)*, Philadelphia, PA, USA, Mar. 2005.

133. S. Venkatesan, S. H. Simon, and R. A. Valenzuela. Capacity of a Gaussian MIMO channel with nonzero mean. *Proc. 2003 IEEE Vehicular Technology Conf. (VTC'03)*, Oct. 2003.

134. S. Verdú. Capacity region of Gaussian CDMA channels: The symbol synchronous case. In *Proc. Allerton Conf. on Communication, Control and Computing*, pages 1025–1034, Monticello, IL, Oct. 1986.

135. S. Verdú. *Multiuser Detection*. Cambridge University Press, Cambridge, UK, 1998.

136. S. Verdú. Spectral efficiency in the wideband regime. *IEEE Trans. on Information Theory*, 48(6):1319–1343, June 2002.

137. S. Verdú and S. Shamai. Spectral efficiency of CDMA with random spreading. *IEEE Trans. on Information Theory*, 45(2):622–640, Mar. 1999.

138. S. Vishwanath, N. Jindal, and A. Goldsmith. Duality, achievable rates and sum rate capacity of the Gaussian MIMO broadcast channel. *IEEE Trans. on Info. Theory*, 49(10):2658–2668, Oct. 2003.

139. E. Visotsky and U. Madhow. Space-time transmit precoding with imperfect feedback. *IEEE Trans. on Information Theory*, 47:2632–2639, Sep. 2001.

140. Z. Wang and G. Giannakis. Outage mutual information of space-time MIMO channels. *IEEE Trans. on Information Theory*, 50(4):657–663, Apr. 2004.

141. S. Wei and D. Goeckel. On the minimax robustness of the uniform transmission power strategy in MIMO systems. *IEEE Communications Letters*, 7(11):523–524, Nov. 2003.

142. W. Weichselberger, M. Herdin, H. Ozcelik, and E. Bonek. Stochastic MIMO channel model with joint correlation of both link ends. *IEEE Trans. on Wireless Communications*, 5:90–100, Jan. 2006.

143. J. H. Winters, J. Salz, and R. D. Gitlin. The impact of antenna diversity on the capacity of wireless communication systems. *IEEE Trans. Commun.*, 42(2/3/4):1740–1751, Feb/Mar/Apr 1994.

144. K. K. Wong, R. D. Murch, and K. B. Letaief. A joint-channel diagonalization for multi-user mimo antenna systems. *IEEE Trans. on Wireless Communications*, 2(4):773–786, July 2003.

145. X. Wu, A. Das, J. Li, and R. Laroia. Fractional power reuse in cellular networks. In *Proc. Allerton Conf. on Comm., Control, and Computing*, Oct. 2006.

146. R. Xu and F. C. M. Lau. Performance analysis for MIMO systems using zero forcing detector over fading channels. *IEE Proc.-Commun.*, 153(1):74–80, Feb. 2006.

147. Y. G. Li. Simplified channel estimation for OFDM systems with multiple transmit antennas. *IEEE Trans. on Wireless Communications*, 1:67–75, Jan. 2002.

148. T. Yoo, N. Jindal, and A. Goldsmith. Multi-antenna broadcast channels with limited feedback and user selection. *IEEE Journal on Sel. Areas in Communications*, 25(7):1478–1491, Sep. 2007.

149. W. Yu and J. Cioffi. Sum capacity of gaussian vector broadcast channels. *IEEE Trans. on Info. Theory*, 50(9):1875–1892, Sept. 2004.

150. H. Zhang and H. Dai. Co-channel interference mitigation and cooperative processing in downlink multicell multiuser MIMO networks. *European Journal on Wireless Communications and Networking*, 4th Quarter 2004.

151. Q. T. Zhang, X. W. Cui, and X. M. Li. Very tight capacity bounds for MIMO-correlated Rayleigh-fading channels. *IEEE Trans. Wireless Commun.*, 4(2):681–688, Mar. 2005.

152. L. Zheng and D. N. C. Tse. Diversity and multiplexing: A fundamental tradeoff in multiple-antenna channels. *IEEE Trans. on Information Theory*, 49(5):1073–1096, May 2003.

7

INTERFERENCE AVOIDANCE FOR CDMA SYSTEMS

Dimitrie C. Popescu, Sennur Ulukus, Christopher Rose, and Roy Yates

7.1 INTRODUCTION

Though the success of cellular telephony heralds the wireless age, wireless networking by the masses is a relatively new and exciting phenomenon with potentially even larger impact on our society. Accelerating sales of wireless LANs (802.11) for home and small office use, and even small business as a customer lure, indicate that wireless connectivity is increasingly seen as an important, if not vital, part of modern life. In addition, there is demand for other wireless networking technology geared toward different applications such as wiring replacement (Bluetooth, HomeRF), paging networks, and wide area email access (Blackberry). Furthermore, as wireless networks become more ubiquitous and less expensive, sensor/actuator networks, now primarily the commercial domain of burglar/fire alarms and climate control, could become much more prevalent.

However, the increasing demand for various wireless services is tempered by simple economics. Usable spectrum is scarce and therefore expensive. Thus, applications which cannot be deployed almost full-blown with a predictably stable revenue base (i.e., cellular systems) cannot usually afford spectrum licenses and must therefore share use of various unlicensed bands—for example, the Unlicensed National Information Infrastructure (UNII [9]) at 5 GHz and other unlicensed bands at 900 MHz and 2.4 GHz.

Advances in Multiuser Detection. Edited by Michael L. Honig
Copyright © 2009 John Wiley & Sons, Inc.

Unfortunately, shared spectrum use implies mutual interference between systems, often co-located, whose owners and/or traffic types and/or service objectives may be completely different. Thus, the prospect of spending development dollars for equipment and services that may be rendered worthless by perfectly legal interference from another system has an appropriately chilling effect on technology and service development. This was clearly articulated by equipment manufacturers in meetings on unlicensed band technology [60,62].

Nonetheless, the success of in-home/small office wireless LAN products such as 802.11 provides an existence proof for an incrementally developing wireless mass market. In addition, as microelectronic technology advances along with Moore's law, it is becoming increasingly economical to incorporate complex modulation, coding and protocols into inexpensive devices. This immediately prompts the question of whether appropriate communications methods and algorithms will make it possible to lower the development cost entry barrier for potential equipment manufacturers and service providers by conferring some degree of immunity from crippling mutual interference between systems. Thus, the ultimate aim of unlicensed wireless system design is to provide robust methods and algorithms which promote peaceful and efficient coexistence.

Unfortunately, even the simplest problem of mutual interference—the "interference channel"—has defied complete solution for over fifty years [2,7,10,11,18]). This analytic gap hampers quantitative evaluation of modern wireless network alternatives, and prompts a search for heuristic approaches that can nudge wireless systems toward efficient use. Interference avoidance, loosely defined, is exactly such a heuristic.

The basic idea behind interference avoidance is to iteratively place signal energy in the least-occupied portion of a signal space until convergence. It has been shown that over a wide range of communications scenarios such iterative greedy interference avoidance provides maximum resource utilization for a variety of system metrics including Signal-to-Interference Ratio (SIR), sum capacity, or total squared correlation [3,49,61,63,73]. Obviously, this notion of optimality under rational behavior (personal greed) is extremely attractive in unlicensed bands which are the wireless equivalent of the "wild west" with no central authority to exercise control.

Though it has been shown that in the presence of non-adaptive interference (non-agile users), interference avoidance by agile users is beneficial for everyone [63], in general interference avoidance is more an organizing principle than a panacea. Interference avoidance has been developed almost exclusively for single receiver systems—that is, where all users are decoded jointly. It clearly does not account for independently operated interference avoiding systems where information is not shared between receivers. In fact, if naively applied to such multi-receiver scenarios, even simple convergence of various interference avoidance algorithms cannot be guaranteed.

Nonetheless, jointly decoded interference avoidance provides an obvious outer bound on performance and offers the possibility of circumventing "Tragedy of the Commons" type problems [5,6,41]. Furthermore, a performance comparison to the usual interference channel scenario is telling—joint decoding often achieves much higher sum capacity than independent decoding owing both to diversity and to the greater amount of usable energy incident on the distributed receiver [50,52,54]. That is, cooperative collection and combining of signals from all receivers on which transmitted energy is incident both increases the amount of signal energy captured and in addition confers a multiple receiver diversity benefit as well. In short, the interference channel capacity region is often a significantly smaller subset of the joint collection/decoding capacity region.

These two benefits strongly motivate a system architecture where information is shared among receivers, even of otherwise independent systems. With increasingly available high bandwidth land line connections coupled to the cost of siting radio transceivers, the notion of a distributed transceiver facility on the uplink seems prudent and inevitable rather than far-fetched. And interference avoidance still provides some degree of autonomy since modulation waveforms need not be jointly chosen—users simply iterate greedy actions and an efficient operating point is reached automatically. All that is needed is a common shared measurement of interference levels at the receiver(s).

In the following sections, we will describe basic interference avoidance principles, different types of interference avoidance algorithms and how such algorithms might be implemented. We then extend basic interference avoidance to cover a variety of typical wireless scenarios including slowly fading channels, rapidly fading channels, MIMO channels, asynchronous systems, power control and finally the dynamically changing subsets of active users sure to be a feature of real unlicensed wireless systems.

We will find that these extended versions of interference avoidance usually lead to mutually water-filled spectra [52,87] for the mutually interfering users. This observation in some ways brings us back on the original motivation of completely independent systems with completely independent receivers [12,18,50–53,55]. If interfering users are treated as noise, mutual waterfilling allows us to identify fixed points of greedy behavior and examine when they are near-optimal or strongly suboptimal. The design of procedures to move between or eradicate certain of these fixed points is beyond the scope of this chapter. However, a number of interesting ideas have surfaced [12,18,55] that flow naturally from a general interference avoidance perspective.

We close the chapter with a summary of results and a selection of open problems.

7.2 INTERFERENCE AVOIDANCE BASICS

We consider the uplink of a synchronous CDMA communication system with K users having signature waveforms $\{S_\ell(t)\}_{\ell=1}^{K}$ of finite duration T and equal received power at a common receiver (base station). Without loss of generality we assume unit received power for each user. The received signal is [78]:

$$R(t) = \sum_{\ell=1}^{K} b_\ell S_\ell(t) + n(t) \tag{7.1}$$

where b_ℓ is the information symbol sent by user ℓ with unit-energy signature $S_\ell(t)$, and $n(t)$ is an additive Gaussian noise process that corrupts the signal at the receiver. We assume that all signals are representable in an arbitrary N-dimensional signal space, in which each user's signature waveform $S_\ell(t)$ is equivalent to an N-dimensional unit-norm codeword vector $\mathbf{s}_\ell$[1] and Gaussian noise process $n(t)$ is equivalent to the noise vector $\mathbf{n}$ with correlation matrix $E[\mathbf{nn}^\mathsf{T}] = \mathbf{W}$. The equivalent received signal vector $\mathbf{r}$ at the base station is then given by the vector equation:

$$\mathbf{r} = \sum_{\ell=1}^{K} b_\ell \mathbf{s}_\ell + \mathbf{n} = \mathbf{Sb} + \mathbf{n} \tag{7.2}$$

where $\mathbf{b} = [b_1 \cdots b_\ell \cdots b_K]^\mathsf{T}$ contains the symbols sent by users 1 through K and:

$$\mathbf{S} = \begin{bmatrix} | & & | & & | \\ \mathbf{s}_1 & \cdots & \mathbf{s}_\ell & \cdots & \mathbf{s}_K \\ | & & | & & | \end{bmatrix} \tag{7.3}$$

is an $N \times K$ matrix with unit norm columns that are the user codewords $\mathbf{s}_\ell$. The auto-correlation matrix of the received signal in equation (7.2) is:

$$\mathbf{R} = E[\mathbf{rr}^\mathsf{T}] = \mathbf{SS}^\mathsf{T} + \mathbf{W} \tag{7.4}$$

A unit norm receiver filter, $\mathbf{c}_k$, is used to estimate the symbol transmitted by a given user k. This estimate is computed as:

$$\hat{b}_k = \mathbf{c}_k^\mathsf{T} \mathbf{r} \tag{7.5}$$

so the signal-to-interference plus noise-ratio (SINR) for user k becomes:

$$\gamma_k = \frac{(\mathbf{c}_k^\mathsf{T} \mathbf{s}_k)^2}{\sum_{\ell=1, \ell \neq k}^{K} (\mathbf{c}_k^\mathsf{T} \mathbf{s}_\ell)^2 + E[(\mathbf{c}_k^\mathsf{T} \mathbf{n})^2]} \tag{7.6}$$

Interference avoidance provides distributed algorithms by which individual users increase/maximize their SINR through adaptation of their codewords. We note that even though interference avoidance is an egocentric procedure of selfish codeword updates, it converges to a socially optimal solution in which global system metrics like the information theoretic sum capacity or total squared correlation (TSC) are

---

[1]For simplicity of exposition we have assumed unit energy codewords. However, it should be noted that all the results we present hold even if unequal codeword energies $|\mathbf{s}_k|^2 = p_k$ are assumed [63].

optimized. Sum capacity is expressed as:

$$C_s = \frac{1}{2}\log(\det \mathbf{R}) - \frac{1}{2}\log(\det \mathbf{W}) \tag{7.7}$$

and was used in the context of optimizing CDMA codewords in [17,64,79–81]. We note that for fixed user power, sum capacity is upper bounded by the value corresponding to the optimal CDMA codewords [79–81].

The total squared correlation (TSC), that is the sum of squared correlations of user codewords, is expressed as:

$$\text{TSC} = \sum_{i=1}^{K}\sum_{j=1}^{K}(\mathbf{s}_i^{\mathsf{T}}\mathbf{s}_j)^2 = \text{tr}\left[(\mathbf{SS}^{\mathsf{T}})^2\right] \tag{7.8}$$

and was used in the context of optimizing CDMA codewords in [64,81]. The TSC is lower bounded, and its lower bound [85]:

$$\text{TSC} \geq \frac{K^2}{N} \tag{7.9}$$

is achieved for the same optimal CDMA codewords that maximize sum capacity [64,81], which are also referred to as Welch Bound Equality (WBE) sequences.

A more general expression of the TSC, called the general squared correlation (GSC), is used in the context of interference avoidance algorithms [48,63]. This is defined as:

$$\text{GSC} = \text{tr}\left[\mathbf{R}^2\right] = \text{tr}\left[\left(\mathbf{SS}^{\mathsf{T}} + \mathbf{W}\right)^2\right] \tag{7.10}$$

In white noise, when the noise covariance is a scaled identity matrix $\mathbf{W} = \sigma\mathbf{I}$, this is related to TSC by an additive constant:

$$\begin{aligned}\text{GSC} &= \text{tr}\left[(\mathbf{SS}^{\mathsf{T}})^2\right] + \sigma^2\text{tr}\left[(\mathbf{I})^2\right] + 2\sigma\,\text{tr}\left[(\mathbf{SS}^{\mathsf{T}})\right]\\ &= \text{tr}\left[(\mathbf{SS}^{\mathsf{T}})^2\right] + \sigma^2 N + 2\sigma K \end{aligned} \tag{7.11}$$

The GSC is also lower bounded, and its lower bound is achieved for generalized WBE sequences (GWBE)[2] [80,81].

---

[2] According to [80, 81], in a GWBE codeword ensemble oversized users—with input power constraints large relative to the input power constraints of the other users—have orthogonal codewords that span the dimensions of the signal space with minimum noise energy, while non-oversized users have codewords that satisfy an aggregate water filling solution over the remaining signal dimensions.

7.2.1 Greedy Interference Avoidance: The Eigen-Algorithm

Assuming simple matched filters at the receiver for all users, the SINR for user k in equation (7.6) is expressed as:

$$\gamma_k = \frac{(\mathbf{s}_k^\mathsf{T}\mathbf{s}_k)^2}{\sum_{\ell=1,\ell\neq k}^K (\mathbf{s}_k^\mathsf{T}\mathbf{s}_\ell)^2 + E[(\mathbf{s}_k^\mathsf{T}\mathbf{n})^2]} = \frac{1}{\mathbf{s}_k^\mathsf{T}\left(\sum_{\ell=1,\ell\neq k}^K \mathbf{s}_\ell\mathbf{s}_\ell^\mathsf{T} + E[\mathbf{n}\mathbf{n}^\mathsf{T}]\right)\mathbf{s}_k} \tag{7.12}$$

We define the autocorrelation matrix of the interference-plus-noise seen by user k as:

$$\mathbf{R}_k = \sum_{\ell=1,\ell\neq k}^K \mathbf{s}_\ell\mathbf{s}_\ell^\mathsf{T} + \mathbf{W} = \mathbf{R} - \mathbf{s}_k\mathbf{s}_k^\mathsf{T} \tag{7.13}$$

and rewrite equation (7.12) as:

$$\gamma_k = \frac{1}{\mathbf{s}_k^\mathsf{T}\mathbf{R}_k\mathbf{s}_k} \tag{7.14}$$

Maximizing the SINR γ_k is equivalent to minimizing the inverse SINR:

$$\beta_k = \frac{1}{\gamma_k} = \mathbf{s}_k^\mathsf{T}\mathbf{R}_k\mathbf{s}_k \tag{7.15}$$

Note that for unit norm codewords, equation (7.15) represents the Rayleigh quotient for matrix $\mathbf{R}_k$, and recall from linear algebra [69, p. 348] that this is minimized by the eigenvector corresponding to the minimum eigenvalue of the given matrix.[3] Thus, the SINR for user k can be greedily maximized by having user k replace its current codeword $\mathbf{s}_k$ with the minimum eigenvector $\mathbf{x}_k$ of the autocorrelation matrix $\mathbf{R}_k$ of the interference-plus-noise seen by user k. We call this procedure *greedy interference avoidance* since by replacing its current codeword with the minimum eigenvector of the interference-plus-noise correlation matrix, user k avoids interference by placing its transmitted energy in that region of the signal space with minimum interference-plus-noise energy and greedily maximizes SINR without paying attention to potentially negative effects this action may have on other users in the system. Applied iteratively by all users in the system, this procedure defines *the eigen-algorithm for interference avoidance* [63] that is formally presented in Figure 7.1. Numerically, a fixed point of the eigen-algorithm is defined with respect to a stopping criterion. That is, we say that a fixed point is reached when the difference between two consecutive values of the stopping criterion is within a specified tolerance ε. The stopping criterion can be an individual one, like the codeword SINR or the Euclidian distance between codewords and their corresponding replacements, or a global one like sum capacity or GSC. We note that in the case of individual stopping criteria all values corresponding to all codewords must be within the specified tolerance for the algorithm to stop.

---

[3]This is also referred to as the minimum eigenvector.

1. Start with a randomly chosen codeword ensemble $\{s_\ell\}_{\ell=1}^K$.

2. For $k = 1, \ldots, K$

    (a) Compute the autocorrelation matrix R_k of the interference-plus-noise seen by user k

    (b) Find the minimum eigenvalue $\lambda_N^{(k)}$ and associated unit eigenvector x_k of R_k.

    (c) If user k's codeword s_k is not already a suitable eigenvector of R_k, replace it by x_k.

3. Repeat step 2 until a fixed point is reached.

Figure 7.1. The Eigen-Algorithm.

Convergence of the eigen-algorithm to a fixed point is ensured by the fact that the algorithm monotonically increases sum capacity which is upper bounded, respectively monotonically decreases GSC which is lower bounded. We note first from equation (7.13) that:

$$\mathbf{R} = \mathbf{R}_k + s_k s_k^T \tag{7.16}$$

To show that sum capacity C_s in equation (7.7) is monotonically increased by application of greedy interference avoidance, we will verify that:

$$\Delta_{C_s} = \frac{1}{2} \log \det\left(\mathbf{R}_k + x_k x_k^T\right) - \frac{1}{2} \log \det\left(\mathbf{R}_k + s_k s_k^T\right) \tag{7.17}$$

is non-negative. We note that $\mathbf{R}_k$ is always invertible due to the presence of the non-singular noise covariance matrix $\mathbf{W}$, and we can factor it out in equation (7.17) to obtain:

$$\Delta_{C_s} = \frac{1}{2} \det\left(\mathbf{I} + \mathbf{R}_k^{-\frac{1}{2}} x_k x_k^T \mathbf{R}_k^{-\frac{1}{2}}\right) - \frac{1}{2} \det\left(\mathbf{I} + \mathbf{R}_k^{-\frac{1}{2}} s_k s_k^T \mathbf{R}_k^{-\frac{1}{2}}\right) \tag{7.18}$$

Since the matrices $\mathbf{R}_k^{-\frac{1}{2}} s_k s_k^T \mathbf{R}_k^{-\frac{1}{2}}$ and $\mathbf{R}_k^{-\frac{1}{2}} x_k x_k^T \mathbf{R}_k^{-\frac{1}{2}}$ each have rank one, equation (7.18) further reduces to:

$$\Delta_{C_s} = \frac{1}{2}(1 + x_k^T \mathbf{R}_k^{-1} x_k) - \frac{1}{2}(1 + s_k^T \mathbf{R}_k^{-1} s_k) = x_k^T \mathbf{R}_k^{-1} x_k - s_k^T \mathbf{R}_k^{-1} s_k \tag{7.19}$$

If x_k is chosen as the minimum eigenvector of $\mathbf{R}_k$, then x_k is also the maximum eigenvector of $\mathbf{R}_k^{-1}$, and it follows that $\Delta_{C_s} \geq 0$ from the properties of Rayleigh quotient [69, p. 348]. Thus, the eigen-algorithm monotonically increases sum capacity.

To show that GSC in equation (7.10) is monotonically decreased by application of greedy interference avoidance, we need to show that the difference:

$$\Delta_{GSC} = \text{tr}\left[(\mathbf{R}_k + \mathbf{s}_k \mathbf{s}_k^T)^2\right] - \text{tr}\left[(\mathbf{R}_k + \mathbf{x}_k \mathbf{x}_k^T)^2\right] \tag{7.20}$$

in GSC, before and after greedy interference avoidance is applied, is non-negative.

After canceling similar terms and replacing the traces by the corresponding quadratic forms, the GSC reduction becomes:

$$\Delta_{GSC} = 2(\mathbf{s}_k^T \mathbf{R}_k \mathbf{s}_k - \mathbf{x}_k^T \mathbf{R}_k \mathbf{x}_k) \tag{7.21}$$

When $\mathbf{x}_k$ is chosen to be the minimum eigenvector of $\mathbf{R}_k$, it follows that $\Delta_{GSC} \geq 0$ and that the eigen-algorithm monotonically decreases GSC.

We note that in the context of the eigen-algorithm maximizing sum capacity and minimizing GSC are equivalent procedures, although this may not be true in general. In particular, Δ_{C_s} and Δ_{GSC} are not identical functions of $\mathbf{x}_k$ and there can be algorithms that monotonically improve one metric but yield fluctuations in the other. We also note that maximizing sum capacity and/or minimizing GSC imply optimizing the overall system performance, although they may not imply fairness to individual users in the system. More specifically, the codeword update of a given user maximizes its corresponding SINR, but may negatively affect other user(s) in the system whose SINRs may decrease after the update. Nevertheless, empirical studies [48] have shown that the minimum SINR is not decreased by codeword updates, and if the user with the worst SINR had an acceptable connection when the iterations started, then usually no other user's connection will be any worse than this.

To summarize, the eigen-algorithm increases sum capacity and decreases GSC at each step. Since both sum capacity and GSC are bounded from above and below respectively, the eigen-algorithm must converge to a fixed point. However, such convergence may not imply convergence of the eigen-algorithm to extremal values of sum capacity and GSC, since it may get trapped in suboptimal fixed points. Nevertheless, it has been shown that there is only a finite number of suboptimal fixed points, which can always be escaped by using various tactics [61]. This ensures convergence of the eigen-algorithm to a class of GWBE codewords for which sum capacity is maximized; respectively, GSC is minimized.

7.2.2 MMSE Interference Avoidance

An alternative interference avoidance procedure can be defined using the minimum mean square error (MMSE) receiver filter as the codeword replacement vector. The MMSE receiver filter for user k is obtained [37] by minimizing the mean squared error (MSE) between the filter output and the transmitted information symbol corresponding to user k:

$$\text{MSE}_k = E[(\mathbf{c}_k^T \mathbf{r} - b_k)^2] = \mathbf{c}_k^T \mathbf{R}_k \mathbf{c}_k + (\mathbf{c}_k^T \mathbf{s}_k - 1)^2 \tag{7.22}$$

The MMSE filter for user k is then:

$$\mathbf{c}_k = \arg \min_{\mathbf{c}_k} \text{MSE}_k \tag{7.23}$$

Since MSE_k is a quadratic function in $\mathbf{c}_k$, the necessary and sufficient condition for optimal $\mathbf{c}_k$ is:

$$\frac{\partial}{\partial \mathbf{c}_k}(\text{MSE}_k) = 0 \tag{7.24}$$

This implies:

$$2\mathbf{R}_k\mathbf{c}_k + 2\mathbf{s}_k(\mathbf{s}_k^\mathsf{T}\mathbf{c}_k - 1) = 0 \tag{7.25}$$

and thus:

$$\mathbf{c}_k = (\mathbf{R}_k + \mathbf{s}_k\mathbf{s}_k^\mathsf{T})^{-1}\mathbf{s}_k = \mathbf{R}^{-1}\mathbf{s}_k \tag{7.26}$$

Using the matrix inversion lemma [24, p. 19] equation (7.26) can be rewritten as:

$$\mathbf{c}_k = \frac{\mathbf{R}_k^{-1}\mathbf{s}_k}{1 + \mathbf{s}_k^\mathsf{T}\mathbf{R}_k^{-1}\mathbf{s}_k} \tag{7.27}$$

We note that $\mathbf{c}_k$ in equation (7.26) does not have unit norm. With the appropriate normalization, we obtain:

$$\mathbf{c}_k = \frac{\mathbf{R}_k^{-1}\mathbf{s}_k}{(\mathbf{s}_k^\mathsf{T}\mathbf{R}_k^{-2}\mathbf{s}_k)^{1/2}} \tag{7.28}$$

the unit norm MMSE receiver filter for user k.

We also note that the MMSE receiver filter maximizes the SINR [37] in equation (7.6). Thus, when MMSE receiver filters are assumed, the SINR for user k can be maximized by having user k replace its codeword with the normalized MMSE receiver filter corresponding to $\mathbf{s}_k$. Therefore, this method was dubbed the *MMSE update* in [73] and defines an alternative interference avoidance procedure. Applied iteratively by all users in the system this procedure defines *the MMSE algorithm for interference avoidance*, which is formally given in Figure 7.2.

As is the case with the eigen-algorithm, convergence of the MMSE algorithm to a fixed point is also defined with respect to a stopping criterion, and we will use the same global criteria as in the case of the eigen-algorithm: sum capacity and GSC.

To show the MMSE algorithm increases sum capacity, we go back to equation (7.19) and replace $\mathbf{x}_k$ by the expression for $\mathbf{c}_k$ in equation (7.28). This yields the

1. Start with a randomly chosen codeword ensemble $\{s_\ell\}_{\ell=1}^{K}$.

2. For $k = 1, \ldots, K$

   (a) Compute the normalized MMSE receiver filter c_k for
       user k using equation (7.28)

   (b) Replace the user k codeword s_k by c_k

3. Repeat step 2 until a fixed point is reached.

Figure 7.2. The MMSE Algorithm.

one-step increase:

$$\Delta'_{C_s} = \frac{s_k^T R_k^{-3} s_k}{s_k^T R_k^{-2} s_k} - s_k^T R_k^{-1} s_k \tag{7.29}$$

in sum capacity for the MMSE update. In order to show that Δ'_{C_s} is non-negative, we rewrite s_k, in terms of the matrix of eigenvectors Φ of R_k^{-1}, as $s_k = \Phi z_k$ where:

$$R_k^{-1} = \Phi \Lambda^{-1} \Phi^T \tag{7.30}$$

The change in sum capacity becomes:

$$\Delta'_{C_s} = \frac{z_k^T \Lambda^{-3} z_k}{z_k^T \Lambda^{-2} z_k} - z_k^T \Lambda^{-1} z_k \tag{7.31}$$

Rearranging, we have:

$$\Delta'_{C_s} = \frac{z_k^T \Lambda^{-3} z_k - z_k^T \Lambda^{-2} z_k z_k^T \Lambda^{-1} z_k}{z_k^T \Lambda^{-2} z_k} \tag{7.32}$$

and factoring yields:

$$\Delta'_{C_s} = \frac{z_k^T [\Lambda^{-2} (I - z_k z_k^T) \Lambda^{-1}] z_k}{z_k^T \Lambda^{-2} z_k} \tag{7.33}$$

Since $I - z_k z_k^T$ is positive semi-definite and Λ is positive definite, Δ'_{C_s} must be non-negative, which implies that the MMSE update monotonically increases sum capacity.

To show that GSC is monotonically decreased by application of the MMSE update, we evaluate the GSC reduction Δ_{GSC} in (21) with x_k replaced by the MMSE update c_k.

In this case, the reduction is GSC is:

$$\Delta'_{\text{GSC}} = 2(\mathbf{s}_k^\mathsf{T}\mathbf{R}_k\mathbf{s}_k - \mathbf{c}_k^\mathsf{T}\mathbf{R}_k\mathbf{c}_k) = 2\mathbf{s}_k^\mathsf{T}\mathbf{R}_k\mathbf{s}_k - 2\frac{\mathbf{s}_k^\mathsf{T}\mathbf{R}_k^{-1}\mathbf{s}_k}{\mathbf{s}_k^\mathsf{T}\mathbf{R}_k^{-2}\mathbf{s}_k} \tag{7.34}$$

The covariance matrix $\mathbf{R}_k$ of the interference plus noise seen by user k is symmetric and positive definite. Therefore, it is invertible and since $\mathbf{s}_k$ is unit norm we can write:

$$1 = \|\mathbf{s}_k\|^2 = \mathbf{s}_k^\mathsf{T}\mathbf{R}_k^{-1/2}\mathbf{R}_k^{1/2}\mathbf{s}_k \tag{7.35}$$

Applying the Schwarz inequality [69, p. 147] we have:

$$1 = \left(\mathbf{s}_k^\mathsf{T}\mathbf{R}_k^{-1/2}\mathbf{R}_k^{1/2}\mathbf{s}_k\right)^2 \le \|\mathbf{R}_k^{-1/2}\mathbf{s}_k\|^2 \|\mathbf{R}_k^{-1/2}\mathbf{s}_k\|^2 \tag{7.36}$$

$$= \left(\mathbf{s}_k^\mathsf{T}\mathbf{R}_k^{-1}\mathbf{s}_k\right)\left(\mathbf{s}_k^\mathsf{T}\mathbf{R}_k\mathbf{s}_k\right) \tag{7.37}$$

Furthermore, using the Schwarz inequality again, we obtain:

$$\left(\mathbf{s}_k^\mathsf{T}\mathbf{R}_k^{-1}\mathbf{s}_k\right)^2 \le \|\mathbf{s}_k\|^2 \|\mathbf{R}_k^{-1}\mathbf{s}_k\|^2 = \mathbf{s}_k^\mathsf{T}\mathbf{R}_k^{-1}\mathbf{s}_k \tag{7.38}$$

Applying the inequality (7.37) to (7.34) yields:

$$\Delta'_{\text{GSC}} \ge 2\mathbf{s}_k^\mathsf{T}\mathbf{R}_k\mathbf{s}_k - 2\frac{(\mathbf{s}_k^\mathsf{T}\mathbf{R}_k^{-1}\mathbf{s}_k)^2(\mathbf{s}_k\mathbf{R}_k\mathbf{s}_k)}{\mathbf{s}_k^\mathsf{T}\mathbf{R}_k^{-2}\mathbf{s}_k} \tag{7.39}$$

$$= 2\mathbf{s}_k^\mathsf{T}\mathbf{R}_k\mathbf{s}_k\left[1 - \frac{(\mathbf{s}_k^\mathsf{T}\mathbf{R}_k^{-1}\mathbf{s}_k)^2}{\mathbf{s}_k^\mathsf{T}\mathbf{R}_k^{-2}\mathbf{s}_k}\right] \tag{7.40}$$

It follows from (7.38) that $\Delta'_{\text{GSC}} \ge 0$, which proves that GSC is monotonically decreased for the MMSE update as well.

Thus, as it was the case with the eigen-algorithm, convergence of the MMSE algorithm is guaranteed by the monotonic increase of sum capacity and decrease of GSC coupled to the upper and lower bounds, respectively, of sum capacity and GSC. And once again, convergence in these metrics *does not* guarantee convergence to extremal values of the metrics. Nevertheless, it has been shown that a noisy version of the MMSE update always converges to a class of codewords for which GSC is minimized [3]. These codewords form GWBE sets that correspond to maximum sum capacity as well.

7.2.3 Other Algorithms for Interference Avoidance

In the previous sections we presented two different interference avoidance algorithms, both of which increase sum capacity and decrease GSC at each step, and for which convergence to the optimal point that maximizes sum capacity and minimizes GSC has been established.

A general class of interference avoidance algorithms can be defined by any replacement procedure for which the new codeword $\mathbf{x}_k$ satisfies either:

$$\mathbf{x}_k^\mathsf{T} \mathbf{R}_k^{-1} \mathbf{x}_k \geq \mathbf{s}_k^\mathsf{T} \mathbf{R}_k^{-1} \mathbf{s}_k \tag{7.41}$$

or:

$$\mathbf{x}_k^\mathsf{T} \mathbf{R}_k \mathbf{x}_k \leq \mathbf{s}_k^\mathsf{T} \mathbf{R}_k \mathbf{s}_k \tag{7.42}$$

When iterated over all users, any such procedure will result in a convergent algorithm, since it either increases sum capacity as equation (7.41) does, or decreases GSC as equation (7.42) does. We note that only convergence to a fixed point is guaranteed for such procedures, but not to the optimal fixed point. Among the many choices that satisfy equations (7.41) and (7.42), one choice is useful from a practical perspective: to adjust the current codeword incrementally in either the direction of the optimal codeword, or in such a way that sum capacity is increased or GSC is decreased. We now introduce two such procedures: *lagged interference avoidance* and *gradient descent interference avoidance*.

Lagged interference avoidance is the more obvious of these incremental procedures and corresponds to the codeword update equation:

$$\mathbf{s}_k(t+1) = \frac{\alpha \mathbf{s}_k(t) + m\beta \mathbf{x}_k(t)}{||\alpha \mathbf{s}_k(t) + m\beta \mathbf{x}_k(t)||} \tag{7.43}$$

where $\alpha, \beta \in \mathbb{R}^+$, $m = \mathrm{sgn}[\mathbf{s}_k^\mathsf{T}(t)\mathbf{x}_k(t)]$, and $\mathbf{x}_k(t)$ is the minimum eigenvector of $\mathbf{R}_k(t)$, the interference-plus-noise covariance seen by user k at time step t. To ensure that codewords change incrementally, we require $|\alpha| \gg |\beta|$, and we explicitly include a time index t to emphasize the incremental nature of the process. It can be easily shown that this lagged interference avoidance procedure monotonically increases sum capacity as well as decreases GSC [48,68].

The second procedure for interference avoidance with incremental codeword adaptation is based on a gradient descent technique. More precisely, in this case we would like to reduce the expression $\mathbf{s}_k^\mathsf{T} \mathbf{R}_k \mathbf{s}_k$ with each iteration, while maintaining unit norm codewords and excluding the trivial case $\mathbf{s}_k = 0$. Using the Rayleigh quotient:

$$\rho_k = \frac{\mathbf{s}_k^\mathsf{T} \mathbf{R}_k \mathbf{s}_k}{\mathbf{s}_k^\mathsf{T} \mathbf{s}_k} \tag{7.44}$$

and taking its gradient with respect to the codeword components $\{s_{kj}\}$ we obtain:

$$\nabla \rho_k = \frac{2[s_k^T s_k R_k s_k - (s_k^T R_k s_k)s_k]}{(s_k^T s_k)^2} \tag{7.45}$$

such that the iteration $s_k(t + 1) = s_k(t) - \mu \nabla \rho_k(t)$ with μ a suitably small constant, would increase the SINR. Since we require unit norm codewords, we normalize the codeword as:

$$s_k(t + 1) = \frac{s_k(t) - \mu \nabla \rho_k(t)}{\|s_k(t) - \mu \nabla \rho_k(t)\|} \tag{7.46}$$

This iteration decreases $s_k^T R_k s_k$ while maintaining a unit norm s_k. In addition, this iteration also monotonically decreases the GSC [48,68]. We note that gradient descent interference avoidance does not explicitly require calculation of the minimum eigenvector, which is a distinct computational advantage. However, as with any incremental method, convergence will proceed more slowly than if the optimal codeword were chosen straight away at each step [48,68].

7.3 INTERFERENCE AVOIDANCE OVER TIME-INVARIANT CHANNELS

Interference avoidance algorithms were introduced in Section 7.2 in a basic CDMA scenario in which individual users are assigned a single codeword for transmission, and communication channels are characterized by an ideal response with only additive Gaussian noise corrupting the received signal at the base station receiver. In this section, we extend application of interference avoidance to more general scenarios in which users are assigned multiple codewords for transmission, and communication channels are no longer ideal [42,43].

Specifically, in this section we consider the uplink of a synchronous CDMA system with K users in which each user sends frames of data symbols using a multicode CDMA approach as described schematically in Figure 7.3. Each symbol in a given user's frame is assigned a distinct N-dimensional codeword such that the CDMA symbol transmitted by the user is:

$$x_\ell = \sum_{m=1}^{K_\ell} b_m^{(\ell)} s_m^{(\ell)} = S_\ell b_\ell, \qquad \ell = 1, \ldots, K \tag{7.47}$$

where $b_\ell = [b_1^{(\ell)} \cdots b_{K_\ell}^{(\ell)}]^T$ is the vector containing the data symbols in user ℓ's frame, and $S_\ell = \left[s_1^{(\ell)} \cdots s_m^{(\ell)} \cdots s_{K_\ell}^{(\ell)} \right]$ is the $N \times K_\ell$ codeword matrix corresponding to user ℓ whose columns $s_m^{(\ell)}$, $m = 1, \ldots, K_\ell$, are the codewords assigned to each of the K_ℓ symbols in user ℓ's frame.

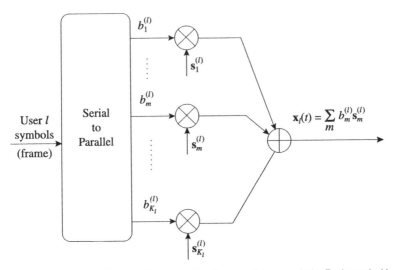

Figure 7.3. Multicode CDMA scenario for sending frames of data symbols. Each symbol in user ℓ's frame is assigned a distinct codeword and the resulting CDMA symbol $\mathbf{x}_\ell$ is a superposition of all codewords scaled by their corresponding data symbols.

The channel between a given user ℓ and the base station is described by the channel matrix $\mathbf{H}_\ell$ which incorporates propagation effects like attenuation and/or multipath. The dimension of the channel matrix $\mathbf{H}_\ell$ and the values of its elements are determined by specific time and/or frequency domain representation of the channel employed. For example, in the case of DS-CDMA systems with multipath $\mathbf{H}_\ell$ is an $N \times N$ circulant matrix containing the coefficients of the discrete-time channel impulse response [43,58], while for multicarrier (MC) CDMA systems $\mathbf{H}_\ell$ is a $N \times N$ diagonal matrix containing the channel frequency gains along the main diagonal [42,43]. We assume that all channels are time-invariant, and that each user knows its corresponding channel matrix, which is fixed. The case of time-varying fading channels is discussed in Section 7.4.

For this multicode CDMA scenario with non-ideal user channels, the received signal vector at the base station is then expressed as:

$$\mathbf{r} = \sum_{\ell=1}^{K} \mathbf{H}_\ell \mathbf{S}_\ell \mathbf{b}_\ell + \mathbf{n} \tag{7.48}$$

where $\mathbf{n}$ is the additive Gaussian noise vector that corrupts the received signal at the base station receiver, with the same covariance matrix $E[\mathbf{n}\mathbf{n}^\mathsf{T}] = \mathbf{W}$ as in previous sections. Assuming that symbols in a given user's frame are uncorrelated with unit energy such that $E[\mathbf{b}_\ell \mathbf{b}_\ell^\mathsf{T}] = \mathbf{I}_{K_\ell}$, the covariance matrix of the received signal in equation (7.48) is:

$$\mathbf{R} = E[\mathbf{r}\mathbf{r}^\mathsf{T}] = \sum_{\ell=1}^{K} \mathbf{H}_\ell \mathbf{S}_\ell \mathbf{S}_\ell^\mathsf{T} \mathbf{H}_\ell^\mathsf{T} + \mathbf{W} \tag{7.49}$$

7.3.1 Interference Avoidance with Diagonal Channel Matrices

Application of greedy interference avoidance for multicode CDMA systems with diagonal channel matrices extends in a straightforward way. We start by rewriting the received signal in equation (7.48) from the perspective of a given user k as:

$$\mathbf{r} = \mathbf{H}_k \mathbf{S}_k \mathbf{b}_k + \sum_{\ell=1, \ell \neq k}^{K} \mathbf{H}_\ell \mathbf{S}_\ell \mathbf{b}_\ell + \mathbf{n} \tag{7.50}$$

in which the first term is the desired signal corresponding to user k while the rest represents interference coming from other users and noise. We note that all the $\mathbf{H}_\ell$ matrices are assumed invertible, although some of their elements may be of $O(\varepsilon)$. Nevertheless, as discussed in [48,49], this does not restrict application of greedy interference avoidance since those dimensions corresponding to very small gains will be completely avoided.

From the perspective of user k, interference avoidance must now be applied to optimizing its codewords in the presence of combined noise and interference from other users. In order to do this, we define an equivalent "inverse-channel" observation for user k by pre-multiplying with the corresponding inverse channel matrix $\mathbf{H}_k^{-1}$ in equation (7.50) to obtain:

$$\mathbf{r}_k = \mathbf{S}_k \mathbf{b}_k + \mathbf{H}_k^{-1} \left(\sum_{\ell \neq k} \mathbf{H}_\ell \mathbf{S}_\ell \mathbf{b}_\ell + \mathbf{n} \right) \tag{7.51}$$

The covariance matrix of the inverse-channel received signal $\mathbf{r}_k$ is:

$$\mathbf{R}^{(k)} = \mathbf{S}_k \mathbf{S}_k^\mathsf{T} + \mathbf{H}_k^{-1} \left(\sum_{\ell \neq k} \mathbf{H}_\ell \mathbf{S}_\ell \mathbf{S}_\ell^\mathsf{T} \mathbf{H}_\ell + \mathbf{W} \right) \mathbf{H}_k^{-1} \tag{7.52}$$

and is related to the original received signal covariance matrix in equation (7.49) by:

$$\mathbf{R}^{(k)} = \mathbf{H}_k^{-1} \mathbf{R} \mathbf{H}_k^{-1} \tag{7.53}$$

Based on the inverse-channel observation, the greedy interference avoidance for user k consists of replacing codeword m of user k, $\mathbf{s}_m^{(k)}$, with the minimum eigenvector of the corresponding interference-plus-noise covariance matrix under channel k inversion, which is given by:

$$\mathbf{R}_m^{(k)} = \mathbf{R}^{(k)} - \mathbf{s}_m^{(k)} \mathbf{s}_m^{(k)\mathsf{T}} \tag{7.54}$$

Application of greedy interference avoidance in this multiuser/multicode CDMA context also monotonically increases sum capacity [48,49], as it was the case in the basic CDMA scenario discussed in Section 7.2. We note that, in this case, sum capacity is given by an expression identical to that in equation (7.7), but in which $\mathbf{R}$ has the more complex expression of equation (7.49).

Numerous interference avoidance algorithms can be formulated based on repeated application of the greedy interference avoidance procedure. These are defined by the various ways in which user codewords are selected for replacement. For example, one algorithm could be defined by replacement at a given step of one codeword of a given user, followed by replacement of a randomly selected codeword of a randomly selected user. Alternatively, at a given step of the algorithm, one could replace the codeword, which will yield the maximum increase in sum capacity. We note that, regardless of the order in which codewords are selected for update, convergence of all such algorithms to a fixed point is guaranteed by the monotonic increase in sum capacity implied by greedy interference avoidance, along with the fact that sum capacity is an upper-bounded measure. With respect to this measure, the optimal point with maximum sum capacity corresponds to simultaneous water filling by all users in their respective signal spaces, provided that users' transmit covariance matrices are allowed to span the entire transmit signal space [87]. In the multicode CDMA framework, this requirement translates to the requirement $K_\ell \geq N$ that the number of codewords assigned for transmission to any given user ℓ be equal to or larger than the dimension N of the signal space.

Empirically, it was observed that when users have at least as many codewords as signal space dimensions, algorithms based on repeated application of greedy interference avoidance with random initializations and various codeword updates yield an optimal codeword ensemble that corresponds to simultaneous water filling by all users in their inverted channel problems. An analytical proof of this result in general, for any possible order of codeword updates is not available, although it has been proved for some specific cases [48,49]. One of these cases consists of updating all codewords of a given user sequentially until convergence to a fixed point is reached, and then iterating for all users in the system. This is an extension of the eigen-algorithm in Section 7.2.1 to the multiuser/multicode scenario, and represents an instance of the "iterative water filling" procedure in [87]. We summarize this algorithm in Figure 7.4 An alternative algorithm based on greedy interference avoidance, which is not iterative water filling, but for which convergence to a simultaneous water filling solution was also proved [48,49], is presented in Figure 7.5. We note that, while this alternative algorithm may not seem attractive from a practical implementation point of view due to the extra complexity required for finding that codeword in the ensemble, which implies maximum increase in sum capacity, it is important from a theoretical perspective since, even though it is *not a water filling procedure*, it still converges to a simultaneously water-filled solution.

7.3.2 Interference Avoidance with General Channel Matrices

In the previous section, we presented the extension of interference avoidance to channels characterized by diagonal matrices, which is applicable to multicarrier CDMA systems. In this section, we discuss how application of greedy interference avoidance generalizes to any type of channel matrices. To ensure utmost generality we will approach the considered system from a very general perspective and assume that

1. Start with a randomly chosen codeword ensemble specified by the user codeword matrices $\mathbf{S}_1, \ldots, \mathbf{S}_K$

2. For each user $k = 1 \ldots K$,

    (a) Define the ''inverse-channel observation'' for user k as in equation (7.51).

    (b) Adjust user k's codewords sequentially such that the codeword corresponding to symbol m is replaced by the minimum eigenvector of matrix $\mathbf{R}_m^{(k)}$ in equation (7.54).

    (c) Repeat step (b) iteratively for each user until a fixed point is reached for which further modification of codewords will bring no additional improvement.

    (d) If a suboptimal point is reached use escape methods [61] and repeat steps (b)-(c).

3. Repeat step 2 iteratively for each user until a fixed point is reached for which further modification of codewords will bring no additional improvement.

Figure 7.4. The multiuser Eigen-Algorithm for dispersive channels.

different users reside in different signal spaces, with different dimensions and potential overlap among them, and all being subspaces of the receiver signal space. Each user's signal space, as well as the receiver signal space, are of finite dimension, and are implied by a finite signaling interval $\mathcal{T}$ and finite bandwidths W_ℓ for each user ℓ, respectively and W (which includes all W_ℓ corresponding to all users) for the receiver [30]. We denote by N_ℓ the dimension of user ℓ's signal space, and by N, $N \geq N_\ell$ for all ℓ, the dimension of the receiver signal space.

The transmitted signal by user ℓ is expressed by the same equation (7.47) but in which user ℓ's codeword matrix $\mathbf{S}_\ell$ is of dimensions $N_\ell \times K_\ell$ in this case. The received signal vector at the base station will be expressed in this case by the same equation (7.48) but in which channel matrices $\mathbf{H}_\ell$ can be any $N \times N_\ell$ matrices.

In order to apply greedy interference avoidance in this case, we start again by rewriting the received signal in equation (7.48) from the perspective of a given user

1. Start with a randomly chosen codeword ensemble
   specified by the user codeword matrices $\mathbf{S}_1, \dots, \mathbf{S}_K$

2. Define the ''inverse-channel observations'' for all
   users k as in equation (7.51)

3. Identify the codeword $\mathbf{s}_m^{(k)}$ whose replacement will
   maximally increase sum capacity. If no codeword will
   increase sum capacity, and suboptimal maxima escape
   methods [61] are ineffective for improvement, then
   STOP. Otherwise,

   (a) adjust $\mathbf{s}_m^{(k)}$ by replacing it with the minimum
       eigenvector of matrix $\mathbf{R}_m^{(k)}$ in equation (7.54)

   (b) Return to step 2

Figure 7.5. The maximum capacity increase algorithm for interference avoidance.

k as in equation (7.50):

$$\mathbf{r} = \mathbf{H}_k \mathbf{S}_k \mathbf{b}_k + \mathbf{z}_k \tag{7.55}$$

where $\mathbf{z}_k$ denotes the interference-plus-noise seen by user k. That is:

$$\mathbf{z}_k = \sum_{\ell=1,\ell \neq k}^{K} \mathbf{H}_\ell \mathbf{S}_\ell \mathbf{b}_\ell + \mathbf{n} \tag{7.56}$$

which has covariance matrix:

$$\mathbf{Z}_k = E[\mathbf{z}_k \mathbf{z}_k^\mathsf{T}] = \sum_{\ell=1,\ell \neq k}^{K} \mathbf{H}_\ell \mathbf{S}_\ell \mathbf{S}_\ell^\mathsf{T} \mathbf{H}_\ell^\mathsf{T} + \mathbf{W} \tag{7.57}$$

Since $\mathbf{Z}_k$ is symmetric, it can be diagonalized as:

$$\mathbf{Z}_k = \mathbf{E}_k \boldsymbol{\Delta}_k \mathbf{E}_k^\mathsf{T} \tag{7.58}$$

Furthermore, because $\mathbf{Z}_k$ is a positive definite covariance matrix, we can define the whitening transformation:

$$\mathbf{T}_k = \boldsymbol{\Delta}_k^{-1/2} \mathbf{E}_k^\mathsf{T} \tag{7.59}$$

In transformed coordinates, equation (7.55) is equivalent to:

$$\tilde{\mathbf{r}} = \mathbf{T}_k \mathbf{r} = \mathbf{T}_k \mathbf{H}_k \mathbf{S}_k \mathbf{b}_k + \mathbf{T}_k \mathbf{z}_k = \tilde{\mathbf{H}}_k \mathbf{S}_k \mathbf{b}_k + \mathbf{w}_k \tag{7.60}$$

where $\tilde{\mathbf{H}}_k = \mathbf{T}_k \mathbf{H}_k$ is the channel matrix seen by user k in the new coordinates and $\mathbf{w}_k = \mathbf{T}_k \mathbf{z}_k$ is the equivalent "white noise" with covariance matrix $E[\mathbf{w}_k \mathbf{w}_k^\mathsf{T}] = \mathbf{T}_k \mathbf{Z}_k \mathbf{T}_k^\mathsf{T} = \mathbf{I}$, the identity matrix. We note that the received signal covariance matrix in the transformed coordinates is related to the original signal covariance matrix by:

$$\tilde{\mathbf{R}} = E[\tilde{\mathbf{r}} \tilde{\mathbf{r}}^\mathsf{T}] = \mathbf{T}_k \mathbf{R} \mathbf{T}_k^\mathsf{T} \tag{7.61}$$

We now apply the singular value decomposition (SVD) [69, p. 442] to the transformed channel matrix corresponding to user k, yielding:

$$\tilde{\mathbf{H}}_k = \mathbf{U}_k \mathbf{D}_k \mathbf{V}_k^\mathsf{T} \tag{7.62}$$

The $N \times N$ matrix $\mathbf{U}_k$ has as columns the eigenvectors of $\tilde{\mathbf{H}}_k \tilde{\mathbf{H}}_k^\mathsf{T}$, the $N_k \times N_k$ matrix $\mathbf{V}_k$ has as columns the eigenvectors of $\tilde{\mathbf{H}}_k^\mathsf{T} \tilde{\mathbf{H}}_k$, and the $N \times N_k$ matrix $\mathbf{D}_k$ contains the singular values of $\tilde{\mathbf{H}}_k$ on the main diagonal and zero elsewhere. We note that because $\mathbf{T}_k$ is invertible, the rank of $\tilde{\mathbf{H}}_k$ will be equal to that of $\mathbf{H}_k$. Without loss of generality, we assume that $\mathbf{H}_k$ has full rank[4] N_k, so that the singular value matrix $\mathbf{D}_k$ can be partitioned as:

$$\mathbf{D}_k = \begin{bmatrix} \tilde{\mathbf{D}}_k \\ \mathbf{0} \end{bmatrix} \tag{7.63}$$

with $\tilde{\mathbf{D}}_k$ an $N_k \times N_k$ diagonal matrix containing the non-zero singular values along the diagonal and zeros in the rest. The left inverse of $\mathbf{D}_k$ is defined as:

$$\mathbf{D}_k^\dagger = \begin{bmatrix} \tilde{\mathbf{D}}_k^{-1} & \mathbf{0} \end{bmatrix} \tag{7.64}$$

with:

$$\mathbf{D}_k^\dagger \mathbf{D}_k = \mathbf{I}_{N_k} \tag{7.65}$$

Returning to equation (7.60) in which the SVD for transformed channel matrix $\tilde{\mathbf{H}}_k$ has been applied, we have:

$$\tilde{\mathbf{r}} = \mathbf{U}_k \mathbf{D}_k \mathbf{V}_k^\mathsf{T} \mathbf{S}_k \mathbf{b}_k + \mathbf{w}_k \tag{7.66}$$

We pre-multiply by $\mathbf{U}_k^\mathsf{T}$ to obtain:

$$\mathbf{r}_k = \mathbf{U}_k^\mathsf{T} \tilde{\mathbf{r}} = \mathbf{D}_k \mathbf{V}_k^\mathsf{T} \mathbf{S}_k \mathbf{b}_k + \mathbf{U}_k^\mathsf{T} \mathbf{w}_k \tag{7.67}$$

and define $\tilde{\mathbf{S}}_k = \mathbf{V}_k^\mathsf{T} \mathbf{S}_k$ and $\tilde{\mathbf{w}}_k = \mathbf{U}_k^\mathsf{T} \mathbf{w}_k$. Note that because both $\mathbf{U}_k$ and $\mathbf{V}_k$ are orthogonal matrices they preserve norms of vectors. Thus, columns of $\tilde{\mathbf{S}}_k$ are also unit norm as were the columns of $\mathbf{S}_k$. Also, because the equivalent noise term $\mathbf{w}_k$ is white, then $\tilde{\mathbf{w}}_k$ will remain white with the same covariance matrix equal to the identity matrix.

---

[4]This is not a restriction since if $\tilde{\mathbf{H}}_k$ is not full rank then in some dimensions the user k signal space will have zero projection on the output space. Therefore we can redefine a reduced codeword matrix $\tilde{\mathbf{S}}_k$ which uses only dimensions with nonzero projections on the output space.

With these additional definitions, we have:

$$\mathbf{r}_k = \mathbf{D}_k \tilde{\mathbf{S}}_k \mathbf{b}_k + \tilde{\mathbf{w}}_k \tag{7.68}$$

At this point, we define an equivalent observation for user k by pre-multiplying with the left inverse of $\mathbf{D}_k$ to obtain:

$$\tilde{\mathbf{r}}_k = \mathbf{D}_k^\dagger \mathbf{r}_k = \tilde{\mathbf{S}}_k \mathbf{b}_k + \tilde{\mathbf{z}}_k \tag{7.69}$$

with $\tilde{\mathbf{z}}_k = \mathbf{D}_k^\dagger \tilde{\mathbf{w}}_k$. For the equivalent observation equation (7.69) application of greedy interference avoidance consists of replacing codeword m of user k by the minimum eigen-vector of the corresponding interference-plus-noise covariance matrix in the transformed problem:

$$\mathbf{R}_m^{(k)} = \tilde{\mathbf{S}}_k \tilde{\mathbf{S}}_k^\mathsf{T} - \tilde{\mathbf{S}}_k^{(k)} \tilde{\mathbf{S}}_k^{(k)\mathsf{T}} + \tilde{\mathbf{D}}_k^{-2} \tag{7.70}$$

We note that, similar to the diagonal channel matrices case, application of greedy interference avoidance in this general scenario also monotonically increases sum capacity and reaches a fixed point [48]. In addition, when users have at least as many codewords as signal space dimensions, repeated application of greedy interference avoidance with various codeword replacement procedures with respect to codewords/users yields a simultaneous water filling solution for all users in their respective signal spaces, which is identical to that in [87] and corresponds to maximum sum capacity. This is just like the case of diagonal channel matrices, and similar algorithms for interference avoidance can be defined in this case as well. However, due to the very general framework used in this case, these algorithms will be applicable to a wide variety of wireless scenarios. We note the special case of multiuser systems with multiple transmit and receive antennas, which is discussed in [45,47].

7.4   INTERFERENCE AVOIDANCE IN FADING CHANNELS

An important characteristic of wireless communication systems is the unavoidable presence of fading, caused by the nature of the wireless communication channel. To maximize the overall network capacity, one should exploit the variations in the channel fade levels while allocating the available resources. In this section, the objective of resource allocation is to maximize the information theoretic ergodic (expected) sum capacity. We consider allocating both powers and the signature sequences of the users as functions of the channel state information (CSI) in order to achieve this objective. The presentation in this section is based on the work published in [28].

In the presence of fading and AWGN, the received signal vector is given by [78]:

$$\mathbf{r} = \sum_{i=1}^K \sqrt{p_i h_i} b_i \mathbf{s}_i + \mathbf{n} \tag{7.71}$$

where $\mathbf{s}_i = [s_{i1}, \ldots, s_{iN}]\mathsf{T}$, p_i, h_i, b_i are the unit energy signature sequence, transmit power, channel gain and information symbol, respectively, of user i, and $\mathbf{n}$ is a zero-mean Gaussian random vector with covariance $\sigma^2 \mathbf{I}_N$. The information symbol b_i is assumed to have unit energy, i.e., $E[b_i^2] = 1$. We assume that the receiver and all of the transmitters have perfect knowledge of the channel states of all users represented as a vector $\mathbf{h} = [h_1, \ldots, h_K]^\mathsf{T}$.

For a given set of signature sequences and a fixed set of channel gains, $\mathbf{h}$, the sum capacity $C_s(\mathbf{h}, \bar{\mathbf{p}}, \mathbf{S})$ is [76]:

$$C_s(\mathbf{h}, \bar{\mathbf{p}}, \mathbf{S}) = \frac{1}{2} \log \det \left(\mathbf{I}_N + \sigma^{-2} \sum_{i=1}^{K} h_i \bar{p}_i \mathbf{s}_i \mathbf{s}_i^\mathsf{T} \right) \tag{7.72}$$

where $\bar{p}_i$ is the average power of user i, $\bar{\mathbf{p}} = [\bar{p}_1, \ldots, \bar{p}_K]$, and $\mathbf{s} = [\mathbf{s}_1, \ldots, \mathbf{s}_K]$. In the presence of fading, if the channel state is modeled as a random vector, the quantity $C_s(\mathbf{h}, \bar{\mathbf{p}}, \mathbf{S})$ is random as well, and the ergodic sum capacity is found as the expected value of $C_s(\mathbf{h}, \bar{\mathbf{p}}, \mathbf{S})$. Instead of keeping the transmit power of user i fixed to $\bar{p}_i$ as in (7.72), we can choose the transmit powers of the users $p_i(\mathbf{h})$, $i = 1, \ldots, K$, as a function of the channel state $\mathbf{h}$ with the aim of maximizing the ergodic sum capacity of the system subject to average transmit power constraints for all users. Similarly, we can choose the signature sequences $\mathbf{S}$ to be a function of the channel state as well; let us denote it by $\mathbf{S}(\mathbf{h})$ to show the dependence on the channel state. Therefore, our problem is to solve for the jointly optimum transmit powers and signature sequences as functions of the channel state in order to maximize the ergodic sum capacity of the system in the presence of fading. The problem can be stated as the maximization of the ergodic sum capacity:

$$E_{\mathbf{h}} \left[\frac{1}{2} \log \det \left(\mathbf{I}_N + \sigma^{-2} \sum_{i=1}^{K} h_i p_i(\mathbf{h}) \mathbf{s}_i(\mathbf{h}) \mathbf{s}_i(\mathbf{h})^\mathsf{T} \right) \right] \tag{7.73}$$

subject to average power constraints for all users:

$$E_{\mathbf{h}}[p_i(\mathbf{h})] \leq \bar{p}_i, \qquad i = 1, \ldots, K \tag{7.74}$$

over the powers $\mathbf{p}(\mathbf{h})$ and signature sequences $\mathbf{S}(\mathbf{h})$, at all channel states $\mathbf{h}$.

In order to jointly optimize the powers and the signature sequences, we first fix the power distributions of all users over all fading states, $\mathbf{p}(\mathbf{h})$. Then, the corresponding optimal signature sequence set at every channel state will consist of a combination of orthogonal and GWBE sequences [79]. This is due to the fact that the signature sequences at a fading state $\mathbf{h}$ can be chosen independently of the signature sequences at any other state, since once the powers are fixed, there are no constraints relating $\mathbf{S}(\mathbf{h})$ to $\mathbf{S}(\bar{\mathbf{h}})$ for $\mathbf{h} \neq \bar{\mathbf{h}}$. Since the optimum signature sequences at each channel state depend only on powers $\mathbf{p}(\mathbf{h})$ and the channel state $\mathbf{h}$, we can express the capacity at each channel state only as a function of the powers, and optimize the ergodic capacity in terms of the power allocations of all users.

First consider the case when $K \leq N$. For any fixed channel state, the optimal choice of signature sequences for a given power control policy $\mathbf{p}(\mathbf{h})$ is an orthogonal set [64,79]. Then, our problem is equivalent to solving K independent Goldsmith-Varaiya problems [13] (see also [27]), the solution to which is a single user waterfilling for each user. More precisely, the optimal solution $\mathbf{p}^*(\mathbf{h})$ is the unique solution satisfying the Karush-Kuhn-Tucker (KKT) conditions, and is given by:

$$p_i^*(\mathbf{h}) = \left(\frac{1}{\lambda_i} - \frac{\sigma^2}{h_i} \right)^+, \qquad i = 1, \ldots, K \tag{7.75}$$

where λ_i is solved by applying (7.75) to (7.74).

When $K > N$, it has been shown in [79], for a non-fading channel, that given the power constraints of all users, one can group the users into two sets L and $\bar{L}$, of oversized and non-oversized users, respectively. Users $i \in L$ are assigned orthogonal sequences, and users $i \in \bar{L}$ are assigned GWBE sequences. For a channel with fading, at a certain channel state $\mathbf{h}$, and for a certain arbitrary power distribution of users which assigns powers $p_1, \ldots, p_K$ to channel state $\mathbf{h}$, let us define the matrix $\mathbf{D} = \text{diag}(p_1 h_1, \ldots, p_K h_K)$, and define μ_i to be the eigenvalues of the matrix $\mathbf{SDS}^\mathsf{T}$. Then the signature sequences that maximize the sum capacity for any fixed $\mathbf{h}$ satisfy [73]:

$$\mathbf{SDS}^\mathsf{T} \mathbf{s}_i = \mu_i \mathbf{s}_i, \qquad i = 1, \ldots, K \tag{7.76}$$

with repetitions of some of the μ_i (since there are only N eigenvalues of $\mathbf{SDS}^\mathsf{T}$), where the optimal μ_i s are given by [79]:

$$\mu_i(\mathbf{h}) = \begin{cases} \dfrac{\sum_{j \in \bar{L}(\mathbf{h})} p_j h_j}{N - |L(\mathbf{h})|}, & i \in \bar{L}(\mathbf{h}) \\ p_i h_i, & i \in L(\mathbf{h}) \end{cases} \tag{7.77}$$

Using the optimum eigenvalue assignment in (7.77) at each channel state, using (7.73), the signature-sequence-optimized sum capacity can be expressed as:

$$E_{\mathbf{h}} \left[\frac{1}{2} \sum_{i \in L(\mathbf{h})} \log \left(1 + \frac{p_i(\mathbf{h}) h_i}{\sigma^2} \right) + \frac{1}{2} (N - |L(\mathbf{h})|) \log \left(1 + \frac{\sum_{i \in \bar{L}(\mathbf{h})} p_i(\mathbf{h}) h_i}{\sigma^2 (N - |L(\mathbf{h})|)} \right) \right] \tag{7.78}$$

Next, we need to solve for the optimum power distributions $p_i(\mathbf{h})$, $i = 1, \ldots, K$ and the number of oversized users at each channel state $L(\mathbf{h})$. To this end, we will first determine the number of users who transmit with nonzero powers at any given channel state. For a given channel state $\mathbf{h}$, let the set of users that will transmit with non-zero powers be $\bar{K}(\mathbf{h})$. Then, we will prove that, for power control policies that maximize (7.78), the number of users in $\bar{K}(\mathbf{h})$ cannot exceed N.

First, we note that the function in (7.78) is concave, and the maximization (7.78) is subject to the affine set of average power constraints (7.74). Therefore, a power vector $\mathbf{p}^*(\mathbf{h})$ achieves the global optimum of the maximization problem if and only if it satisfies the KKT conditions. Then, writing the KKT conditions for the objective function in (7.78), it is easy to show that:

$$\frac{h_i}{\mu_i(\mathbf{h}) + \sigma^2} \leq \lambda_i \tag{7.79}$$

where $\mu_i(\mathbf{h})$ is given by (7.77), and equality holds if $p_i(\mathbf{h}) > 0$. Now, let us assume that the number of non-zero components in $\mathbf{p}^*(\mathbf{h})$ is $|\bar{K}(\mathbf{h})| > N$, for a given $\mathbf{h}$. Then, some users must share some of the available dimensions, i.e., not all users can be made orthogonal to each other. In fact, we can find at most $N - 1$ sequences that are orthogonal to all other sequences in the system, or equivalently, at least $|\bar{K}(\mathbf{h})| - N + 1$ users will have the same $\mu_i(\mathbf{h}) = \sum_{j \in L(\mathbf{h})} h_j p_j / (N - |L(\mathbf{h})|)$. Then, substituting this into (7.79), we get $h_i/\lambda_i = h_j/\lambda_j$ for $i \neq j$, $i, j \in \bar{K}(\mathbf{h})$ for at least $|\bar{K}(\mathbf{h})| - N + 1$ users. Note that as the channel fading is assumed to be a continuous random variable, this event has zero probability, and at most one user with GWBE sequences (one with highest h_i/λ_i ratio, as in [29]) may transmit, with probability 1. But this contradicts the assumption that $|\bar{K}(\mathbf{h})| > N$, which establishes our desired result, i.e., $|\bar{K}(\mathbf{h})| \leq N$ almost surely. This result may be viewed as a generalization of [29] to a vector channel with a unit rank constraint on the covariance matrices of the inputs; [29] showed that in scalar MAC (i.e., when $N = 1$), at most one user may transmit at a channel state with probability 1.

An important implication of this result is that, since the optimal power allocation dictates that at most N users transmit with positive powers at any given channel state, orthogonal sequences should be assigned to those users that are transmitting with positive powers. That is, although we allowed for allocating GWBE sequences to some of the users, the solution implies that there is at most one such user, and the problem reduces to the orthogonal case. The optimal power allocation is again single user waterfilling, similar to the solution given in (7.75), i.e.:

$$p_i*(\mathbf{h}) = \begin{cases} 1/\lambda_i - \sigma^2/h_i, & i \in \bar{K}(\mathbf{h}) \\ 0, & \text{otherwise} \end{cases} \tag{7.80}$$

Here, one needs to be careful about the transmit regions. Unlike the case where the actual number of users is $K \leq N$, the users in the set $\bar{K}(\mathbf{h})$ change with $\mathbf{h}$; thus a channel adaptive allocation of the orthogonal sequences is necessary. Our convention is that we assign a sequence from an orthogonal set to a user wherever its power is positive.

To specify the optimal power allocation completely, let us define $\gamma_i = h_i/\lambda_i$. Then, the probability that $\gamma_i = \gamma_j$, for $i \neq j$ is zero. Therefore, we can always find a unique order statistics $\{\gamma_{[i]}\}_{i=1}^{K}$ such that $\gamma_{[1]} > \cdots > \gamma_{[K]}$, for each given $\mathbf{h}$. Let us now place σ^2 in that ordering, assuming that at least one of the $\gamma_{[i]}$s is larger than σ^2. Define

$\gamma_{[K+1]} = 0$. Then, for some $n \in \{1, \ldots, K\}$, let:

$$\gamma_{[1]} \geq \cdots \geq \gamma_{[n]} > \sigma^2 \geq \gamma_{[n+1]} \geq \cdots \geq \gamma_{[K+1]} \tag{7.81}$$

where the equalities are included for the sake of consistency of the indices, and do not affect the solution (note the strict inequality just before σ^2).

First, let $n \leq N$. Then, we see that (7.80) gives positive powers for all n users, and thus all n users with highest γ_is will transmit with the non-zero powers given in (7.80). When $n > N$, there are more than N users satisfying the positivity constraints $\gamma_i > \sigma^2$. However, we know from our derivation that only the user with the highest γ_i from the set we intend to assign GWBE sequences may transmit. Therefore, a total of N users with the highest γ_is transmit at this channel state.

Finally, we can summarize the jointly optimal power and signature sequence allocation policy as:

$$p_i^*(\mathbf{h}) = \begin{cases} 1/\lambda_i - \sigma^2/h_i, & i \in \Omega \\ 0, & \text{otherwise} \end{cases} \tag{7.82}$$

$$\mathbf{s}_i^*(\mathbf{h})^{\mathsf{T}} \mathbf{s}_j^*(\mathbf{h}) = 0, \qquad i, j \in \Omega, \ i \neq j \tag{7.83}$$

$$\Omega = \left\{ i \leq \min(K, N) \middle| \gamma_{[i]} > \sigma^2 \right\} \tag{7.84}$$

7.4.1 Iterative Power and Sequence Optimization in Fading

We found in the previous section that the optimal power control strategy is a waterfilling over some favorable channel states for each user. However, in order to obtain the optimal power levels one should also compute the Lagrange multipliers λ_i, from the average power constraints. It turns out that the power allocation of each user still depends in a complicated fashion on those of the other users through these λ_i. In this section, we provide an iterative method to obtain the jointly optimal power and signature sequence allocation, and hence the λ_i.

In [27], we have shown that for fixed signature sequences $\mathbf{S}$, the optimal single-user update that maximizes the sum capacity as a function of $p_k(\mathbf{h})$ is given by:

$$p_k(\mathbf{h}, \mathbf{S}) = \left(\frac{1}{\lambda_k} - \frac{1}{h_k \mathbf{s}_k^{\mathsf{T}} \mathbf{A}_k^{-1} \mathbf{s}_k} \right)^+ \tag{7.85}$$

where the *interference covariance matrix* $\mathbf{A}_k$ is defined as:

$$\mathbf{A}_k = \sigma^2 \mathbf{I}_N + \sum_{i \neq k} h_i p_i(\mathbf{h}) \mathbf{s}_i \mathbf{s}_i^{\mathsf{T}} = \sigma^2 \mathbf{I} + \mathbf{SDS}^{\mathsf{T}} - h_k p_k(\mathbf{h}) \mathbf{s}_k \mathbf{s}_k^{\mathsf{T}} \tag{7.86}$$

We can find and fix the optimal signature sequences at each state for a given power allocation using results of [79]. Then, plugging these sequences in (7.86), multiplying both sides by the optimal signature sequence $\mathbf{s}_k^*$, and noting that the signature sequences that maximize the sum capacity for a fixed set of power constraints satisfy

(7.76), we obtain:

$$\mathbf{A}_k \mathbf{s}_k^* = (\sigma^2 + \mu_k - h_k p_k) \mathbf{s}_k^* \tag{7.87}$$

where the μ_k are given by (7.77). Therefore:

$$\mathbf{s}_k^{*\mathrm{T}} \mathbf{A}_k^{-1} \mathbf{s}_k^* = \frac{1}{\sigma^2 + \mu_k - h_k p_k} \tag{7.88}$$

This shows that we can represent the base level for the waterfilling in (7.85) as a function of the power levels in the previous iteration. Substituting this in (7.85), we get the optimal power allocation at the $n + 1$st step, $p_k^{n+1}(\mathbf{h})$ for user k, with optimal sequences and fixed powers $\{p_i(\mathbf{h})\}_{i \neq k}$ from the previous iteration:

$$p_k^{n+1}(\mathbf{h}) = \left(\frac{1}{\lambda_k^{n+1}} - \frac{\sigma^2 + \mu_k^n(\mathbf{h}) - h_k p_k^n(\mathbf{h})}{h_k} \right)^+ \tag{7.89}$$

where we use $\{p_1^{n+1}(\mathbf{h}), \ldots, p_{k-1}^{n+1}(\mathbf{h}), p_k^n(\mathbf{h}), \ldots, p_K^n(\mathbf{h})\}$ to compute $\mu_k^n(\mathbf{h})$. Combining this with (7.77) gives us the power update at each step. It is easy to observe that, once the eigenvalues $\mu_k^n(\mathbf{h})$ are determined using the power levels from the previous iteration, we can use (7.89) to solve for kth user's power by waterfilling. Note that, the Lagrange multiplier λ_k^{n+1} is chosen to satisfy the average power constraint of user k at each iteration, and can be obtained by plugging (7.89) into the constraint in (7.74). The waterfilling algorithm automatically obtains the value of λ_k^{n+1} as it is the inverse of the "water level."

The proposed algorithm may be interpreted in two ways. First, it may be seen as an iteration from a set of powers to another set of powers as given by (7.89). Therefore, one may run this algorithm starting with an arbitrary power distribution, to obtain the capacity maximizing power distribution when the algorithm converges. The signature sequences may then be assigned to the users after the algorithm converges: at each channel state, the users that have non-zero powers (there will be at most N such users) are assigned signature sequences from an orthogonal set. Second, the algorithm may be seen as an iteration from powers to signature sequences, and then back to powers again. Specifically, for a given set of powers, the optimal sequences may be found using (7.76) and (7.77), i.e., as in [79]; corresponding to these sequences, base levels for the waterfilling in (7.85) can be computed using (7.87) and (7.88), and new powers may be found using (7.85) as in [27].

7.5 INTERFERENCE AVOIDANCE IN ASYNCHRONOUS SYSTEMS

We will formulate the problem of designing optimum signature sequences in asynchronous CDMA systems from two different angles as in the synchronous case: First, we will find the optimum signature sequences that maximize the user capacity when we constrain ourselves to one-shot matched filter receivers. We will observe, somewhat similar in spirit to the synchronous case, that if we used M-shot optimum linear filtering, i.e., M-shot MMSE filters, with the identified optimum signature sequences, such complicated filters would reduce to one-shot matched filters. Therefore, we will conclude that there is no loss in constraining ourselves to matched filters. This approach, which we present in Section 7.5.1, is based on our work published in [74].

Secondly, we will find the optimum signature sequences that maximize the information theoretic sum capacity. In this case, the entire received signal is observed over the entire time axis and the optimum receivers are used. The optimum signature sequences that maximize the information theoretic sum capacity turn out to be the same as those that maximize the user capacity. We present this approach, which is based on our work published in [36], in Section 7.5.2. The optimum signature sequences in the asynchronous case minimize a quantity called the total squared asynchronous correlation (TSAC). The existence of the optimum signature sequences that minimize the TSAC for arbitrary user delay profiles and powers has been shown through an explicit construction algorithm in [36]. We will not discuss that construction algorithm here, however, we will present iterative asynchronous interference avoidance algorithms that are based on decreasing the TSAC at every iteration by the update of a single user's signature sequence, in Section 7.5.3.

7.5.1 Interference Avoidance for User Capacity Maximization

We consider a single-cell symbol-asynchronous (but chip-synchronous) CDMA system with K users and processing gain N. The received signal in the nth symbol interval of user k is given as (see Figure 7.6):

$$\mathbf{r}_k(n) = \sqrt{p_k} b_k(n) \mathbf{s}_k + \sum_{l \neq k} \sqrt{p_l} \left[b_l(n) T_L^{d_{kl}} \mathbf{s}_l + b_l(n+1) T_R^{d_{kl}} \mathbf{s}_l \right] + \mathbf{n}_k \qquad (7.90)$$

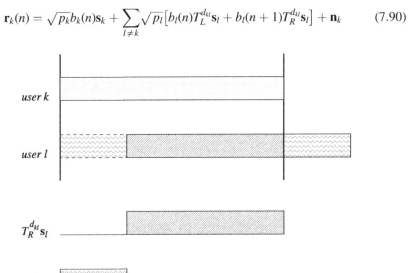

Figure 7.6. Asynchronous interference calculation.

where p_k, $b_k(n)$ and $\mathbf{s}_k$ are the received power, nth transmitted symbol and signature sequence of user k, respectively, and $\mathbf{n}_k$ is a zero-mean Gaussian random vector with $E[\mathbf{n}_k \mathbf{n}_k^\mathsf{T}] = \sigma^2 \mathbf{I}_N$. The signature sequences of all users are of unit energy, i.e., $\mathbf{s}_k^\mathsf{T} \mathbf{s}_k = 1$, for all k. For users k and l, d_{kl} represents the relative time delay of user l with respect to the time delay of user k. That is, $d_{kl} = d_l - d_k$, where d_k and d_l are the time delays of users k and l, respectively. Symbols T_R^d and T_L^d denote the operations of shifting, to right and left, respectively, of a vector by d and $N - d$ chips (components). For both operators, the vacated positions in the vector are filled with zeros. That is, for a vector $\mathbf{x} = [x_1, \ldots, x_N]^\mathsf{T}$ and integer $d \geq 0$, we define:

$$T_L^d \mathbf{x} = [x_{N-d+1}, \ldots, x_N, 0^{N-d}]^\mathsf{T} \quad \text{and} \quad T_R^d \mathbf{x} = [0^d, x_1, \ldots, x_{N-d}]^\mathsf{T} \quad (7.91)$$

where 0^d denotes d consecutive zeros.

We will use one-shot matched filters as the receivers. The decision statistics for the kth user in the nth symbol interval is $y_k(n) = \mathbf{s}_k^\mathsf{T} \mathbf{r}_k(n)$, where we do assume that the matched filter receiver of each user is perfectly aligned with the symbol interval of the user. Since $\mathbf{s}_k^\mathsf{T} \mathbf{s}_k = 1$, the SIR of the kth user is then given by:

$$\text{SIR}_k = \frac{p_k}{\sum_{l \neq k} A_{kl} p_l + \sigma^2} \quad (7.92)$$

where we define the $K \times K$ matrix $\mathbf{A}$ with the entries:

$$A_{kl} = \begin{cases} (\mathbf{s}_k^\mathsf{T} T_L^{d_{kl}} \mathbf{s}_l)^2 + (\mathbf{s}_k^\mathsf{T} T_R^{d_{kl}} \mathbf{s}_l)^2, & k \neq l \\ 0, & k = l \end{cases} \quad (7.93)$$

The common SIR target β is said to be feasible iff one can find non-negative powers $\{p_k\}_{k=1}^K$ such that $\text{SIR}_k \geq \beta$ for all k, which can be written in an equivalent matrix form as:

$$\mathbf{p} \geq \beta(\mathbf{A}\mathbf{p} + \sigma^2 \mathbf{1}) \quad (7.94)$$

where $\mathbf{1}$ is the vector of all ones. It is well-known that if the common SIR target β is feasible, then the optimum power vector, i.e., the componentwise smallest feasible power vector, is found by solving (7.94) with equality [86]. Furthermore, the power control problem is feasible if [66]:

$$\beta < \frac{1}{\rho_A} \quad (7.95)$$

where ρ_A is the largest (also called the Perron-Frobenius) eigenvalue of the symmetric non-negative matrix $\mathbf{A}$. We define the matrix $\mathbf{R} = \mathbf{A} + \mathbf{I}$ so that $R_{kk} = (\mathbf{s}_k^\mathsf{T} \mathbf{s}_k)^2 = 1$ and $\mathbf{R}$ represents the squared asynchronous cross correlations of the signature sequences. The Perron–Frobenius eigenvalue of $\mathbf{R}$ satisfies $\rho_R = \rho_A + 1$, and the feasibility condition in (7.95) can also be expressed as:

$$\beta < \frac{1}{\rho_R - 1} \tag{7.96}$$

That is, for a single cell CDMA system, the range of common achievable SIR values is determined only by the Perron–Frobenius eigenvalue of the squared asynchronous cross correlation matrix **R** which depends only on the signature sequences of the users and their relative time delays. For a given signature sequence set $\{s_k\}_{k=1}^K$ and a set of time delays $\{d_k\}_{k=1}^K$, the supremum of common achievable SIR targets equals $1/(\rho_R - 1)$. Our aim is to choose the signature sequences of the users, for any given set of time delays, such that the common achievable SIR is maximized. Therefore, we seek the signature sequence set that maximizes $1/(\rho_R - 1)$, or, equivalently, minimizes ρ_R.

We note that it is hard to characterize the dependence of ρ_R on individual signature sequences. Instead, our approach is to tie the Perron–Frobenius eigenvalue of **R**, ρ_R, to another parameter of **R** which can be related to the signature sequences in a more direct way. By this approach, we will be able to characterize the optimum signature sequences in a closed form expression in addition to being able to devise an iterative and distributed signature sequence update algorithm that will construct progressively better signature sequence sets. To this end, we start our derivation with the following bounds on the Perron–Frobenius eigenvalue of **R** in terms of its row-sums [66]:

$$\min_k \sum_{l=1}^K R_{kl} \le \rho_R \le \max_k \sum_{l=1}^K R_{kl} \tag{7.97}$$

Similar bounds that can be obtained using column-sums of **R** are identical to (7.97) since **R** is symmetric. We also have the following bound from a simple application of the Rayleigh quotient [69]:

$$\frac{1}{K} \sum_{k=1}^K \sum_{l=1}^K R_{kl} \le \rho_R \tag{7.98}$$

which is equivalent to $(\mathbf{1}^\mathsf{T}\mathbf{R}\mathbf{1})/(\mathbf{1}^\mathsf{T}\mathbf{1}) \le \rho_R$. Combining (7.97) and (7.98) and the fact that the minimum row-sum lower bounds the average of the row-sums yields:

$$\min_k \sum_{l=1}^K R_{kl} \le \frac{1}{K} \sum_{k=1}^K \sum_{l=1}^K R_{kl} \le \rho_R \le \max_k \sum_{l=1}^K R_{kl} \tag{7.99}$$

We define the total squared asynchronous correlation (TSAC) as:

$$\mathrm{TSAC} = \sum_{k=1}^K \sum_{l=1}^K R_{kl} \tag{7.100}$$

Note that the TSAC is equal to the sum of the entries of the asynchronous squared correlation matrix **R**. Since we want to minimize ρ_R, and since ρ_R is lower bounded by

$TSAC/K$, it is reasonable to try to minimize the TSAC over the space of all possible signature sequences. Although it is not clear that ρ_R decreases as TSAC decreases, we will show that the signature sequence sets that achieve a particular lower bound on TSAC are precisely those that minimize ρ_R.

Next, our aim is to prove two facts: that TSAC is lower bounded by K^2/N, just as in the Welch bound (7.9) for the synchronous case, and secondly, that when the signature sequences achieve this lower bound, the squared asynchronous correlation (SAC) "seen" by each users is the same (the uniformly good property). These are Theorems 3 and 4 in [74]; however, here we will present a much simpler development for their proofs.

In the sequel, we will concentrate on a time duration that is equal to one symbol interval, i.e., N chip intervals. However, this symbol interval is not assumed to be time aligned to any particular user's symbol period. This symbol interval is depicted in Figure 7.7, where each box represents a chip interval. For each user, the white and gray chips in Figure 7.7 correspond to symbols with time stamps n and $n + 1$ of that user. In particular, for user k, the white chips on the left represent the last d_k chips in the signature $\mathbf{s}_k$ used to transmit symbol n, and the gray chips on the right are the first $N - d_k$ chips of $\mathbf{s}_k$ used to transmit symbol $n + 1$. We will call the gray chips preceeded by d_k zeros the "left signatures" of the users and the white chips followed by $N - d_k + 1$ zeros the "right signatures." We denote the left signature of user k by $\mathbf{s}_k^L$ and the right signature of user k by $\mathbf{s}_k^R$. We define two $N \times K$ matrices, $\mathbf{S}_L$ and $\mathbf{S}_R$ that contain all the left signatures and the right signatures, respectively.

With this new representation, we can obtain a more concise expression for the squared asynchronous cross correlation terms A_{kl} in (7.93). As in (7.93), the

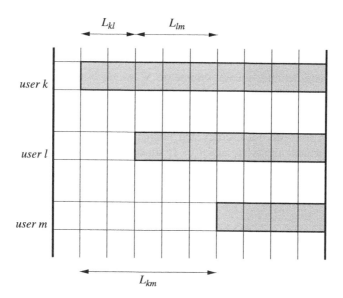

Figure 7.7. Asynchronous system representation.

interference a pair of users create to each other has two components: these two components can be expressed as correlations of two vectors restricted to two sets of chip indices. For users k and l, let L_{kl} denote the set of chip indices for which users k and l transmit symbols with different time stamps. In terms of Figure 7.7, L_{kl} is the set of columns (chip indices) for which rows (users) k and l have different colors (e.g., white and gray). With this new representation, $R_{kl} = A_{kl} + 1$ can be expressed as:

$$R_{kl} = \left([\mathbf{S}_L^\mathsf{T}\mathbf{S}_L]_{kl} + [\mathbf{S}_R^\mathsf{T}\mathbf{S}_R]_{kl} \right)^2 + \left([\mathbf{S}_L^\mathsf{T}\mathbf{S}_R]_{kl} + [\mathbf{S}_R^\mathsf{T}\mathbf{S}_L]_{kl} \right)^2 \qquad (7.101)$$

Therefore, using (7.100) and (7.101), the TSAC can be expressed in terms of $\mathbf{S}_L$ and $\mathbf{S}_R$ as:

$$\begin{aligned}
\text{TSAC} &= \text{tr}\left[\mathbf{S}_L\mathbf{S}_L^\mathsf{T}\mathbf{S}_L\mathbf{S}_L^\mathsf{T}\right] + 2\text{tr}\left[\mathbf{S}_L\mathbf{S}_R^\mathsf{T}\mathbf{S}_R\mathbf{S}_L^\mathsf{T}\right] + \text{tr}\left[\mathbf{S}_R\mathbf{S}_R^\mathsf{T}\mathbf{S}_R\mathbf{S}_R^\mathsf{T}\right] \\
&\quad + 2\text{tr}\left[\mathbf{S}_L\mathbf{S}_L^\mathsf{T}\mathbf{S}_R\mathbf{S}_R^\mathsf{T}\right] &(7.102) \\
&\geq \text{tr}\left[\mathbf{S}_L\mathbf{S}_L^\mathsf{T}\mathbf{S}_L\mathbf{S}_L^\mathsf{T}\right] + 2\text{tr}\left[\mathbf{S}_L\mathbf{S}_L^\mathsf{T}\mathbf{S}_R\mathbf{S}_R^\mathsf{T}\right] + \text{tr}\left[\mathbf{S}_R\mathbf{S}_R^\mathsf{T}\mathbf{S}_R\mathbf{S}_R^\mathsf{T}\right] &(7.103) \\
&= \text{tr}\left[\left(\mathbf{S}_L\mathbf{S}_L^\mathsf{T} + \mathbf{S}_R\mathbf{S}_R^\mathsf{T}\right)\left(\mathbf{S}_L\mathbf{S}_L^\mathsf{T} + \mathbf{S}_R\mathbf{S}_R^\mathsf{T}\right)^\mathsf{T}\right] &(7.104) \\
&\geq \frac{1}{N}\left(\text{tr}\left[\mathbf{S}_L\mathbf{S}_L^\mathsf{T} + \mathbf{S}_R\mathbf{S}_R^\mathsf{T}\right]\right)^2 &(7.105) \\
&= \frac{1}{N}\left(\text{tr}\left[(\mathbf{S}_L + \mathbf{S}_R)(\mathbf{S}_L + \mathbf{S}_R)^\mathsf{T}\right]\right)^2 &(7.106) \\
&= \frac{K^2}{N} &(7.107)
\end{aligned}$$

where, in getting (7.102), we used the fact that $\mathbf{S}_L^\mathsf{T}\mathbf{S}_R$ is strictly lower triangular and $\mathbf{S}_R^\mathsf{T}\mathbf{S}_L$ is strictly upper triangular; (7.103) follows because we ignored non-negative term $2\text{tr}[\mathbf{U}\mathbf{U}^\mathsf{T}]$ where $\mathbf{U} = \mathbf{S}_L\mathbf{S}_R^\mathsf{T}$; (7.105) follows because $\text{tr}[\mathbf{V}\mathbf{V}^\mathsf{T}] \geq \frac{1}{N}(\text{tr}[\mathbf{V}])^2$ for positive semi-definite, square, symmetric $\mathbf{V}$, which is true because $\sum_{i=1}^N \lambda_i^2 \geq \frac{1}{N}\left(\sum_{i=1}^N \lambda_i\right)^2$, for non-negative λ_i, $i = 1, \ldots, N$; and (7.106) follows because $\text{tr}\left[\mathbf{S}_L^\mathsf{T}\mathbf{S}_R\right] = 0$ since, by definition, $[\mathbf{S}_L^\mathsf{T}\mathbf{S}_R]_{ii} = 0$ for all i, as the left and right signatures, for each user, do not overlap; and finally, (7.107) follows because $\mathbf{S}_L + \mathbf{S}_R$ has columns whose norms are equal to 1, since these columns correspond to the rotated (in time) versions of the signature sequences, which are all unit-norm.

Therefore, we have proved that:

$$\text{TSAC} = \sum_{k=1}^{K}\sum_{l=1}^{K} R_{kl} \geq \frac{K^2}{N} \qquad (7.108)$$

We first note that the lower bound on the TSAC is exactly the same as the lower bound on the TSC. Secondly, we observe that in order for the lower bound on the TSAC to be achieved, we need both inequalities in (7.103) and (7.105) to be satisfied with equality.

This happens when $\mathbf{U}$ is equal to zero, and $\mathbf{V}$ is a multiple of identity matrix, i.e., all λ_i are equal. That is, the lower bound on the TSAC is achieved if:

$$\mathbf{S}_L \mathbf{S}_R^T = 0 \qquad (7.109)$$

$$\mathbf{S}_L \mathbf{S}_L^T + \mathbf{S}_R \mathbf{S}_R^T = \frac{K}{N} \mathbf{I}_N \qquad (7.110)$$

These are exactly the same conditions as in Theorem 3 in [74]. Finally, we note that if the signature sequences are such that the lower bound on the TSAC is achieved, then:

$$\sum_{l=1}^{K} R_{kl} = \frac{K}{N}, \qquad k = 1, \ldots, K \qquad (7.111)$$

This is the asynchronous version of the "uniformly good property" that was proved for the synronous case [38]. This property means that if the signature sequences achieve the lower bound on the TSAC, then all users "see" the same amount of interference from each other.

We recall that our aim is to minimize ρ_R in order to maximize the achievable SIR, β. Note that (7.108) combined with (7.99) says that:

$$\rho_R \geq \frac{K}{N} \qquad (7.112)$$

Since our aim is to minimize ρ_R, we note that we cannot do better than to choose signature sequences that achieve (7.112) with equality. Meanwhile, (7.111) says that when the signature sequences can be chosen such that the TSAC lower bound is achieved with equality, then all of the row-sums equal K/N. By (7.99), the row-sums sandwich ρ_R, and so (7.112) is satisfied with equality, yielding the lowest possible ρ_R: $\rho_R = K/N$. Therefore, using (7.96), the bound on the common achievable SIR target in this asynchronous case is:

$$\beta < \frac{1}{K/N - 1} \qquad (7.113)$$

which is the same as the bound in the synchronous case, which can also be written in terms of an upper bound on the number of users that can be supported at SIR level β as [81]:

$$\frac{K}{N} < 1 + \frac{1}{\beta} \qquad (7.114)$$

In conclusion, we observe that the optimum signature sequences minimize the TSAC, and therefore, satisfy the conditions in (7.109) and (7.110). In addition, we observe that the user capacity of the asynchronous system is the same as the user capacity of the synchronous system, that is, there is no loss in capacity due to asynchrony when the signature sequences are chosen optimally.

7.5.2 Interference Avoidance for Sum Capacity Maximization

Consider the first M symbol durations. The chip matched filter output of the receiver can be denoted by an $MN \times 1$ real-valued column vector $\mathbf{y}$ that satisfies the system model:

$$\mathbf{y} = \sum_{k=1}^{K} S_k \mathbf{x}_k + \mathbf{n} \tag{7.115}$$

Here $\mathbf{y} = \left[\mathbf{y}(1)^\mathsf{T}, \mathbf{y}(2)^\mathsf{T}, \ldots, \mathbf{y}(M)^\mathsf{T} \right]^\mathsf{T}$, where $\mathbf{y}(m)$ is the chip-matched filter output vector of the mth symbol duration; $\mathbf{x}_k = [x_k(1) \ x_k(2) \ \cdots \ x_k(M)]^\mathsf{T}$ is the source symbol vector of user k; $\mathbf{n}$ is the Gaussian noise with zero mean and covariance matrix $E[\mathbf{n}\mathbf{n}^\mathsf{T}] = \sigma^2 I_{MN}$; and S_k is the signature matrix of user k, which is given by:

$$S_k = \begin{bmatrix} s_k^L & 0 & \ddots & 0 \\ s_k^R & s_k^L & 0 & \ddots \\ 0 & \ddots & \ddots & 0 \\ \ddots & 0 & s_k^R & s_k^L \end{bmatrix} \tag{7.116}$$

The average power of the normalized source signal of the kth user is restricted to:

$$\mathrm{tr}\left(E\left[\mathbf{x}_k \mathbf{x}_k^\mathsf{T}\right] \right) \leq M P_k \tag{7.117}$$

where P_k is the average power per symbol of user k. Given J as an arbitrary group of users, the mutual information $I(\mathbf{x}_{k \in J}; \mathbf{y} | \mathbf{x}_{k \notin J})$ is given by [76]:

$$I(\mathbf{x}_{k \in J}; \mathbf{y} | \mathbf{x}_{k \notin J}) \leq \frac{1}{2} \log \det \left(I_{MN} + \sum_{k \in J} \frac{S_k E[\mathbf{x}_k \mathbf{x}_k^\mathsf{T}] S_k^\mathsf{T}}{\sigma^2} \right) \tag{7.118}$$

with equality if the signals of users $k \in J$ are Gaussian.

Assume that the user delay profile is known to both the transmitters and the receiver. The capacity region of the system is given by the convex closure, over independent random vectors $\mathbf{x}_k$ satisfying (7.117), of the union of the following heptagons:

$$\lim_{M \to \infty} \bigcap_{J \subset \{1,2,\ldots,K\}} \left\{ (R_1, \ldots, R_K) : 0 \leq \sum_{k \in J} R_k \leq \frac{1}{M} I(\mathbf{x}_{k \in J}; \mathbf{y} | \mathbf{x}_{k \notin J}) \right\} \tag{7.119}$$

Suppose that the signature sequences are given. In the situation when user signals are symbol synchronous, it is shown in [77] that the rate constraints are maximized if the source signals have white power spectra, i.e., $E[x_k x_k^T] = P_k I_M$, for all k. However, in the symbol asynchronous case, in general, there is no unique power spectrum that can maximize the rate upper bounds in (7.119) simultaneously [77].

Combining (7.118) and (7.119), the sum capacity per symbol of the system satisfies:

$$C_s \leq \lim_{M \to \infty} \max_{\mathrm{tr}\left(E\left[x_k x_k^T\right]\right) \leq MP_k} \frac{1}{2M} \log \det \left(I_{MN} + \sum_{k=1}^{K} \frac{S_k E[x_k x_k^T] S_k^T}{\sigma^2} \right) \qquad (7.120)$$

where equality holds when the input signals are stationary Gaussian.

Define the block circulant signature matrix of user k by:

$$\hat{S}_k = \begin{bmatrix} s_k^L & 0 & \ddots & 0 & s_k^R \\ s_k^R & s_k^L & 0 & \ddots & 0 \\ 0 & \ddots & \ddots & \ddots & \ddots \\ \ddots & \ddots & \ddots & \ddots & 0 \\ 0 & \ddots & 0 & s_k^R & s_k^L \end{bmatrix} \qquad (7.121)$$

Then for any set of users J, the two matrices:

$$\sum_{k \in J} \hat{S}_k E[x_k x_k^T] \hat{S}_k^T \qquad \text{and} \qquad \sum_{k \in J} S_k E[x_k x_k^T] S_k^T$$

are asymptotically equivalent [16,36]. Consequently, we can replace the signature matrices with their circulant versions, and rewrite the sum capacity in (7.120) as:

$$C_s \leq \lim_{M \to \infty} \max_{\mathrm{tr}\left(E\left[x_k x_k^T\right]\right) \leq MP_k} \frac{1}{2M} \log \det \left(I_{MN} + \sum_{k=1}^{K} \frac{\hat{S}_k E[x_k x_k^T] \hat{S}_k^T}{\sigma^2} \right) \qquad (7.122)$$

Note that all circulant matrices of the same dimension have the identical eigen-vector set. Define the $M \times 1$ vector q_m and the $M \times M$ Fourier transform matrix Q_M as:

$$q_m = \left[1 \quad e^{j\frac{2\pi(m-1)}{M}} \quad \cdots \quad e^{j\frac{2\pi(m-1)(M-1)}{M}} \right]^H \qquad (7.123)$$

$$Q_M = \frac{1}{\sqrt{M}} [q_1 \quad q_2 \quad \cdots \quad q_M] \qquad (7.124)$$

where the superscript H denotes the conjugate transpose. Denote the mth component of vector $\boldsymbol{q}_n$ by q_{nm}. According to [16,83], we can decompose the block circulant matrix $\hat{\boldsymbol{S}}_k$ as:

$$\hat{\boldsymbol{S}}_k = \begin{bmatrix} q_{11}\boldsymbol{I}_N & \cdots & q_{M1}\boldsymbol{I}_N \\ \cdots & \cdots & \cdots \\ q_{1M}\boldsymbol{I}_N & \cdots & q_{MM}\boldsymbol{I}_N \end{bmatrix} \boldsymbol{\Phi}_k^* \boldsymbol{Q}_M^H \tag{7.125}$$

where the superscript $*$ denotes the conjugate operation; $\boldsymbol{\Phi}_k$ is a block diagonal matrix, defined as:

$$\boldsymbol{\Phi}_k = \begin{bmatrix} \phi_{k1} & 0 & \cdots & 0 \\ 0 & \phi_{k2} & \cdots & \cdots \\ \cdots & \cdots & \cdots & \cdots \\ 0 & \cdots & 0 & \phi_{kM} \end{bmatrix} \tag{7.126}$$

and ϕ_{km} in equation (7.126) is an $N \times 1$ column vector given by:

$$\phi_{km} = s_k^L + s_k^R e^{-j\frac{2\pi(m-1)}{M}} \tag{7.127}$$

Substituting (7.125) into (7.122), we obtain:

$$\left| \boldsymbol{I}_{MN} + \sum_{k=1}^{K} \frac{\hat{\boldsymbol{S}}_k E[\boldsymbol{x}_k \boldsymbol{x}_k^T] \hat{\boldsymbol{S}}_k^T}{\sigma^2} \right| = \left| \boldsymbol{I}_{MN} + \sum_{k=1}^{K} \frac{\boldsymbol{\Phi}_k^* \boldsymbol{Q}_M^H E[\boldsymbol{x}_k \boldsymbol{x}_k^T] \boldsymbol{Q}_M \boldsymbol{\Phi}_k^T}{\sigma^2} \right|$$

$$\leq \left| \boldsymbol{I}_{MN} + \sum_{k=1}^{K} \frac{\boldsymbol{\Phi}_k^* \boldsymbol{P}_k \boldsymbol{\Phi}_k^T}{\sigma^2} \right| \tag{7.128}$$

where $\boldsymbol{P}_k$ is an $M \times M$ diagonal matrix, whose diagonal entries are equal to those of $\boldsymbol{Q}_M^H E[\boldsymbol{x}_k \boldsymbol{x}_k^T] \boldsymbol{Q}_M$. The last inequality in (7.128) is due to the generalized Hadamard inequality [4,77], which indicates that the determinant of a positive-definite matrix is upper-bounded by the product of the determinants of its diagonal blocks.

Denote the mth diagonal entry of $\boldsymbol{P}_k$ by p_{km}. Since equation (7.128) holds with equality when $\boldsymbol{Q}_M^H E[\boldsymbol{x}_k \boldsymbol{x}_k^T] \boldsymbol{Q}_M$ is diagonal, substituting (7.128) into (7.122), we obtain:

$$C_s \leq \lim_{M \to \infty} \max_{\sum_{m=1}^{M} p_{km} \leq MP_k} \frac{1}{2M} \sum_{m=1}^{M} \log \det \left(\boldsymbol{I}_N + \sum_{k=1}^{K} \frac{\phi_{km}^* p_{km} \phi_{km}^T}{\sigma^2} \right) \tag{7.129}$$

When the signature sequences of the users are given, the sum capacity can be found by solving (7.129) with an iterative water filling algorithm [77], whose convergence is studied in [87]. It can be seen that, given an arbitrary signature set and user delay profile, the sum capacity is usually *not* achieved by input signals with white spectra (i.e., $P_k = P_k I_M$).

Without loss of generality and to simplify the presentation, let us assume from this point on, in this section that the user powers are all equal, $P_k = P$. The case of arbitrary user powers was treated in [36] with appropriate definition of oversized users, etc., as in the synchronous case [81]. Since S_k is replaced by $\hat{S}_k$, $\forall k$, we can view the K-user M-symbol asynchronous system as an MK-user symbol synchronous system. Therefore, we can see that an upper bound on the sum capacity is [81]:

$$
C_s \leq \frac{N}{2} \log\left(1 + \frac{KP}{N\sigma^2}\right) \tag{7.130}
$$

Now, using the definition in (7.126) and choosing $p_{km} = P$, which satisfies the power constraint, we can develop an equivalent representation for the term inside the log det expression in (7.129):

$$
\sum_{k=1}^{K} \phi_{km}^* p_{km} \phi_{km}^T = P \sum_{k=1}^{K} \left(s_k^L + s_k^R e^{j\frac{2\pi(m-1)}{M}}\right)\left(s_k^L + s_k^R e^{-j\frac{2\pi(m-1)}{M}}\right)^T
$$

$$
= P\left(S_L S_L^T + S_R S_R^T\right) + P e^{-j\frac{2\pi(m-1)}{M}} S_L S_R^T
$$

$$
+ P e^{j\frac{2\pi(m-1)}{M}} S_R S_L^T \tag{7.131}
$$

Now, we can see that the conditions in (7.109)–(7.110) are necessary and sufficient in order to achieve the upper bound in (7.130) on the sum capacity.

In conclusion, we observe that the optimum signature sequences that maximize the information theoretic sum capacity are the same as those that maximize the user capacity (as in the synchronous case), and therefore, that there is no loss in capacity due to asynchrony when the signature sequences are chosen optimally.

7.5.3 TSAC Reduction: Iterative Algorithms

Following the closed-form expressions for the signature sequence sets maximizing the information theoretic sum capacity [64,79] and user capacity [81], references [72,73] introduced the iterative adaptation of signature sequences for synchronous CDMA systems. Since the optimum signature sequences minimize the TSC in the synchronous case, the algorithms in [63,72,73] were designed to decrease (more precisely, not to increase) the TSC at each iterative step. Here, we will design algorithms that decrease the TSAC at each iteration. To this end, we first separate the terms that depend on the signature sequence of the kth user in the TSAC. From the TSAC definition (7.100), the definition of $\mathbf{R} = \mathbf{A} + \mathbf{I}$, and the fact that $\mathbf{A}$ in (7.93) is symmetric, we can write:

$$\text{TSAC} = (\mathbf{s}_k^{\mathsf{T}}\mathbf{s}_k)^2 + 2\mathbf{s}_k^{\mathsf{T}}\mathbf{B}_k\mathbf{s}_k + \gamma_k \tag{7.132}$$

where $\gamma_k = \sum_{i \neq k}\sum_{j \neq k} R_{ij}$ denotes the squared asynchronous correlation terms that do not depend on $\mathbf{s}_k$ and where in terms of $\bar{\mathbf{s}}_{kl} = T_L^{d_{kl}}\mathbf{s}_l$ and $\tilde{\mathbf{s}}_{kl} = T_R^{d_{kl}}\mathbf{s}_l$, the left and right signatures of the lth asynchronous user with respect to the kth user:

$$\mathbf{B}_k = \sum_{l \neq k} \left(\bar{\mathbf{s}}_{kl}\bar{\mathbf{s}}_{kl}^{\mathsf{T}} + \tilde{\mathbf{s}}_{kl}\tilde{\mathbf{s}}_{kl}^{\mathsf{T}} \right) \tag{7.133}$$

In order to minimize the TSAC, we are looking for updates of the signature sequence of the kth user from $\mathbf{s}_k$ to some $\mathbf{c}_k$ that is guaranteed to decrease (not to increase) the TSAC. Let us denote the TSAC after the $\mathbf{s}_k \rightarrow \mathbf{c}_k$ update as $\overline{\text{TSAC}}$. Then:

$$\overline{\text{TSAC}} = (\mathbf{c}_k^{\mathsf{T}}\mathbf{c}_k)^2 + 2\mathbf{c}_k^{\mathsf{T}}\mathbf{B}_k\mathbf{c}_k + \gamma_k \tag{7.134}$$

Restricting the new (updated) signature sequence of the kth user to be of unit energy as well, i.e., $\mathbf{c}_k^{\mathsf{T}}\mathbf{c}_k = 1$, we note that $\overline{\text{TSAC}} \leq \text{TSAC}$ if:

$$\mathbf{c}_k^{\mathsf{T}}\mathbf{B}_k\mathbf{c}_k \leq \mathbf{s}_k^{\mathsf{T}}\mathbf{B}_k\mathbf{s}_k \tag{7.135}$$

Although there are many possible $\mathbf{s}_k \rightarrow \mathbf{c}_k$ updates that would guarantee that (7.135) holds, we will propose two of them here. The two similar updates used in the synchronous CDMA context were given in [72,73] and in [63]. We call the first update the *asynchronous MMSE update* which we define as:

$$\mathbf{c}_k = \frac{\left(\mathbf{B}_k + a^2\mathbf{I}_N \right)^{-1}\mathbf{s}_k}{\left[\mathbf{s}_k^{\mathsf{T}}\left(\mathbf{B}_k + a^2\mathbf{I}_N \right)^{-2}\mathbf{s}_k \right]^{1/2}} \tag{7.136}$$

and we call the second update the *asynchronous eigen-update*, which we define as the normalized eigen-vector of $\mathbf{B}_k$ corresponding to its smallest eigen-value. Note that, in the asynchronous MMSE update, the new signature sequence of user k, $\mathbf{c}_k$, is the normalized one-shot MMSE receiver filter for that user when the signature sequences of all other users are fixed. Similar to the synchronous MMSE update [72,73], the new signature sequence can be obtained using an adaptive [1,37,39,59] or a blind [20] implementation of the one-shot MMSE filter.

One can also devise algorithms as in [75], where both the signature sequences and the receiver filters are updated in an on-line fashion. The proof that the asynchronous eigen update decreases the TSAC follows from the Rayleigh quotient applied to the matrix $\mathbf{B}_k$ [69]. The proof that the asynchronous MMSE update decreases the TSAC can be carried out in a very similar fashion to the proof that the MMSE update decreases the TSC [72,73].

7.6 FEEDBACK REQUIREMENTS FOR INTERFERENCE AVOIDANCE

In distributed implementations of interference avoidance methods, user codewords and corresponding receiver filters are adapted iteratively using feedback information received from the base station in order to improve performance. As a consequence, for practical implementation one must consider the feedback channel between the base station receiver, which has the information required for codeword adaptation, and the transmitter, which uses this information to perform the actual codeword adaptation. Ideally, with unlimited feedback capacity, the transmitter and receiver have access to the same information, but in practice only limited feedback is available [31,40,65].

In particular, compact representation of codewords is extremely important for systems that employ interference avoidance since as opposed to current CDMA systems where uniform-amplitude codeword chips are used, interference avoidance employs real-valued "chips"—real-valued coefficients for a set of orthonormal basis functions of the signal space used by the transmitter and receiver.

In numerical studies [44], it was found that optimal codeword ensembles can be represented using approximately 4 bits per dimension per codeword on average. Thus, describing a codeword with 128 real-valued "chips" would require 512 bits. If codewords are computed at the receiver and fed back, this seems a reasonably large feedback burden considering that a single codeword probably conveys relatively few bits. Of course, this still may be acceptable if codewords need not be changed very frequently or the downlink channel from receiver to transmitter is very large relative to the uplink multiple access channel.

Nonetheless, developing parsimonious codeword representations or ways of delivering the minimum information necessary for codeword construction at the transmitter is an important part of producing practical interference avoidance methods. In what follows, we describe two approaches, both empirical and therefore not completely satisfying, but both seeming to offer significant reductions in the necessary amount of codeword feedback.

7.6.1 Codeword Tracking for Interference Avoidance

Early work [44] implicitly assumes that codewords are in effect made from whole cloth, and then fed back to the transmitter. That is, codewords are derived anew during each iteration of the algorithm and are completely unrelated to previous codewords. In a slowly varying environment, however, the *entropy rate* associated with each codeword may be much less than the implicit entropy of, say, 512 bits per update postulated above for a 128-chip codeword. The direct approach of quantifying codeword update entropy rates seems difficult. In addition, the aggregate codeword feedback rate must necessarily scale linearly with the number of codewords which must be fed back. This can be particularly onerous for overloaded (more users than signal dimensions) systems.

More recent work on CDMA codeword adaptation with feedback [65] considers feedback limited to B bits in the context of a large system limit where the number of users K, signal dimensions N, and feedback bits B tend to infinity with fixed ratios of the system load K/N and feedback bits per codeword element B/N, and studies the performance of random vector quantization scheme in which codebook entries are independent and isotropically distributed.

One alternative to codeword feedback is interference covariance feedback. This is grossly inefficient when the signal space dimension is large and the number of users small. That is, the number of entries in the covariance matrix goes as the number of signal dimensions N squared, so if only $K \ll N$ codewords need be fed back, it may be simpler to simply feed back the codewords themselves. However, as the number of users approaches and exceeds the number of signal space dimensions, covariance matrix feedback becomes much more attractive.

Furthermore, covariance feedback offers a simple means to quantify the necessary amount of feedback information. Specifically, the codeword updates themselves are dependent on interference covariance matrix feedback. Thus, we can via the information processing theorem [11] obtain a bound on the necessary feedback level relatively simply.

In [67] a noisy covariance feedback channel was considered and its capacity calculated. Thus, the rate at which update information could be delivered to the transmitter was limited. By adjusting the noise level on the feedback channel and noting its effect on performance, the upper bound on the amount of information necessary to produce nearly optimal transmitted codewords was found to be on the order of one bit per dimension. This is a significant reduction from four bits per dimension codeword quantization results.

More careful analytic treatment of covariance feedback performance should be the subject of future work.

7.6.2 Reduced-Rank Signatures

Another approach to reducing the amount of codeword feedback information is to in effect directly quantize the codeword space in a sort of principal component approach where the codeword dimensions along which the most improvement will be had are identified and fed back. That is, one might identify small set of codeword *subspaces* and feed back information only about these principal directions.

Paraphrasing from [56,58], one might constrain the $N \times 1$ signature to lie in a D-dimensional subspace $\mathcal{S}_D$, where $D < N$. That is, the signature for user k is:

$$\mathbf{s}_k = \mathbf{F}_k \mathbf{c}_k \tag{7.137}$$

where $\mathbf{F}_k$ is $N \times D$ and the columns span $\mathcal{S}_D$, and $\mathbf{c}_k$ is the $D \times 1$ vector of combining coefficients. Each matrix $\mathbf{F}_k$ is chosen differently for each user, so that the associated signatures lie in different subspaces. Optimizing $\mathbf{s}_k$ with respect to equation (7.137) is termed *reduced-rank* signature optimization. This is analogous to reduced-rank, or subspace, techniques, which have been considered for receiver optimization [14,15,21–23,84].

The combining coefficients $\mathbf{c}_k$ are estimated at the receiver, individually quantized, and transmitted back to the transmitter. The parameter D has an intuitively pleasing interpretation: $D = 1$ corresponds to conventional power control, and $D = N$ corresponds to full signature adaptation. As D increases, the degrees of freedom for avoiding interference increase; however, there are more coefficients to quantize. In the presence of fixed interference the optimized reduced-rank signature is the projection of the full-rank optimized signature onto the subspace spanned by the columns of $\mathbf{F}_k$ [58].

In the absence of *a priori* information about where the signatures for the interferers lie in $\mathbb{R}^N$, we can select the columns of $\mathbf{F}_k$ to be isotropic. Another possibility, which simplifies the computation of the optimized signatures, is to choose the first column of $\mathbf{F}_k$ as a signature with *i.i.d.* elements, and generate the remaining columns with different orthogonal masks so that $\mathbf{F}_k^T \mathbf{F}_k = \mathbf{I}$. (For example, the columns might be non-overlapping segments of a randomly chosen signature [58].)

The performance of the optimized signature (i.e., with unlimited feedback bits) can be estimated by a large system analysis in which $(D, K, N) \to \infty$ with fixed D/N and K/N [58]. (A large system analysis of reduced-rank receivers with random signatures is presented in [23].) Results presented in [65] indicate that there is a critical $D < N$, which offers a significant improvement in performance relative to taking $D = N$. Reduced-rank signature adaptation can also be applied to multi-code CDMA. In that case, power and rate can be allocated among the optimized reduced-rank signatures for a particular user [57].

7.7 RECENT RESULTS ON INTERFERENCE AVOIDANCE

7.7.1 Interference Avoidance and Power Control

The uplink CDMA system considered in Section 7.2 assumed that all users have equal received power at the base station, and in this case interference avoidance algorithms yield a WBE codeword ensemble that implies uniform SINR for all users [63,73,79]. The algorithms extend in a straightforward manner to a system in which users are received with different powers at the base station. In this case, the received signal at the base station is given by:

$$\mathbf{r} = \sum_{\ell=1}^{K} b_\ell \sqrt{p_\ell} \mathbf{s}_\ell + \mathbf{n} = \mathbf{S}\mathbf{P}^{1/2}\mathbf{b} + \mathbf{n} \qquad (7.138)$$

with $\mathbf{P} = \text{diag}(p_1, p_2, \ldots, p_K)$ containing received powers for all users.

When matched filters are used at the receiver for all users, the SINR for a given user k becomes:

$$\gamma_k = \frac{(\sqrt{p_k}\mathbf{s}_k^T \mathbf{s}_k)^2}{\sum_{\ell=1, \ell \neq k}^{K} (\mathbf{s}_k^T \mathbf{s}_\ell \sqrt{p_\ell})^2 + E[(\mathbf{s}_k^T \mathbf{n})^2]} = \frac{p_k}{\mathbf{s}_k^T \mathbf{R}_k \mathbf{s}_k} \qquad (7.139)$$

with the correlation matrix of the interference-plus-noise seen by user k being expressed in this case as:

$$\mathbf{R}_k = \sum_{\ell=1, \ell \neq k}^{K} p_\ell \mathbf{s}_\ell \mathbf{s}_\ell^{\mathsf{T}} + \mathbf{W} = \mathbf{R} - p_k \mathbf{s}_k \mathbf{s}_k^{\mathsf{T}} \qquad (7.140)$$

where

$$\mathbf{R} = \mathbf{SPS}^{\mathsf{T}} + \mathbf{W} \qquad (7.141)$$

is the correlation matrix of the received signal $\mathbf{r}$ in equation (7.138).

In order to maximize user k's SINR through codeword adaptation one may still replace the current codeword of user k with the minimum eigenvector of $\mathbf{R}_k$. Thus, when user powers are assumed fixed, greedy interference avoidance and the eigen-algorithm in Section 7.2.1 apply with no changes, and in that case the algorithm yields an ensemble of GWBE codewords with eventual oversized users [63,80,81]. Oversized users have large powers relative to the other users in the system and get exclusive use of signal dimensions with minimum noise energy. We note that users achieve maximum possible SINRs corresponding to their powers and cross-correlations of GWBE codewords. We also note that by allowing users to change their power in addition to their codeword, we provide an extra degree of freedom and allow more flexibility to users in achieving SINRs that match more closely their quality of service requirements.

In this section, we present an algorithm that combines codeword adaptation through greedy interference avoidance with a power control mechanism in a two-stage code-word and power update. In the first stage the algorithm decreases the effective inter-ference seen by a given user k through greedy interference avoidance, by replacing its current codeword with the minimum eigenvector of $\mathbf{R}_k$. If, after the first stage, the SINR of the given user is below a specified target SINR, then in the second stage the given user increases its power attempting to meet the specified target. The new user power is the minimum between the value that matches the specified target SINR and the maximum allowed user power $p_k^{\max}$. We note that the target SINRs $\{\gamma_1^*, \ldots, \gamma_k^*, \ldots, \gamma_K^*\}$ must satisfy the admissibility condition:

$$\sum_{k=1}^{K} \frac{\gamma_k^*}{1 + \gamma_k^*} < N \qquad (7.142)$$

in order for a valid codeword and power allocation to exist [80,81]. We present the algorithm formally in Figure 7.8. Convergence of the eigen-algorithm with power control to a fixed point is defined with respect to sum capacity C_s, which is given by the expression in equation (7.7) but with $\mathbf{R}$ in equation (7.141), and which is upper bounded by the sum capacity of the K-user Gaussian multiple access channel with the corresponding power constraints on total user power and noise [64,79]. Following the same line of reasoning as in equations (7.17)–(7.19) in Section 7.2, we obtain that C_s is monotonically increased by the eigen-algorithm with power con-trol, and because it is upper bounded the algorithm will always reach a fixed point.

1. Start with a random set of user codewords and powers
   specified by matrices S and P respectively.

2. Specify a set of desired target SINRs $\gamma_1^*, \ldots, \gamma_K^*$
   satisfying the condition in equation (7.142)

3. For each user $k = 1, \ldots, K$,

   (a) Compute $\mathbf{R}_k$ using equation (7.140) and determine
       the minimum eigenvalue λ_k and eigenvector $\mathbf{x}_k$

   (b) Minimize the effective interference for user k by
       replacing its current codeword $\mathbf{s}_k$ with the minimum
       eigenvector $\mathbf{x}_k$ of $\mathbf{R}_k$

   (c) If user k's SINR after codeword replacement is
       below the specified target γ_k^*, increase user k
       power to meet the target SINR:

   $$p_k = \min\{p_k^{\max}, \gamma_k^* \lambda_k\}.$$

   Otherwise, leave p_k unchanged.

4. Repeat step 3 until a fixed point is reached.

Figure 7.8. The Eigen-Algorithm with power control.

Among all fixed points of the eigen-algorithm with power control, an optimal fixed point corresponds to a GWBE codeword ensemble with eventual oversized users [80,81].

We note that although an analytical convergence proof of the eigen-algorithm with power control to optimal GWBE codeword ensembles is not available, extensive simulations have shown that this is reached when the algorithm is initialized with random user codewords [46], provided that the specified target SINRs are admissible as defined by equation (7.142). This is consistent with empirical observations made on the eigen-algorithm which show that with random codeword initialization this always converges to a GWBE codeword ensemble [61,63].

7.7.2 Adaptive Interference Avoidance Algorithms

Interference avoidance algorithms presented so far are static in the number of users and do not allow variable target SINRs. Each time these change, the algorithms must be reiterated in order to determine a new optimal solution for the new number of users and/or target SINRs. Other algorithms for codeword adaptation for uplink CDMA systems [17,25,26,70,71,79–81] have the same characteristic and are not adaptable to changing numbers of active users/target SINRs in the system. In order to overcome this limitation, recent research [19] proposes using Grassmannian signatures in dynamic systems with variable number of users. These are designed to support a maximum number of active users in the system subject to a given interference level, and have the nice property that interference among users does not change when less users are active in the system. As noted in [19] the disadvantage associated with equiangular Grassmannian signatures is that they may not exist for any desired system configuration specified by a given number of users and signal space dimensions.

Recently, an alternative approach to dealing with variable number of active users and/or target SINRs in the uplink of a CDMA system has been proposed. This uses an adaptive algorithm with incremental updates similar to the ones proposed for joint incremental codeword and power adaptation based on interference avoidance [32,33]. The algorithm moves the system incrementally from an optimal configuration with a given number of active users and/or target SINRs, to a new optimal configuration with a different number of active users/target SINRs [34]. The transition between the two optimal configurations is based on an adaptive interference avoidance procedure: when a change in the system status occurs this translates to a change of the SINR of active users, which will employ a greedy gradient-based technique to optimize their corresponding spectral efficiency subject to constraints on the SINR. The spectral efficiency function used in deriving the adaptive interference avoidance algorithm is expressed in terms of the user SINR as:

$$\eta_k = \ln(1 + \gamma_k) \quad [\text{nats/s/Hz}], \qquad k = 1, \dots, K \qquad (7.143)$$

This expression corresponds to the spectral efficiency of a single-user bandlimited AWGN channel [82], and is a reasonable optimization criterion for individual users in the system who have access only to their corresponding SINR, with no knowledge of the other user SINRs. When replacing the expression of γ_k from equation (7.139) we write user k's spectral efficiency as a function of its codeword and power:

$$\eta_k = \ln\left(1 + \frac{p_k}{s_k^T R_k s_k}\right) \quad [\text{nats/s/Hz}], \quad k = 1, \dots, K \qquad (7.144)$$

In the adaptive interference avoidance algorithm each user k will perform joint codeword and power adaptation to maximize its corresponding spectral efficiency subject to target SINR constraint $\gamma_k = \gamma_k^*$ and to unit norm constraint on codewords $s_k^T s_k = 1$. Thus, the equations of the codeword and power updates for the adaptive algorithm are obtained from solving the constrained optimization problem:

$$\max_{s_k, p_k} \eta_k \quad \text{subject to} \quad \begin{cases} \dfrac{p_k}{s_k^{\mathsf{T}} \mathbf{R}_k s_k} = \gamma_k^* \\ s_k^{\mathsf{T}} s_k = 1 \end{cases} \tag{7.145}$$

We define the Lagrange multipliers λ_k and ξ_k associated with user k constraints in (7.145) such that the user k Lagrangian function is:

$$L_k(s_k, p_k, \lambda_k, \xi_k)$$

$$= \eta_k(s_k, p_k) + \lambda_k(\gamma_k - \gamma_k^*) + \xi_k(s_k^{\mathsf{T}} s_k - 1) \tag{7.146}$$

$$= \ln\left[1 + \frac{p_k}{s_k^{\mathsf{T}} \mathbf{R}_k s_k}\right] + \lambda_k\left[\frac{p_k}{s_k^{\mathsf{T}} \mathbf{R}_k s_k} - \gamma_k^*\right] + \xi_k(s_k^{\mathsf{T}} s_k - 1) \tag{7.147}$$

The necessary conditions for maximizing the Lagrangian in equation (7.147) are obtained by taking its partial derivatives with respect to the corresponding variables. Equating the partial derivative of the Lagrangian with respect to the codeword s_k to zero, we obtain an eigenvalue/eigenvector equation corresponding to matrix $\mathbf{R}_k$. That is:

$$\frac{\partial L_k}{\partial s_k} = 0 \quad \Rightarrow \quad \mathbf{R}_k s_k = \nu_k s_k \tag{7.148}$$

where ν_k is expressed in terms of the Lagrange multipliers, as well as user power p_k and codeword s_k. We note that the exact expression of ν_k is not relevant, and that for any eigenvector of $\mathbf{R}_k$ we have that $\partial L_k / \partial s_k = 0$ which satisfies the necessary condition in equation (7.148). A good choice for user k's codeword that satisfies the necessary condition in equation (7.148) is the minimum eigen-vector x_k of $\mathbf{R}_k$: for given power p_k this maximizes user k's SINR and implicitly its spectral efficiency, by minimizing the effective interference that corrupts user k's signal at the receiver. This choice defines the greedy interference avoidance procedure, and may generate a sudden change in the user's codeword as the minimum eigenvector may be far away in signal space from the current codeword employed by the user for transmission. If the receiver does not get immediate feedback on the new codeword, tracking sudden changes may generate errors at the receiver since it is not realistic to assume that the corresponding matched filter receiver changes instantaneously to the new user codeword.

A more desirable approach for an adaptive algorithm is to change the user codeword in small increments as suggested in Section 7.2.3, with a corresponding incremental change of the receiver filter. This way the receiver is capable of following codeword changes, and can continue to detect transmitted symbols correctly. We will therefore use an incremental update that adapts the codeword in the direction of the minimum eigen-vector $\mathbf{x}_k$ defined by:

$$\mathbf{s}_k(i+1) = \frac{\mathbf{s}_k(i) + m\beta\mathbf{x}_k(i)}{\|\mathbf{s}_k(i) + m\beta\mathbf{x}_k(i)\|} \tag{7.149}$$

where $m = \text{sgn}(\mathbf{s}_k^T\mathbf{x}_k)$, and β is a parameter that limits how far in terms of Euclidian distance the updated codeword can be from the old codeword. This is an incremental interference avoidance codeword update, which for given power p_k implies an increase in user k's SINR [68], and implicitly in its spectral efficiency. We note that since the update in equation (7.149) always generates new codewords that have unit norm, the value of the Lagrange multiplier ξ is irrelevant, and does not need to be obtained from the associated necessary condition for maximum $\partial L_k/\partial\xi_k = 0$.

User power will be adapted incrementally as well, to avoid sudden changes in the system. Since the Lagrangian is a concave function of user power, incremental adaptation in the direction of the corresponding gradient provides maximum increase in the spectral efficiency and implies the power update equation:

$$p_k(i+1) = p_k(i) + \mu_p\frac{\partial L_k}{\partial p_k}\bigg|_{\mathbf{s}_k=\mathbf{s}_k(i+1)} \tag{7.150}$$

where $0 < \mu_p < 1$ and $\partial L_k/\partial p_k$, after the user codeword has been updated as specified by equation (7.149), is:

$$\frac{\partial L_k}{\partial p_k}\bigg|_{\mathbf{s}_k=\mathbf{s}_k(i+1)} = \frac{1}{\mathbf{s}_k(i+1)^T\mathbf{R}_k(i)\mathbf{s}_k(i+1) + p_k(i)}$$
$$+ \frac{\lambda_k(i)}{\mathbf{s}_k(i+1)^T\mathbf{R}_k(i)\mathbf{s}_k(i+1)} \tag{7.151}$$

The Lagrange multiplier $\lambda_k(i)$ is adapted incrementally as well. Given the constant $\mu_\lambda > 0$:

$$\lambda_k(i) = -\mu_\lambda\frac{\partial L_k}{\partial\lambda_k}\bigg|_{\mathbf{s}_k=\mathbf{s}_k(i+1)} \tag{7.152}$$

$$= -\mu_\lambda\left[\frac{p_k(i)}{\mathbf{s}_k(i+1)^T\mathbf{R}_k(i)\mathbf{s}_k(i+1)} - \gamma_k^*\right] \tag{7.153}$$

We note that the Lagrangian L_k is a linear function of λ_k with slope determined by $\partial L_k/\partial\lambda_k$, and is increased by moving λ_k in the corresponding direction indicated by the slope. This implies that the update in equation (7.153) is essentially a steepest ascent gradient update. We also note that this term acts as an extra correction factor in the power update equation (7.150), having more or less influence depending on

Initial Data:

- Codeword matrix $\mathbf{S}$, power matrix $\mathbf{P}$, target SINRs $\gamma_1^*, \ldots, \gamma_K^*$.

- Noise covariance matrix $\mathbf{W}$

- Constants μ_p, μ_λ, β, and tolerance ϵ.

Triggering Events:

- The SINR of an active user differs from the target SINR.

- New users are admitted: their codewords, powers, and target SINRs are added to the system by augmenting the corresponding matrices and increasing K.

- Inactive users are dropped: their codewords, powers, and target SINRs are removed from the system, and K is decreased.

Admissibility check:

- IF the admissibility condition (7.142) is satisfied, GO TO *Adaptation Stage*; ELSE STOP: the system became infeasible.

Adaptation Stage:

1. IF change in spectral efficiency is bigger than ϵ for any user GO TO Step 2, ELSE STOP: an optimal configuration has been reached.

2. FOR each user $k = 1, \ldots, K$, DO

    (a) Compute current $\mathbf{R}_k(i)$ using equation (7.140) and determine its minimum eigenvector $\mathbf{x}_k(i)$.

    (b) Replace the current codeword $\mathbf{s}_k(i)$ using codeword update equation (7.149).

    (c) Update user k's Lagrange multiplier using equation (7.153).

    (d) Update user k's power using equation (7.150).

3. GO TO Step 1.

Figure 7.9. Adaptive interference avoidance algorithm.

how close the SINR:

$$\gamma_k(i) = \frac{p_k(i)}{s_k(i+1)^{\mathsf{T}}\mathbf{R}_k(i)s_k(i+1)} \tag{7.154}$$

after codeword adaptation is to the target SINR γ_k^*.

The adaptive interference avoidance algorithm consists of two distinct stages performed sequentially by active users in the system: one in which users adapt incrementally the codeword, followed by incremental adaptation of their power. The algorithm is distributed, and may be run independently by active users to adapt to changes in the system configuration as reflected by changes of their SINRs and corresponding spectral efficiencies. We note that a change in the system configuration may occur as a result of various events like for example admitting new active users into the system, dropping idle/inactive users, or changing the target SINRs of active users. The algorithm is formally given in Figure 7.9.

Convergence of this algorithm was investigated using a game-theoretic framework in [35]. In addition, extensive simulations have shown that the algorithm reaches a GWBE ensemble of codewords and powers for users [80,81], for which the sum of allocated powers among all valid power allocations for the given target SINRs is minimum. This optimal configuration is reached within a specified tolerance, provided that the target SINRs of active users satisfy the admissibility condition in equation (7.142). The tolerance and speed of convergence of the algorithm can be adjusted through parameters μ_λ, μ_p, β, and ε as is the case in general with gradient-based algorithms.

7.8 SUMMARY AND CONCLUSIONS

In the previous sections, we have explored the concept of iterative interference avoidance for wireless systems. An emphasis was placed on unlicensed systems, but the basic concepts apply to any communications medium where efficient multiple access is an issue. The driving force behind (and surprise of) interference avoidance is personal greed by individual users—rather than leading to system collapse, greediness leads to optimally efficient use of the shared resource. This is a fortuitous result, especially in an unlicensed environment. We have also seen that interference avoidance is robust under the usual wireless system "impairments" such as frequency/ space selective channels, fading channels, asynchronous users and a dynamic palette of users entering and leaving the system.

The current major obstacle to implementation of distributed interference avoidance is the amount of information which must be fed back to transmitters for codeword adaptation. Individual codeword feedback can be onerous owing to the number of degrees of freedom per codeword so various work-arounds have been proposed including global covariance feedback and reduced rank methods.

However, depending on the application, the issue could also be moot. That is, in wireless systems, the downlink (to the mobile) and uplink (from the mobile) can often be grossly asymmetric with the downlink being much much faster than the multiple access uplink. In this case, codeword computation at a common receiver and downlink dispersal to users would be relatively simple and iterative interference avoidance might be one of many potential optimal codeword computation algorithms used.

Another interesting observation regarding the potential utility of distributed interference avoidance is that the high cost of siting for wireless antennas has caused carriers to co-locate equipment. Taking this trend farther, one could even imagine, as the cost of computing hardware plummets with Moore's Law but the cost of specialized antenna/front-end hardware increases, co-location of equipment leading to an even more intimate receiver co-location where carriers share the cost of expensive multi-antenna, mixed-circuit front end transceiver hardware. Multiple "commodity" (and proprietary) backend processing units could then be attached to compose and decode the necessary waveforms. In such a scenario, implicitly competitive and certainly administratively decoupled, one could imagine what might be called "interference avoidance in a box" whereby different carriers would adjust their waveforms to avoid one another on the common transceiver hardware, but without explicit software coordination. That is, all would have access to the front end receiver signals (and hence covariance information) so that the issue over-the-air feedback bandwidth becomes unimportant. However, the lack of shared/coordinated algorithms across carriers would make iterative avoidance of interference extremely important.

Regardless of what the future holds, the mechanics of iterative interference avoidance methods have been studied and the basic idea behind the various forms of the algorithm—greed—seems to stand up to analytic scrutiny as a means to achieve better performance. We find this result particularly interesting in light of the fact that optimality is an "emergent" property where no individual user is seeking to maximize its capacity through waterfilling—at least not directly since a single codeword simply *cannot* waterfill across a multidimensional space. The robustness of the result suggests we do well to examine other joint optimization problems in communications from this sort of "mole's eye view."

REFERENCES

1. M. Abdulrahman, A. U. H. Sheikh, and D. D. Falconer. Decision feedback equalization for CDMA in indoor wireless communications. *IEEE Journal on Selected Areas in Communications*, 12(4):698–706, May 1994.

2. R. Ahlswede. The capacity region of a channel with two senders and two receivers. *Annals of probability*, 2:805–814, 1974.

3. P. Anigstein and V. Anantharam. Ensuring Convergence of the MMSE Iteration for Interference Avoidance to the Global Optimum. *IEEE Transactions on Information Theory*, 49(4):873–885, April 2003.

4. R. Bellman. *Introduction to Matrix Analysis*. New York: McGraw-Hill, 1960.

5. Y. Benkler. Overcoming Agoraphobia: Building the Commons of the Digitally Networked Environment. *Harvard Journal of Law and Technology*, 11:287, 1998.

6. T. J. Brennan. The spectrum as a commons: Tomorrow's vision, not today's prescription. *Journal of Law and Economics*, 41(2):791–803, 1998.

7. A. B. Carleial. Interference Channels. *IEEE Transaction on Information Theory*, 24(1):60–70, January 1978.

8. N. Clemens and C. Rose. Intelligent Power Allocation Strategies in an Unlicensed Spectrum. In *Proceedings First IEEE International Symposium on New Frontiers in Dynamic Spectrum Access Networks—DySPAN 2005*, pages 37–42, Baltimore, MD, 2005.

9. Federal Communications Commission. FCC Report and Order 97-5: Amendment of the commission's rules to provide for operation of unlicensed NII devices in the 5 GHz frequency range. ET Docket No. 96-102, 1997.

10. M. Costa. On the Gaussian Interference Channel. *IEEE Transactions on Information Theory*, 31(5):607–615, September 1985.

11. T. M. Cover and J. A. Thomas. *Elements of Information Theory*. Wiley-Interscience, 1991.

12. R. Etkin, A. Parehk, and D. Tse. Spectrum Sharing for Unlicensed Bands. *IEEE Journal on Selected Areas in Communications*, 25(3):517–528, April 2007.

13. A. J. Goldsmith and P. P. Varaiya. Capacity of fading channels with channel side information. *IEEE Transactions on Information Theory*, 43(6):1986–1992, November 1997.

14. J. S. Goldstein and I. S. Reed. Reduced-rank adaptive filtering. *IEEE Transactions on Signal Processing*, 45(2):492–496, Feb 1997.

15. J. S. Goldstein, I. S. Reed, and L. L. Scharf. A multistage representation of the Wiener filter based on orthogonal projections. *IEEE Transactions on Information Theory*, 44(7):2943–2959, November 1998.

16. R. Gray. On the asymptotic eigenvalue distribution of toeplitz matrices. *IEEE Transactions on Information Theory*, IT-18(6):725–730, November 1972.

17. T. Guess. Optimal Sequences for CDMA with Decision-Feedback Receivers. *IEEE Transactions on Information Theory*, 49(4):886–900, April 2003.

18. T. S. Han and K. Kobayashi. A new achievable Rate Region for the interference channel. *IEEE Transaction on Information Theory*, 27(1):49–60, January 1981.

19. R. W. Heath Jr., J. A. Tropp, I. Dhillon, and T. Strohmer. Construction of Equiangular Signatures for Synchronous CDMA Systems. In *Proceedings 8th IEEE International Symposium on Spread Spectrum Techniques and Applications—ISSSTA'04*, volume 1, pages 708–712, Sydney, Australia, August 2004.

20. M. Honig, U. Madhow, and S. Verdú. Blind adaptive multiuser detection. *IEEE Transactions on Information Theory*, 41(4):944–960, July 1995.

21. M. L. Honig. A Comparison of Subspace Adaptive Filtering Techniques for DS-CDMA Interference Suppression. In *Proceedings 1997 IEEE Military Communications Conference—MILCOM'97*, volume 2, pages 836–840, Monterey, CA, November 1997.

22. M. L. Honig and J. S. Goldstein. Adaptive Reduced-Rank Interference Suppression Based on the Multi-Stage Wiener Filter. *IEEE Transactions on Communications*, 50(6):986–994, June 2002.

23. M. L. Honig and W. Xiao. Performance of Reduced-Rank Linear Interference Suppression. *IEEE Transactions on Information Theory*, 47(5):1928–1946, July 2001.

24. R. A. Horn and C. A. Johnson. *Matrix Analysis*. Cambridge University Press, Cambridge, United Kingdom, 1985.

25. G. N. Karystinos and D. A. Pados. New Bounds on the Total Squared Correlation and Optimum Design of DS-CDMA Binary Signature Sets. *IEEE Transactions on Communications*, 51(1):48–51, January 2003.

26. G. N. Karystinos and D. A. Pados. The Maximum Squared Correlation, Sum Capacity, and Total Asymptotic Efficiency of Minimum Total Squared Correlation Binary Signature Sets. *IEEE Transactions on Information Theory*, 51(1):351–354, January 2005.

27. O. Kaya and S. Ulukus. Optimum power control for CDMA with deterministic sequences in fading channels. *IEEE Transactions on Information Theory*, 50(10):2449–2458, October 2004.

28. O. Kaya and S. Ulukus. Ergodic sum capacity maximization for CDMA: Optimum resource allocation. *IEEE Transactions on Information Theory*, 51(5):1831–1836, May 2005.

29. R. Knopp and P. A. Humblet. Information Capacity and Power Control in Single-Cell Multiuser Communications. In *Proceedings 1995 IEEE International Conference on Communications—ICC'95*, volume 1, pages 331–335, Seattle, WA, June 1995.

30. H. J. Landau and H. O. Pollack. Prolate Spheroidal Wave Functions, Fourier Analysis and Uncertainty—III: The Dimension of the Space of Essentially Time- and Band-Limited Signals. *The Bell System Technical Journal*, 41(4):1295–1335, July 1962.

31. D. J. Love, R. W. Heath Jr., W. Santipach, and M. L. Honig. What Is the Value of Limited Feedback for MIMO Channels? *IEEE Communications Magazine*, 42(10):54–59, October 2004.

32. C. Lăcătuş and D. C. Popescu. Interference Avoidance With Incremental Power Updates for Uplink CDMA Systems. In *Proceedings 2006 IEEE Wireless Communications and Networking Conference—WCNC 2006*, volume 4, pages 1842–1847, Las Vegas, NV, April 2006.

33. C. Lăcătuş and D. C. Popescu. Joint Incremental Codeword and Power Adaptation in CDMA Systems. In *Proceedings 2006 IEEE Region 5 Technology and Science Conference*, pages 185–189, San Antonio, TX, April 2006.

34. C. Lăcătuş and D. C. Popescu. Adaptive Interference Avoidance for Dynamic Wireless Systems. In *2007 IEEE Consumer Communications and Networking Conference—CCNC'07*, pages 150–154, Las Vegas, NV, January 2007.

35. C. Lăcătuş and D. C. Popescu. Adaptive Interference Avoidance for Dynamic Wireless Systems: A Game-Theoretic Approach. *IEEE Journal on Selected Topics in Signal Processing*, 1(1):189–202, June 2007. Special issue on adaptive waveform design for agile sensing and communications.

36. J. Luo, S. Ulukus, and A. Ephremides. Optimal sequences and sum capacity of symbol asynchronous CDMA systems. *IEEE Transactions on Information Theory*, 51(8):2760–2769, August 2005.

37. U. Madhow and M. L. Honig. MMSE Interference Suppression for Direct-Sequence Spread-Spectrum CDMA. *IEEE Transactions on Communications*, 42(12):3178–3188, December 1994.

38. J. L. Massey and T. Mittelholzer. Welch's bound and sequence sets for code-division multiple-access systems. In R. Capocelli, A. De Santis, and U. Vaccaro, editors, *Sequences II: Methods in Communication, Security and Computer Science*. Springer-Verlag, 1991.

39. S. L. Miller. An adaptive direct-sequence code-division multiple-access receiver for multiuser interference rejection. *IEEE Transactions on Communications*, 43(2/3/4): 1746–1755, February/March/April 1995.

40. K. K. Mukkavilli, A. Sabharwal, E. Erkip, and B. Aazhang. On Beamforming with Finite Rate Feedback in Multiple Antenna Systems. *IEEE Transactions on Information Theory*, 49(10):2562–2579, October 2003.

41. E. Ostrom. *Governing the Commons: The Evolution of Institutions for Collective Action.* Cambridge University Press, Cambridge, 1990.

42. M. J. M. Peacock, I. B. Collings, and M. L. Honig. Asymptotic Analysis of MMSE Multiuser Receivers for Multisignature Multicarrier CDMA in Rayleigh Fading. *IEEE Transactions on Communications*, 52(6):964–972, June 2004.

43. M. J. M. Peacock, I. B. Collings, and M. L. Honig. Asymptotic Spectral Efficiency of Multiuser Multisignature CDMA in Frequency-Selective Channels. *IEEE Transactions on Information Theory*, 52(3):1113–1129, March 2006.

44. D. C. Popescu and C. Rose. Codeword Quantization for Interference Avoidance. In *Proceedings 2000 IEEE International Conference on Acoustics, Speech, and Signal Processing—ICASSP 2000*, volume 6, pages 3670–3673, Istanbul, Turkey, June 2000.

45. D. C. Popescu and C. Rose. Interference Avoidance and Multiuser MIMO Systems. *International Journal of Satellite Communications and Networking*, 21(1):143–161, January 2003. Invited paper for special issue on interference suppression techniques for satellite systems.

46. D. C. Popescu and C. Rose. Interference Avoidance and Power Control for Uplink CDMA Systems. In *Proceedings 58th IEEE Vehicular Technology Conference—VTC 2003 Fall*, volume 3, pages 1473–1477, Orlando, FL, October 2003.

47. D. C. Popescu and C. Rose. Multiuser MIMO Systems and Interference Avoidance. In *Proceedings 2003 IEEE International Conference on Acoustics, Speech, and Signal Processing—ICASSP 2003*, volume 4, pages IV-828–IV-831, Hong Kong, P. R. China, April 2003. Invited paper.

48. D. C. Popescu and C. Rose. *Interference Avoidance Methods for Wireless Systems.* Kluwer Academic Publishers, New York, NY, 2004.

49. D. C. Popescu and C. Rose. Codeword Optimization for Uplink CDMA Dispersive Channels. *IEEE Transactions on Wireless Communications*, 4(4):1563–1574, July 2005.

50. O. Popescu. *Interference Avoidance for Wireless Systems with Multiple Receivers.* PhD thesis, Rutgers University, Department of Electrical and Computer Engineering, 2004. Thesis Director: Prof. C. Rose. Available online at http://www.winlab.rutgers.edu/ ~otilia/thesis.pdf.

51. O. Popescu, D. C. Popescu, and C. Rose. Greedy Interference Avoidance in Non-Collaborative Multi-Base Wireless Systems. In *Proceedings 39th Conference on Information Sciences and Systems—CISS'05*, The Johns Hopkins University, Baltimore, MD, March 2005.

52. O. Popescu, D. C. Popescu, and C. Rose. Simultaneous Water Filling in Mutually Interfering Systems. *IEEE Transactions on Wireless Communications*, 6(3):1102–1113, March 2007.

53. O. Popescu and C. Rose. Water Filling May Not Good Neighbors Make. In *Proceedings 2003 IEEE Global Telecommunications Conference—GLOBECOM'03*, volume 3, pages 1766–1770, San Francisco, CA, December 2003.

54. O. Popescu and C. Rose. Sum Capacity and TSC Bounds in Collaborative Multi-Base Wireless Systems. *IEEE Transactions on Information Theory*, 50(10):2433–2438, October 2004.

55. O. Popescu, C. Rose, and D. C. Popescu. Strong Interference and Spectrum Warfare. In *Proceedings 38th Conference on Information Sciences and Systems—CISS 2004*, pages 83–88, Princeton, NJ, March 2004.

56. G. Rajappan and M. L. Honig. Multi-dimensional amplitude control for DS-CDMA. In *Proceedings 49th IEEE Vehicular Technology Conference—VTC'99 Spring*, volume 2, pages 1256–1260, Houston, TX, May 1999.

57. G. Rajappan and M. L. Honig. Spreading Code Adaptation for DS-CDMA with Multipath. In *Proceedings 21st Century Military Communications Conference—MILCOM 2000*, volume 2, pages 1164–1168, Los Angeles, CA, October 2000.

58. G. S. Rajappan and M. L. Honig. Signature Sequence Adaptation for DS-CDMA with Multipath. *IEEE Journal on Selected Areas in Communications*, 20(2):384–395, February 2002.

59. P. B. Rapajic and B. S. Vucetic. Adaptive receiver structures for asynchronous CDMA systems. *IEEE Journal on Selected Areas in Communications*, 12(4):685–697, May 1994.

60. C. Rose. WINLAB Focus'99 on Radio Networks for Everything. Available on line http://www.winlab.rutgers.edu/pub/symposiums/Focus~99/Index.html/focus9%9, May 1999. New Brunswick, NJ.

61. C. Rose. CDMA Codeword Optimization: Interference Avoidance and Convergence Via Class Warfare. *IEEE Transactions on Information Theory*, 47(6):2368–2382, September 2001.

62. C. Rose and A. T. Ogielski. WINLAB Focus'98 on the U-NII. Available on line http://www.winlab.rutgers.edu/pub/symposiums/foc-org/Index.html, June 1998. Long Branch, NJ.

63. C. Rose, S. Ulukus, and R. Yates. Wireless Systems and Interference Avoidance. *IEEE Transactions on Wireless Communications*, 1(3):415–428, July 2002.

64. M. Rupf and J. L. Massey. Optimum Sequence Multisets for Synchronous Code-Division Multiple-Access Channels. *IEEE Transactions on Information Theory*, 40(4):1226–1266, July 1994.

65. W. Santipach and M. L. Honig. Signature Optimization for CDMA with Limited Feedback. *IEEE Transactions on Information Theory*, 51(10):3475–3492, October 2005.

66. E. Seneta. *Non-negative Matrices and Markov Chains*. Springer Verlag, 1981. 2nd edition.

67. J. Singh and C. Rose. Codeword Adaptation and Tracking for Distributed Interference Avoidance. In *WINLAB Technical Memorandum*, volume 231, Piscataway, NJ, March 2003.

68. J. Singh and C. Rose. Distributed Incremental Interference Avoidance. In *Proceedings 2003 IEEE Global Telecommunications Conference—Globecom 2003*, volume 1, pages 415–419, San Francisco, CA, December 2003.

69. G. Strang. *Linear Algebra and Its Applications*. Harcourt Brace Jovanovich College Publishers, San Diego, CA, 1988.

70. J. A. Tropp, I. S. Dhillon, and R. W. Heath. Finite-Step Algorithms for Constructing Optimal CDMA Signature Sequences. *IEEE Transactions on Information Theory*, 50(11):2916–2921, November 2004.

71. J. A. Tropp, I. S. Dhillon, R. W. Heath, and T. Strohmer. Designing Structured Tight Frames Via An Alternating Projection Method. *IEEE Transactions on Information Theory*, 51(1):188–209, January 2005.

72. S. Ulukus. *Power Control, Multiuser Detection and Interference Avoidance in CDMA Systems*. PhD thesis, Rutgers University, Department of Electrical and Computer Engineering, 1998. Available at http://www.ece.umd.edu/~ulukus.

73. S. Ulukus and R. Yates. Iterative Construction of Optimum Signature Sequence Sets in Synchronous CDMA Systems. *IEEE Transactions on Information Theory*, 47(5):1989–1998, July 2001.

74. S. Ulukus and R. D. Yates. User capacity of asynchronous CDMA systems with matched filter receivers and optimum signature sequences. *IEEE Transactions on Information Theory*, 50(5):903–909, May 2004.

75. S. Ulukus and A. Yener. Iterative transmitter and receiver optimization for CDMA networks. *IEEE Transactions on Wireless Communications*, 3(6):1879–1884, November 2004.

76. S. Verdú. Capacity region of Gaussian CDMA channels: The symbol-synchronous case. In *24th Allerton Conference on Communication, Control and Computing*, pages 1025–1034, Monticello, IL, October 1986.

77. S. Verdú. The capacity region of the symbol-asynchronous gaussian multiple-access channel. *IEEE Transactions on Information Theory*, 35(4):733–751, July 1989.

78. S. Verdú. *Multiuser Detection*. Cambridge University Press, Cambridge, United Kingdom, 1998.

79. P. Viswanath and V. Anantharam. Optimal Sequences and Sum Capacity of Synchronous CDMA Systems. *IEEE Transactions on Information Theory*, 45(6):1984–1991, September 1999.

80. P. Viswanath and V. Anantharam. Optimal Sequences for CDMA Under Colored Noise: A Schur-Saddle Function Property. *IEEE Transactions on Information Theory*, 48(6):1295–1318, June 2002.

81. P. Viswanath, V. Anantharam, and D. Tse. Optimal Sequences, Power Control and Capacity of Spread Spectrum Systems with Multiuser Linear Receivers. *IEEE Transactions on Information Theory*, 45(6):1968–1983, September 1999.

82. P. Viswanath and D. N. C. Tse. Sum Capacity of the Vector Gaussian Broadcast Channel and Uplink-Downlink Duality. *IEEE Transactions on Information Theory*, 49(8):1912–1921, August 2003.

83. A. Viterbi and J. Omura. *Principles of Digital Communication and Coding*. New York: McGraw-Hill, 1979.

84. X. Wang and H. V. Poor. Blind Multiuser Detection: A Subspace Approach. *IEEE Transactions on Information Theory*, 44(2):677–690, March 1998.

85. L. R. Welch. Lower Bounds on the Maximum Cross Correlation of Signals. *IEEE Transactions on Information Theory*, IT-20(3):397–399, May 1974.

86. R. D. Yates. A framework for uplink power control in cellular radio systems. *IEEE Journal on Selected Areas in Communications*, 13(7):1341–1347, September 1995.

87. W. Yu, W. Rhee, S. Boyd, and J. M. Cioffi. Iterative Water-Filling for Gaussian Vector Multiple-Access Channels. *IEEE Transactions on Information Theory*, 50(1):145–152, January 2004.

8

CAPACITY-APPROACHING MULTIUSER COMMUNICATIONS OVER MULTIPLE INPUT/MULTIPLE OUTPUT BROADCAST CHANNELS

Uri Erez and Stephan ten Brink

8.1 INTRODUCTION

Previous chapters have focused on the Multiple-Access (MAC) channel, or the uplink in a typical communication scenario. In this chapter, we consider communication on the downlink. In multiuser broadcast (BC) channels, a transmitter communicates with a multiplicity of users simultaneously. While it is possible to restrict transmission to schemes such that signals arriving at different users are orthogonal, this is in general suboptimal. To approach the capacity of the Gaussian broadcast channel, we need to use transmission schemes that result in interference between users. Thus, for a given user, we are now faced with a detection problem of recovering the desired signal in the presence of interfering (undesired) signals. In other words, we are faced with a multiuser detection problem.

Advances in Multiuser Detection. Edited by Michael L. Honig
Copyright © 2009 John Wiley & Sons, Inc.

In this chapter, we shall see that this multiuser detection problem may be circumvented using *precoding* techniques. In effect, we will see how (capacity-achieving) detection at each receiver may be performed without joint decoding of the messages of the interfering users. This insight hinges on the "dirty paper" coding result [1] which will be described.

We will outline recent advances in the field and discuss multiuser detection versus multiuser precoding for the Gaussian MIMO broadcast scenario. In fact, the latter approach allows us to achieve the capacity of the Gaussian MIMO broadcast channel. The underlying building block will be an effective dirty paper precoder for coding for channels with known interference. We will describe how such a precoder may be designed as well as the appropriate detection techniques needed when such coding is used.

8.2 MANY-TO-ONE MULTIPLE ACCESS VERSUS ONE-TO-MANY SCALAR BROADCAST CHANNELS

Consider the communication scenario depicted in Figure 8.1 between a central node (for instance, a basestation in a cellular setting) and a set of "users." For simplicity, we consider a system with two users. The uplink is a multiple access channel as considered in previous chapters. It is mathematically described by:

$$y = h_1 x_1 + h_2 x_2 + n \tag{8.1}$$

where x_1, x_2, h_1, and h_2 are the users' transmit symbols and the corresponding channel gains, respectively, y is the received signal and $n \sim \mathcal{N}(0, P_n)$ is *i.i.d.* circularly symmetric complex Gaussian noise. The transmitters must satisfy the power constraint $E[|x_i|^2] \leq P_{x,i}$. It is well known [4] that capacity can be achieved by *superposition*

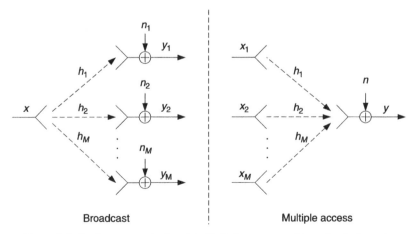

Figure 8.1. Scalar communication link: Multiple access and broadcast channels.

coding (and time-sharing) of codewords from AWGN codes and successive decoding at the receiver. We refer to such coding as SSD (superposition with successive decoding) in the sequel. That is, user i draws its codebook according to an *i.i.d.* complex Gaussian distribution with power $P_{x,i}$. Specifically, the two corner points of the MAC capacity region correspond to one of the following rate pairs:

1.

$$R_1 = \log_2\left(1 + \frac{|h_1|^2 P_{x,1}}{P_n}\right) \tag{8.2}$$

$$R_2 = \log_2\left(1 + \frac{|h_2|^2 P_{x,2}}{|h_1|^2 P_{x,1} + P_n}\right) \tag{8.3}$$

2.

$$R_1 = \log_2\left(1 + \frac{|h_1|^2 P_{x,1}}{|h_2|^2 P_{x,2} + P_n}\right) \tag{8.4}$$

$$R_2 = \log_2\left(1 + \frac{|h_2|^2 P_{x,2}}{P_n}\right) \tag{8.5}$$

In the first case, the receiver first decodes the message of User 2, treating x_1 as AWGN noise. It then subtracts off x_2 from the received signal y and then decodes x_1. In the second case, the order of decoding is reversed.

Consider now communication on the downlink. Here we have:

$$y_i = h_i x + n_i$$

where $n_i \sim \mathcal{N}(0, P_n)$, $i = 1, 2$, the noises n_1 and n_2 are statistically independent, and where the transmitter is subject to the power constraint $E[|x|^2] \leq P_x$.

This is a *degraded* scalar Gaussian channel. Without loss of generality we assume that User 1 has a stronger channel, i.e., $|h_1| \geq |h_2|$. It is well known [4] that—just as for the MAC channel—capacity may again achieved using superposition coding with successive decoding at the receiver. That is, we may send $x = x_1 + x_2$ where x_1 and x_2 are taken from *i.i.d.* Gaussian codebooks with powers $P_{x,1}$ and $P_{x,2}$, respectively, and such that $P_{x,1} + P_{x,2} = P_x$. Therefore, when using superposition coding the received signal is given by:

$$y_i = h_i x_1 + h_i x_2 + n_i \tag{8.6}$$

Comparing (8.6) and (8.1), we see that when superposition coding is used, the Gaussian MAC and BC channels bear a great resemblance. Specifically, the received signal in both cases is a linear combination of the transmitted codewords.[1] This in turn means that similar coding and detection approaches are relevant to both setups.

---

[1] Precise notions of uplink-downlink duality between Gaussian MAC and BC channels have been explored in depth in [5,6,7].

We turn now to the *capacity region* of the scalar Gaussian BC channel. As User 1 can decode whatever message is sent to User 2, the following rate region is achievable:

$$
\left\{ (R_1, R_2): \quad
\begin{aligned}
R_1 &\leq \log_2\left(1 + \frac{|h_1|^2 P_{x,1}}{P_n}\right) \\
R_2 &\leq \log_2\left(1 + \frac{|h_2|^2 P_{x,2}}{|h_2|^2 P_{x,1} + P_n}\right)
\end{aligned}
, \quad P_{x,1} + P_{x,2} = P_x \right\}
\tag{8.7}
$$

where we go over all possible power allocations $P_{x,1}$ and $P_{x,2}$ satisfying $P_{x,1} + P_{x,2} = P_x$. Indeed, this is the capacity region of the channel.

The SSD scheme for the BC channel is not quite dual to the SSD scheme for the MAC channel, as for the BC channel the message of User 2 must be decoded first. Thus, a particular ordering is imposed based on the strengths of the users. Furthermore, successive decoding is still done *at the receiver* as before, even though the roles of encoder/decoder are swapped.

8.3   ALTERNATIVE APPROACH: DIRTY PAPER CODING

An alternative approach which, in a sense, is a "closer" dual to the MAC SSD scheme is based on dirty paper coding (DPC). We start with a brief review of the dirty paper coding result.

8.3.1   The Dirty Paper Coding Result

Consider the channel:

$$
y = x + s + n
\tag{8.8}
$$

where n is *i.i.d.* $\mathcal{N}(0, P_n)$ as before and where s is arbitrary interference known to the transmitter.

The channel model is illustrated in Figure 8.2. A result of Costa [1] states that the capacity of the dirty paper channel (8.8) is $\log_2(1 + P_x/P_n)$, so that interference does not incur any loss in capacity if it is known to the transmitter.[2] An important property is that with dirty paper coding, the transmitted signal x is statistically independent[3] of the interference s (although x is a function of s). Furthermore, the transmitted signal (when optimal coding is used) has the statistical characteristics of an *i.i.d.* Gaussian codebook of power P_x, just like an optimal code for an interference-free AWGN channel. We will further elaborate on the dirty paper result as well as describe applicable coding strategies in subsequent sections.

---

[2] Costa proved the result for *i.i.d.* Gaussian interference. The result was extended to arbitrary interference in [2,3].

[3] The independence holds when the interference is Gaussian and optimal coding is used. For more general interference (or when suboptimal coding is used), statistical independence may still be guaranteed by using additive dithering as will be described in the sequel.

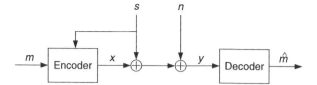

Figure 8.2. The dirty paper channel model.

8.3.2 DPC vs. SSD Approach for a Coded Interference Signal

We now highlight an inherent advantage of the DPC approach over the SSD approach when coding in the presence of known interference. Suppose that the interference signal s in (8.8) is a codeword (possibly a message intended to some other user) of an *i.i.d.* Gaussian codebook of power P_s. Consider an encoder that uses an *i.i.d.* Gaussian codebook with power P_x and rate (arbitrarily close to) $\log_2(1 + P_x/P_n)$ to generate the signal x. That is, we wish to send at a rate close to the capacity of the interference-free channel.

If the *receiver* knows the codebook of the interference signal (no knowledge about the interference is required of the transmitter), then the SSD approach is applicable. Thus, the receiver should be able to first decode the interference message in the presence of its intended signal x. This implies that the rate of the interference codebook should satisfy:

$$R_S < \log_2\left(1 + \frac{P_s}{P_x + P_n}\right)$$

This constraint severely restricts the applicability of the SSD approach as will become clear when we discuss the Gaussian MIMO broadcast channel. With the DPC approach (which is applicable when the interference is known to the transmitter, and assumes no knowledge of the nature of the interference at the receiver), the rate of the interference message is irrelevant and furthermore it makes no difference whether it is a coded signal or not. In fact the interference can be arbitrary.

8.3.3 Scalar Broadcast Using the DPC Approach

We can derive an alternative coding scheme based on precoding at the transmitter that achieves precisely the capacity region (8.7). The encoder can first encode the message intended for User 2, treating the code of User 1 as AWGN noise. Next, the encoder can regard the message of User 2 as *known interference* and use dirty paper coding to send the message of User 1. This results in precisely the same rate region that we obtained previously. In fact the transmitted signal is still a superposition of x_1 and x_2 as given in (8.6) with the two signals having the same statistical characterization as before. The only difference is that $\mathbf{x}_1$, rather than being a codeword from an AWGN code, is now precoded "against" $\mathbf{x}_2$. The resulting rate region (obtained by varying P_1 and P_2) is depicted in Figure 8.3. We denote this rate region by $\mathcal{R}_{21}$ where the order of

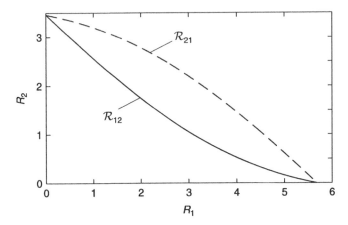

Figure 8.3. Rate regions corresponding to different encoding ordering for scalar BC channel. Here $P_x = 10$, $P_n = 1$, $|h_1|^2 = 5$, $|h_2|^2 = 1$.

the subscripts corresponds to the encoding order of users at the transmitter. As noted earlier, $\mathcal{R}_{21}$ is the capacity region of the channel.

An important feature of the precoding approach is that roles of User 1 and 2 can be reversed now, resulting in the following rate region:

$$\mathcal{R}_{12} = \left\{ (R_1, R_2): \begin{array}{l} R_1 \leq \log_2\left(1 + \dfrac{|h_1|^2 P_{x,1}}{|h_1|^2 P_{x,2} + P_n}\right) \\ R_2 \leq \log_2\left(1 + \dfrac{|h_2|^2 P_{x,2}}{P_n}\right) \end{array}, \quad P_{x,1} + P_{x,2} = P_x \right\} \qquad (8.9)$$

This rate region is depicted in Figure 8.3 alongside the rate region $\mathcal{R}_{21}$. Note that although any rate pair in the region $\mathcal{R}_{12}$ may be achieved with SSD coding (since $\mathcal{R}_{12} \subset \mathcal{R}_{21}$), the roles of User 1 and User 2 *cannot* be reversed when using the SSD approach, since, while User 1 can always decode User 2's message, the converse is in general not true. Thus, the DPC approach for the BC channel bears a greater resemblance to SSD coding for the Gaussian MAC channel than does the successive decoding at receiver approach (as observed in [8]).

As the scalar Gaussian channel is a degraded channel, this greater flexibility in user ordering of DPC does not result in an advantage as far as the capacity region is concerned, and both the DPC and the SSD approach can be used to achieve capacity. There are, however, important differences between the two schemes, among which are:

- SSD coding uses standard AWGN codes. As we shall see below, DPC requires more sophisticated coding techniques to achieve comparable performance.

- To apply SSD coding to the scalar Gaussian broadcast channel, the transmitter has to know only the rate pair to be used. On the other hand, in DPC, the transmitter needs to know the channel coefficients with good precision.
- When using SSD coding, the better users must know the codebooks of the weaker users and decode the latter users' messages before they can decode the message intended to them. In contrast, in DPC, each user decodes only its own message and the complexity of the decoder does not grow with the number of users. In this sense, DPC transfers some of the computational load from the receivers to the transmitter.
- SSD decoding may suffer from error propagation and this may degrade performance as the number of users grows. DPC does not suffer from any error propagation as all messages are perfectly known to the transmitter.
- Dirty paper coding also has many other applications that we do not discuss here. See, e.g., [9].

We next turn to more general Gaussian MIMO broadcast channels. In this case, as the channel is no longer degraded in general, there is no natural ordering of users corresponding to channel quality (which would hold for *any* transmitted signal). This, in turn, limits the applicability of the SSD approach to such channels. Thus, the wider applicability of DPC will prove to be of key importance.

Before considering the general MIMO case, we consider an example to illustrate the role of DPC in a simple setup.

8.4 A SIMPLE 2 × 2 EXAMPLE

Consider a broadcast link between a basestation equipped with two transmit antennas and two users, each equipped with one antenna. See Figure 8.4. The received signals are:

$$\begin{bmatrix} y_1 \\ y_2 \end{bmatrix} = \begin{bmatrix} h_{11} & h_{12} \\ h_{21} & h_{22} \end{bmatrix} \begin{bmatrix} x_1 \\ x_2 \end{bmatrix} + \begin{bmatrix} n_1 \\ n_2 \end{bmatrix} \tag{8.10}$$

where x_i, y_i and n_i are the input, output and AWGN noise respectively, and where $h_{i,j}$ are the channel gain coefficients. The input is subject to the power constraint $E[\|\mathbf{x}\|^2] \leq P_x$, and the noises n_1 and n_2 are complex Gaussian and statistically independent, each with zero mean and power P_n.

If the two users could cooperate, the system would be equivalent to a single user 2×2 MIMO system, and coding could be handled by standard techniques. In particular, applying unitary transformations at the transmitter and the (joint) receiver (i.e., using the singular-value decomposition), we could arrive at an equivalent channel where the channel matrix is *diagonal*. Thus, communication to the users could be

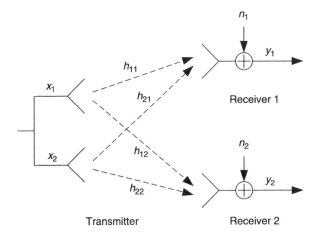

Figure 8.4. Communication link from a two-antenna transmitter to two users equipped with one antenna each.

made orthogonal, avoiding mutual interference, and the problem would be reduced to coding over parallel scalar AWGN channels.

The problem is that diagonalization entails both pre- and post-processing. Since the receivers cannot perform joint processing, diagonalization is precluded. In fact, the capacity of this seemingly simple channel was only recently found [14].

Nonetheless, one can still gain structural simplification using simple preprocessing only. Specifically, applying an appropriate unitary transformation at the transmitter, we can obtain an equivalent channel with a *triangular* channel matrix.

Let us rewrite (8.10) in matrix form:

$$\mathbf{y} = \mathbf{Hx} + \mathbf{n}$$

Let $\mathbf{H} = \mathbf{LQ}$ be the LQ factorization [10] of $\mathbf{H}$, where $\mathbf{L}$ is a lower triangular matrix and $\mathbf{Q}$ is a unitary matrix.[4] We then have:

$$
\begin{aligned}
\mathbf{y} &= \mathbf{Hx} + \mathbf{n} \\
&= \mathbf{LQx} + \mathbf{n} \\
&= \mathbf{Lx}' + \mathbf{n}
\end{aligned}
\tag{8.11}
$$

where $\mathbf{x} = \mathbf{Q}^*\mathbf{x}'$. More explicitly, we arrive at the following channel:

$$
\begin{bmatrix} y_1 \\ y_2 \end{bmatrix} = \begin{bmatrix} l_1 & 0 \\ l_{21} & l_2 \end{bmatrix} \begin{bmatrix} x_1' \\ x_2' \end{bmatrix} + \begin{bmatrix} n_1 \\ n_2 \end{bmatrix}
\tag{8.12}
$$

---

[4]The LQ decomposition may be obtained by performing the Gramm-Schmidt orthogonalization process proceeding from the first row of $\mathbf{H}$ to the last. If we require that the elements on the diagonal be non-negative, the factorization is unique.

Note that since $\mathbf{Q}$ is unitary, we have $\|\mathbf{x}'\|^2 = \|\mathbf{x}\|^2$ and hence the power constraint remains unaffected. Clearly User 1 now sees an interference-free AWGN channel:

$$y_1 = l_1 x_1' + n_1$$

User 2, on the other hand, suffers from interference from the signal sent to User 1:

$$y_2 = l_2 x_2' + s + n_2$$

where we denote the interference (or "crosstalk") by $s = l_{21} x_1'$. More generally (with more users in the system), using an LQ decomposition, each user will experience interference only from "previous" users.

Consider now transmission based on the SSD approach. In order for User 2 to be able to decode and strip off the interference, we would have to limit the transmission rate of User 1 to satisfy:

$$R_1 \leq \log_2 \left(1 + \frac{|l_{21}|^2 P_{x,1}}{|l_2|^2 P_{x,2} + P_n} \right) \tag{8.13}$$

This constraint would need to be imposed *in addition* to the condition:

$$R_1 \leq \log_2 \left(1 + \frac{|l_1|^2 P_{x,1}}{P_n} \right) \tag{8.14}$$

In contrast, if we use the DPC approach, the interference may be (pre-)cancelled, regardless of the magnitude of l_{21} and the rate R_1. The following DPC transmission scheme [11,12] may thus be used.

Transmitter:

- First use an AWGN code to encode the information to be sent to User 1, resulting in the signal x_1'.
- Next, apply dirty paper precoding to encode the message of User 2 treating User 1's signal as known interference. This results in the signal x_2'.
- Apply the unitary transformation $\mathbf{Q}^*$ to the resulting signals so that the transmitted signal is $\mathbf{x} = \mathbf{Q}^* \mathbf{x}'$. In other words, the messages to User 1 and 2 are sent over the "directions" $\mathbf{q}_1^*$ and $\mathbf{q}_2^*$, where $\mathbf{q}_1^*$ and $\mathbf{q}_2^*$ are the first and second column of $\mathbf{Q}^*$ respectively.

Receivers:

- Receiver 1 decodes its message using a standard decoder since an AWGN code is used for this user.
- Receiver 2 uses a dirty paper decoder to be discussed in later sections.

- Note that in general (when there are more users), we will use standard AWGN coding and decoding only for one (the "first") user.[5]

The resulting achievable rate region is:

$$
\mathcal{R}_{12} = \left\{ (R_1, R_2): \begin{array}{c} R_1 \leq \log_2\left(1 + \dfrac{|l_1|^2 P_{x,1}}{P_n}\right) \\[2mm] R_2 \leq \log_2\left(1 + \dfrac{|l_2|^2 P_{x,2}}{P_n}\right) \end{array}, \ P_{x,1} + P_{x,2} = P_x \right\} \tag{8.15}
$$

Clearly, the roles of User 1 and 2 may be reversed. That is, if we perform the Gramm-Schmidt orthogonalization process starting with the second user, we obtain the following equivalent channel:

$$
\begin{bmatrix} y_1 \\ y_2 \end{bmatrix} = \begin{bmatrix} r_1 & r_{12} \\ 0 & r_2 \end{bmatrix} \begin{bmatrix} x_1' \\ x_2' \end{bmatrix} + \begin{bmatrix} n_1 \\ n_2 \end{bmatrix} \tag{8.16}
$$

Encoding will now be done in reverse order. We first encode User 2's message. User 2 suffers from no interference from User 1. Then we use DPC to encode User 1's message treating User 2's message (i.e., $s = r_{12} x_2'$) as known interference.

It follows that the following region is also achievable:

$$
\mathcal{R}_{21} = \left\{ (R_1, R_2): \begin{array}{c} R_1 \leq \log_2\left(1 + \dfrac{|r_1|^2 P_{x,1}}{P_n}\right) \\[2mm] R_2 \leq \log_2\left(1 + \dfrac{|r_2|^2 P_{x,2}}{P_n}\right) \end{array}, \ P_{x,1} + P_{x,2} = P_x \right\} \tag{8.17}
$$

Numerical Example Let the channel be:

$$
\begin{bmatrix} y_1 \\ y_2 \end{bmatrix} = \begin{bmatrix} 1 & 0.75 \\ 0.75 & 1 \end{bmatrix} \begin{bmatrix} x_1 \\ x_2 \end{bmatrix} + \begin{bmatrix} n_1 \\ n_2 \end{bmatrix} \tag{8.18}
$$

where $\mathbf{n} \sim \mathcal{N}(0, \mathbf{I})$ and where the power constraint is $P = 50$. Then the above mentioned two factorizations of $\mathbf{H}$ are:

$$
\begin{bmatrix} 1 & 0.75 \\ 0.75 & 1 \end{bmatrix} = \begin{bmatrix} 1.25 & 0 \\ 1.2 & 0.35 \end{bmatrix} \begin{bmatrix} 0.8 & 0.6 \\ -0.6 & 0.8 \end{bmatrix} \tag{8.19}
$$

[5]We may, of course, also use dirty paper coding for the first user, where the interference is taken to be zero.

and:

$$\begin{bmatrix} 1 & 0.75 \\ 0.75 & 1 \end{bmatrix} = \begin{bmatrix} 0.35 & 1.2 \\ 0 & 1.25 \end{bmatrix} \begin{bmatrix} 0.8 & -0.6 \\ 0.6 & 0.8 \end{bmatrix} \quad (8.20)$$

The corresponding rate regions are:

$$\mathcal{R}_{12} = \left\{ (R_1, R_2) : \begin{array}{l} R_1 \leq \log_2(1 + 1.25^2 P_1) \\ R_2 \leq \log_2(1 + 0.35^2 P_2) \end{array}, \quad P_1 + P_2 = 50 \right\} \quad (8.21)$$

$$\mathcal{R}_{21} = \left\{ (R_1, R_2) : \begin{array}{l} R_1 \leq \log_2(1 + 0.35^2 P_1) \\ R_2 \leq \log_2(1 + 1.25^2 P_2) \end{array}, \quad P_1 + P_2 = 50 \right\} \quad (8.22)$$

The achievable rate region, shown in Figure 8.5, is then the convex hull of the union of $\mathcal{R}_{12}$ and $\mathcal{R}_{21}$.

Remarks:

- For a fixed preprocessing (linear transformation at the transmitter), the optimal encoding order is the same for all rates. For instance, if we apply the LQ decomposition (8.19) in the example above, we can only lose in rate if we

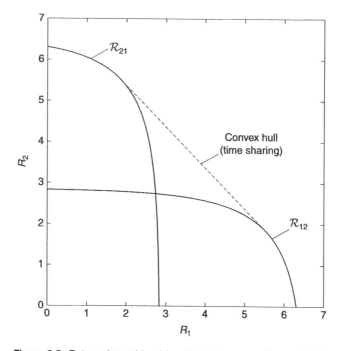

Figure 8.5. Rate region achievable with LQ decomposition and DPC.

encode User 2 first. Thus, the two regions $\mathcal{R}_{12}$ and $\mathcal{R}_{21}$ correspond to both swapping the encoding order and changing the preprocessing matrix.

- The encoding scheme easily generalizes to an arbitrary number of users. We can consider all encoding orderings of users. For each of these, by performing Gramm-Schmidt orthogonalization of **H** in the corresponding order, we arrive at a preprocessing matrix that guarantees zero interference between users after DPC is used. For this reason this scheme is termed Zero-Forcing DPC in [11].

- The Zero-Forcing DPC approach is reminiscent of the Zero-Forcing Decision-Feedback precoding (Tomlinson-Harashima) approach used for intersymbol interference channels. The latter is known to be suboptimal (except in the limit of high SNR). In general, better performance can be achieved by leaving some residual interference. This is also true in the present case. Thus, we need to consider more general linear transformations. It turns out that when this is done, capacity may actually be attained.

8.5 GENERAL GAUSSIAN MIMO BROADCAST CHANNELS

Consider now a transmitter with N_t antennas and N_r receivers (users), where the m-th receiver is equipped with $N_r(m)$ antennas, $m = 1, \ldots M$. Then the received signal of user m is given by:

$$\mathbf{y}_m = \mathbf{H}_m \mathbf{x} + \mathbf{n}_m$$

where $\mathbf{x} = [x_1, x_2, \ldots, x_{N_t}]^T$ is the input vector, $\mathbf{y}_m = [y_{m,1}, y_{m,2}, \ldots, y_{m,N_r(m)}]^T$ is the received vector, $\mathbf{H}_m$ is an $N_r(m) \times N_t$ channel matrix, and the vector $\mathbf{n}_m = [n_{m,1}, n_{m,2}, \ldots, n_{m,N_r(m)}]^T$ is *i.i.d.* $\mathcal{N}(0, \mathbf{I}_{N_r(m) \times N_r(m)})$ noise. The transmitted signal is subject to the power constraint $E[\|\mathbf{x}\|^2] \leq P_x$. For simplicity of notation, we restrict ourselves below to the case of two users.

As before, we may form the transmitted signal as a sum (superposition) of signals $\mathbf{x}_1$ and $\mathbf{x}_2$ conveying the messages to the respective users. Thus, we may write:

$$\mathbf{y}_1 = \mathbf{H}_1(\mathbf{x}_1 + \mathbf{x}_2) + \mathbf{n}_1 \tag{8.23}$$

$$\mathbf{y}_2 = \mathbf{H}_2(\mathbf{x}_1 + \mathbf{x}_2) + \mathbf{n}_2 \tag{8.24}$$

Suppose now that the transmitter has selected the signal $\mathbf{x}_2$ for User 2. Then the term $\mathbf{s} = \mathbf{H}_1\mathbf{x}_2$ may be viewed as known interference suffered by User 1. We wish to use the dirty paper approach to pre-cancel the known interference. Indeed, the scalar dirty paper result of Costa is readily extended to the case of a vector channel as we now state.

8.5.1 Vector Dirty Paper Coding: Reduction to Scalar Case

Consider a single user vector dirty paper channel where the transmitter has N_t antennas and the receiver has N_r antennas:

$$\mathbf{y} = \mathbf{H}\mathbf{x} + \mathbf{s} + \mathbf{n} \tag{5.25}$$

where $\mathbf{s}$ is an interference vector known (non-causally) to the transmitter, the input $\mathbf{x}$ is required to have covariance:

$$E[\mathbf{x}\mathbf{x}^*] = \mathbf{S}_{xx} \tag{8.26}$$

and the noise $\mathbf{n}$ is distributed as $\mathbf{n} \sim N(0, \mathbf{S}_{nn})$ and is *i.i.d.* over time. We assume that the channel matrix $\mathbf{H}$ as well as the noise correlation matrix $\mathbf{S}_{nn}$ are known at both transmission ends. As we next show, the interference $\mathbf{s}$ incurs no loss in capacity. That is, the maximal achievable mutual information when transmitting over the channel (8.25) under the covariance constraint (8.26) is equal to that achievable over the interference–free channel:

$$\mathbf{y} = \mathbf{H}\mathbf{x} + \mathbf{n} \tag{8.27}$$

Specifically, the maximal possible transmission rate is:

$$\log_2\left(\frac{|\mathbf{H}\mathbf{S}_{xx}\mathbf{H}^* + \mathbf{S}_{nn}|}{|\mathbf{S}_{nn}|}\right) \tag{8.28}$$

We turn now to the derivation of the result. We will see that by linear pre/post operations, the vector dirty paper channel may be reduced to parallel (independent) scalar dirty paper channels. Beyond yielding the vector DPC result, this method also reduces the coding problem for the vector dirty paper channel to that of coding for the scalar dirty paper channel. Our exposition follows that of [13].

We start by diagonalizing the covariance matrices of the input $\mathbf{x}$ and the noise $\mathbf{n}$. Let $\mathbf{Q}_n$ be such that $\mathbf{S}_{nn} = \mathbf{Q}_n\Lambda_n\mathbf{Q}_n^*$ and let $\mathbf{Q}_x$ be such that $\mathbf{S}_{xx} = \mathbf{Q}_x\Lambda_x\mathbf{Q}_x^*$. Then we have:

$$\mathbf{y} = \mathbf{H}\mathbf{Q}_x\Lambda_x^{1/2}\mathbf{x}' + \mathbf{s} + \mathbf{Q}_n\Lambda_n^{1/2}\mathbf{n}' \tag{8.29}$$

$$= \mathbf{H}'\mathbf{x}' + \mathbf{s} + \mathbf{Q}_n\Lambda_n^{1/2}\mathbf{n}' \tag{8.30}$$

where $\mathbf{x}' = \Lambda_x^{-1/2}\mathbf{Q}_x^*\mathbf{x}$, $\mathbf{n}' = \Lambda_n^{-1/2}\mathbf{Q}_n^*\mathbf{n}$ and $\mathbf{H}' = \mathbf{H}\mathbf{Q}_x\Lambda_x^{1/2}$. Notice that $\mathbf{x}' \sim \mathcal{N}(0, \mathbf{I}_{N_t \times N_t})$ and $\mathbf{n}' \sim \mathcal{N}(0, \mathbf{I}_{N_r \times N_r})$.

Applying the linear transformation $\Lambda_n^{-1/2}\mathbf{Q}_n^*$ to the received vector $\mathbf{y}$ we thus have:

$$\mathbf{y}' = \Lambda_n^{-1/2}\mathbf{Q}_n^*\mathbf{y} \tag{8.31}$$

$$= \Lambda_n^{-1/2}\mathbf{Q}_n^*\mathbf{H}'\mathbf{x}' + \Lambda_n^{-1/2}\mathbf{Q}_n^*\mathbf{s} + \mathbf{n}' \tag{8.32}$$

$$= \mathbf{H}''\mathbf{x}' + \mathbf{s}' + \mathbf{n}' \tag{8.33}$$

where

$$\mathbf{H}'' = \Lambda_n^{-1/2}\mathbf{Q}_n^*\mathbf{H}' \tag{8.34}$$

$$= \Lambda_n^{-1/2}\mathbf{Q}_n^*\mathbf{H}\mathbf{Q}_x\Lambda_x^{1/2} \tag{8.35}$$

and

$$s' = \Lambda_n^{-1} Q_n^* s \tag{8.36}$$

Finally, let $H'' = U\Sigma_{H''}V^*$ be the SVD decomposition [10] of H'' where U is an $N_r \times N_r$ unitary matrix, V is an $N_t \times N_t$ unitary matrix, and $\Sigma_{H''}$ is an $N_t \times N_t$ matrix with non-negative entries on the diagonal and zero elsewhere. Applying U^* to y' (to form y'') it follows from (8.33) that:

$$y'' = U^*y' \tag{8.37}$$
$$= \Lambda_{H''}V^*x' + U^*s' + U^*n' \tag{8.38}$$
$$= \Lambda_{H''}x'' + s'' + n'' \tag{8.39}$$

where we define $x'' = V^*x'$, $s'' = U^*s'$ and $n'' = U^*n'$. Note that n'' has the same distribution as n' since U^* is unitary, i.e., $S_{n''n''} = I_{N_r \times N_r}$. Also, since V is unitary, it follows that $S_{x''x''} = S_{x'x'} = I_{N_t \times N_t}$.

Since $\Sigma_{H''}$ is diagonal and since $S_{x''x''}$ and $S_{n''n''}$ are also diagonal, the channel (8.39) represents parallel scalar channels for which the standard Costa result holds.[6] That is, the interference s'' incurs no loss in mutual information.

Since all operations leading from (8.25) to (8.39) are invertible, it follows that the two channels are equivalent. Therefore, for a given covariance matrix S_{xx} the channel (8.25) has the same achievable mutual information as the interference-free channel (8.27), and the achievable rate is given in (8.28). The transmitted signal is:

$$x = Q_x\Lambda_x^{1/2}Vx''$$

That is, after DPC is applied to produce the signal x_i'', the latter is multiplied by the i-th column of the matrix $Q_x\Lambda_x^{1/2}V$.

8.5.2 DPC Rate Region

We now apply the vector DPC result to the two user Gaussian MIMO broadcast channel (8.23, 8.24). Suppose the transmitter first selects the codeword for User 2, treating User 1's signal as noise. Thus, when coding for User 1, the term $s = H_1x_2$ is known interference. Applying now the vector dirty paper result it follows that the mutual information of the channel (8.23) is equal to that of the interference-free channel

$$y_1 = H_1x_1 + n_1$$

Thus, interference from previously encoded messages may be completely eliminated. Furthermore, as will be seen in the next sections, the precoded signal of User 1 will be statistically equivalent to Gaussian noise.

---

[6]Note that there will be no more than $\min(N_r, N_t)$ non-trivial scalar channels.

It remains to choose the covariance matrices $\mathbf{S}_i = E[\mathbf{x}_i \mathbf{x}_i^*]$ of the two users. In order to satisfy the power constraint, we require:

$$\text{tr}(\mathbf{S}_1) + \text{tr}(\mathbf{S}_2) \le P_x$$

Going over all possible choices for $\mathbf{S}_1$, $\mathbf{S}_2$, we arrive at the following rate regions:

$$\mathcal{R}_{12} = \left\{ (R_1, R_2): \begin{array}{l} R_1 \le \log_2 \dfrac{|\mathbf{I} + \mathbf{H}_1(\mathbf{S}_1 + \mathbf{S}_2)\mathbf{H}_1^*|}{|\mathbf{I} + \mathbf{H}_1 \mathbf{S}_2 \mathbf{H}_1^*|} \\ R_2 \le \log_2 |\mathbf{I} + \mathbf{H}_2 \mathbf{S}_2 \mathbf{H}_2^*| \end{array}, \quad \text{tr}(\mathbf{S}_1 + \mathbf{S}_2) \le P_x \right\} \quad (8.40)$$

and

$$\mathcal{R}_{21} = \left\{ (R_1, R_2): \begin{array}{l} R_1 \le \log_2 |\mathbf{I} + \mathbf{H}_1 \mathbf{S}_1 \mathbf{H}_1^*| \\ R_2 \le \log_2 \dfrac{|\mathbf{I} + \mathbf{H}_2(\mathbf{S}_1 + \mathbf{S}_2)\mathbf{H}_2^*|}{|\mathbf{I} + \mathbf{H}_2 \mathbf{S}_1 \mathbf{H}_2^*|} \end{array}, \quad \text{tr}(\mathbf{S}_1 + \mathbf{S}_2) \le P_x \right\} \quad (8.41)$$

The dirty paper rate region is given by the convex hull of the union of $\mathcal{R}_{12}$ and $\mathcal{R}_{21}$. Furthermore, the derivation can readily be extended to any number of users. In has been shown [14] that, in fact, the dirty paper region is the true capacity region of the Gaussian MIMO broadcast channel. Algorithms for the calculation of the capacity region and the optimal covariance matrices (for a given precoding order) can be found in [6,15].

It remains to show how a scalar dirty paper coding system may be designed. This will be our focus in the remaining sections.

8.6 CODING WITH SIDE INFORMATION AT THE TRANSMITTER

As we have seen in the previous sections, the capacity of the vector broadcast channel can be achieved by appropriate precoding at the transmitter. Precoding for a *scalar* channel was shown to be sufficient. The result of section 8.3, Figure 8.2 can be briefly summarized as follows: Consider a generalization of the AWGN channel $y = x + n$ with an interference s, in addition to the Gaussian noise n. Assume that s is Gaussian distributed and *known* to the transmitter. Costa showed [1] that the scalar channel:

$$y = x + s + n \quad (8.42)$$

has the same capacity as $y = x + n$, which is:

$$C = \frac{1}{2} \log_2(1 + P_x / P_n)$$

per real dimension. This result was later extended to interference with arbitrary distributions in [2]. Thus, the interference s does not cause any loss in capacity provided that we can implement a "dirty paper" coding scheme: The transmitted signal x can be viewed as the "ink," which is chosen dependent on the interference s ("dirt") that is present at the encoder; we can say: "x writes on the dirty paper s."

In the following sections, we discuss ways of realizing dirty paper coding to achieve the promised capacities. We start with one-dimensional quantization and discuss its losses and limitations. We then move on to multi-dimensional vector quantization, and describe how to design a combined channel coding and precoding scheme that can approach capacity closely.

8.6.1 A Naive Attempt

Let the input constellation $\mathcal{C}$ be restricted to the interval $\mathcal{V} = [-2, 2)$; for instance, $\mathcal{C}$ could be an M-PAM constellation conveying $M' = \log_2 M$ bits per symbol, in which case:

$$\mathcal{C} = \left\{ \left. \frac{-2(M - 1 - 2 \cdot i)}{M} \right| i = 0, \ldots, M - 1 \right\}$$

Let m_j be a sequence of symbols chosen for transmission; for instance, m_j may be uncoded symbols or possibly coded symbols from an outer channel code. For brevity of notation, we drop the discrete-time index j in the following. Since the transmitter knows s, it could simply pre-subtract it: instead of transmitting $x = m$, it could send:

$$x = m - s$$

The received signal is:

$$y = x + s + n = m - s + s + n = m + n$$

and thus, the interference would be eliminated, reducing the problem back to coding for an AWGN channel. However, rather than $P_m = E[|m|^2]$ the transmitter now (with m, s assumed to be independent) consumes the power:

$$E[|m - s|^2] = P_m + P_s$$

Since the interference power $P_s = E[|s|^2]$ can be arbitrarily big and much larger than the power P_m of the desired signal, the simple pre-cancellation approach is likely to suffer a severe *power loss*. A precoding technique that can overcome this power loss while still allowing the use of virtually standard AWGN codes is *Tomlinson-Harashima (TH) precoding* [16,17].

8.6.2 Scalar Quantization: Tomlinson-Harashima Precoding

The basic idea of Tomlinson-Harashima precoding is to limit the power of the interference by applying a one-dimensional (scalar) modulo operation before transmitting the signal. Denote by Λ the set of all integers divisible by 4:

$$\Lambda = \{\ldots, -8, -4, 0, 4, 8, \ldots\}$$

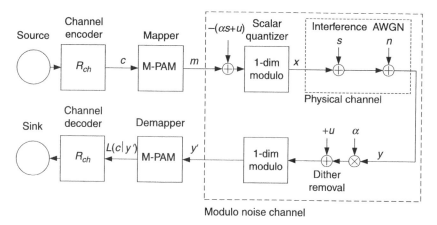

Figure 8.6. Dirty paper coding using Tomlinson-Harashima precoding.

or simply:

$$\Lambda = 4\mathbb{Z}$$

Rather than transmitting $x = m-s$, as done in the previous section, the transmitter sends:

$$x = m - s + \lambda \qquad (8.43)$$

where λ is defined to be the (unique) element of Λ such that $(m - s + \lambda) \in \mathcal{V}$, i.e., the transmitted signal x is brought back into the modulo interval[7] $[-2, 2)$. This step can be viewed as a *quantization* operation. Equation (8.43) may equivalently be written as:

$$\begin{aligned} x &= [m - s] \bmod \Lambda \\ &= [m - s] - Q[m - s] \end{aligned} \qquad (8.44)$$

that is, the *difference* (or quantization error) between $m-s$ and $Q[m-s]$ is transmitted, where the quantization operator $Q[m-s] = -\lambda$ denotes that element of Λ which is closest to $m-s$. In the same way, the receiver applies a modulo operation to obtain:

$$\begin{aligned} y' &= y \bmod \Lambda \\ &= [(m - s + \lambda) + s + n] \bmod \Lambda \\ &= [m + n] \bmod \Lambda \end{aligned}$$

Thus, TH precoding transforms the dirty paper channel (8.42) into a modulo-additive noise channel. An overview of the communication link is given in Figure 8.6. (Note that, for now, no MMSE scaling is used, i.e., we assume $\alpha = 1$ in the figure.) Next we discuss the dither signal u that is required to make the power of x independent of the distribution of s.

---

[7]Contrary to the conventional definition, $x \bmod \mathbb{Z}$ shall reduce x to the interval $[-\frac{1}{2}, \frac{1}{2})$ rather than $[0, 1)$ throughout this chapter.

8.6.2.1 Dither Signal The transmit power P_x of (8.44) now depends on both the symbol m and the interference s. To make x independent of the distribution of s we add a *dither* u that is uniformly distributed over $\mathcal{V}$ and assumed to be known at *both* transmission ends (e.g., generated by a pseudo-random sequence). Transmission takes on the form:

$$x = [m - (s + u)] \bmod \Lambda \tag{8.45}$$

and the receiver obtains:

$$
\begin{aligned}
y' &= [x + (s + u) + n] \bmod \Lambda \\
&= [(m - (s + u) + \lambda) + (s + u) + n] \bmod \Lambda \\
&= [m + n] \bmod \Lambda
\end{aligned}
\tag{8.46}
$$

Obviously, the dither signal u cancels out at the receiver and has no further impact on the structure of the modulo-additive noise channel. Besides ensuring that the transmitted power is independent of the distribution of s, the dither signal u is also essential to make x independent of m for the case of MMSE (α-)scaling, as will be introduced in the later Section 8.6.2.3.

8.6.2.2 Losses of Tomlinson-Harashima Precoding The capacity of the resulting mod-Λ channel is achieved by taking m to be uniformly distributed over $\mathcal{V}$. This results in an output y' that is also uniform over the interval $\mathcal{V}$, leading to a mutual information (per real dimension) satisfying:

$$
\begin{aligned}
I(m; y') &= h(y') - h(n \bmod \Lambda) \\
&= \frac{1}{2}\log_2(2\pi e P) - h(n \bmod \Lambda) - \frac{1}{2}\log_2\left(\frac{2\pi e}{12}\right) \\
&\geq \frac{1}{2}\log_2(2\pi e P) - h(n) - \frac{1}{2}\log_2\left(\frac{2\pi e}{12}\right) \\
&= \frac{1}{2}\log_2(\mathrm{SNR}) - \frac{1}{2}\log_2\left(\frac{2\pi e}{12}\right) \\
&= \frac{1}{2}\log_2(1 + \mathrm{SNR}) - \underbrace{\frac{1}{2}\log_2\left(\frac{1 + \mathrm{SNR}}{\mathrm{SNR}}\right)}_{\text{modulo loss}} - \underbrace{\frac{1}{2}\log_2\left(\frac{2\pi e}{12}\right)}_{\text{shaping loss}}
\end{aligned}
$$

The differential entropy $h(x)$ of the continuous random variable x is defined as $h(x) = -\int p(\xi)\log_2 p(\xi)d\xi$, with probability density function (PDF) $p(\xi)$ of x. Neglecting the gain in having $(n \bmod \Lambda)$ rather than n (which is a good approximation at high SNR, with $\mathrm{SNR} = P_x/P_n$), we find the following losses:

1. *Modulo Loss*: Due to the modulo operation, the output is restricted to the interval $\mathcal{V}$ and has the same power as the input. This corresponds to having $\log_2(\mathrm{SNR})$ rather than $\log_2(1 + \mathrm{SNR})$.

2. *Shaping Loss*: The output is uniform rather than Gaussian. This corresponds to a loss of $\log_2\left(\frac{2\pi e}{12}\right) \approx 1.53\,\mathrm{dB}$ at high SNR, and a loss of more than 4 dB at low SNR (see, e.g., [24]).

3. *Power Loss*: A third loss, sometimes attributed to TH precoding, is referred to as "the power loss" [18]. This loss, however, is an artifact coming from using uncoded *discrete input-constrained* PAM. For an M-PAM constellation, the transmit power is $(1 - 1/M^2) \cdot 4/3$. As we have seen before, using the same constellation with TH precoding, the transmit power is $4/3$. Thus, there is a loss of $4/(3M^2)$ in extra power needed for transmission. This loss stems from the fact that, while transmitting symbols restricted to the interval $[-2(M-1)/M, 2(M-1)/M)$, we are "paying" for the larger interval $\mathcal{V} = [-2, 2)$. To maximize the mutual information, however, we should use a constellation that is uniform over the entire region of $\mathcal{V}$. Obviously, in the limit of $M \to \infty$, this power loss goes to zero [18].

Figure 8.7 shows the effect of these losses in the spectral efficiency chart. The rightmost curve plots the lower bound on the mutual information of the scalar quantizer, $\log_2(\mathrm{SNR}) - \log_2\left(\frac{2\pi e}{12}\right)$, as a function of E_b/N_0, with $E_b/N_0 = P_x/(M'R_{ch}2\sigma_n^2)$, σ_n^2 being the noise power per real dimension (for PAM modulation, $\sigma_n^2 = P_n$), and R_{ch} being the code rate of the channel code. The loss is more than 4 dB at 1 bit/s/Hz.

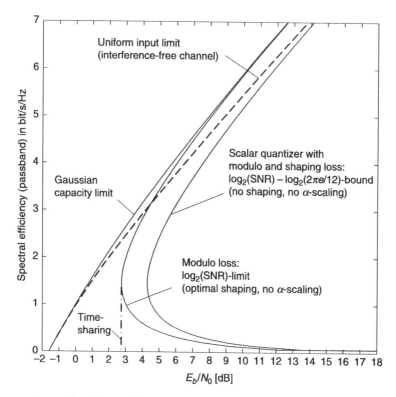

Figure 8.7. Capacity and mutual information limits relevant for Tomlinson-Harashima precoding.

Even with optimal shaping, the modulo loss is significant at low SNR. The uniform input limit (dashed lines) further illustrates that for an interference-free AWGN channel $y = x + n$ shaping does not play a role at low SNR. For a given SNR, we can further use *time-sharing* between lattice transmission schemes designed for different SNRs, as long as the *average* power consumed is the designated power. This, as indicated by the vertical dashed line in Figure 8.7 provides better performance than the direct approach for very low SNRs: the lower legs of the mutual information limit curves, that bend away to high E_b/N_0-values, can always be overcome by time-sharing.

8.6.2.3 *MMSE Scaling*

The performance of Tomlinson-Harashima precoding can be improved at low SNR by using minimum-mean-squared-error (MMSE) scaling ("α-scaling"). It *removes the modulo loss* that was described in the previous section. Rather than (8.45), the transmitter sends:

$$x = [m - (\alpha s + u)] \bmod \Lambda \qquad (8.47)$$

with $0 < \alpha \le 1$. After adding interference and noise on the channel, the receiver obtains:

$$y = x + s + n = [m - (\alpha s + u)] \bmod \Lambda + s + n \qquad (8.48)$$

The receiver first scales the channel output by α, adds the dither signal u, and then proceeds as before:

$$
\begin{aligned}
y' &= [\alpha y + u] \bmod \Lambda \\
&= [\alpha x + \alpha s + \alpha n + u] \bmod \Lambda
\end{aligned}
\qquad (8.49)
$$

By adding:

$$0 = (1 - \alpha)x - (1 - \alpha)x = [m - (\alpha s + u)] \bmod \Lambda - \alpha x - (1 - \alpha)x$$

we can write (8.49) as:

$$
\begin{aligned}
y' &= [m - (\alpha s + u) - \alpha x - (1 - \alpha)x + \alpha x + \alpha s + \alpha n + u] \bmod \Lambda \\
&= [m - (1 - \alpha)x + \alpha n] \bmod \Lambda
\end{aligned}
\qquad (8.50)
$$

This formulation allows an insightful interpretation of the effective modulo-noise channel:

1. The additive noise term $-(1 - \alpha)x$ is uniformly distributed over the interval $[-2(1 - \alpha), 2(1 - \alpha))$. A value $\alpha < 1$ can be thought of *blurring* the transmitted constellation symbol m. Note that x is made independent of m through the dither u, see (8.47).
2. The term αn is scaled Gaussian noise. A value $\alpha < 1$ *cools* the noise.

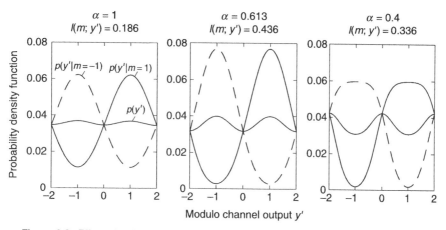

Figure 8.8. Effect of noise cooling (α-scaling) for a BPSK constellation at SNR = 2 dB.

This is illustrated in Figure 8.8 for the case of a BPSK constellation, $m \in \{-1, +1\}$. For $\alpha = 1$, there is no constellation blur, but noise folding, owing to the modulo operation. The more we "cool" the noise, i.e., the smaller we make α, the less noise folding we have, but the more constellation blurring occurs: there is a trade-off between cooling and blurring. The optimal α that maximizes the mutual information depends on the SNR. It can be computed by minimizing the entropy of the effective folded noise $n' = [-(1-\alpha)x + \alpha n] \bmod \Lambda$ in (8.50). Neglecting the modulo, this is appoximately the case for:

$$\alpha = P_x/(P_x + P_n) = \text{SNR}/(1 + \text{SNR}) \qquad (8.51)$$

As we will later see, this choice of α is optimal for a shaping lattice with a dimensionality tending to infinity; it is also a good approximation of the optimal α for Tomlinson-Harashima precoding. For SNR = 2 dB, we obtain $\alpha = 0.613$, as depicted in the center of Figure 8.8.

Figure 8.9 provides an overview of the different capacity limits when α-scaling is applied. At high SNRs, the gap to capacity is dominated by the lack of shaping, and thus approaches 1.53 dB; the MMSE scaling does not help to overcome the shaping loss. For low SNR, however, the benefits of α-scaling can clearly be seen. The mutual information limits of BPSK modulation per dimension (i.e., with a maximum value of 2 bit/s/Hz) are shown for different values of α as dashed lines, indicating that the right choice of α is important. The "α-envelope" was obtained using (8.51). Still, even with MMSE scaling, the gap to capacity is more than 4 dB at zero spectral efficiency, and the lowest possible E_b/N_0-operating point is at about 2.4 dB. In Section 8.6.3 we will see how the shaping loss can be overcome by going from one-dimensional to multi-dimensional precoding.

8.6.2.4 *One-Dimensional Soft-Symbol Metric* For close-to-capacity operation, we have to apply channel coding. At the transmitter, a convolutional code or

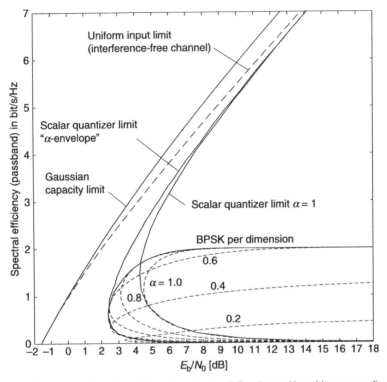

Figure 8.9. Capacity and mutual information limits of Tomlinson-Harashima precoding with α-scaling.

a powerful turbo code of rate R_{ch} can be used to generate coded bits $c = c_{ch}$ that are mapped onto an M-PAM constellation, before performing the scalar quantization and transmission over the communication channel (Figure 8.6). At the receiver, an appropriate demapping has to be performed to provide soft reliability values to the outer channel decoder. The modulo-channel can be taken into account by using log-likelihood ratio values (L–values, e.g., [19]). With reference to (8.50), let us first consider the likelihood functions that are involved.

Likelihood Functions With noise power $\sigma_n^2 = P_n$ (assuming M-PAM transmission) the PDF of the *i.i.d.* Gaussian noise αn is:

$$p_G(\xi) = \frac{1}{\alpha \sigma_n \sqrt{2\pi}} \cdot \exp\left[-\frac{\xi^2}{2\alpha^2 \sigma_n^2}\right] \tag{8.52}$$

Correspondingly, the PDF of the uniformly distributed term $(1 - \alpha)x$ in (8.50) is:

$$p_U(\xi) = \begin{cases} \dfrac{1}{4(1 - \alpha)}; & -2(1 - \alpha) \le \xi \le 2(1 - \alpha) \\[2mm] 0 & \text{else} \end{cases} \tag{8.53}$$

The convolution $p_{UG}(\xi) = p_U(\xi) * p_G(\xi)$ yields the PDF of the sum $-(1 - \alpha)x + \alpha n$ as:

$$
p_{UG}(\xi) = \begin{cases} \dfrac{\mathrm{erf}\left(\dfrac{\xi + 2 \cdot (1 - \alpha)}{\alpha \sigma_n \sqrt{2}}\right) - \mathrm{erf}\left(\dfrac{\xi - 2 \cdot (1 - \alpha)}{\alpha \sigma_n \sqrt{2}}\right)}{8 \cdot (1 - \alpha)}; & 0 < \alpha < 1 \\[2em] p_G(\xi); & \alpha = 1 \end{cases} \quad (8.54)
$$

After the one-dimensional modulo (noise folding) at the receiver, we obtain an approximation of the likelihood-function for a hypothesized transmit amplitude m as:

$$
p'(y'|m) \approx \frac{1}{7} \cdot \sum_{l=-3}^{3} p_{UG}(y' - m + 4l) \tag{8.55}
$$

Only a few neighboring modulo-intervals need be considered in practice, corresponding to l running from -3 to 3 in (8.55). Moreover, an approximation of p_{UG} using a Gaussian distribution with effective unfolded noise power $\sigma_{n'}^2 = (1 - \alpha)^2 P_x + \alpha^2 P_n$ is sufficient.

Log-Likelihood Ratio Values For M-PAM demapping, with $M' = \log_2 M$ outer channel coded bits $\mathbf{c}_{ch} = (c_{ch,1}, \ldots, c_{ch,M'})$ per PAM symbol amplitude $m = m(\mathbf{c}_{ch})$, the *a posteriori* L-values are:

$$
L_D(c_{ch,i}|y') = \ln \frac{P[c_{ch,i} = 0|y']}{P[c_{ch,i} = 1|y']} \tag{8.56}
$$

With:

$$
P[c_{ch,i}|y'] = P[c_{ch,i}] \cdot p(y'|c_{ch,i})/p(y')
$$

this writes as:

$$
L_D(c_{ch,i}|y') = L_A(c_{ch,i}) + L_E(c_{ch,i}|y') \tag{8.57}
$$

with *a priori* knowledge:

$$
L_A(c_{ch,i}) = \ln \frac{P[c_{ch,i} = 0]}{P[c_{ch,i} = 1]} \tag{8.58}
$$

as provided by the outer channel decoder in the "turbo" feedback loop, and *extrinsic* information:

$$
L_E(c_{ch,i}|y') = \ln \frac{\displaystyle\sum_{\substack{\forall m \\ \text{with } c_{ch,i}=0}} p'(y'|m) \cdot e^{\frac{1}{2}\mathbf{c}_{ch,[i]}^T \cdot \mathbf{L}_{A,[i]}}}{\displaystyle\sum_{\substack{\forall m \\ \text{with } c_{ch,i}=1}} p'(y'|m) \cdot e^{\frac{1}{2}\mathbf{c}_{ch,[i]}^T \cdot \mathbf{L}_{A,[i]}}} \tag{8.59}
$$

as passed on to the channel decoder (see, e.g., [19]). The vector $\mathbf{c}_{ch,[i]}$ carries $M' - 1$ bits $c_{ch,j}$, with $j = 1, \ldots, M'$ and $j \neq i$. Correspondingly, $\mathbf{L}_{A,[i]}$ is the vector of $M' - 1$ *a priori* L-values according to (8.58), with the ith entry for $c_{ch,i}$ skipped. Note that the amplitude m, as used for enumerating the summations in numerator and denominator, is a function of the hypothesized coded bits $\mathbf{c}_{ch}$.

The L-value computation for BPSK demapping is particularly simple and writes as:

$$L_E(c_{ch}|y') = \ln \frac{p'(y'|m = +1)}{p'(y'|m = -1)}$$

With the log-likelihood ratio values based on the modulo-metric, and M-PAM mapping using Gray-labeling, simple AWGN coding is sufficient to get close to the mutual information limits of Tomlinson-Harashima precoding (Figure 8.9). In Gray-labeling, the bit labels of neighboring signal amplitudes differ by only a single digit.

8.6.3 Vector Quantization: Sign-Bit Shaping

We have removed the modulo loss by applying MMSE scaling in section 8.6.2.3, and are left with the shaping loss. In *one-dimensional* (TH) precoding, we chose the symbol that is closest to the interference, and transmitted the difference using a scalar modulo operation; the transmitted output was *uniformly* distributed. Now, in *multi-dimensional* precoding, we choose the *sequence* that is closest to the interference, and transmit the difference *vector*, using the Viterbi algorithm to find the minimum energy sequence; the transmitted output is *Gaussian-like* distributed, which provides the desired reduction in transmit power, commonly referred to as *shaping gain*.

8.6.3.1 *Lattices*   A *lattice* Λ is a discrete subgroup of the Euclidean space $\mathbb{R}^n$ with the ordinary vector addition operation. If λ_1, λ_2 are in Λ, then their sum and difference are also in Λ. The translated set $\mathbf{x} + \Lambda$ is referred to as a *coset* of Λ for any $\mathbf{x} \in \mathbb{R}^n$. The fundamental Voronoi region $\mathcal{V}$ of $\Lambda \subset \mathbb{R}^n$ is the set of minimum Euclidean norm coset representatives of the cosets of Λ. Every $\mathbf{x} \in \mathbb{R}^n$ can be uniquely written as $\mathbf{x} = \lambda + \mathbf{r}$ with $\lambda \in \Lambda, \mathbf{r} \in \mathcal{V}$, where $\lambda = Q_\mathcal{V}(\mathbf{x})$ is a nearest neighbor of $\mathbf{x}$ in Λ, and $\mathbf{r} = \mathbf{x} \mod \Lambda$ is the apparent error $\mathbf{x} - Q_\mathcal{V}(\mathbf{x})$. We may thus write $\mathbb{R}^n = \Lambda + \mathcal{V}$ and $\mathcal{V} = \mathbb{R}^n \mod \Lambda$. A comprehensive introduction to lattices can be found in [20].

8.6.3.2 *Shaping Gain*   Assuming optimal MMSE scaling to overcome the modulo loss, (8.47) can be written as:

$$I(m; y') \geq \frac{1}{2}\log_2(1 + \text{SNR}) - \underbrace{\frac{1}{2}\log_2[2\pi e G(\Lambda)]}_{\text{shaping loss}} \tag{8.60}$$

with $G(\Lambda) = 1/12$ being the normalized second moment of the lattice $\Lambda = \mathbb{Z}$, where, for any lattice, $G(\Lambda)$ is defined as:

$$G(\Lambda) = \frac{1}{n|\mathcal{V}|^{1+2/n}} \int_{\mathcal{V}} \|\mathbf{x}\|^2 \, d\mathbf{x} \tag{8.61}$$

The volume of a Voronoi region is denoted as $|\mathcal{V}|$. By definition (8.61), $G(\Lambda)$ is invariant under scaling (and isometry), and, for any dimension n of the cubic lattice $\mathbb{Z}^n$, we have $G(\mathbb{Z}^n) = 1/12$. The region that has the smallest normalized second moment is the n-hypersphere, which, in the limit of $n \to \infty$ approaches:

$$\lim_{n \to \infty} G(n - \text{sphere}) = \frac{1}{2\pi e} \approx \frac{1}{17}$$

The *shaping gain* of a lattice Λ is defined as:

$$g_s|_{dB} = 10 \log_{10} \frac{G(\mathbb{Z}^n)}{G(\Lambda)} = 10 \log_{10} \frac{1}{12G(\Lambda)} \tag{8.62}$$

It is a geometric property of Λ and measures how much more power is needed when using an input uniformly distributed over a hypercube $[-\frac{1}{2}, \frac{1}{2})^n$ (such as in Tomlinson-Harashima precoding), rather than a distribution uniform over the Voronoi region $\mathcal{V}$ of a multi-dimensional lattice Λ, in order to obtain the same volume (or entropy). The ultimate shaping gain is achieved using the n-hypersphere (and $n \to \infty$):

$$g_s(\Lambda)|_{dB} \text{ (optimal shaping)} = 10 \log_{10} \frac{2\pi e}{12} \approx 1.53 \text{ dB} \tag{8.63}$$

in which case the shaping loss in (8.60) goes to zero ("optimal shaping"), recovering the capacity of the interference-free AWGN channel. Note that the lattice $\mathbb{Z}$ as used for TH precoding has no shaping gain.

8.6.3.3 Communication Using Lattices

Generalizing (8.47)–(8.50) for the vector case, let Λ denote an n-dimensional lattice and let $\mathcal{V}$ denote its fundamental Voronoi region. Also, let $\mathbf{u}$ be a random vector variable (dither) uniformly distributed over $\mathcal{V}$. At the *transmitter*, the input alphabet is restricted to $\mathcal{V}$. For any $\mathbf{m} \in \mathcal{V}$, the encoder sends:

$$\mathbf{x} = [\mathbf{m} - (\alpha\mathbf{s} + \mathbf{u})] \bmod \Lambda \tag{8.64}$$

We say that the lattice Λ performs a "vector quantization." After adding the interference vector $\mathbf{s}$ and noise vector $\mathbf{n}$ on the channel, the receiver computes:

$$\mathbf{y}' = [\alpha\mathbf{y} + \mathbf{u}] \bmod \Lambda \tag{8.65}$$

and the effective modulo-Λ noise channel writes as:

$$\mathbf{y}' = [\mathbf{m} - (1 - \alpha)\mathbf{x} + \alpha\mathbf{n}] \bmod \Lambda \tag{8.66}$$

For a good shaping lattice and the dimensionality n towards infinity (and optimal α), the effective folded noise $\mathbf{n}' = [-(1 - \alpha)\mathbf{x} + \alpha\mathbf{n}] \bmod \Lambda$ becomes Gaussian distributed (per dimension), and we obtain:

$$\mathbf{y}' = [\mathbf{m} + \mathbf{n}'] \bmod \Lambda \tag{8.67}$$

With a sequence length of n, the normalized (per complex dimension) mutual information of (8.67) is given by:

$$\frac{1}{n} I(\mathbf{m}; \mathbf{y}') = \frac{1}{n} h(\mathbf{y}') - \frac{1}{n} h(\mathbf{n}')$$

$$\geq \log_2(2\pi e P_m) - \log_2\left(2\pi e \frac{P_m P_n}{P_m + P_n}\right)$$

$$= \log_2(1 + \text{SNR})$$

where the first equation follows since the channel is modulo additive, and the inequality follows since the second moment of the effective noise is no greater (and asymptotically equal) to $P_m P_n / (P_m + P_n)$, and the entropy of a Gaussian random vector is maximal for a given second moment. Since $\mathbf{m}$ and $\mathbf{x}$ are uniformly distributed over $\mathcal{V}$, we have $P_x = P_m$.

Compared to one-dimensional precoding, where we were left with a uniformly distributed noise component, now the interference $\mathbf{s}$ has been completely "absorbed" in n-dimensional space by the shaping lattice. As $\mathbf{n}'$ becomes Gaussian, minimizing the power:

$$P_{n'} = (1 - \alpha)^2 P_x + \alpha^2 P_n \tag{8.68}$$

also minimizes the entropy of the effective noise, and thus, maximizes mutual information. The choice of α in (8.51) is optimal in this case. In the following sections, we will show how vector quantization can be realized at both sides, transmitter and receiver. Lattices based on linear codes will play an important role.

8.6.3.4 Lattice Precoding at the Transmitter

Let us first reconsider Tomlinson-Harashima coding in view of lattice precoding, before moving on to the multi-dimensional generalization described in the next sections. The set:

$$\Lambda = \{\ldots, -8, -4, 0, 4, 8, \ldots\} = 4\mathbb{Z}$$

of section 8.6.2 can be regarded as a one-dimensional lattice. Transmission can be thought of as follows: Rather than deciding on transmitting m, we may imagine that

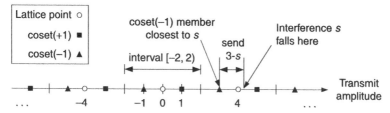

Figure 8.10. Coset view of Tomlinson-Harashima transmission of the symbol " -1."

m now represents any point in the set $m + \lambda$ where $\lambda \in \Lambda$. Such a set is called the *coset* of m. Choosing a symbol m now corresponds to choosing an entire coset.

Figure 8.10 shows an example for a BPSK constellation $m \in \{\pm 1\}$. The modulo-interval around the origin is called the *fundamental Voronoi region* $\mathcal{V}$ of the lattice Λ, and the elements $-1, +1$ within $\mathcal{V}$ are referred to as *coset leaders*. The two induced cosets are:

$$\text{coset}(-1) = \{\ldots, -9, -5, -1, 3, 7, \ldots\}$$

and

$$\text{coset}(1) = \{\ldots, -7, -3, 1, 5, 9, \ldots\}$$

That *member* of the coset to be transmitted is picked which is closest to the interference s (transmitted is the difference, i.e., the quantization error). This scalar precoding scheme extends to the vector case by constructing multi-dimensional lattices based on linear codes, which allow us to achieve shaping gains.

8.6.3.5 *High-Dimensional Lattices from Linear Codes* Let the code $\mathcal{C}$

denote the set of codewords $\mathbf{c}_{vq}$ satisfying $\mathbf{H}\mathbf{c}_{vq}^T = 0$. The $(n - k) \times n$-matrix $\mathbf{H}$ is referred to as parity-check matrix, while the $k \times n$-matrix $\mathbf{G}$ is referred to as generator matrix, with $\mathbf{G}\mathbf{H}^T = 0$. Assume, for now, that the entries of $\mathbf{c}_{vq}$, $\mathbf{u}_{vq}$, $\mathbf{H}$ and $\mathbf{G}$ are elements from GF(2), i.e., can take on values 0 and 1. An information bit sequence $\mathbf{u}_{vq}$ is mapped to a codeword sequence by $\mathbf{c}_{vq} = \mathbf{u}_{vq}\mathbf{G}$. However, we will *not* use this mapping in precoding, as not the *codewords* $\mathbf{c}_{vq}$, but rather, the *cosets* carry the information.

To gain further insight into the coset view of a linear code, we write the codewords $\mathbf{c} = \mathbf{c}_{vq}$ and all possible correctable error patterns $\mathbf{e}_j$ (over, e.g., a binary symmetric channel) in a "standard array," which covers all 2^n binary sequences. This is depicted in Figure 8.11. The top row shows the 2^k codewords, with the all-zero codeword in the top left corner. The leftmost column shows all 2^{n-k} correctable error pattens $\mathbf{e}_1, \ldots, \mathbf{e}_{2^{n-k}}$ ($\mathbf{e}_1 + 0$ stands for the "zero-error" pattern). Each row represents a coset($\mathbf{e}_j$), with $\mathbf{e}_j$ being its coset leader within the fundamental Voronoi region around the all-zero codeword. Observe that the members of any coset($\mathbf{e}_j$) can be

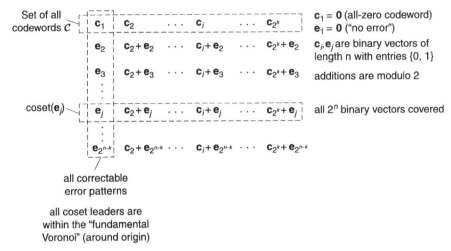

Figure 8.11. Coset view of linear code $\mathcal{C}$ of rate k/n.

enumerated by going through all possible codewords $\mathbf{c}_i$. All members of the same coset have the same *syndrome* $\mathbf{s}_j$, which is defined as $(\mathbf{e}_j + \mathbf{c}_i)\mathbf{H}^T = \mathbf{e}_j\mathbf{H}^T = \mathbf{s}_j$.

Figure 8.12 (left) provides another visualization of the linear binary code $\mathcal{C}$, with 2^k codewords $\mathbf{c}_i$, 2^{n-k} cosets $\mathbf{e}_j$, and, as an example, a coset($\mathbf{e}_4$). Note that the particular shape of the Voronoi region, as well as the positions of the n-dimensional binary vectors $\mathbf{c}_i$, $\mathbf{e}_j$, as depicted in two dimensions, is just illustrative. We can think of $\mathcal{C}$ as a "finite binary lattice," and, so far, we are just covering a subspace of $\mathbb{R}^n$. To associate with $\mathcal{C}$ a lattice $\Lambda(\mathcal{C})$ in $\mathbb{R}^n$ that can be used for shaping, we need to extend the binary

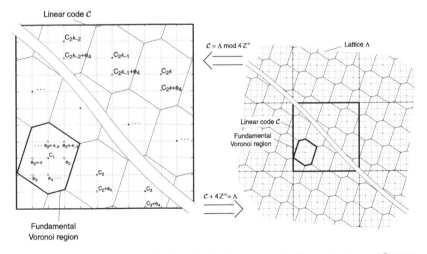

Figure 8.12. Construction A: "Finite lattice" of linear code (left) is extended to $\mathbb{R}^n$ (right).

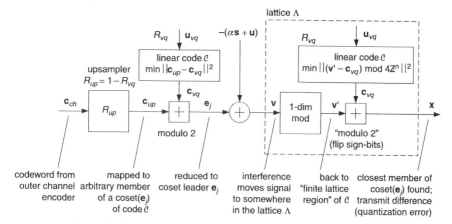

Figure 8.13. Precoding using a lattice Λ based on a linear code $\mathcal{C}$.

code into $\mathbb{R}^n$ by the cubic lattice $\mathbb{Z}^n$, as done by "Construction A" [20]:

$$\Lambda = \mathcal{C} + 2\mathbb{Z}^n \tag{8.69}$$

Thus, any point of the lattice can be obtained by adding a sequence of even integer components to a codeword $\mathbf{c}_{vq} \in \mathcal{C}$.

From now on, let us assume that $\mathcal{C}$ has elements 0, 2 rather than 0, 1; i.e., we multiply the binary lattice by two. This allows us to use the same modulo interval $[-2, 2)$ per dimension as in the TH case, and we obtain:

$$\Lambda = \mathcal{C} + 4\mathbb{Z}^n \tag{8.70}$$

The tessellating of $\mathbb{R}^n$ by $\mathcal{C}$ is illustrated in Figure 8.12 (right).

We have arrived at a convenient construction of multi-dimensional lattices. The next question is how the precoding is actually done. If the code $\mathcal{C}$ was used in classic channel coding, the encoding would be using the generator matrix, $\mathbf{c}_{vq} = \mathbf{u}_{vq}\mathbf{G}$, and the 2^k codewords $\mathbf{c}_{vq}$ would carry the information. However, in lattice precoding, we do not map the information sequence to the codewords, but rather, to the 2^{n-k} available cosets (or syndromes). Thus, lattice precoding can also be referred to as "coset coding" (or syndrome coding). For an unconstrained binary sequence of length n, each symbol carries 1 bit; as there are only 2^{n-k} coset leaders to choose from (for a sequence length n), the shaped sequence carries $1 - R_{vq}$ bits per symbol, with $R_{vq} = k/n$. Note that it is the redundancy $1 - R_{vq}$ that allows us to choose the minimum energy sequence out of a set of 2^k "equivalent" sequences in the coset, i.e., to perform the shaping.

With reference to Figure 8.13, precoding involves the following steps:

1. Mapping of channel codeword sequence to coset member of shaping code $\mathcal{C}$: The outer codeword $\mathbf{c}_{ch}$ of length n_{ch} is mapped to *any member* $\mathbf{c}_{up}$ of the coset($\mathbf{e}_j$) of length n_{vq} by an unambiguous mapping ("upsampler") where

$n_{vq} = n_{ch}/R_{up}$, and $R_{up} = 1 - R_{vq}$. For example, if $R_{vq} = 1/2$, then a simple repetition code of rate $R_{up} = 1/2$ is sufficient to perform this task. Thus, we can say that the channel encoder of rate R_{ch} chooses a coset from the "shaping code" C of rate R_{vq}.

2. Reduce $\mathbf{c}_{up}$ modulo the shaping lattice Λ, to arrive at the coset leader $\mathbf{e}_j$. We assume an outer binary code, and thus, the upsampled sequence $\mathbf{c}_{up}$ is part of the standard-array of C, member of some coset $\mathbf{e}_j$, and can be written as $\mathbf{c}_{up} = \mathbf{c}_{vq,i} + \mathbf{e}_j$ (mod-2 addition). The search over the "finite binary lattice" of C suffices to find the coset leader $\mathbf{e}_j$, which is done by enumeration over all 2^k codewords of C, minimizing the *Hamming distance* between $\mathbf{c}_{up}$ and C (i.e., we need to find the codeword $\mathbf{c}_{vq,i}$)

$$d^2(\mathbf{c}_{up}, C) = \min_{\mathbf{c}_{vq} \in C} \left\| \mathbf{c}_{up} - \mathbf{c}_{vq} \right\|^2$$

3. Pre-subtract the scaled interference and dither sequence, $-(\alpha\, \mathbf{s} + \mathbf{u})$. This takes the signal to somewhere in the lattice over $\mathbb{R}^n$.

4. Repeat step 2, i.e., reduce the input $\mathbf{v} = \mathbf{e}_j - (\alpha\mathbf{s} + \mathbf{u})$. modulo the shaping lattice Λ. This time, however, as $\mathbf{v}$ is somewhere in $\mathbb{R}^n$, it would mean a search over the *infinite* lattice Λ. Luckily, due to the periodicity of Λ in $\mathbb{Z}^n$, applying a one-dimensional quantizer to Λ results in C, and thus, enumeration over the *finite* lattice C suffices:

$$\Lambda \bmod 4\mathbb{Z}^n = C \tag{8.71}$$

Denote the output of the one-dimensional modulo by $\mathbf{v}' = \mathbf{v} \bmod 4\mathbb{Z}^n$. As $\mathbf{v}'$ is continuous-valued (unlike the discrete binary vectors of the first stage), we need to minimize a *modulo-Euclidean distance* (rather than Hamming distance) over C by enumeration over the coset members:

$$d^2(\mathbf{v}', C) = \min_{\mathbf{c}_{vq} \in C} \left\| (\mathbf{v}' - \mathbf{c}_{vq}) \bmod 4\mathbb{Z}^n \right\|^2 \tag{8.72}$$

which is equivalent to performing a minimization over the infinite lattice Λ

$$d^2(\mathbf{v}, \Lambda) = \min_{\lambda \in \Lambda} \left\| \mathbf{v} - \lambda \right\|^2$$

The result of this minimization is the member of the coset($\mathbf{e}_j$) that is closest to $\mathbf{e}_j - (\alpha\mathbf{s} + \mathbf{u})$. Transmitted is the difference vector. The distribution of the components of the difference vector is no longer uniform, but Gaussian-like, which provides the desired power reduction.

Note, again, that the actual value of the information sequence $\mathbf{u}_{vq}$ is of no particular interest; rather, it is just used for enumeration over the coset members. To emphasize that the binary entries of $\mathbf{u}_{vq}$ do not carry any information, we could refer to them as "virtual" information bits. Note also that steps 2 and 4 can be combined; they were presented separately just for the convenience of explanation. Performing a single modulo-Λ operation after adding the interference and dither sequences is sufficient.

8.6.3.6 *Sign-Bit Shaping* The enumeration over the coset members is done by going through all possible codewords ("complete decoder"). For convolutional codes, a particularly efficient implementation of a complete decoder is available through the Viterbi algorithm, operating on a trellis structure, which is, thus, also referred to as "trellis shaping" [21]. We can easily obtain a linear block code from a convolutional code by properly terminating the trellis.

Recall that the classic shaping problem for (interference-free) AWGN channels is concerned with high SNR, i.e., high spectral efficiency; Figure 8.7 showed, indeed, that there is hardly any penalty from using a uniform input distribution at low SNR. Thus, rather than using BPSK, as we did in the previous section, an M-PAM constellation is used per dimension, carrying $M' = \log_2 M$ bits per symbol, to obtain higher spectral efficiency. Each amplitude level m is associated with a bit label $\mathbf{b} = (b_1, b_2, \ldots, b_{M'})$, i.e., $m = m(\mathbf{b})$. For M-PAM, the basic lattice construction stays the same; however, the upsampled coset member sequence $\mathbf{c}_{up}$ now only represents the *sign-bits* $b_{M'}$ of the mapping, and we can regard the other (unsigned) bits $(b_1, b_2, \ldots, b_{M'-1})$ as just traveling "piggybacking" with the signed bits. Accounting for the redundancy added in the upsampler, there are $M' - 1 + (1 - R_{vq})$ bits conveyed per symbol. To obtain a valid lattice structure, we can only use bit labels that fulfill:

$$
\begin{aligned}
m(b_1, b_2, \ldots, b_{M'-1}, b_{M'} = 0) & \\
= [m(b_1, b_2, \ldots, b_{M'-1}, b_{M'} = 1) + 2] & \bmod 4\mathbb{Z}
\end{aligned}
\tag{8.73}
$$

Thus, Gray-labeling, which is a popular choice in bit-interleaved coded modulation (BICM), is not allowed. Other labelings, like natural labeling, or a Gray-like labeling within the unsigned bits $(b_1, b_2, \ldots, b_{M'-1})$ have to be used. Figure 8.14 only shows a single modulo-Λ reduction, as steps 2 and 4 of the previous section are combined. The Viterbi algorithm just influences (flips) the sign-bits of the M-PAM symbols, while, owing to the labeling constraint of (8.73), the unsigned bits remain unchanged. This is the reason why it is also referred to as "sign-bit shaping."

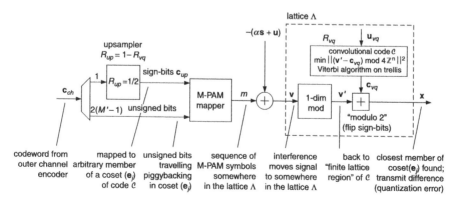

Figure 8.14. Precoding using sign-bit shaping.

We can further extend the scheme by allowing the shaping code to choose from *several* amplitude levels, rather than just flipping the sign of the amplitude [22]. For example, with two bits per *M*-PAM symbol coming from a shaping code, it can choose one out of four different amplitude levels. This way, for the same code memory, a bigger shaping gain can be achieved. However, we will only use simple sign-bit shaping in the following.

8.6.3.7 Coset Decoding at the Receiver

For detection at the receiver, another modulo-Λ operation is performed, to find the most probable coset, and thus to recover the transmitted information ("coset decoding"). Assuming sign-bit shaping, the Viterbi algorithm could be used to perform this task. However, we have separated the operations of channel coding and precoding (shaping) at the transmitter. Thus, from the perspective of the outer channel decoder, the lattice decoder is just an inner processing stage which should provide *soft reliability information* on the outer channel coded bits. As the Viterbi algorithm performs a maximum-likelihood sequence estimation (MLSE), it cannot provide soft outputs, but rather, has to be replaced by a BCJR A Posteriori Probability (APP) decoder [23]. The BCJR provides the initial APP (log-likelihood ratio) values and may be viewed as a bitwise *quantization detector*. In turn, the outer channel decoder refines the estimates of the bit probabilities using the redundancy R_{ch} in the coded bit stream. These are then fed back to the BCJR quantizer, performing *iterative detection and decoding* to closely approach capacity. Different channel codes can be used, such as low-density parity-check (LDPC) codes or repeat-accumulate (RA) codes (a subclass of LDPC codes with linear encoding complexity). LDPC and RA codes allow close-to-capacity operation and can be "matched" to the quantization detector to optimize the convergence behavior of the concatenated scheme. Details of the code design procedure by matching extrinsic information transfer curves can be obtained from [24].

Figure 8.15 shows a block diagram of the inner quantization detector. In the same way as at the transmitter, a one-dimensional modulo is used to bring the received

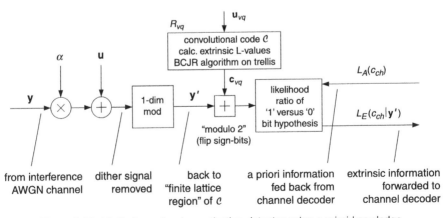

Figure 8.15. Multi-dimensional quantization detector using a priori knowledge.

signal back into the region of $\mathcal{C}$, and an enumeration over the finite number of coset members using the virtual information bits $\mathbf{u}_{vq}$ suffices. The received signal after MMSE scaling, dither addition and one-dimensional modulo is:

$$\mathbf{y}' = (\alpha\mathbf{y} + \mathbf{u}) \bmod 4\mathbb{Z}^n$$

Rather than just finding the closest coset, corresponding to minimizing a metric like (8.72), i.e., search over the "finite binary lattice" of $\mathcal{C}$:

$$d^2(\mathbf{y}', \mathcal{C}) = \min_{\mathbf{c}_{vq} \in \mathcal{C}} \left\| (\mathbf{y}' - \mathbf{c}_{vq}) \bmod 4\mathbb{Z}^n \right\|^2$$

we have to perform a log-likelihood ratio value calculation on the coded bits of the outer channel code (compare to section 8.6.2.4), by *summation* of log-likelihood values over the coset. The L-value computation is now performed over a sequence $\mathbf{y}'$, and no longer over individual symbols y', as the PAM symbols are connected through the sign-bits which were chosen at the transmitter according to the codebook of the underlying shaping code $\mathcal{C}$.

Let $\mathbf{c}_{ch}$ denote the vector of coded bits from the outer channel decoder (length n_{ch}), and $\mathbf{u}_{vq}$ the vector of all k_{vq} information bits of the APP detector (or "vector quantizer"). In analogy to (8.56), the APP vector quantizer computes the *a posteriori* L-values on the outer channel coded bits as:

$$L_D(c_{ch,i}|\mathbf{y}') = \ln \frac{P[c_{ch,i} = 0|\mathbf{y}']}{P[c_{ch,i} = 1|\mathbf{y}']}$$

which can be written as:

$$L_D(c_{ch,i}|\mathbf{y}') = L_A(c_{ch,i}) + L_E(c_{ch,i}|\mathbf{y}') \tag{8.74}$$

$L_A(c_{ch,i})$ provides the interface for feeding back soft information from the outer channel decoder to the inner quantization detector in an iterative turbo detection loop. It is set to zero for the first pass through the detector. After replacing the simple likelihood function $p'(y'|m)$ of (8.55) by $p'(\mathbf{y}'|\mathbf{c}_{vq}, \mathbf{u}_{vq})$, the demapping of signed and unsigned bits is performed by enumerating over the coset using $\mathbf{u}_{vq}$ (and summation of log-likelihood values in numerator and denominator):

$$L_E(c_{ch,i}|\mathbf{y}') = \ln \frac{\displaystyle\sum_{\substack{\mathbf{c}_{ch} \in \mathbb{C}_{i,0} \\ \forall \mathbf{u}_{vq}}} p'(\mathbf{y}'|\mathbf{c}_{ch}, \mathbf{u}_{vq}) \cdot \exp\left(\frac{1}{2}\mathbf{c}_{ch,[i]}^T \cdot \mathbf{L}_{A,[i]}\right)}{\displaystyle\sum_{\substack{\mathbf{c}_{ch} \in \mathbb{C}_{i,1} \\ \forall \mathbf{u}_{vq}}} p'(\mathbf{y}'|\mathbf{c}_{ch}, \mathbf{u}_{vq}) \cdot \exp\left(\frac{1}{2}\mathbf{c}_{ch,[i]}^T \cdot \mathbf{L}_{A,[i]}\right)} \tag{8.75}$$

where $\mathbf{c}_{ch,[i]}$ denotes the sub-vector of $\mathbf{c}_{ch}$ obtained by omitting its ith element $c_{ch,i}$, and $\mathbf{L}_{A,[i]}$ denotes the vector of all L_A-values, also omitting its ith element. The set $\mathbb{C}_{i,0}$

contains all 2^{n-1} bit vectors $\mathbf{c}_{ch}$ having $c_{ch,i} = 0$, i.e., $\mathbb{C}_{i,0} = \{\mathbf{c}_{ch}|c_{ch,i} = 0\}$, and $\mathbb{C}_{i,1} = \{\mathbf{c}_{ch}|c_{ch,i} = 1\}$. The evaluation of (8.75) is efficiently done by exploiting the underlying trellis structure [23] of C; for this, $p'(\mathbf{y}'|\mathbf{c}_{ch}, \mathbf{u}_{vq})$ is broken up into logarithmic metric increments that are based on the one-dimensional modulo metric:

$$\ln p'(y_1', y_2'|m_1, m_2) = \ln p'(y_1'|m_1) + \ln p'(y_2'|m_2)$$

$$\approx \ln \sum_{l=-3}^{3} \exp\left[-\frac{(y_1' - m_1 + 4l)^2}{2\sigma_{vq}^2}\right] + \ln \sum_{l=-3}^{3} \exp\left[-\frac{(y_2' - m_2 + 4l)^2}{2\sigma_{vq}^2}\right] \quad (8.76)$$

where the two M-PAM symbols m_1, m_2 per metric increment ($R_{vq} = 1/2$) are dependent on $c_{ch,1}, c_{ch,2}, \ldots, c_{ch,2M'-2}$, the information bit u_{vq} of the shaping code, and the state transition in the trellis. The effective noise power per real dimension is $\sigma_{vq}^2 = P_{n'}$ (assuming M-PAM transmission), with $P_{n'}$ according to (8.68).

8.6.3.8 Mutual Information Limits   Figure 18.16 shows the mutual information of an equivalent channel between precoder input $\mathbf{m}$ and quantizer detector soft output $\mathbf{L}_E$. Any operating point located on the right side of the respective

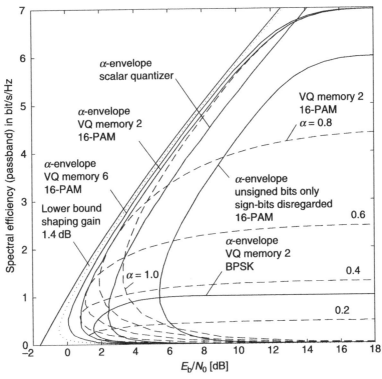

Figure 8.16. Mutual information limits of sign-bit shaping with BPSK and 16-PAM per dimension, vector quantizer of memory 2 and memory 6.

mutual information curve is attainable provided that an appropriate channel coding scheme is applied [24]. BPSK and 16-PAM mappings are used per dimension, with a vector quantizer of rate $R_{vq} = 1/2$, corresponding to a maximal spectral efficiency of $1\,\mathrm{bit/s/Hz}$ and $7\,\mathrm{bit/s/Hz}$, respectively. All curves are "α-envelopes," i.e., $\alpha = \mathrm{SNR}/(1 + \mathrm{SNR})$ was used for computing the mutual information at each SNR. The curve of the scalar quantizer is given as a reference. As can be seen, the gains of vector quantization are biggest for low SNR. Specifically, we observe the following:

- BPSK per dimension is too small: Although a vector quantizer (VQ) of memory 2 gains about 1 dB over the scalar quantizer at $0.25\,\mathrm{bit/s/Hz}$, we can further improve performance by just going into more amplitude levels, like 16-PAM; using 4-PAM (not shown) would already be sufficient.

- A VQ of memory 2, 16-PAM (shown for different α-values) improves more than 1.5 dB over the scalar quantizer at low SNR, and about 1 dB at high SNR. For spectral efficiencies greater than $4\,\mathrm{bit/s/Hz}$, no MMSE scaling is needed, and $\alpha = 1$ suffices, while for lower spectral efficiency the right choice of α is crucial.

- For a VQ of memory 6, the advantage is about 2.5 dB over scalar quantization, and about 0.8 dB over a memory 2 VQ at $0.5\,\mathrm{bit/s/Hz}$. This, however, comes at the expense of a 16-times increase in complexity over memory 2.

- Based on (8.60), (8.62), we computed a lower bound for a shaping gain of $g_s(\Lambda)|_{\mathrm{dB}} = 1.4\,\mathrm{dB}$, which is shown on the very left side of the chart as a dotted curve. The notion of "lower bound" in this case means that the actual curve is even more to the left (while, of course, staying below the AWGN capacity curve). By simulation experiments we determined that a convolutional code of memory greater than 16 is required to obtain this shaping gain, which is too complicated for implementation. It is worth mentioning that the lower bound (8.60) accurately predicts the performance of the memory 2 and memory 6 vector quantizers with measured shaping gains of 0.98 dB and 1.22 dB, respectively.

- A simple quantizer detector which only performs an M-PAM demapping per dimension while completely disregarding the constraints imposed by the sign-bits has the worst performance, shown on the very right side of the figure. The sign-bits are only used for shaping at the transmitter, but neither for communicating information, nor for improving the quality of the soft output in a trellis at the receiver.

The complexity of a memory 2 vector quantizer appears to be reasonable in view of the gains over the scalar quantizer. A higher quantizer memory is only justified if operating points at very low SNR are expected to be frequently used.

8.6.4 The Role of Channel Knowledge

8.6.4.1 Single User vs. Multiuser MIMO

Channel knowledge is particulary important in order to approach capacity of the multiuser MIMO Gaussian broadcast channel. While channel knowledge is essential to achieve capacity even over a single user MIMO link, its role is even more prominent in a multiuser setting. In essence, precise channel knowledge in a single user setting is needed in order to determine the optimal input covariance matrix, i.e., in order to determine the water-filling power allocation over the channel eigenmodes. Nonetheless, at high SNR, using a (scaled) identity covariance matrix results in little loss. Thus, one can work in a mode where only the desired transmission rate is fed back from the receiver, or even in an "open loop" mode with no feedback.[8]

In contrast, in multiuser MIMO broadcast, obtaining a reliable channel estimate is indispensable. While we have seen that the DPC transmission scheme in a multiuser setting hinges on having channel knowledge at the transmitter, high-precision channel knowledge is essential even in suboptimal approaches such as linear precoding. As the users are separated and cannot perform joint processing, imperfect channel estimates will necessarily result in residual interference that will play the role of noise. For this reason, a multiuser broadcast system needs to operate in a "closed loop" mode where considerable resources are spent to inform the transmitter of the channel.

8.6.4.2 *Obtaining Channel Knowledge*   In a practical setting, obtaining good estimates of the channel at the transmitter may be quite challenging. In a fixed wired system (such as in DSL systems), the channel is typically nearly constant in time, and reliable channel estimation may be fed back to the transmitter with small overhead, or may be inferred indirectly when working in a time-division mode.

Obtaining reliable channel estimates at the transmitter in a mobile wireless environment is a much more challenging task. Typically, the channel variation in time necessitates frequent updates of the channel estimation. If the channel estimate is to be obtained by feedback from the users (receivers), providing reliable estimates may result in considerable feedback rate as well as impose strict latency constraints. Effective feedback mechanisms and quantization schemes for channel information are a topic of current research.

8.7   SUMMARY

While the first decade of multiuser communications has focused on many-to-one links, the wireless channel has brought new communication scenarios, like the MIMO broadcast channel, where a given user experiences interference from signals intended for other users. Such interference is well suited to be treated by recently developed methods for coding (or precoding) in the presence of known interference. In this chapter, we have provided an overview of such communication scenarios, and presented a method for achieving capacity using coding with side information at the transmitter, also referred to as "dirty paper coding." We have shown that the scalar dirty paper coding scheme is sufficient as a basic building block. Vector quantization has been identified as an important concept to realize dirty paper coding. Capacity can be closely approached

---

[8]In this case an outage event may occur and feedback at the MAC layer will provide a "no-acknowledge" (NACK) indication.

at any spectral efficiency using iterative quantization detection and decoding. While several challenges for practical implementation over wireless channels remain, like fast and reliable feedback of channel state information, dirty paper coding is readily applicable to various physical instantiations, such as cable-bound systems.

REFERENCES

1. M. H. M. Costa. "Writing on dirty paper," *IEEE Trans. Inf. Theory*, IT-29, pp. 439–441, May 1983.

2. U. Erez, S. Shamai (Shitz), and R. Zamir. "Capacity and lattice-strategies for cancelling known interference," *Proc. Internat. Symp. Inf. Theory and Its Appl.*, pp. 681–684, Nov. 2000.

3. A. S. Cohen and A. Lapidoth. "Generalized writing on dirty paper," *Proc. Internat. Symp. Inf. Theory*, p. 227, Jun./Jul. 2002.

4. T. M. Cover and J. A. Thomas. *Elements of Information Theory*. Wiley, New York, 1991.

5. D. N. C. Tse and P. Viswanath. "Sum capacity of the vector Gaussian broadcast channel and uplink-downlink duality," *IEEE Trans. Inf. Theory*, vol. 49, no. 8, pp. 1912–1921, Aug. 2003.

6. S. Vishwanath, N. Jindal, and A. Goldsmith, "Duality, achievable rates, and sum-rate capacity of Gaussian MIMO broadcast channels," *IEEE Trans. Inf. Theory*, vol. 49, no. 10, pp. 2658–2668, Oct. 2003.

7. W. Yu and T. Lan. "Minimax duality of Gaussian vector broadcast channels," *Proc. Internat. Symp. Inf. Theory*, p. 177, July 2004.

8. W. Yu and J. Cioffi. "Sum Capacity of Gaussian Vector Broadcast Channels," *IEEE Trans. Inf. Theory*, vol. 50, no. 9, pp.1875–1892, Sept. 2004.

9. R. Zamir, S. Shamai (Shitz), and U. Erez. "Nested linear/lattice codes for structured multiterminal binning," *IEEE Trans. Inf. Theory*, vol. 48, no. 6, pp. 1250–1276, June 2002.

10. G. Strang. "Linear Algebra and its Applications," Third Edition, San Diego, Harcourt Brace Jovanovich Publishers, 1988.

11. G. Caire and S. Shamai (Shitz). "On achievable rates in a multi-antenna broadcast down-link," Proc. 38th Annual Allerton Conference on Commun. Control and Computing, pp. 1188–1193, Oct. 2000.

12. G. Ginis and J. M. Cioffi. "A Multi-user Precoding Scheme Achieving Crosstalk Cancellation with Application to DSL Systems," *Proc. 34th Asilomar Conference*, pp. 1627–1631, Oct. 2000.

13. T. Pedro, W. Utschick, G. Bauch, and J. Nossek. "Efficient Implementation of Successive Encoding Schemes for the MIMO OFDM Broadcast Channel," *Proc. Internat. Conf. on Commun.*, June 2006.

14. H. Weingarten, Y. Steinberg, and S. Shamai (Shitz). "The Capacity Region of the Gaussian Multiple-Input Multiple-Output Broadcast Channel," *IEEE Trans. Inf. Theory*, vol. 52, no. 9, pp. 3936–3964, Sept. 2006.

15. H. Viswanathan, S. Venkatesan, and H. Huang. "Down-link capacity evaluation of cellular networks with known-interference cancellation," *IEEE Journ. on Select. Areas in Commun.*, vol. 21, no. 5, pp. 802–811, 2003.

16. M. Tomlinson. "New automatic equalizer employing modulo arithmetic," *Electronic Lett.*, vol. 7, pp. 138–139, Mar. 1971.

17. H. Harashima and H. Miyakawa. "Matched-transmission technique for channels with intersymbol interference," *IEEE Trans. Commun.*, pp. 774–780, Aug. 1972.

18. S. Shamai (Shitz) and R. Laroia. "The intersymbol interference channel: lower bounds on capacity and channel precoding loss," *IEEE Trans. Inf. Theory*, IT-42, pp. 1388–1404, Sept. 1996.

19. J. Hagenauer, E. Offer, and L. Papke. "Iterative decoding of binary block and convolutional codes," *IEEE Trans. Inf. Theory*, IT-42, no. 2, pp. 429–445, Mar. 1996.

20. J. H. Conway and N. J. A. Sloane. Sphere Packings, Lattices and Groups. New York: Springer–Verlag, 1988.

21. G. D. Forney, Jr. "Trellis shaping." *IEEE Trans. Inf. Theory*, vol. 38, pp. 281–300, Mar. 1992.

22. M. W. Marcellin, T. R. Fischer. "Trellis coded quantization of memoryless and Gauss-Markov sources," *IEEE Trans. Commun.*, vol. 38, no. 1, pp. 82–93, Jan. 1990.

23. L. Bahl, J. Cocke, F. Jelinek, and J. Raviv. "Optimal decoding of linear codes for minimizing symbol error rate," *IEEE Trans. Information Theory*, IT-20, pp. 284–287, Mar. 1974.

24. S. ten Brink and U. Erez. "A close-to-capacity dirty paper coding scheme," *Proc. Internat. Symp. on Inf. Theory*, p. 533, July 2004. As well as in U. Erez and S. ten Brink, *IEEE Trans. Inf. Theory*, vol. 51, no. 10, pp. 3417-3432, Oct. 2005.

INDEX

Advances in Multiuser Detection. Edited by Michael L. Honig
Copyright © 2009 John Wiley & Sons, Inc.